T0180218

# Computational Materials Science

Kaoru Ohno · Keivan Esfarjani
Yoshiyuki Kawazoe

# Computational Materials Science

## From Ab Initio to Monte Carlo Methods

Second Edition

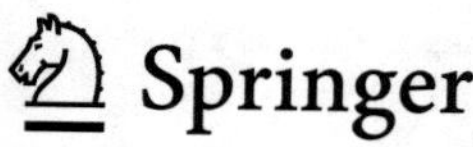

Kaoru Ohno
Department of Physics
Yokohama National University
Yokohama
Japan

Yoshiyuki Kawazoe
New Industry Creation Hatchery Center
Tohoku University
Sendai
Japan

Keivan Esfarjani
Department of Mechanical
and Aerospace Engineering
Materials Science and Engineering
and Physics
University of Virginia
Charlottesville, VA
USA

ISBN 978-3-662-58587-0 ISBN 978-3-662-56542-1 (eBook)
https://doi.org/10.1007/978-3-662-56542-1



Printed on acid-free paper

This Springer imprint is published by the registered company Springer-Verlag GmbH, DE part of Springer Nature
The registered company address is: Heidelberger Platz 3, 14197 Berlin, Germany

# Preface to the Second Edition

During these 18 years after this book was first published, there has been great progress in computational materials science, in particular in ab initio approaches. In density functional theory (DFT), the so-called DFT + $U$ approach and hybrid functionals are now very widely used as well as the standard approaches such as the local density approximation (LDA) or the generalized gradient approximation (GGA) by many researchers to study properties of a large variety of materials. There have been many other advances in these years. For example, calculations of the electron–phonon interaction have recently become possible with much interest in many practical applications. These new calculation methods are explained in detail in this second edition. In addition, we gave a brief explanation of dynamical mean field theory (DMFT), density matrix renormalization group (DMRG) theory, and the tensor network as well. Instead, we deleted some old results on the ab initio molecular dynamics simulations of $C_{60}$ and a section that described the optimized effective potential (OEP) method. Similarly, we could not include the subjects related to Dirac cone, topological insulators, or superconductivity in this book, because we wanted to put more emphasis on the computation of the properties rather than the physics of materials.

Another important progress was made in quasiparticle theory and Green's function methods, which are directly related to the photoelectron spectroscopy experiments. Now, the accuracy of the self-consistent $GW\Gamma$ calculations including vertex corrections becomes comparable to quantum chemistry methods such as the multireference single- and double-configuration interaction (MRDCI). Using these methods, one may treat arbitrary excited eigenstates of the Hamiltonian. Since it is not easy to learn the detail of quantum field theory, we decided to explain the essence of these new theories without introducing standard field theoretical many-body perturbation theory. So, we hope that the readers who do not have theoretical physics background can easily understand these new sections only from the knowledge of creation and annihilation operators. We also added a brief explanation of parquet diagrams, which are expected to yield accurate results for materials with strong electron correlations.

The ab initio calculations can be also used to construct parameters of model Hamiltonians which are used to perform various multiscale simulations. For example, calculations of the thermal conductivity, which require large-scale molecular dynamics simulations or are based on a lattice dynamics (LD) model, have been recently performed using ab initio methods. DFT calculations are used in this case to compute parameters of a LD model with high accuracy. We have accordingly included a new section "Thermal Conductivity" in Chap. 2. Similarly, DFT calculations are nowadays widely used to determine parameters of an electronic tight-binding model by the so-called maximally localized Wannier function method. Thus, produced Wannier functions can then be used to treat correlated systems using more sophisticated methods such as DMFT, or calculate Berry phase applied to newly developed topological materials, or used to compute electron transmission and conductance or electron–phonon interaction in various materials and devices.

As another example, in order to perform lattice Monte Carlo simulations for alloys, one may use the renormalized local free energies determined by an ab initio total energy calculation as an input. An example of such a calculation of FePt alloys is newly added in Chap. 4. There, we briefly explained how to introduce the method to treat a FePt alloy cluster in the vacuum and the phase field model in a sense to make a bridge between ab initio calculations and coarse-grained models. It is explained in Sect. 5.4.6.

In the Appendix, we revised some sections and added three new chapters: "Appendix A. Fourier Interpolation in 1 and 3 Dimensions" which gives explicit formulas for Fourier series and interpolations used in a crystal; "Appendix F. Partitioning Formula, Downfolding and Green's Functions," which is a very powerful and handy formula to project the Green's function of an infinite system onto a finite subsystem, without loss of spectral information; and "Appendix G. Statistical Data Analysis," which shows how statistical data generated from a molecular dynamics or a Monte Carlo run should be processed and analyzed, especially in the context of correlations needed to compute response functions or dynamical properties. We also largely revised chapter "Appendix C. Vibrational Properties" and added an explanation of "Block Davidson Method," which is now widely used as an iterative matrix diagonalization scheme, in chapter "Appendix E. Optimization Methods Used in Materials Science." We hope all these modifications and inclusion of new topics and methods are beneficial to the readers.

Finally, we would like to express our sincere thanks once again to Dr. Claus Ascheron. Without his patience and encouragement, this revision could never be completed.

Yokohama, Japan
Charlottesville, USA
Sendai, Japan
October 2017

Kaoru Ohno
Keivan Esfarjani
Yoshiyuki Kawazoe

# Preface to the First Edition

There has been much progress in the computational approaches in the field of materials science during the past two decades. In particular, computer simulation has become a very important tool in this field since it is a bridge between theory, which is limited by its too simplified modelings, and experiment which is limited by the physical parameters. Computer simulation, on the other hand, can partially fulfill both of these paradigms, since it is based on theories and is in fact performing experiment but under any arbitrary, however, unphysical conditions.

This progress is indebted to the advances in computational physics and chemistry. Ab initio methods are being used widely and frequently in order to determine the electronic and/or atomic structures of different materials. The ultimate goal is to be able to predict various properties of a material just from its atomic coordinates and, also in some cases, to even predict the stable atomic positions of a given material. However, at this time, the applications of ab initio methods are very much limited with respect to the number of particles and the timescale of a dynamical simulation. This is one extremum of the methodology on the basis of the very accurate electronic-level calculations.

When one does not require such ab initio calculations, there are many alternative powerful semiempirical techniques such as tight-binding method. More empirical methods, e.g., classical molecular dynamics and Monte Carlo simulations, offer also powerful ways which enable one to treat even larger-scale systems. This is the other extremum where one uses methods such as the lattice gas model which has a very simple interaction function with few parameters only.

How to bridge between these two completely different levels of calculations is also one big theme in the field of computational materials science, and we have addressed this issue in the fourth chapter.

The basic theories and their applications are, thus, quite widespread and usually not always very easy for students, scientists, and engineers to survey and understand correctly. In particular, what has been a recent progress in solid-state physics and chemistry is now used in materials research and even biology.

Because of the very rapid progress in computational methods in materials research nowadays, there has been a strong demand to have such a comprehensive and introductory book which covers most of the important computational methodologies and basic theories behind them, covering from very rigorous methods such as ab initio methods to more empirical methods such as those using classical potentials or lattice models. In this book, we have attempted to put together all these methods in a unified description and in a systematic fashion, from the basics to very advanced methods, so that students and researchers can acquire an overview of this quickly evolving field and its applications. Several references have also been supplied in case one needs to get a deeper insight into more advanced problems or details of implementations of some of the methods.

In recent years, physicists, chemists, materials scientists, and even biochemists have been using the same computational tools in order to solve their very different problems in nature. The progress in quantum chemistry has been very quickly reflected onto quantum physics and materials science and vice versa. This book is intended to introduce many basic theories behind the recent methodologies adopted in computational materials science. We started from small-scale behaviors and highly accurate treatments and finished with very large-scale behaviors using simple models in order to introduce different levels of methodologies more systematically and more easily.

Some emphasis is also put on points which are not fully established but seem important since they might have important applications in the future. Such examples are the quantum Monte Carlo method which is the most accurate treatment of the many-fermion problem possible up to now, beyond the Born-Oppenheimer approximation and the GW approximation. All these fields are becoming very active because of their importance and the recent advances in computer hardware which make simulations of their level possible.

We would like to express our sincere thanks to Dr. Claus Ascheron for his suggesting and encouraging us writing this book from the very beginning until publishing this as a new volume in Springer's series in Solid-State Sciences and also for his very patient cooperation in completing this book. We also would like to thank Dr. Angela Lahee for her very good assistance of improving the style and English of this book. We think that this book would have never been appeared without their kind and patient cooperations.

Sendai, Japan
March 1999

Kaoru Ohno
Keivan Esfarjani
Yoshiyuki Kawazoe

# Contents


**1 Introduction** . . . 1
1.1 Computer Simulation as a Tool for Materials Science . . . 1
1.2 Modeling of Natural Phenomena . . . 2

**2 Ab Initio Methods** . . . 7
2.1 Introduction . . . 7
2.2 Hartree–Fock and Density Functional Approaches . . . 8
2.2.1 Quantum Mechanics of Identical Particles . . . 9
2.2.2 Virial Theorem . . . 12
2.2.3 The Hartree–Fock (HF) Approximation . . . 13
2.2.4 Electron Correlations Beyond the HF Approximation . . . 18
2.2.5 Density Functional Theory . . . 21
2.2.6 Local Density Approximation (LDA) . . . 26
2.2.7 Generalized Gradient Approximation (GGA) . . . 32
2.2.8 Self-interaction Correction (SIC) . . . 36
2.2.9 LDA + $U$, GGA + $U$ . . . 38
2.2.10 Hybrid Functionals . . . 39
2.3 Explicit Methods to Calculate Electronic Structures . . . 43
2.3.1 Periodic Systems . . . 43
2.3.2 Group Theory . . . 49
2.3.3 LCAO, OPW and Mixed-Basis Approaches . . . 56
2.3.4 Pseudopotential Approach . . . 63
2.3.5 APW Method . . . 65
2.3.6 KKR, LMTO and ASW Methods . . . 69
2.3.7 Some General Remarks . . . 72
2.3.8 Ab Initio $O(N)$ and Related Methods . . . 75
2.4 Perturbation and Linear Response . . . 81
2.4.1 Effective-Mass Tensor . . . 81
2.4.2 Dielectric Response . . . 83
2.4.3 Magnetic Susceptibility . . . 87

2.4.4 Chemical Shift . . . . . . . . . . . . . . . . . . . . . . . . . . . . . 89
2.4.5 Phonon Spectrum . . . . . . . . . . . . . . . . . . . . . . . . . . . . 91
2.4.6 Electrical Conductivity . . . . . . . . . . . . . . . . . . . . . . . . . 96
2.4.7 Thermal Conductivity . . . . . . . . . . . . . . . . . . . . . . . . . 99
2.4.8 Time-Dependent Density Functional Theory . . . . . . . . . . 108
2.4.9 Dynamical Mean Field Theory, Density Matrix Renormalization Theory, and Tensor Network . . . . . . . . . 109
2.5 Ab Initio Molecular Dynamics . . . . . . . . . . . . . . . . . . . . . . . 117
2.5.1 Car–Parrinello Method . . . . . . . . . . . . . . . . . . . . . . . . . 117
2.5.2 Steepest Descent, Conjugate Gradient, and Block Davidson Methods . . . . . . . . . . . . . . . . . . . . . . . . . . . 120
2.5.3 Formulation with Plane Wave Basis . . . . . . . . . . . . . . . . 121
2.5.4 Formulation with Other Bases . . . . . . . . . . . . . . . . . . . . 123
2.5.5 Application to Fullerene Systems . . . . . . . . . . . . . . . . . . 125
2.5.6 Application to Point Defects in Crystals . . . . . . . . . . . . . 129
2.5.7 Application to Other Systems . . . . . . . . . . . . . . . . . . . . 131
2.5.8 Coherent Potential Approximation (CPA) . . . . . . . . . . . . 134
2.6 Beyond the Born . . . . . . . . . . . . . . . . . . . . . . . . . . . . . . . . 136
2.7 Quasiparticle Theory and Green's Function Methods . . . . . . . . . . 141
2.7.1 Quasiparticle (QP) Theory . . . . . . . . . . . . . . . . . . . . . . . 141
2.7.2 Green's Function Formalism . . . . . . . . . . . . . . . . . . . . . . 148
2.7.3 GW Approximation . . . . . . . . . . . . . . . . . . . . . . . . . . . 157
2.7.4 Exchange and Coulomb Holes . . . . . . . . . . . . . . . . . . . . 169
2.7.5 Vertex Correction . . . . . . . . . . . . . . . . . . . . . . . . . . . . 170
2.7.6 Inclusion of Ladder Diagrams . . . . . . . . . . . . . . . . . . . . . 173
2.7.7 Further Remarks: Cusp Condition and Parquet Diagrams . . . . . . . . . . . . . . . . . . . . . . . . . . . . . . . . . 177
References . . . . . . . . . . . . . . . . . . . . . . . . . . . . . . . . . . . . . 181

**3 Tight-Binding Methods** . . . . . . . . . . . . . . . . . . . . . . . . . . . . 199
3.1 Introduction . . . . . . . . . . . . . . . . . . . . . . . . . . . . . . . . . . . 199
3.2 Tight-Binding Formalism . . . . . . . . . . . . . . . . . . . . . . . . . . . 200
3.2.1 Tight-Binding Parametrization . . . . . . . . . . . . . . . . . . . . 201
3.2.2 Calculation of the Matrix Elements . . . . . . . . . . . . . . . . . 203
3.2.3 Total Energy . . . . . . . . . . . . . . . . . . . . . . . . . . . . . . . . 204
3.2.4 Forces . . . . . . . . . . . . . . . . . . . . . . . . . . . . . . . . . . . . 205
3.3 Methods to Solve the Schrödinger Equation for Large Systems . . . . . . . . . . . . . . . . . . . . . . . . . . . . . . . . . . . . 207
3.3.1 The Density Matrix $O(N)$ Method . . . . . . . . . . . . . . . . 207
3.3.2 The Recursion Method . . . . . . . . . . . . . . . . . . . . . . . . . 211
3.3.3 Recursive Green's Function Method . . . . . . . . . . . . . . . . 213

3.4 Self-consistent Tight-Binding Formalism . . . . . . 214
3.4.1 Parametrization of the Coulomb Integral $\mathcal{U}$ . . . . . . 217
3.5 Applications to Fullerenes, Silicon and Transition-Metal Clusters . . . . . . 219
3.5.1 Fullerene Collisions . . . . . . 220
3.5.2 $C_{240}$ Doughnuts and Their Vibrational Properties . . . . . . 221
3.5.3 IR Spectra of $C_{60}$ and $C_{60}$ Dimers . . . . . . 221
3.5.4 Simulated Annealing of Small Silicon Clusters . . . . . . 225
3.5.5 Titanium and Copper Clusters . . . . . . 227
3.6 Conclusions . . . . . . 228
References . . . . . . 229

**4 Empirical Methods and Coarse-Graining** . . . . . . 231
4.1 Introduction . . . . . . 231
4.2 Reduction to Classical Potentials . . . . . . 232
4.2.1 Polar Systems . . . . . . 233
4.2.2 Van der Waals Potential . . . . . . 235
4.2.3 Potential for Covalent Bonds . . . . . . 237
4.2.4 Embedded-Atom Potential . . . . . . 240
4.3 Cluster Expansion . . . . . . 241
4.3.1 Lattice Gas Model . . . . . . 242
4.3.2 The Connolly–Williams Approximation . . . . . . 242
4.4 Potential Renormalization . . . . . . 244
4.4.1 Basic Idea: Two-Step Renormalization Scheme . . . . . . 244
4.4.2 The First Step . . . . . . 246
4.4.3 The Second Step . . . . . . 248
4.4.4 Application to Si . . . . . . 250
4.4.5 Application to FePt Alloys . . . . . . 252
References . . . . . . 259

**5 Monte Carlo Methods** . . . . . . 261
5.1 Introduction . . . . . . 261
5.2 Basis of the Monte Carlo Method . . . . . . 262
5.2.1 Stochastic Processes . . . . . . 263
5.2.2 Markov Process . . . . . . 265
5.2.3 Ergodicity . . . . . . 269
5.3 Algorithms for Monte Carlo Simulation . . . . . . 273
5.3.1 Random Numbers . . . . . . 273
5.3.2 Simple Sampling Technique . . . . . . 276
5.3.3 Importance Sampling Technique . . . . . . 278
5.3.4 General Comments on Dynamic Models . . . . . . 281
5.4 Applications . . . . . . 283

5.4.1 Systems of Classical Particles . . . . . 283
5.4.2 Modified Monte Carlo Techniques . . . . . 286
5.4.3 Percolation . . . . . 290
5.4.4 Polymer Systems . . . . . 292
5.4.5 Classical Spin Systems . . . . . 297
5.4.6 Nucleation . . . . . 311
5.4.7 Crystal Growth . . . . . 316
5.4.8 Fractal Systems . . . . . 319
References . . . . . 332

**6 Quantum Monte Carlo (QMC) Methods** . . . . . 339
6.1 Introduction . . . . . 339
6.2 Variational Monte Carlo (VMC) Method . . . . . 340
6.3 Diffusion Monte Carlo (DMC) Method . . . . . 342
6.4 Path-Integral Monte Carlo (PIMC) Method . . . . . 346
6.5 Quantum Spin Models . . . . . 349
6.6 Other Quantum Monte Carlo Methods . . . . . 350
References . . . . . 351

**Appendix A: Fourier Interpolation in 1 and 3 Dimensions** . . . . . 353

**Appendix B: Molecular Dynamics and Mechanical Properties** . . . . . 357

**Appendix C: Vibrational Properties** . . . . . 371

**Appendix D: Calculation of the Ewald Sum** . . . . . 387

**Appendix E: Optimization Methods Used in Materials Science** . . . . . 393

**Appendix F: Partitioning Formula, Downfolding and Green's Functions** . . . . . 403

**Appendix G: Statistical Analysis of Data** . . . . . 409

**Index** . . . . . 415

# Chapter 1
# Introduction

## 1.1 Computer Simulation as a Tool for Materials Science

Although categorized as an engineering field which normally studies artificial materials, metallurgy has also treated natural objects in its long history. Metallurgy and Alchemy, which are the basis of present-day materials science, have also produced all the natural sciences, including chemistry and physics. The field of materials science is expanding rapidly to include amorphous, ceramic, polymer, and nanoscale materials. Alchemy was not successful, but its methods still contribute to industry, because they introduce new useful materials.

Since materials are complex in nature, it has never been possible to treat them directly by theory; experimental studies have always been the main method in materials research. In the past, physical concepts could only be applied to pure crystals, and real materials with defects and grain boundaries could not be treated theoretically. Recent progress in computer technology is now opening a new era in which one can treat real, complex materials as they are. The first dramatic change has happened in chemistry. In the past, the most important work of chemists has been to categorize chemicals, and for this purpose, computer databases have been used as a new tool. In the 1980s, molecular-orbital calculations were applied to calculate the properties of chemicals which were of interest in research and in industry. At present, chemicals, including drugs, are designed first by applying molecular-orbital calculations, and then experimental tests confirm the theoretical results, before the molecules are realized. This means new chemicals with desired properties can now be designed by using such calculations. The physical and chemical properties of large molecules such as DNA are also studied by using supercomputers.

In molecular orbital (MO) calculations, the MO is assumed to be a linear combination of atomic orbitals. Although it seems sufficient to express the molecular orbital in this way, a large number of basis functions might be needed in order to accurately represent the electronic wavefunction and, consequently, the charge density. In this case, usually, the Hartree–Fock approximation is used to describe electron-electron interaction effects. This approximation is thought to be rough, since it does not


K. Ohno et al., *Computational Materials Science*,
https://doi.org/10.1007/978-3-662-56542-1_1

include correlation effects. A better approximation, the local-density approximation (LDA), will give ground-state properties almost perfectly and more efficiently. The LDA is, of course, an approximation, and cannot be applied to excited states and highly correlated systems. The latter are among the hot topics in theoretical physics.

The application of ab initio methods to real materials is still limited. At the moment, even if we employ the fastest supercomputer with the largest memory, we can treat at most a few hundred atoms, and it is not possible to compute all of the properties of real materials. Periodic boundary conditions serve in the study of crystals, but do not really increase the number of atoms included in the system. Therefore, other methods, such as the tight-binding method or classical molecular dynamics, are very important in order to overcome the complexities of some materials. In this case, we assume potential parameters that are extracted from experiment or ab initio calculations. The latter are, however, only valid in the region where they were fitted, and not beyond. It is, nevertheless, useful to study complex materials on the basis of these approximations. Popular potentials such as Lennard–Jones or Finnis–Sinclair are known to be good enough for computing some of the properties of various materials under various conditions and gain insights and learn about trends rather than making quantitative predictions.

The above treatments, ab initio molecular dynamics, tight binding, and classical molecular dynamics, are all "deterministic". The state of the system is determined completely by the initial condition. This approach to materials research is useful in understanding stable structures, vibrations, and growth at the atomistic level. Another method for treating complex systems is the Monte Carlo method. It assumes very idealized or simplified interaction parameters and can treat larger number of atoms. Physical and chemical properties of large systems can be obtained as a statistical average over randomly moved particles. It is, therefore, a "stochastic" method. In Chap. 5, we introduce the Monte Carlo method, starting with some concepts from statistical physics. The standard Ising model, classical spin systems, percolation and quantum Monte Carlo, all using this method, are described.

## 1.2 Modeling of Natural Phenomena

How do we understand Nature's behavior? Except for philosophical sayings such as "I think therefore I exist", we should consider the methods of human understanding. Philosophy, in each era, has been based on the physical understanding of that time. (This statement might be inverted by philosophers!) In the basic conceptual framework of physics, we try to understand the world by discarding meanings. In other words, the science of physics expresses the complex world by numerical equations. For example, in Newtonian mechanics, the acceleration is proportional to the external force, and the proportionality constant is called the mass. It is an amazing fact that by only this equation, the motion of planets or the trajectory of the Space Shuttle can be computed very accurately. However, since the velocity cannot exceed the speed of light, this approximation cannot hold for very fast motions close to the light velocity.

Einstein's principle of relativity is not even relevant to advanced technology in space, let alone daily life, but it predicts the motion of light or light particles passing close to a heavy object such as the sun.

Another standard method of modeling is "linearization". The Newtonian equation is the linear approximation of the relativistic equation of motion at speeds much smaller than the light velocity. Other well-known examples are Ohm's law and Hooke's law. For these, a moderate voltage or force is assumed. The reason why the linear approximation holds for complex systems is that the potential around the stable position can be expressed by a quadratic equation. It is also important to study the situation outside the range of linearity. It is not easy to solve nonlinear equations by analytic methods, and normally, numerical solutions are needed. One of the recent fields in which computer simulation has been applied recently is Chaos theory, where small variations of some parameters cause drastic changes in the solution.

Human knowledge has expanded greatly, and the "ultimate elements of matter" were found first to be protons and neutrons, and then quarks and gluons. However, in the 20th century, this reductionist approach of separating matter into small elements reached its end. As a typical example, the Superconducting Super Collider (SSC) project in the USA was stopped. It is not possible to make an accelerator surrounding the earth, either physically or economically. The search for the ultimate constituents of matter is still important, even though it may now be no more than the dream of monkey removing the successive skins of an onion in a fruitless attempt to find what is inside.

The subject covered in this book is, fundamentally, the many-body problem, which has long been known to be difficult to solve analytically. The two-body equation in classical dynamics is reduced to a one-body equation as follows. The equations for two particles are expressed as

$$m_1\ddot{r}_1 = f(r_1 - r_2), \tag{1.1}$$

$$m_2\ddot{r}_2 = f(r_2 - r_1). \tag{1.2}$$

Introducing the relative coordinate

$$r = r_1 - r_2, \tag{1.3}$$

the reduced equation for $r$ is obtained as

$$\frac{m_1 m_2}{m_1 + m_2}\ddot{r} = f(r). \tag{1.4}$$

This equation can be solved as a one-body equation. What will happen when we increase the number of particles? This reduction method can not be applied to systems of more than three particles. The study of three-body systems was established as a field of research long ago. Except for linear, or in general, symmetric systems where the number of degrees of freedom can be reduced, the following general three-body equation can not be solved analytically:

$$m_1\ddot{r}_1 = f(r_1, r_2, r_3), \tag{1.5}$$

$$m_2\ddot{r}_2 = f(r_1, r_2, r_3), \tag{1.6}$$

$$m_3\ddot{r}_3 = f(r_1, r_2, r_3). \tag{1.7}$$

Numerical treatment is the only way to solve many-body problems in general. The basic equation in molecular dynamics is an extension of this scheme to include a large number of particles. It has been extended to quantum systems by Faddeev.

In solid-state physics, the term of "many-body problem" is used in a different way. That is to say, materials are assumed to be composed of an infinite number of atoms (actually $10^{23}$ for one mole) and such a material is referred to as "bulk". Because of the high degree of symmetry present in the system, namely the invariance under translations of the unit cell, it is possible to use powerful theorems such as Bloch's theorem to solve this many-(almost infinite) body system. The symmetry can be deduced from the observation of X-ray diffraction patterns; without actually seeing the atoms, it was possible to discover the periodicity of crystals!

Therefore, it is important to model the physical systems that occur in nature. The surface of water, like a mirror, is very rough at an atomistic level. Basically we can understand nature in relation to our size (in units of meters and seconds). It is not easy to recognize objects with different scales and they have no meaning in daily life. Amounts of water or electricity are not measured by their number of molecules or electrons. We measure these amounts in proper units such as litres or amperes. And even during a period of high inflation, we will not use money in units of $10^{23}$!

How do we aim to understand nature by first-principles calculations? We will start with the categorization of forces. That is,

- (1) strong interaction (nuclear force)
- (2) Coulomb force
- (3) weak interaction (the force related to $\beta$ decay)
- (4) gravitational force.

These are distinguished by their orders of magnitude. That is, the strength of the force is strongest for (1) and weakest for (4). In units of $\sqrt{\hbar c} = 1.778 \times 10^{13}\,\mathrm{J}^{1/2}\,\mathrm{m}^{1/2}$, their strengths are $10, 0.1, 10^{-3}, 10^{-19}$, respectively. The interactions are long-ranged, as $1/r$ for (4); usually shielded with a range of molecular size, for (2); and very short-ranged, with a range of order $10^{-15}$ m, for (1) and (3). These are the four basic kinds of force, which characterize present-day science.

One of the main research areas in physics is the unification of these forces. For example, the magnetic force is unified with the Coulomb force in Maxwell's equations, and is expressed in terms of $c$ (the speed of light) and $e$ (the electron charge).

The electromagnetic and weak interactions have already been unified by Weinberg and Salam. Although larger-scale unification is being studied by many researchers, it is a very difficult task to include all the forces into a single one. There are other considerations claiming that a fifth or even sixth category of force exists! These studies belong in the domain of elementary-particle physics.

The forces we talk about in this book originate from the Coulomb force only. Although people categorize them as ionic, covalent, molecular and van der Waals forces, they are all different manifestations of the Coulomb force combined with Heisenberg uncertainty and Pauli exclusion principles. The material properties studied here for molecules, crystals and polymers are determined by the fundamental forces between their constituents. Since a complete ab initio calculation is not possible, determination of the properties is strongly dependent on the establishment of interaction potentials and on experimental observations. At present, materials cannot be directly designed by ab initio methods in a computer used as a virtual laboratory. However, because of the dramatic progress of computer power, it will be possible in the future to realize our dream of predicting new materials by computer. This is the reason why we study computer simulation methods such as molecular dynamics and Monte Carlo methods; the aim is to design materials.

# Chapter 2
# Ab Initio Methods

## 2.1 Introduction

In fields such as materials science, electronics, mechanical engineering and bio-science, not to mention physics and chemistry, the keywords "first principles" and "ab initio" have been widely used recently. The principal idea of these keywords is to regard many-atom systems as many-body systems composed of electrons and nuclei, and to treat everything on the basis of first principles of quantum mechanics, without introducing any empirical parameters.

Except for very light elements such as hydrogen and helium, the quantum behavior of nuclei is not very important for most purposes, and one may consider nuclei to be structureless points with mass and charge and treat them classically. On the other hand, electrons cannot be treated as classical particles in any situation, since, if one did so, all atoms would shrink rapidly to points by emitting photons. In nature, if atoms emit photons, they go to and then stay in their ground state without further shrinkage because of Heisenberg's uncertainty principle in quantum mechanics. When the positions of all nuclei are held fixed in a system, the electrons reach a steady state and form an "electron cloud" with a stable charge density distribution determined quantum mechanically.

First-principles (or ab initio) methods enable one to study the electronic structure of solids, surfaces, or clusters as accurately as possible with moderate computational effort. Moreover, using ab initio molecular-dynamics methods, which calculate the forces exerted on atoms at each time step, one can simulate the evolution of atomic and electronic motions without assuming empirical parameters. Apart from particular types of electronic motion independent of atomic motion, such as plasma oscillations, ionization-neutralization processes, electronic transition processes with photoemission or photoabsorption, electron tunneling, electric currents and so forth, electrons move so as to follow very quickly the nuclear motion. This is because the electron mass is much smaller than that of nuclei. In such a case, where the change in the electronic state occurs very rapidly compared to the nuclear motion, the assumption that the electrons are always in a steady state holds generally. This is called



K. Ohno et al., *Computational Materials Science*,
https://doi.org/10.1007/978-3-662-56542-1_2

the Born–Oppenheimer adiabatic approximation. Using this approximation, one can completely separate the calculation of the electronic structure from that of the ionic motion and perform the two calculations separately at each step. This is the basis of the usual ab initio molecular-dynamics methods.

However, to treat rigorously the quantum states of many-electron systems is not easy, even with the most sophisticated methods of modern physics and chemistry, and one has to rely on some approximate theories. That is, only God can use the true methodology represented by the term, "first-principle methods"; humans have to use a methodology which is fairly reliable but not exact.

In this chapter, the basics of first principles calculations are described in the next two sections, after which ab initio molecular-dynamics methods based on the Born–Oppenheimer (BO) adiabatic approximation are introduced. Then, as a practical application of ab initio molecular-dynamics simulation, the recently found carbon cluster, $C_{60}$ fullerene, is treated, for example. Finally, the treatment of electron correlations is mentioned. The assumption that the steady state is realized for the electronic system for any atomic positions at any time is only an assumption. More details about the BO approximation can be found in Sect. 2.6. In the following, atomic units ($e = \hbar = m = 1$) are used unless mentioned otherwise. In atomic units, which we abbreviate to a.u., the length unit is the Bohr radius ($0.529 \times 10^{-10}$ m), and the energy unit is 1 Hartree ($= 2$ Rydberg $= 27.2$ eV $= 4.36 \times 10^{-18}$ J).

## 2.2 Hartree–Fock and Density Functional Approaches

There are many first-principles approaches for determining the ground state and excited states of many-atom systems. These approaches are classified into four main groups: the first is quantum chemistry approach starting from the Hartree–Fock approximation, which offers a rigorous one-electron approximation. These are, for example, Møller–Plesset (MP) perturbation theory, configuration interaction (CI), and coupled cluster (CC) theory. The second is an approach based on the density functional theory which offers an exact background for the many-body problem but can be solved only approximately. In both cases, the resulting equation is one-electron-like (rigorously in the case of the Hartree–Fock theory and effectively in the case of the density functional theory). The correction from the the Hartree–Fock approximation is often referred to as the electron correlation. The third approach is the Green' function approach on the basis of many-body perturbation theory (MBPT). The fourth approach is quite different using quantum Monte Carlo methods, which will be described in Chap. 6. In the present chapter, we describe in detail the formulation starting from the quantum mechanics of many identical particles and leading to the density functional theory, and its widespread applications and several problems.

### 2.2.1 Quantum Mechanics of Identical Particles

For quantum mechanics, one may refer to many good text books, for example, those by Messiah [1], Schiff [2], Dirac [3] and Landau and Lifshitz [4]. Let us write the Hamiltonian of a quantum mechanical system composed of $N$ identical particles as $\mathcal{H}$ and its $\xi$th eigenfunction (wavefunction) as $\Psi_\xi(1, 2, \ldots, N)$. The particle coordinates $(1, 2, \ldots, N)$ include in general a coordinate associated with the internal spin degree of freedom (spin coordinate) $\{s_i\}$ as well as the spatial coordinate $\{\boldsymbol{r}_i\}$. The Schrödinger equation reads

$$\mathcal{H}\Psi_\xi(1, 2, \ldots, N) = E_\xi \Psi_\xi(1, 2, \ldots, N), \tag{2.1}$$

where $E$ denotes the energy eigenvalue of the $\xi$th state. Since $\mathcal{H}$ is Hermitian, all eigenvalues are real numbers.

If one applies a permutation operator $P$ which interchanges several particles simultaneously among each other, the wavefunction does not change, apart from a phase factor, because the system is composed of identical particles. Writing the phase factor as $\kappa_P$, one has

$$\mathrm{P}\Psi(1, \ldots, i, \ldots, j, \ldots, N) = \kappa_P \Psi(1, \ldots, i, \ldots, j, \ldots, N). \tag{2.2}$$

Notice that the wavefunction goes back to the original one when we apply the same permutation operator $P$ twice. Thus we have $\kappa_P^2 = 1$, i.e. $\kappa_P = \pm 1$. Thus the wavefunction must be symmetric or antisymmetric under one interchange operation. (Here the word "interchange" means a single permutation between two particles.) A particle is called a boson if the wavefunction is symmetric, and $\kappa_P$ is unity. On the other hand, it is called a fermion if the wavefunction is antisymmetric, and $\kappa_P = (-)^P$, i.e. $\kappa_P = 1$ for an even number of interchanges (even permutation) and $\kappa_P = -1$ for an odd number of interchanges (odd permutation). According to quantum mechanics, only an integer multiple of $\frac{1}{2}$ is allowed for the spin angular momentum (we call it simply the "spin") of a particle. And, according to relativistic quantum theory [5], it is found that particles having an integer $(0, 1, 2, \ldots)$ spin are bosons and particles having a half-integer $(\frac{1}{2}, \frac{3}{2}, \ldots)$ spin are fermions.

Because of the antisymmetry of the fermion wavefunction, one may readily show that if two fermions come on the same position, the fermion wavefunction becomes identically zero. That is, two fermions cannot occupy the same one-particle state and cannot come close to each other. This property is called Pauli's exclusion principle [6]. On the other hand, there is no such rule for bosons, because the boson wavefunction is symmetric for permutations.

Now we turn our attention to electron systems. Electrons interact with each other and also with nuclei by the Coulomb interaction. Rigorously speaking, electrons obey the relativistic Dirac equation [5], having the spin-$\frac{1}{2}$ angular momentum as an internal degree of freedom. Since their spin is half-integer, electrons are fermions. Now we consider the case where the kinetic energies of the electrons are relatively small and relativistic effects can be ignored. Moreover, nuclear motion is usually much slower than electron motion and nuclei can be treated classically as a point charge due to their large mass, although there are some exceptional cases in particular for light elements like hydrogen. Here we assume that electrons follow perfectly the nuclear motions and always at the steady state at the time. This is equivalent to the assumption that the position of every nucleus is held fixed. Roughly speaking, we assume that electrons follows adiabatically the nuclei motion without changing their state at all. this assumption is called the Born–Oppenheimer (BO) approximation. For more detail, see Sect. 2.6. The $N$-electron system, for such a case, is described by the Schrödinger equation with the Hamiltonian

$$\mathcal{H} = -\frac{1}{2}\sum_{i=1}^{N} \nabla_i^2 + \sum_{i>j}^{N} \frac{1}{|\boldsymbol{r}_i - \boldsymbol{r}_j|} + \sum_{i=1}^{N} v(\boldsymbol{r}_i), \tag{2.3}$$

where the first term on the right-hand side represents the electron kinetic energy, the second term is the electron–electron Coulomb interaction and the third term, $v(\boldsymbol{r}_i)$, denotes the Coulomb potential caused by the nuclei.

Each electron has a spin coordinate equal to either $+1/2$ (upwards) or $-1/2$ (downwards). In the nonrelativistic case, since the Hamiltonian does not depend on the spin coordinate, spins are good quantum numbers, i.e. the spin state of each electron is either upwards or downwards. In other words, if the spin wavefunction is nonzero for the spin coordinate $s_i = +1/2$, the wavefunction is zero for $s_i = -1/2$. Such a spin state is called the upward ($\uparrow$) spin state and represented conveniently by $s_i = \uparrow$. Conversely, if it is zero for $s_i = +1/2$, it is nonzero for $s_i = -1/2$. Such a state is called the downward ($\downarrow$) spin state and represented by $s_i = \downarrow$. Whether the spin state of the $i$th electron is in the $\uparrow$ state or the $\downarrow$ state will be indicated by the index $\xi$ of the eigenstate of the total system. In the present nonrelativistic case, in the absence of an external magnetic field, the eigenstates that are different only in the spin states are energetically degenerate. On the other hand, when the spin–orbit coupling, which is one of the important relativistic effects (see the end of Sect. 2.2.10) is not negligible, the spin state of each electron becomes unclear and the eigenstate energies split according to a more complex symmetry such as a singlet or triplet spin state. That is, even in this case, the total spin angular momentum is still a good quantum number.

Let us consider the simplest system, composed of $N$ electrons moving independently. In this case, if we write the Hamiltonian for the $n$th electron as $\mathcal{H}_i$, the total Hamiltonian reads

$$\mathcal{H} = \mathcal{H}_1 + \mathcal{H}_2 + \cdots + \mathcal{H}_N. \tag{2.4}$$

The wavefunction of $N$ independent particles is in general written as the product of $N$ one-particle wavefunctions. In other words, since the $\mathcal{H}_i$ commute with one another, the eigenstate can be constructed as a product of one-particle eigenstates. For the eigenvalue equation

$$\mathcal{H}\Psi_{\lambda_1,\lambda_2,\ldots,\lambda_N}(1,2,\ldots,N) = E_{\lambda_1,\lambda_2,\ldots,\lambda_N}\Psi_{\lambda_1,\lambda_2,\ldots,\lambda_N}(1,2,\ldots,N), \tag{2.5}$$

one finds the antisymmetric solution to be

$$\begin{aligned}\Psi_{\lambda_1,\lambda_2,\ldots,\lambda_N}(1,2,\ldots,N) &= \frac{1}{\sqrt{N!}}\sum_P (-)^P P\psi_{\lambda_1}(1)\psi_{\lambda_2}(2)\cdots\psi_{\lambda_N}(N) \\ &= \frac{1}{\sqrt{N!}}\begin{vmatrix} \psi_{\lambda_1}(1) & \psi_{\lambda_2}(1) & \cdots & \psi_{\lambda_N}(1) \\ \psi_{\lambda_1}(2) & \psi_{\lambda_2}(2) & \cdots & \psi_{\lambda_N}(2) \\ \vdots & \vdots & & \vdots \\ \psi_{\lambda_1}(N) & \psi_{\lambda_2}(N) & \cdots & \psi_{\lambda_N}(N) \end{vmatrix},\end{aligned} \tag{2.6}$$

and

$$E_{\lambda_1,\lambda_2,\ldots,\lambda_N} = \varepsilon_{\lambda_1} + \varepsilon_{\lambda_2} + \cdots + \varepsilon_{\lambda_N}, \tag{2.7}$$

where $(-)^P$ represents the permutation factor equal to $+1$ or $-1$ for an even or odd number of interchanges, respectively. This form of the wavefunction is exactly the same as the determinant of a matrix and is called the Slater determinant [7]. The $\varepsilon_{\lambda_i}$ in (2.7) are the eigenvalues for the eigenstates $\lambda_i$ of the one-electron eigenvalue equation $\mathcal{H}_i\psi_{\lambda_i}(i) = \varepsilon_{\lambda_i}\psi_{\lambda_i}(i)$.

Owing to the antisymmetry of the wavefunction, it is not allowed for two electrons to occupy the same quantum state. This can be understood from, for example, the properties of the Slater determinant (2.6). This fact manifests Pauli's exclusion principle concerning quantum states, and is one of the most important consequences of this antisymmetry for quantum states in many-fermion (many-electron) systems. For systems composed of many bosons, no such principle exists, because of the symmetry of the Bose wavefunction. Therefore, many identical bosons can occupy simultaneously one quantum state. This rule concerning the occupation of quantum states for fermions and bosons affects significantly the statistics of quantum systems especially at low temperatures. The statistics obeyed by fermions and bosons are called, respectively, Fermi–Dirac statistics [8, 9] and Bose–Einstein statistics [10, 11].

There are many quantum states $\lambda_i$ having different energy eigenvalues; each of these states is called an energy level. In the ground state, the quantum states of the system are occupied by electrons one by one, from energetically lower levels to higher levels, because of Pauli's exclusion principle. The highest occupied level is called the Fermi level and the eigenvalue energy at this level is called the Fermi energy. All the levels below the Fermi level are occupied and all the levels above the Fermi level are empty.

When electrons occupy levels one by one from lower to higher energies, often one spin state, say the $\uparrow$ spin state, is occupied first over several levels, which are similar both energetically and symmetrically (for example, 3d levels in the case of transition metal elements), and then the opposite ($\downarrow$) spin state is occupied later in these levels. Such an order of electron occupation is energetically more favorable than a homogeneous order in which each level is occupied doubly by $\uparrow$ and $\downarrow$ electrons regardless of the spin. This is because electrons with the same spin cannot come near each other because of Pauli's exclusion principle and hence the energy loss due to the electron–electron Coulomb interaction among these electrons (when they come close) is significantly reduced. Therefore, open-shell atoms and molecules, i.e. those having a degenerate level which is only partially occupied, generally have a "spin magnetic moment". The rule stating that states of same spin are occupied first is called "Hund's rule" [12].

### 2.2.2 *Virial Theorem*

Here we comment on the virial theorem.

$$\mathcal{H}^{\text{total}}\{\boldsymbol{r}_i, \boldsymbol{R}_k\} = \mathcal{H} + \sum_{k>l} \frac{Z_k Z_l}{|\boldsymbol{R}_k - \boldsymbol{R}_l|} \tag{2.8}$$

stands for the total Hamiltonian including the Coulomb interaction between nuclei, where $\boldsymbol{R}_k$ and $\boldsymbol{R}_l$ denote the positions of nuclei, and $\mathcal{H}$ is given by (2.3) with $v(\boldsymbol{r}_i) = -\sum_k Z_k/|\boldsymbol{r}_i - \boldsymbol{R}_k|$. Now we multiply all electron and nuclear coordinates ($\{\boldsymbol{R}_k\}$ and $\{\boldsymbol{r}_i\}$) by a factor $\alpha$, i.e., all lengths are scaled by $\alpha$. Then, since the $N$-electron wavefunction $\Psi_\gamma\{\boldsymbol{r}_i\}$ of an eigenstate $\gamma$ of this Hamiltonian is normalized as

$$\int \Psi_\gamma^*\{\alpha\boldsymbol{r}_i\}\Psi_\gamma\{\alpha\boldsymbol{r}_i\}\prod_i^N d(\alpha\boldsymbol{r}_i) = 1, \tag{2.9}$$

the total energy of this eigenstate $\gamma$ at the equilibrium atomic positions $\{\boldsymbol{R}_k\}$,

$$E_\gamma^N = \int \Psi_\gamma^*\{\alpha\boldsymbol{r}_i\}\mathcal{H}^{\text{total}}\{\alpha\boldsymbol{r}_i, \alpha\boldsymbol{R}_k\}\Psi_\gamma\{\alpha\boldsymbol{r}_i\}\prod_i^N d(\alpha\boldsymbol{r}_i), \tag{2.10}$$

has the $\alpha$-dependence only in the form of $\langle\hat{T}\rangle/\alpha^2 + \langle\hat{V}\rangle/\alpha$, where $\mathcal{H}\{\boldsymbol{r}_i, \boldsymbol{R}_k\}$ is given by (2.8), $\langle\hat{T}\rangle$ and $\langle\hat{V}\rangle$ denote the kinetic and potential contributions to $E_\gamma^N$. If the atomic configuration is fully relaxed at this excited state, the total energy must be stationary with respect to the change in $\alpha$. Therefore, differentiating $E_\gamma^N$ with respect to $\alpha$ and setting $\alpha$ at unity, the virial theorem $2\langle\hat{T}\rangle + \langle\hat{V}\rangle = 0$ is readily derived for any (either ground or excited) eigenstate of the Hamiltonian (2.8) [13]. A formulation for calculating arbitrary electronic excited eigenstates is given in Sect. 2.7.

### 2.2.3 *The Hartree–Fock (HF) Approximation*

In order to treat the electronic states of many-atom systems or crystals correctly, it is necessary to perform calculations for complex interacting electron systems on the basis of quantum mechanics. However, it is, in general, not an easy task, even using some approximations. As the simplest method for calculating the electronic states of many-electron systems, there is a one-electron approximation. According to quantum mechanics, one can determine the ground state of the Hamiltonian $\mathcal{H}$ by means of the "variational principle" for the normalized wavefunction $\Psi(1, 2, \ldots, N)$:

$$\begin{aligned}\langle\Psi|\mathcal{H}|\Psi\rangle &= \sum_{s_1}\sum_{s_2}\cdots\sum_{s_N}\int \Psi^*(1, 2, \ldots, N)\mathcal{H}\Psi(1, 2, \ldots, N)d\boldsymbol{r}_1 d\boldsymbol{r}_2\cdots d\boldsymbol{r}_N \\ &= \text{minimum} = E_0. \end{aligned} \tag{2.11}$$

According to this variational principle, one may try to find an approximate wavefunction $\Psi(1, 2, \ldots, N)$ so as to minimize the expectation value $\langle\Psi|\mathcal{H}|\Psi\rangle$, since we know that the expectation value is always greater than the true ground-state energy $E_0$.

Now we suppose the electrons to be independent and the trial function $\Psi(1, 2, \ldots, N)$ to be a Slater determinant of single-particle wavefunctions:

$$\Psi(1, 2, \ldots, N) = \frac{1}{\sqrt{N!}}\begin{vmatrix} \psi_1(1) & \psi_2(1) & \cdots & \psi_N(1) \\ \psi_1(2) & \psi_2(2) & \cdots & \psi_N(2) \\ \vdots & \vdots & & \vdots \\ \psi_1(N) & \psi_2(N) & \cdots & \psi_N(N) \end{vmatrix}. \tag{2.12}$$

Here $\psi_\lambda(i)$ denotes the one-electron wavefunction of the $\lambda$th level, which depends on the coordinates indicating both the spatial position $\boldsymbol{r}_i$ and the spin direction $s_i$ of the $i$th electron, and forms an orthonormal set:

$$\langle\psi_\lambda|\psi_\nu\rangle = \sum_{s_i}\int \psi_\lambda^*(i)\psi_\nu(i)d\boldsymbol{r}_i = \delta_{\lambda\nu}. \tag{2.13}$$

Using (2.12) as a trial function, we calculate the expectation value $\langle\Psi|\mathcal{H}|\Psi\rangle$ and get

$$\begin{aligned}\langle\Psi|\mathcal{H}|\Psi\rangle = &\sum_{\lambda=1}^{N}\langle\psi_\lambda|\mathcal{H}_0|\psi_\lambda\rangle + \frac{1}{2}\sum_{\lambda,\nu}\langle\psi_\lambda\psi_\nu|U|\psi_\lambda\psi_\nu\rangle \\ &-\frac{1}{2}\sum_{\lambda,\nu}\langle\psi_\lambda\psi_\nu|U|\psi_\nu\psi_\lambda\rangle, \end{aligned} \tag{2.14}$$

where we divided the Hamiltonian into the one-electron part $\mathcal{H}_0$ and the electron–electron Coulomb interaction $U$ as follows:

$$\mathcal{H} = \sum_i \mathcal{H}_0(i) + \frac{1}{2}\sum_{i,j} U(i,j). \tag{2.15}$$

The one-electron part is composed of the kinetic energy and the nucleus Coulomb potential and is given by

$$\mathcal{H}_0(i) = -\frac{1}{2}\nabla_i^2 + v(\boldsymbol{r}_i) \tag{2.16}$$

with

$$v(\boldsymbol{r}_i) = -\sum_j \frac{Z_j}{|\boldsymbol{r}_i - \boldsymbol{R}_j|}. \tag{2.17}$$

Here $Z_j$ is the nucleus charge of the $j$th atom. The electron–electron Coulomb interaction is given by

$$U(i,j) = \frac{1}{|\boldsymbol{r}_i - \boldsymbol{r}_j|}. \tag{2.18}$$

Using a Lagrange multiplier to guarantee the normalization condition (2.13), we solve the variational problem (2.14) to obtain

$$\begin{aligned}\sum_{\lambda=1}^{N}\langle\delta\psi_\lambda|\mathcal{H}_0|\psi_\lambda\rangle + \sum_{\lambda,\nu}\langle\delta\psi_\lambda\psi_\nu|U|\psi_\lambda\psi_\nu\rangle - \sum_{\lambda,\nu}\langle\delta\psi_\lambda\psi_\nu|U|\psi_\nu\psi_\lambda\rangle \\ - \sum_\lambda \varepsilon_\lambda\langle\delta\psi_\lambda|\psi_\lambda\rangle = 0,\end{aligned} \tag{2.19}$$

where $\varepsilon_\lambda$ ($\lambda = 1, 2, \ldots, N$) denotes the Lagrange multiplier. To satisfy (2.19) for an arbitrary variation $\langle\delta\psi_\lambda|$, the one-electron wavefunction $\psi_\lambda$ should satisfy

$$\begin{aligned}\mathcal{H}_0\psi_\lambda(i) + \left[\sum_{\nu=1}^{N}\sum_{s_j}\int \psi_\nu^*(j)U(i,j)\psi_\nu(j)d\boldsymbol{r}_j\right]\psi_\lambda(i) \\ -\left[\sum_{\nu=1}^{N}\sum_{s_j}\int \psi_\nu^*(j)U(i,j)\psi_\lambda(j)d\boldsymbol{r}_j\right]\psi_\nu(i) = \varepsilon_\lambda\psi_\lambda(i).\end{aligned} \tag{2.20}$$

This equation and the approximation used to derive it (i.e. the use of a single Slater determinant (2.12) to express the many-electron wavefunction) are called, respectively, the "Hartree–Fock equation" and the "Hartree–Fock approximation". Thus, the Hartree–Fock equation is no less than the rigorous one-electron approximation for interacting fermions which takes into account the antisymmetry of the wavefunction. In the Hartree–Fock equation, there are two contributions from the electron–electron Coulomb interaction $U(i, j)$. One is the "Coulomb term" (or the "direct term" or the "Hartree term" [14]) which is independent of the symmetry of the wavefunction and is given by the second term on the left-hand side of (2.20),

$$I(i)\psi_\lambda(i) = \sum_{\nu=1}^{N}\sum_{s_j}\int \frac{\psi_\nu^*(j)\psi_\nu(j)}{|\boldsymbol{r}_i - \boldsymbol{r}_j|} d\boldsymbol{r}_j \psi_\lambda(i). \tag{2.21}$$

The operator $I(i)$ is called the Hartree operator. The other is the "exchange term" (or "Fock term" [15]), which is given by the third term on the left-hand side of the same equation,

$$J(i)\psi_\lambda(i) = -\sum_{\nu=1}^{N}\sum_{s_j}\int \frac{\psi_\nu^*(j)\psi_\lambda(j)}{|\boldsymbol{r}_i - \boldsymbol{r}_j|} d\boldsymbol{r}_j \psi_\nu(i). \tag{2.22}$$

The operator $J(i)$ is called the Fock operator. The exchange term stems from the antisymmetry of the wavefunction with respect to two-particle permutation (exchange), because of Pauli's exclusion principle. Using the above notations and introducing the Hartree–Fock operator

$$\mathcal{H}_{\mathrm{HF}}(i) = \mathcal{H}_0(i) + I(i) + J(i), \tag{2.23}$$

we can rewrite the Hartree–Fock equation (2.20) as

$$\mathcal{H}_{\mathrm{HF}}(i)\psi_\lambda(i) = \varepsilon_\lambda \psi_\lambda(i). \tag{2.24}$$

In (2.20), the Coulomb term and the exchange term exactly cancel each other for $\nu = \lambda$, and therefore the summation with respect to $\nu$ also includes $\lambda$. We call a similar approximation ignoring the antisymmetry which led to to the exchange term the "Hartree approximation". The Hartree equation is given by (2.20) with the Coulomb term (without the self-interaction $\nu = \lambda$, to be precise) but without the exchange term. Since, in (2.20), $U$ does not depend on spin, the summations with respect to the spin coordinates in the Coulomb and exchange terms can be performed easily. Introducing the spin function $\theta_\lambda(s_i)$, we may write the one-electron wavefunction as

$$\psi_\lambda(i) = \psi_\lambda(\boldsymbol{r}_i, s_i) = \phi_\lambda(\boldsymbol{r}_i)\theta_\lambda(s_i). \tag{2.25}$$

Then, replacing $\boldsymbol{r}_i$ and $\boldsymbol{r}_j$ by $\boldsymbol{r}$ and $\boldsymbol{r}'$, (2.20) can be rewritten as

$$\mathcal{H}_0\phi_\lambda(\boldsymbol{r}) + \sum_{\nu=1}^{N} \int \phi_\nu^*(\boldsymbol{r}')U(\boldsymbol{r}-\boldsymbol{r}')\phi_\nu(\boldsymbol{r}')d\boldsymbol{r}'\phi_\lambda(\boldsymbol{r})$$
$$- \sum_{\nu=1(\lambda,\nu \text{ spin-parallel})}^{N} \int \phi_\nu^*(\boldsymbol{r}')U(\boldsymbol{r}-\boldsymbol{r}')\phi_\lambda(\boldsymbol{r}')d\boldsymbol{r}'\phi_\nu(\boldsymbol{r}) = \varepsilon_\lambda\phi_\lambda(\boldsymbol{r}), \tag{2.26}$$

where we put $U(\boldsymbol{r}-\boldsymbol{r}') = 1/|\boldsymbol{r}-\boldsymbol{r}'|$. The term which has been omitted in the rigorous Hartree–Fock approximation is called the "electron correlation" in the very strict sense of the term. We use the term "exchange–correlation" to express the electron correlation together with the exchange contribution.

Multiplying both sides of (2.26) by $\phi_\lambda^*(\boldsymbol{r})$ from the left-hand side and integrating with respect to the spatial coordinate $\boldsymbol{r}$ (and summing up with respect to the spin coordinate $s$), one obtains

$$\varepsilon_\lambda = \int \phi_\lambda^*(\boldsymbol{r})\mathcal{H}_0(\boldsymbol{r})\phi_\lambda(\boldsymbol{r})d\boldsymbol{r} + I_\lambda + J_\lambda, \tag{2.27}$$

where $I_\lambda$ and $J_\lambda$ denote the Coulomb and exchange integrals which are given, respectively, by

$$I_\lambda = \sum_{\nu=1}^{N} \int\int \frac{\phi_\lambda^*(\boldsymbol{r})\phi_\nu^*(\boldsymbol{r}')\phi_\lambda(\boldsymbol{r})\phi_\nu(\boldsymbol{r}')}{|\boldsymbol{r}-\boldsymbol{r}'|} d\boldsymbol{r}d\boldsymbol{r}'. \tag{2.28}$$

and

$$J_\lambda = - \sum_{\nu=1(\lambda,\nu \text{ spin-parallel})}^{N} \int\int \frac{\phi_\lambda^*(\boldsymbol{r})\phi_\nu^*(\boldsymbol{r}')\phi_\nu(\boldsymbol{r})\phi_\lambda(\boldsymbol{r}')}{|\boldsymbol{r}-\boldsymbol{r}'|} d\boldsymbol{r}d\boldsymbol{r}'. \tag{2.29}$$

In the Hartree–Fock approximation, the total potential energy is given by

$$\langle \hat{V} \rangle = \sum_{\lambda}^{N} \left[ \frac{1}{2}(I_\lambda + J_\lambda) + \int \phi_\lambda^*(\boldsymbol{r})v(\boldsymbol{r})\phi_\lambda(\boldsymbol{r})d\boldsymbol{r} \right], \tag{2.30}$$

and the total kinetic energy is given by

$$\langle \hat{T} \rangle = -\frac{1}{2}\sum_{\lambda}^{N} \int \phi_\lambda^*(\boldsymbol{r})\nabla^2\phi_\lambda(\boldsymbol{r})d\boldsymbol{r}. \tag{2.31}$$

Then, it is easy to show that the virial theorem $2\langle \hat{T} \rangle + \langle \hat{V} \rangle = 0$ exactly holds in a similar way explained in the previous section using (2.9) and (2.10), if the nuclear positions are fully relaxed at the stable geometry.

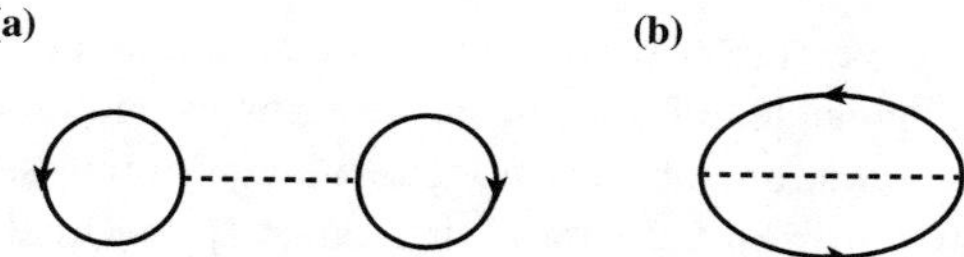


**Fig. 2.1** Diagrammatic representations for **a** the Coulomb integral $I_\lambda$ and **b** the exchange integral $J_\lambda$ in the Hartree–Fock approximation

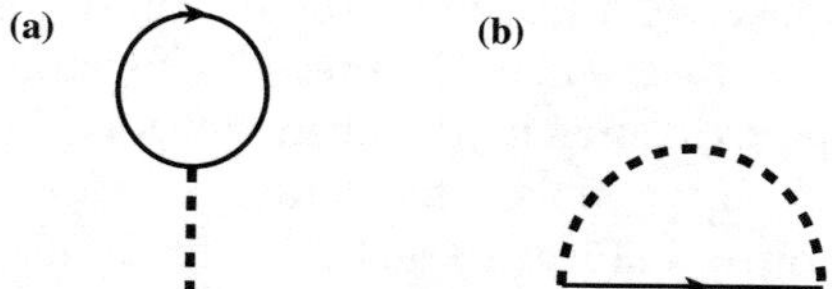


**Fig. 2.2** Diagrammatic representations for **a** the Coulomb operator $I(i)$ and **b** the exchange operator $J(i)$ in the Hartree–Fock approximation

The integrals $I_\lambda$ and $J_\lambda$ also appear in the first-order perturbation expansion of the energy eigenvalue of the $\lambda$th level with respect to the electron–electron Coulomb interaction. Perturbation theory in many-body quantum systems can be conveniently formulated in terms of "Feynman diagrams" [16, 17]. When one considers such graphical perturbation theory, it is worth mentioning the existence of the "linked-cluster theorem" [18]. This theorem tells us that, in each order in a perturbation expansion, diagrams composed of more than two isolated, unlinked parts should not appear in the final expression. Only the "linked-cluster diagrams" can give the correct linear-$N$ dependence of extensive physical quantities on the system size $N$. The specific approach of such graphical perturbation theory to the systems of interacting electrons has been considered by many authors [19–21]; Here, the electron-electron Coulomb interaction, i.e., the second term of (2.3), is regarded as a perturbation, and the unperturbed Hamiltonian is comprised of the kinetic energy and the nuclear potential, i.e., the first and third terms of (2.3). So, the unperturbed system is the noninteracting inhomogeneous system. Its ground-state (or excited-state) $N$-electron wavefunction is represented by the Slater determinant made of the one-particle wavefunctions, which are the solutions of $\mathcal{H}_0(i)\psi_\lambda(i) = \varepsilon_\lambda^0\psi_\lambda(i)$. For more details, see Sect. 2.7. Using Feynman diagrams, one may express diagrammatically the Coulomb integral $I_\lambda$ and the exchange integral $J_\lambda$, respectively, by Fig. 2.1a, b. In these figures, the dotted lines denote the bare Coulomb interaction $U(\boldsymbol{r}, \boldsymbol{r}')$, while the solid lines with an arrow denote the one-particle Green's function $G(\boldsymbol{r}, \boldsymbol{r}'; t - t')$, which represents the electron propagation from $\psi_\lambda(\boldsymbol{r}')$ (at time $t'$) to $\psi_\lambda(\boldsymbol{r})$ (at time $t$) (for this reason, the Green's function is also called the propagator). Its Fourier transform from $t - t'$ to $\omega$ is defined as (2.224) in Sect. 2.3.8, (2.447) in Sect. 2.7.2, or (2.474) in Sect. 2.7.3. On the other hand, the Hartree and Fock operators, $I(i)$ and $J(i)$, defined by (2.21) and (2.22) are represented, respectively, by Fig. 2.2a, b. These parts are called the self-energy; see Sect. 2.7 for details.

However, since one-electron wavefunctions which appear in the Hartree–Fock theory are not the eigenstates of $\mathcal{H}_0$ but the self-consistent solutions of (2.26), the Hartree–Fock approximation for inhomogeneous electron systems, where $v(\boldsymbol{r})$ is not uniform, is different from the simple first-order approximation in the perturbation expansion in power series of the electron–electron Coulomb interaction. The only exception is the case of the "jellium model" (or the "electron gas"), where the Coulomb potential caused by nuclei $v(\boldsymbol{r})$ is replaced by a uniform background of positive charge. In this case, the one-electron wavefunction of the Hartree–Fock equation becomes a plane wave, and the Hartree–Fock approximation can be identified as the first-order approximation in perturbation theory. Nevertheless, in general inhomogeneous systems, if we replace the zeroth order Green's function, i.e., the thin solid lines with an arrow in Figs. 2.1 and 2.2, with the clothed Green's function, i.e., the thick solid lines with an arrow as in the first two terms in Figs. 2.21 and 2.22 of Sect. 2.7.2, which is the Green's function constructed by the eigenfunctions $\phi_\lambda(\boldsymbol{r})$ of the Hartree–Fock equation (2.26), Fig. 2.1a, b become the true Hartree and Fock energies ((2.464) in Sect. 2.7.2), and Fig. 2.2a, b become the Hartree and Fock self-energies; see Sect. 2.7.2.

It is possible to evaluate the exchange term in the Hartree–Fock approximation, although it is more time consuming compared to the DFT approaches described in the next section. In any method, it is necessary to solve the basic equation (2.26) self-consistently, and a technique to obtain the self-consistent solution is called the "self-consistent-field (SCF) approach" [27, 28]. For molecules, the Hartree–Fock approximation has been used mainly in the field of quantum chemistry, where each one-electron wavefunction is approximated by a superposition (linear combination) of several "atomic orbitals"; either "Slater-type orbitals" (STO) [22], which have the form of $r^n \exp(-\alpha r)$, "Gaussian-type orbitals" (GTO) [23–25], which have the form of $\exp(-\alpha r^2)$, or "numerical atomic orbitals" (NAO) [26] are used. One possible simplification in terms of STOs has been formulated by Roothaan [29]. In the case of the STO basis, two-center integrals between atomic orbitals centered at different positions or between an atomic orbital and the Coulomb potential can be evaluated analytically [30, 31], while in the case of the GTO basis, even three-center integrals can be evaluated analytically for Coulomb interactions. The method expressing the one-electron wavefunction in terms of a linear combination of atomic orbitals is called the LCAO approach or the molecular-orbital (MO) approach (the term, molecular-orbital, is used mainly for isolated molecules and clusters in chemistry). However, when one says just LCAO or MO approach, it indicates the use of atomic orbitals, but it does not necessary mean the Hartree–Fock method.

### 2.2.4 Electron Correlations Beyond the HF Approximation

Here, we comment on the problem of electron correlations beyond the Hartree–Fock (HF) approximation [32–34]. Let us consider the simplest example of a hydrogen molecule, $H_2$. When the intermolecular distance is small, the two electrons move back and forth between the two atoms so that the Hartree–Fock picture holds well. This picture is called the "itinerant-electron picture" when one talks about the problem of electron correlations. (The Hartree–Fock approach is not necessarily restricted to isolated molecules or clusters but is also applicable to crystals as has been demonstrated by Fulde and Stoll's group [35–38].)

In this case, the 1s bonding orbital is doubly occupied and the 1s antibonding orbital is unoccupied. On the other hand, when the intermolecular distance increases, the electrons tend to localize on each atom so that the above picture of the bonding orbital, in which two electrons move between the two atoms, becomes no longer correct. This localization is mainly due to the cost of the Coulomb repulsive interaction energy for two electrons occupying the same atom, and therefore the direct consequence of the electron correlation. In this case, the Heitler–London picture is more appropriate.

When the distance $r_{AB}$ between the two hydrogen atoms is large enough, all electronic levels of the hydrogen molecule have a twofold degeneracy and the wavefunctions are given by $\phi_{1s_A}(1)\phi_{1s_B}(2)$ and $\phi_{1s_A}(2)\phi_{1s_B}(1)$ or any two independent linear combinations of these two. Here $\phi_{1s_A}(n)$ represents a hydrogenlike 1s wavefunction for electron $n = 1, 2$ about nucleus $A$. This suggests that for large values of $r_{\mathrm{AB}}$ we should use a linear combination of these two product functions as a variational function. Such a treatment of a hydrogen molecule is known as the Heitler–London theory [39, 40]. It is clear that this theory gives the correct description of the system when $r_{\mathrm{AB}}$ becomes infinite, in contrast to the Hartree–Fock theory which does not give this expected behavior.

This picture is called the "localized-electron picture" when we think about electron correlations. An insulator in which the localized electron picture holds is called a "Mott insulator". There are several examples of Mott insulators, the compounds such as NiO containing a transition-metal element, perovskite-structure insulators such as $YTiO_3$ containing a rare-earth-metal element, and semiconductors with impurities at certain concentrations. If one could gradually increase the strength of the electron-electron Coulomb interaction, one would expect a phase transition ("metal-insulator transition") from the metallic phase to the insulating phase. Such a phase transition is called a "Mott transition". The metal-insulator transition occurring in semiconductors when one changes the impurity concentration can be regarded as an example of the Mott transition [41]. NiS undergoes a Mott transition with a 2% volume change (without change in the crystal symmetry) at a certain temperature. In the recently discovered copper oxide series of high-$T_c$ superconductors, a state close to a Mott insulator is realized with a small number of carriers. This fact has attracted much current interest in relation to the mechanism of high-$T_c$ superconductivity [32]. A related topic is Anderson localization [42], which stems from a random configuration of atoms and exists without any correlation effect.

It is possible to take the electron correlation into account by introducing the so-called "configuration interaction" (CI) [43], which tries to represent the many-electron wavefunction in terms of a linear combination of many Slater determinants corresponding to the excited states $\Psi^{(0)}_{\Lambda\to\Gamma}$, in which the one-electron levels $\Lambda$ are excited to the one-electron levels $\Gamma$, as well as the Slater determinant corresponding to the ground state $\Psi_0^{(0)}$. Such a technique is nowadays generally used to improve the result of the Hartree–Fock approximation. In the singly excited CI (SCI), one considers configurations $\Psi_{\lambda\to\gamma}^{(0)}$, in which one empty level $\lambda$ is changed by an empty level $\gamma$. However, these configurations do not couple directly with the ground state $\Psi_0^{(0)}$, because of the Brillouin theorem [44] stating that $\langle\Psi_0^{(0)}|\mathcal{H}|\Psi_{\lambda\to\gamma}^{(0)}\rangle = 0$. Therefore, the first nontrivial CI starts from the doubly excited CI (DCI), in which configurations with two excited levels, $\Psi_{\lambda\nu\to\gamma\delta}$, are considered. However, since the DCI is not based on the linked-cluster theorem in perturbation theory, there is a problem such that the result for the extensive physical quantities obtained by the CI is not necessarily proportional to the number of noninteracting, e.g., He atoms $N$ in the system under consideration; for example, see [27].

If one includes all possible excitations within the Fock space spanned by the basis set, which is called the full CI, such a problem of the size-extensivity or the size-consistency does not occur and the linear scaling can be expected. Another way to remove this problem to confine the "active space" in the one-electron levels. In contrast, in the multi-reference CI (MRCI), one considers excitations not only from the ground state but also from some excited reference states. The most useful multi-reference CI is probably the multi-reference single and double CI (MRDCI) [45]. Although it is still not size-extensive, but the problem is much reduced; the results are enough accurate and reliable. Another possible treatment is to confine "active space" in the one-electron levels and consider all possible excitations in this active space. The active orbitals can be fixed as the same as the HF orbitals or can be relaxed so as to minimize the ground-state total energy. The latter is referred to as the complete active space (CAS) and its self-consistent-field calculation is called the CASSCF calculation [46, 47]. The accuracy of the CASSCF calculation is generally quite high but quite expensive in computation.

There is a different methodology which is based on perturbation theory and considered first by Møller and Plesset [48] in 1934. They assumed the Hartree–Fock equation as an unperturbed system and treated the difference between the Hartree–Fock operator (2.23) and the true Hamiltonian as a perturbation. This special choice of unperturbed Hamiltonian in general (Rayleigh–Schrödinger) perturbation theory is called the Møller–Plesset (MP) perturbation theory, and is used also very popularly among chemists. Any $N \times N$ Slater determinant composed from the solution $\psi_\lambda(i)$ of the Hartree–Fock equation (2.24) is the $N$-body wavefunction in the Hartree-Fock theory. One is the ground-state wavefunction $\Psi_0^{(0)}$ having an energy $E_0^{(0)}$, and all the remaining $\Psi_\Lambda^{(0)}$ $\Psi_\Lambda^{(0)}$ correspond to the excited states, if the ground state is not degenerate. All the excited-state wavefunctions can be used in the perturbation expansion. In the frozen core approximation, one may neglect the excitation from core orbitals. Let us assume that we are calculating the correction to the ground-state total energy. The first-order correction is exactly zero ($E_0^{(1)} = 0$), because the expectation value of the true Hamiltonian $\langle\Psi_0^{(0)}|\mathcal{H}|\Psi_0^{(0)}\rangle$ is the same as that of the Hartree–Fock operator. The second-order correction to the ground-state total energy is given by

$$E_0^{(2)} = \sum_{\Lambda(\neq 0)} \frac{|\langle \Psi_\Lambda^{(0)} | U | \Psi_0^{(0)} \rangle|^2}{E_0^{(0)} - E_\Lambda^{(0)}}$$
$$= \sum_{\lambda' < \nu'}^{\mathcal{E}} \sum_{\lambda < \nu}^{O} \frac{|\langle \psi_{\lambda'} \psi_{\nu'} | U | \psi_\lambda \psi_\nu \rangle - \langle \psi_{\lambda'} \psi_{\nu'} | U | \psi_\nu \psi_\lambda \rangle|^2}{\varepsilon_\lambda + \varepsilon_\nu - \varepsilon_{\lambda'} - \varepsilon_{\nu'}}, \quad (2.32)$$

where the summations with symbols $\mathcal{E}$ and $O$ mean, respectively, those running over the empty states and the occupied states. The calculational scheme using only (2.32) is referred to as the MP2. Similarly, the third-order and the fourth-order schemes which are referred to as the MP3 and MP4, respectively, can be used as well [49, 50]. Note that $E_0^{(2)}$ is negative, because $E_\Lambda^{(0)} > E_0^{(0)}$. Therefore the ground-state energy decreases always by means of the MP2. However, it is not guaranteed that the perturbation series converge rapidly, in particular when the difference between the true ground state and the Hartree–Fock ground state is large. Moreover the MP perturbation theory is not generally applicable to the calculations of excited states [51]; for such a purpose, the CI is commonly used.

There are two other famous, sophisticated theories to be mentioned here which treat also the electron correlation (i.e. go beyond the Hartree–Fock theory) but still guarantee the linear-$N$ dependence of all the extensive physical quantities. One is the independent electron-pair (IEP) approximation (IEPA) [52] and the other is the coupled cluster (CC) theory [53]. These theories can be reformulated by means of the linked-cluster expansion in perturbation theory; see for reviews, [54–56].

The Hartree–Fock approximation is more tractable for a system of a small number of atoms and becomes more complex for crystals. Moreover, one encounters several problems if one uses the Hartree–Fock approximation in crystals. For example, if one uses this approximation in metals, one obtains a zero density of states at the Fermi level. And, if one uses it for semiconductors, the energy gap between the occupied and unoccupied states is overestimated. The reason lies in the fact that the Hartree–Fock approximation cannot describe electronic screening properly; therefore, it is necessary to include correlations, such as in the random-phase approximation (RPA), to recover the correct results. But for inhomogeneous systems, this task becomes quite complicated.

### 2.2.5 *Density Functional Theory*

As an alternative to the Hartree–Fock approximation, one may consider replacing the electron–electron interactions by some effective potential acting on the electrons. At a first glance, such a formulation might seem a very crude approximation. However, as will be discussed in this section, such an alternative method can be formulated quite exactly on the basis of density functional theory, which is usually abbreviated as DFT.

In this theory, the exchange–correlation energy is expressed, at least formally as a functional of the resulting electron charge density distribution, and the electronic states are solved for self-consistently as in the Hartree or Hartree–Fock approximation. The so-called "self-consistent band calculation" in solid-state physics [57–59] corresponds to this method.

In 1964, Hohenberg and Kohn [60] considered the ground state of the electron gas system in the external potential $v(\boldsymbol{r})$, and proved that the following density functional theory (DFT) holds exactly: There is a universal (i.e. independent of the external potential $v(\boldsymbol{r})$) functional $F[\rho(\boldsymbol{r})]$ of the electron charge density distribution $\rho(\boldsymbol{r})$ that defines the total energy of the electronic system by

$$E = \int v(\boldsymbol{r})\rho(\boldsymbol{r})d\boldsymbol{r} + F[\rho(\boldsymbol{r})]. \tag{2.33}$$

The total energy $E$ has a minimum when the charge density $\rho(\boldsymbol{r})$ coincides with the true charge density distribution in the external potential $v(\boldsymbol{r})$. This DFT is exact, if the ground state has no degeneracy.

The proof of this theorem is rather simple and given as follows. The electronic charge density distribution $\rho(\boldsymbol{r})$ is related to the true $N$-body wavefunction $\Psi(\boldsymbol{r}_1, \boldsymbol{r}_2, \ldots, \boldsymbol{r}_N)$ via

$$\begin{aligned}\rho(\boldsymbol{r}) &= \langle\Psi|\sum_{i=1}^{N}\delta(\boldsymbol{r}-\boldsymbol{r}_i)|\Psi\rangle \\ &= \sum_{i=1}^{N}\int \Psi^*(\boldsymbol{r}_1, \boldsymbol{r}_2, \ldots, \boldsymbol{r}_N)\delta(\boldsymbol{r}-\boldsymbol{r}_i)\Psi(\boldsymbol{r}_1, \boldsymbol{r}_2, \ldots, \boldsymbol{r}_N)d\boldsymbol{r}_1, d\boldsymbol{r}_2\cdots d\boldsymbol{r}_N.\end{aligned} \tag{2.34}$$

Then the expectation value for the one-body potential $v(\boldsymbol{r})$ is given by

$$\sum_{i=1}^{N}\langle\Psi|v(\boldsymbol{r}_i)|\Psi\rangle = \int \rho(\boldsymbol{r})v(\boldsymbol{r})d\boldsymbol{r}. \tag{2.35}$$

First, it is easy to conceive from the structure of the Schrödinger equation (2.1) with (2.3) that the true wavefunction $\Psi$ and the charge density distribution $\rho(\boldsymbol{r})$ are, in general, unique functionals of the external potential $v(\boldsymbol{r})$. Conversely, one may ask if the potential $v(\boldsymbol{r})$ is a unique functional of the charge density distribution $\rho(\boldsymbol{r})$. To answer this question, let us suppose that a different potential $v'(\boldsymbol{r})$ gives the same ground-state charge density distribution $\rho(\boldsymbol{r})$. Let the Hamiltonians whose potentials are given by $v(\boldsymbol{r})$ and $v'(\boldsymbol{r})$ be $\mathcal{H}$ and $\mathcal{H}'$, respectively. Similarly, let us write their eigenvalues and eigenfunctions as $E$, $E'$, $\Psi$ and $\Psi'$. We have $\mathcal{H}\Psi = E\Psi$ and $\mathcal{H}'\Psi' = E'\Psi'$. From the variational principle for the ground state, we expect

$$\begin{aligned} E' &= \langle \Psi' | \mathcal{H}' | \Psi' \rangle < \langle \Psi | \mathcal{H}' | \Psi \rangle \\ &= \langle \Psi | \mathcal{H} | \Psi \rangle + \int [v'(\boldsymbol{r}) - v(\boldsymbol{r})] \rho(\boldsymbol{r}) d\boldsymbol{r} \\ &= E + \int [v'(\boldsymbol{r}) - v(\boldsymbol{r})] \rho(\boldsymbol{r}) d\boldsymbol{r} \end{aligned} \tag{2.36}$$

and

$$\begin{aligned} E &= \langle \Psi | \mathcal{H} | \Psi \rangle < \langle \Psi' | \mathcal{H} | \Psi' \rangle \\ &= \langle \Psi' | \mathcal{H}' | \Psi' \rangle + \int [v(\boldsymbol{r}) - v'(\boldsymbol{r})] \rho(\boldsymbol{r}) d\boldsymbol{r} \\ &= E' - \int [v'(\boldsymbol{r}) - v(\boldsymbol{r})] \rho(\boldsymbol{r}) d\boldsymbol{r}. \end{aligned} \tag{2.37}$$

Here we have assumed $\rho(\boldsymbol{r}) = \rho'(\boldsymbol{r})$. Then these two equations lead to the following contradictory result:

$$E + E' < E + E'. \tag{2.38}$$

Thus we find that the initial assumption was wrong and that $v(\boldsymbol{r})$ is a unique functional of $\rho(\boldsymbol{r})$. Since $v(\boldsymbol{r})$ determines the form of the Hamiltonian $\mathcal{H}$ uniquely, then the wavefunction $\Psi$ in the ground state is itself a unique functional of $\rho(\boldsymbol{r})$. If we write the Hamiltonian in the absence of the external field, i.e., the Hamiltonian of the electron gas, as $\mathcal{H}_0$, then the value

$$F[\rho(\boldsymbol{r})] = \langle \Psi | \mathcal{H}_0 | \Psi \rangle \tag{2.39}$$

turns out to be a unique functional of $\rho(\boldsymbol{r})$. Next, how can we prove that the energy given by (2.33) is minimum when $\rho(\boldsymbol{r})$ coincides with the true charge density in the ground state? We know from the variational principle that the energy (2.33) is minimum if $\Psi$ coincides with the true wavefunction in the ground state. As we showed above, $\Psi$ is a unique functional of $\rho(\boldsymbol{r})$, and therefore we come to the conclusion that when $\rho(\boldsymbol{r})$ coincides with the ground-state charge density, the energy (2.33) is minimum. Here we note that the total energy functional (2.33) is a Legendre transform of the universal function $F[\rho]$. This is the other reason why, when we work with the fixed external potential $v(\boldsymbol{r})$, we should minimize $E[\rho]$ to obtain the ground state. $F[\rho(\boldsymbol{r})]$ being universal, one can in principle calculate it for the uniform electron gas, as done by Ceperley and Alder [61] using a quantum Monte Carlo (QMC) method.

Instead of the ground state, we could also consider the grand canonical ensemble with the temperature and the chemical potential held constant. In the grand canonical ensemble, the thermodynamic potential $\Omega$ satisfies a variational principle. This principle states that $\Omega$ is minimum if the density matrix $\hat{\rho}$ used coincides with the true one. That is, if we introduce the thermodynamic potential $\Omega$

$$\Omega[\hat{\rho}] = \mathrm{Tr}\,\hat{\rho}\left(\mathcal{H} - \mu N + \frac{1}{\beta}\ln\hat{\rho}\right) \tag{2.40}$$

as a functional of the density matrix $\hat{\rho}$ (here, $\mu$ is the chemical potential, $\beta = 1/k_B T$, $k_B$ is the Boltzmann constant and $T$ is the temperature), it has its minimum when $\hat{\rho}$ coincides with the true density matrix expected in the grand canonical ensemble,

$$\hat{\rho}_0 = \mathrm{e}^{-\beta(\mathcal{H}-\mu N)} / \mathrm{Tr}\,\mathrm{e}^{-\beta(\mathcal{H}-\mu N)}. \tag{2.41}$$

The potential is then given by

$$\Omega[\hat{\rho}_0] = -\frac{1}{\beta}\ln\mathrm{Tr}\,\mathrm{e}^{-\beta(\mathcal{H}-\mu N)}. \tag{2.42}$$

This variational property at finite temperatures is similar to the variational principle for the ground state. Note that the charge density $\rho(\boldsymbol{r})$ and the density matrix are related via

$$\rho(\boldsymbol{r}) = \mathrm{Tr}\,\hat{\rho}_0\hat{n}(\boldsymbol{r}), \tag{2.43}$$

where $\hat{n}(\boldsymbol{r}) = \psi^\dagger(\boldsymbol{r})\psi(\boldsymbol{r})$ is the number density operator in the real-space representation. In 1965, Mermin [62] used this property to prove the following theory by extending Hohenberg and Kohn's DFT to the case of the grand canonical ensemble: For an electron gas system in an external field $v(\boldsymbol{r})$ in the grand canonical ensemble, there is a universal functional $F[\rho(\boldsymbol{r})]$ of the electron charge density distribution $\rho(\boldsymbol{r})$ (universal means that it does not depend on the form of the external potential $v(\boldsymbol{r})$), such that the thermodynamic potential for the electron system is given by

$$\Omega = \int v(\boldsymbol{r})\rho(\boldsymbol{r})d\boldsymbol{r} + F[\rho(\boldsymbol{r})], \tag{2.44}$$

and this $\Omega$ has a minimum value when $\rho(\boldsymbol{r})$ coincides with the true charge density distribution in the presence of the external potential $v(\boldsymbol{r})$.

DFT can be extended also to the time-dependent theory where the Hamiltonian depends on time $t$ as was first formulated by Runge and Gross [63]. In the time-dependent case, however, the variational principle no longer holds and the Hamiltonian functional depends not only on the charge density $\rho(\boldsymbol{r}, t)$ but also on the wavefunction $\Psi(t_0)$ at the initial time $t_0$. Of course, if the initial state at $t = t_0$ is the ground state, the functional is determined uniquely only by the charge density $\rho(\boldsymbol{r}, t)$, since the wavefunction at $t_0$ is a unique functional of the charge density at $t = t_0$ by virtue of the time-independent density functional theory described above. The proof of the time-dependent density functional theory (TDDFT) is not based on the variational principle but on the fact that both the external potential $v(\boldsymbol{r}, t)$ and the paramagnetic current density operator

$$j(\boldsymbol{r}, t) = \frac{1}{2i} \sum_i \left[ \nabla \delta(\boldsymbol{r} - \boldsymbol{r}_i) + \delta(\boldsymbol{r} - \boldsymbol{r}_i) \nabla \right] \tag{2.45}$$

can be expanded in Taylor series in powers of $t - t_0$ around the initial time $t_0$. Then one may use the continuity equation

$$\frac{\partial}{\partial t} \rho(\boldsymbol{r}, t) = -\nabla \cdot j(\boldsymbol{r}, t) \tag{2.46}$$

to prove the uniqueness of the Hamiltonian as a functional of the density $\rho(\boldsymbol{r}, t)$. Several other theories concerning the time dependence of the exchange–correlation energy,

$$\frac{dE_{\text{xc}}}{dt} = \int d\boldsymbol{r} \dot{\rho}(\boldsymbol{r}, t) \mu_{\text{xc}}(\boldsymbol{r}, t), \tag{2.47}$$

the virial theorem and the scaling of the coupling constant were derived by Hessler et al. [64].

Before ending this section, we comment on several simplified treatments of electronic structure calculation based on DFT.

It may be possible to approximate the density functional of the true electron charge distribution of a many-atom system by that of the superposition of the electron charges calculated rigorously for isolated atoms, if these two charge distributions are not very different. The Harris functional method [65] provides such an approach, based on a considerably simplified treatment of DFT. In this method, one needs to solve the Kohn–Sham equation only once, by using the superposition of the electron charges of isolated atoms as the input density. Therefore, no self-consistent iteration is required. This approximation reduces significantly the amount of computation, and is known to be not too bad for various metals and semiconductors [66], although sometimes differences are found between the results of the Harris functional and self-consistent methods [67–69]. On the basis of this fact, the Harris functional method is often referred to in the explanation of the non-self-consistency of the tight-binding approximation, which will be described in detail in the next chapter.

According to DFT, not only the exchange–correlation potential but also the electron kinetic energy could be expressed, basically, be expressed as a functional of electron density. There have been other recent developments in the field of kinetic-energy functionals [70], although they have not found wide use yet, because of some of their shortcomings. Once a full-functional form of the density functional is obtained for the Hamiltonian, the minimization of the total energy leads to a simple, closed equation for the electron charge density distribution. Thus, if one uses this method, it becomes no longer necessary to solve the Kohn–Sham equation for each wavefunction of each level. This method is very easy to handle, so that it would lead to results with less computational effort, especially in ab initio molecular-dynamics (MD) simulation, as well as in MD simulation based on the tight-binding method which will be described in the next chapter.

### 2.2.6 Local Density Approximation (LDA)

All the material presented above is just the general theory, and nothing has been mentioned about the form of the density functional $F[\rho(\boldsymbol{r})]$ itself in (2.33) and (2.44) for inhomogeneous systems. Kohn and Sham [71] were the first to present a strategy to calculate the electronic states of many-atom systems in the ground state according to DFT, although only approximately. Since then, their method has been called the "local-density approximation" (LDA) and is very widely used nowadays.

Here we derive the LDA according to Kohn and Sham's treatment using the grand canonical ensemble. If we write the electron kinetic energy as $T[\rho(\boldsymbol{r})]$, we have

$$\begin{aligned} F[\rho(\boldsymbol{r})] &= T[\rho(\boldsymbol{r})] + U[\rho(\boldsymbol{r})] + E_{\mathrm{xc}}[\rho(\boldsymbol{r})] - \mu \int \rho(\boldsymbol{r}) d\boldsymbol{r}, \\ U[(\rho(\boldsymbol{r})] &= \frac{1}{2} \int \frac{\rho(\boldsymbol{r})\rho(\boldsymbol{r}')}{|\boldsymbol{r} - \boldsymbol{r}'|} d\boldsymbol{r} d\boldsymbol{r}', \end{aligned} \tag{2.48}$$

where $E_{\mathrm{xc}}[\rho(\boldsymbol{r})]$ represents the exchange–correlation energy functional. Kohn and Sham's local-density approximation is to use the exchange–correlation energy of the homogeneous electron gas, evaluated from the charge density at the point $\boldsymbol{r}$ under consideration. That is, we first put $\rho = \rho(\boldsymbol{r})$ and forget about $\rho(\boldsymbol{r})$; then we replace the exchange–correlation energy at $\boldsymbol{r}$ with the corresponding value for the electron gas system which has a homogeneous charge density $\rho$ everywhere. This approximate treatment is valid whenever the inhomogeneity of the electron density distribution $\rho(\boldsymbol{r})$ due to the existence of the external potential $v(\boldsymbol{r})$ is small. The key approximation of the LDA is to keep this replacement of the exchange–correlation energy even when the inhomogeneity is strong. This treatment is essentially the local approximation for the exchange–correlation energy, and is therefore called the local-density approximation. Thus, in the LDA, the value of the exchange-correlation energy of the homogeneous electron gas of the charge density at that local point is used, as if this charge density is uniform around this area, regardless of the inhomogeneity.

The LDA does not, however, offer a method to study electronic properties in the presence of magnetically ordered structure. In order to treat magnetic ordering, it is necessary to treat the up-spin and down-spin states separately: For this purpose Barth and Hedin [72] formulated for the first time the local-spin-density approximation (LSDA). In 1976, Gunnarsson and Lundqvist [73] extended the LSDA to the grand canonical thermodynamic ensemble. They also pointed out that only the spherical average of the exchange–correlation hole (the hole appearing near the origin of the pair distribution function) influences the exchange–correlation potential, and the local approximation obeys a sum rule stating that the total amount of the exchange–correlation hole is equal to unity. In this sense, the LDA and LSDA are not bad approximations.

According to the variational principle of DFT, we take the variation of (2.33) or (2.44) with respect to the local density to obtain

$$\int \delta\rho(\boldsymbol{r})\left\{\frac{\delta T[\rho]}{\delta\rho(\boldsymbol{r})} + v(\boldsymbol{r}) + \int \frac{\rho(\boldsymbol{r}')}{|\boldsymbol{r}-\boldsymbol{r}'|}d\boldsymbol{r}' + \mu_{\mathrm{xc}}[\rho](\boldsymbol{r}) - \mu\right\}d\boldsymbol{r} = 0. \tag{2.49}$$

In taking the variation, we have to impose the constraint on the total number of electrons

$$\int \rho(\boldsymbol{r})d\boldsymbol{r} = N. \tag{2.50}$$

In (2.49), $\mu_{\mathrm{xc}}[\rho](\boldsymbol{r})$ is the local exchange–correlation potential

$$\mu_{\mathrm{xc}}[\rho](\boldsymbol{r}) = \frac{\delta E_{\mathrm{xc}}}{\delta\rho(\boldsymbol{r})}. \tag{2.51}$$

Without loss of generality, one can write the charge density as

$$\rho(\boldsymbol{r}) = \sum_{\lambda=1}^{N} |\psi_\lambda(\boldsymbol{r})|^2, \tag{2.52}$$

with $\psi_\lambda(\boldsymbol{r})$ satisfying orthonormality, since one is looking for a one-electron theory. Here, a factor of 2 is introduced in front of the summation in (2.52) in order to account for the spin multiplicity. Then, writing the electron kinetic energy as

$$T[\rho(\boldsymbol{r})] = -\frac{1}{2}\sum_{\lambda=1}^{N}\int \psi_\lambda^*(\boldsymbol{r})\nabla^2\psi_\lambda(\boldsymbol{r})d\boldsymbol{r}, \tag{2.53}$$

we find that the solution of (2.49) satisfying (2.50) and (2.52) can be obtained by solving the following effective one-electron Schrödinger equation for $\psi_\lambda$

$$\left\{-\frac{1}{2}\nabla^2 + v(\boldsymbol{r}) + \int \frac{\rho(\boldsymbol{r}')}{|\boldsymbol{r}-\boldsymbol{r}'|}d\boldsymbol{r}' + \mu_{\mathrm{xc}}[\rho](\boldsymbol{r})\right\}\psi_\lambda(\boldsymbol{r}) = \varepsilon_\lambda\psi_\lambda(\boldsymbol{r}), \tag{2.54}$$

where $\varepsilon_\lambda$ is the Lagrange multiplier introduced so as to guarantee the normalization

$$\int \psi_\lambda^*(\boldsymbol{r})\psi_\lambda(\boldsymbol{r})d\boldsymbol{r} = 1. \tag{2.55}$$

The charge density and the kinetic energy are then obtained from the solution via (2.52) and (2.53), respectively. The summation should be taken from the lower energy levels. Equation (2.54) is called the Kohn–Sham equation, and the effective one-electron Hamiltonian associated with (2.54) is called the Kohn–Sham Hamiltonian. Since the Kohn–Sham Hamiltonian is Hermitian, its eigenvalues (Lagrange multipliers) $\varepsilon_\lambda$ are all real.

Here, it is found that the eigenvalue $\varepsilon_\lambda$ of the highest occupied level is identical to the first ionization energy of the $N$-electron system, because $\Delta E/\Delta N =$

$\int(\Delta\rho/\Delta N)(\delta E/\delta\rho(\boldsymbol{r}))d\boldsymbol{r} = \int \psi_\lambda^* \mathcal{H}\psi_\lambda d\boldsymbol{r} = \varepsilon_\lambda$. This is well known as the Koopmans theorem [74]. On the basis of this theorem, the numbers $\varepsilon_\lambda$ are usually identified as the one-electron energy levels, although this statement is not generally correct, because the derivative with respect to $N$ is not equal to the finite difference with respect to $\Delta N = 1$. (see (2.83) and (2.84b) below in connection to Janak's theorem and also the quasiparticle (QP) theory related to the photoelectron spectroscopy explained in Sect. 2.7.) Equation (2.54) might be regarded as a one-electron approximation (or mean-field theory) including the many-particle interaction effectively, since the equation is just for a single electron existing in an effective potential created by the charge density of all the electrons in the system. However, the equation is only an approximation in the sense that we have used the local approximation for the exchange–correlation energy. Once again, one should note that there is no approximation in DFT itself!

Another remark should be made about the wavefunctions. All the above equations should be solved self-consistently. If this is done, it is certainly true that the wavefunction $\psi_\lambda$ is related to the electron charge density and electron kinetic-energy density via the relations (2.52) and (2.53). Nevertheless, there is no theoretical reason to regard the Slater determinant constructed from $\psi_\lambda$ as the true many-electron wavefunction. That is, $\psi_\lambda$ is not the one-electron wavefunction in the sense of the Hartree–Fock approximation, but an object (wavefunction or probability amplitude) which is more directly related to the true electron charge distribution. We will call $\psi_\lambda$ the effective one-electron wavefunction or the Kohn–Sham wavefunction.

The $X\alpha$ method proposed by Slater et al. [75] gives the simplest form for the local-density approximation to the exchange–correlation potential. His method is to assume the exchange–correlation potential to be proportional to the one-third power of the local charge density (from a dimensional analysis of the exchange term of the Hartree–Fock equation), and write it as $\mu_{\mathrm{xc}}[\rho] = -(3\alpha/2)(3\rho/\pi)^{1/3}$. The coefficient $\alpha$ can be determined from the known results for the electron gas, for example. We know that the electron gas system gives exactly the same form for the exchange term. If we consider the pure Fock term in the Hartree-Fock approximation of the electron gas system, we obtain $\alpha = 2/3$. On the other hand, Gaspar [76], followed by Kohn and Sham [71] applied the variational method to Thomas–Fermi atom and derived the same expression with $\alpha = 1$, which is now referred to as the Gaspar–Kohn–Sham formula. Empirically, one has to regard $\alpha$ as a parameter which may be somewhere in the middle between 2/3 and 1. Such a choice of the exchange–correlation potential is called Slater et al.'s $X\alpha$ method. For example, Kmetko [77] optimized the value of $\alpha$ for atoms. Other local-density functionals which are free of parameters have been proposed by Moruzzi, Janak and Williams [78], and Perdew and Zunger [79], by fitting QMC results of Ceperley and Alder [61], and have been used extensively by many people.

In electronic systems, in general, the screening effect becomes important when the electron density increases. When we consider the perturbation series expansion of the long-range electron–electron Coulomb interaction, we find that there are a special type of terms called the "ring diagrams", which become more divergent than the other types of terms (diagrams) in the high-density limit. These terms play a dominant role in the electronic screening, and can be resummed (even in the case of nonuniform systems; see also Sect. 2.7.3) to yield a finite result. This resummation procedure is equivalent to evaluate the inverse of the dielectric matrix. This approximation is often called the "random-phase approximation" (RPA) [17, 20]. In this way, the exchange–correlation energy of a high-density electron gas can be estimated from the RPA. On the other hand, in the low density limit, it is known that the electron gas forms a "Wigner solid", which is a lattice of isolated electrons. When the electron density is not very high and not very low, one needs a formula for the exchange–correlation energy, which interpolates the RPA energy and the energy of the Wigner solid. More accurate values for the exchange–correlation energy of the electron gas have been determined by quantum Monte Carlo simulation [61] and are widely used in treatments within the LDA.

Here we give Perdew–Zunger's interpolation formula for the LSDA [79] fitted to the Ceperley–Alder's QMC results [61] and Gell-Mann and Brueckner's perturbation results [19] in the high-density limit as an example. Let us define the radius $r_s = (3/4\pi\rho)^{1/3}$ (in units of a.u., i.e., Bohr) of the sphere whose volume is equal to the volume per electron, i.e., the total volume divided by the total number of electrons in the system. In a spin polarized system, $\zeta = (\rho_\uparrow - \rho_\downarrow)/\rho$ represents the polarization ratio. Then, the exchange energies for up- and down-spin electrons are given by

$$E_{x\uparrow} = -\frac{3}{2}\left(\frac{3\rho_\uparrow}{4\pi}\right)^{1/3}, \quad E_{x\downarrow} = -\frac{3}{2}\left(\frac{3\rho_\downarrow}{4\pi}\right)^{1/3}, \tag{2.56}$$

which correspond to the Fock energies in the Hartree–Fock approximation of the electron gas system and to $\alpha = 2/3$ in Slater's $X\alpha$ method. For low density regions, $r_s \geq 1$, the unpolarized and polarized parts of the correlation energies are given by

$$E_U = -\frac{0.1423}{1 + 1.0529\sqrt{r_s} + 0.3334 r_s}, \tag{2.57a}$$

$$E_P = -\frac{0.0843}{1 + 1.3981\sqrt{r_s} + 0.2611 r_s}, \tag{2.57b}$$

and those of the correlation potentials are given by

$$V_U = E_U \frac{1 + \frac{7 \times 1.0529}{6}\sqrt{r_s} + \frac{4 \times 0.3334}{3} r_s}{1 + 1.0529\sqrt{r_s} + 0.3334 r_s}, \tag{2.58a}$$

$$V_P = E_P \frac{1 + \frac{7 \times 1.3981}{6}\sqrt{r_s} + \frac{4 \times 0.2611}{3} r_s}{1 + 1.3981\sqrt{r_s} + 0.2611 r_s}. \tag{2.58b}$$

On the other hand, for high density regions, $r_s < 1$, these quantities are given by

$$E_U = 0.03110 \ln r_s - 0.0480 + 0.0020 r_s \ln r_s - 0.0116 r_s, \tag{2.59a}$$

$$E_P = 0.01555\sqrt{r_s} - 0.0269 + 0.0007 r_s^{3/2} - 0.0048 r_s, \tag{2.59b}$$

and

$$V_{\mathrm{U}} = E_{\mathrm{U}} - \frac{0.03110}{3} - \frac{0.0020}{3} r_s \ln r_s + 0.0116 - 0.00203 r_s, \tag{2.60a}$$

$$V_{\mathrm{P}} = E_{\mathrm{P}} - \frac{0.01555}{3} - \frac{0.0007}{3} r_s \ln r_s + \frac{0.0048 - 0.0007}{3} r_s. \tag{2.60b}$$

Using the polarization factors defined by

$$f = ((1+\zeta)^{4/3} + (1-\zeta)^{4/3} - 2)/(2^{4/3} - 2), \tag{2.61a}$$

$$f_z = \frac{4((1+\zeta)^{1/3} - (1-\zeta)^{1/3})}{3(2^{4/3} - 2)}, \tag{2.61b}$$

we have the exchange-correlation potentials

$$V_{\uparrow} = V_{\mathrm{U}} + f(V_{\mathrm{P}} - V_{\mathrm{U}}) + (E_{\mathrm{U}} - E_{\mathrm{U}})(1 - \zeta) f_z + E_{\mathrm{x}\uparrow} - \frac{0.15273333}{r_{s\uparrow}}, \tag{2.62a}$$

$$V_{\downarrow} = V_{\mathrm{U}} + f(V_{\mathrm{P}} - V_{\mathrm{U}}) + (E_{\mathrm{U}} - E_{\mathrm{U}})(-1 - \zeta) f_z + E_{\mathrm{x}\downarrow} - \frac{0.15273333}{r_{s\downarrow}}, \tag{2.62b}$$

and the exchange-correlation energies

$$E_{\mathrm{c}} = E_{\mathrm{U}} + f(E_{\mathrm{P}} - E_{\mathrm{U}}) \tag{2.63}$$

$$E_{\mathrm{xc}\uparrow} = E_{\mathrm{c}} + E_{\mathrm{x}\uparrow}, \quad E_{\mathrm{xc}\downarrow} = E_{\mathrm{c}} + E_{\mathrm{x}\downarrow}, \tag{2.64}$$

for up- and down-spin electrons.

The LDA works best for metals since it has been derived from the uniform electron gas. It usually reproduces the ground-state charge density to within a few percent. Concerning the total energy, the nonmagnetic HCP iron has a lower energy than the magnetic BCC iron, in the LSDA, which is of course in contradiction to experiments. (BCC ($\alpha$) iron is favorable at low temperatures and FCC ($\gamma$) iron becomes favorable at higher temperatures.) It is also known that the local-density approximation fails to include the van der Waals interaction between separated atoms and molecules. A practical procedure for remedying this difficulty was proposed by Kohn et al. [80]. For the van der Waals interaction, several functionals have been proposed so far [81–85].

Here we comment on the calculation of the total energy. Summing the Kohn–Sham energy levels $\varepsilon_{\boldsymbol{k}\lambda}$ over the occupied states, we may easily obtain the band energy:

$$E_{\mathrm{band}} = 2 \sum_{\boldsymbol{k}\lambda}^{O} \varepsilon_{\boldsymbol{k}\lambda}, \tag{2.65}$$

where $O$ means the occupied states and the prefactor 2 comes from the spin multiplicity. (For simplicity, we consider nonpolarized cases only.) However, the total energy is different from this band energy. The reason is as follows: First of all, the Coulomb potential energy between the nuclei is not included in the band energy. Second, not all the contributions which are nonlinear in the charge density $\rho(\boldsymbol{r})$ are given correctly by the band energy. For example, the band energy accounts doubly for the electron–electron Coulomb interaction. This is because the effective one-electron Hamiltonian in the Kohn–Sham equation contains the Coulomb interaction due to the whole electron charge density $\rho(\boldsymbol{r})$ but all the other electrons also feel the same interaction self-consistently. The corresponding contribution to the band energy amounts to

$$2\sum_{\lambda}^{O}\int \psi_{\lambda}^{*}(\boldsymbol{r})\int \frac{\rho(\boldsymbol{r}')}{|\boldsymbol{r}-\boldsymbol{r}'|}d\boldsymbol{r}'\psi_{\lambda}(\boldsymbol{r})d\boldsymbol{r} = \int \frac{\rho(\boldsymbol{r})\rho(\boldsymbol{r}')}{|\boldsymbol{r}-\boldsymbol{r}'|}d\boldsymbol{r}d\boldsymbol{r}', \tag{2.66}$$

while the true contribution to the total energy is just half of this value. Third, for a similar reason, the exchange–correlation energy $E_{\mathrm{xc}}$ is not evaluated correctly in the band energy. That is, because of the nonlinearity, if one defines the exchange–correlation energy density $\varepsilon_{\mathrm{xc}}[\rho(\boldsymbol{r})]$ by

$$E_{\mathrm{xc}} = \int \rho(\boldsymbol{r})\varepsilon_{\mathrm{xc}}[\rho(\boldsymbol{r})]d\boldsymbol{r}, \tag{2.67}$$

one finds generally that $\varepsilon_{\mathrm{xc}}[\rho(\boldsymbol{r})]$ is not equal to $\mu_{\mathrm{xc}}[\rho(\boldsymbol{r})]$, which is given by the derivative of $E_{\mathrm{xc}}$ with respect to $\rho(\boldsymbol{r})$; see (2.51). Thus the total energy is given by

$$\begin{aligned} E = E_{\mathrm{band}} &+ \frac{1}{2}\sum_{j(\neq i)}\frac{Z_i Z_j}{|\boldsymbol{R}_i-\boldsymbol{R}_j|} - \frac{1}{2}\int \frac{\rho(\boldsymbol{r})\rho(\boldsymbol{r}')}{|\boldsymbol{r}-\boldsymbol{r}'|}d\boldsymbol{r}d\boldsymbol{r}' \\ &+ \left(E_{\mathrm{xc}} - \int \mu_{\mathrm{xc}}(\boldsymbol{r})\rho(\boldsymbol{r})d\boldsymbol{r}\right), \end{aligned} \tag{2.68}$$

The first, second, third and fourth terms on the right-hand side of (2.65) represent, respectively, the band energy, the electrostatic energy between nuclei, the correction to the electron–electron Coulomb interaction energy and the correction to the exchange–correlation energy.

The electrostatic energy between the ions (nuclei) in crystals arises either from many positive and negative ions which are alternately and periodically arranged, as in an ionic crystal, or from many positive nuclear charges, also periodically arranged, but in a homogeneous negative (electronic) charge background. In either case, charge neutrality is guaranteed. Since the Coulomb interaction due to point charges is very long-ranged both in real space and in reciprocal space, the evaluation of the sum over only the lattice vectors $\boldsymbol{R}$ or only the reciprocal-lattice vectors $\boldsymbol{G}$ requires much effort. Instead, the evaluation can be performed rather conveniently by decomposing the Coulomb potential into two parts: one is a spatially localized part which can be evaluated rather fast in real space and the other is a (spatially) slowly varying part which can be evaluated faster in reciprocal space. One convenient formulation of such a decomposition was given by Ewald, and the method is called the Ewald sum; see Appendix C for more details.

Apart from the Coulomb interaction between the nuclei, the total energy $E_{\mathrm{el}}$ of the $N$-electron system can be rewritten in the form of a sum of the Hamiltonian $\mathcal{H}$ and the correction $\Delta\mathcal{F}$:

$$E_{\mathrm{el}} = \int \rho(\boldsymbol{r})\mathcal{F}[\rho(\boldsymbol{r})]d\boldsymbol{r}, \tag{2.69}$$

$$\mathcal{F}[\rho(\boldsymbol{r})] = \mathcal{H}[\rho(\boldsymbol{r})] + \Delta\mathcal{F}[(\boldsymbol{r})], \tag{2.70}$$

$$\Delta\mathcal{F}[\rho(\boldsymbol{r})] = -\frac{1}{2}\int \frac{\rho(\boldsymbol{r}')}{|\boldsymbol{r}-\boldsymbol{r}'|}d\boldsymbol{r}' + \{\varepsilon_{\mathrm{xc}}[\rho(\boldsymbol{r})] - \mu_{\mathrm{xc}}[\rho(\boldsymbol{r})]\}, \tag{2.71}$$

where $\mathcal{F}$ is the total energy operator.

Note that the functional derivative of the total energy $E_{\mathrm{el}}$ of the $N$-electron system with respect to the density $\rho(\boldsymbol{r})$ gives the Kohn–Sham Hamiltonian $\mathcal{H}$, if the constraint on the total number of electrons (2.50) is *not* imposed. If this constraint is imposed, the functional derivative becomes zero at the ground state because of the variational principle; see (2.49).

Here we make a brief comment on the virial theorem. The exchange-correlation energy, $E_{\mathrm{xc}} = \int d\boldsymbol{r}\rho(\boldsymbol{r})\varepsilon_{\mathrm{xc}}(\rho(\boldsymbol{r}))$, is divided into a kinetic energy part $T_{\mathrm{xc}}$ and a potential energy part $U_{\mathrm{xc}}$. According to Averill and Painter [86], $T_{\mathrm{xc}}$ is given by

$$T_{\mathrm{xc}} = \int d\boldsymbol{r}\rho(\boldsymbol{r})[3\mu_{\mathrm{xc}}(\rho(\boldsymbol{r})) - 4\varepsilon_{\mathrm{xc}}(\rho(\boldsymbol{r}))]. \tag{2.72}$$

Then the true kinetic-energy and potential-energy contributions to the ground-state total energy is given by $T =$ "$\hat{T}$" $+T_{\mathrm{xc}}$ and $V =$ "$(V_{\mathrm{n-e}} + V_{\mathrm{H}}/2)$" $+E_{\mathrm{xc}} - T_{\mathrm{xc}} =$ "$(V_{\mathrm{n-e}} + V_{\mathrm{H}}/2 - 3\mu_{\mathrm{xc}})$" $+5E_{\mathrm{xc}}$, respectively. Here $V_{\mathrm{n-e}}$ and $V_{\mathrm{H}}$ denote the nuclear-electron Coulomb potential and the Hartree potential, respectively, and "..." is defined by a summation over all occupied states:

$$\text{"..."} = \sum_{\nu}^{\mathrm{occ}} \langle\nu|...|\nu\rangle. \tag{2.73}$$

Then, the virial theorem $V = -2T$ is exactly satisfied within the LDA as demonstrated in [87].

### *2.2.7 Generalized Gradient Approximation (GGA)*

If the electron density becomes nonuniform, the exchange–correlation energy deviates from the result of the uniform electron gas. This deviation may be expressed in terms of the gradient and higher spatial derivatives of the total charge density. The total energy of an electron gas with a slowly varying density gives information about the density functional with gradient corrections. The generalized gradient approximation (GGA) is based on these ideas and is mainly due to Langreth and Perdew [88], Langreth and Mehl [89], Perdew and Wang [90], Perdew [91] and Becke [92]. One form of the GGA exchange–correlation energy and potential is given explicitly in [90, 91], which is often referred to as Perdew–Wang '86. In this form, one has to introduce a cutoff density $\rho_c$, below which the GGA contribution should be discarded. This cutoff density must be determined so as to satisfy the sum rule for the exchange and correlation holes. In order to overcome this complication, Perdew and Wang [93] proposed a new form which does not introduce any cutoff; see also [94]. This new formalism of the GGA is referred to as Perdew–Wang '91 or more simply PW91. The so-called PBE functional was proposed by Perdew, Burke and Ernzorhof [95], which is a little bit more empirical in a sense but made simpler compared to PW91. For example, the exchange part of the functional is given by

$$E_{\mathrm{x}} = -\frac{1}{2}\sum_s \int \rho_s^{4/3} K_s[s] d\boldsymbol{r} \tag{2.74}$$

with

$$s = \frac{|\nabla \rho_s|}{2k_{\mathrm{F}}\rho_s} = \frac{|\nabla \rho_s|}{2(6\pi^2\rho_s)^{1/3}\rho_s} = \frac{x_s}{7.7956}, \quad x_s = \frac{|\nabla \rho_s|}{\rho_s^{4/3}}, \tag{2.75}$$

and

$$K_s^{\mathrm{LDA}} = 3\left(\frac{3}{4\pi}\right)^{1/3}, \tag{2.76}$$

$$K_s^{\mathrm{PW91}} = K_s^{\mathrm{LDA}} \frac{1 + 0.19645 s \sinh^{-1} x_s + (0.2743 - 0.1508 e^{-100 s^2}) s^2}{1 + 0.19645 s \sinh^{-1} x_s + 0.004 s^4}, \tag{2.77}$$

$$K_s^{\mathrm{PBE}} = K_s^{\mathrm{LDA}} \left(1 + \kappa - \frac{\kappa}{1 + \mu s^2/\kappa}\right), \tag{2.78}$$

where $K_s^{\mathrm{LDA}}$ is the Slater's expression for the exchange energy within the LDA, while $K_s^{\mathrm{PW91}}$ and $K_s^{\mathrm{PBE}}$ are those given in the PW91 and PBE functionals; $\kappa = 0.804$ and $\mu = 0.21951$ (older values proposed by Becke [96] were $\kappa = 0.967$ and $\mu = 0.235$). On the other hand, the correlation part of the PBE functional is given by

$$E_{\mathrm{c}}^{\mathrm{PBE}}[\rho_\uparrow, \rho_\downarrow] = \int \rho(\boldsymbol{r}) \left[\varepsilon_{\mathrm{c}}^{\mathrm{LDA}}(r_s, \zeta) + H(r_s, \zeta, t)\right] d\boldsymbol{r}, \tag{2.79}$$

where $\zeta = (\rho_\uparrow - \rho_\downarrow)/\rho$ and $t = |\nabla \rho|/2\phi(\zeta) k_s \rho$ with

$$\phi(\zeta) = \frac{(1+\zeta)^{2/3} + (1-\zeta)^{2/3}}{2}, \quad k_s = \sqrt{\frac{4k_{\mathrm{F}}}{\pi a_{\mathrm{B}}}} = \sqrt{\frac{4(3\pi^2\rho)^{1/3}}{\pi a_{\mathrm{B}}}}. \tag{2.80}$$

The function $H(r_s, \zeta, t)$ in (2.79) is given by

$$H = \frac{e^2}{a_\mathrm{B}} \gamma \phi^3 \ln \left\{ 1 + \frac{\beta}{\gamma} t^2 \left[ \frac{1 + At^2}{1 + At^2 + A^2 t^4} \right] \right\}, \tag{2.81}$$

with

$$A = \frac{\beta}{\gamma} \left[ \exp \left\{ -\frac{\varepsilon_c^{\mathrm{LDA}} a_\mathrm{B}}{\gamma \phi^3 e^2} \right\} - 1 \right]^{-1} \tag{2.82}$$

with $\beta = 0.066725$ and $\gamma = (1 - \ln 2)/\pi^2 = 0.031091$. Now, PBE is most widely used as the GGA. There is also "meta-GGA" proposed by Perdew et al. [97], which requires the orbital kinetic energy density and is free of self-interaction error. More recently, PBEsol [98] was devised to work for solids. It makes slightly better predictions of the lattice parameter and bulk modulus of crystals, but typically is worse for dissociation or cohesive energies.

Anyway, the GGA has a possibility of going beyond the LDA within the framework of DFT. It has been reported that for many molecules the GGA improves significantly the binding energy. For solids, the GGA is known to improve the cohesive energy. For example, in the GGA, the BCC structure of iron having ferromagnetic order has truly the absolute minimum in the total energy and the lattice constant also coincides well with the experimental value [99, 100]. Ab initio MD simulations using the GGA have been performed for an Fe system under pressure [101]. Hamann [102] showed that the hydrogen bonding in an $H_2O$ system is well described by the GGA, the reason being that it was derived using a perturbation expansion in powers of $\nabla \rho$. It is therefore expected to work for systems where there are small density gradients. In other cases, where there are large variations of the charge density, the GGA is not necessarily an improvement on the LDA, since it is not derived from a variational principle. The van der Waals interaction, for example, between argon atoms, between benzene molecules, and between graphite layers [82], cannot be represented by the GGA because of a significant underestimation, and instead more properly represented by the LDA maybe due to an error cancellation; see for example [103] for the interaction between carbon nanotube (CNT) and polyyne $C_{10}H_2$ or hexane ($C_6H_{14}$) as well as between graphite layers and between $C_{60}$ molecules.

It is well known that, for the band gap between the highest occupied states and the lowest unoccupied states of semiconductors and insulators, the LDA and GGA usually give at most 2/3 of the experimental value. If one writes the one-electron spectrum given by DFT for a fractional occupation number $f_\lambda$ as $\varepsilon_\lambda(f_\lambda)$, it is exactly represented as

$$\varepsilon_\lambda(f_\lambda) = \frac{\partial E}{\partial f_\lambda}, \tag{2.83}$$

by Janak's theorem [104] in DFT. Therefore, by integrating (2.83) with respect to $f_\lambda$ from zero to one, one would obtain

$$E^{N}_{f_\lambda=1} - E^{N-1}_{f_\lambda=0} = \int_0^1 \varepsilon_\lambda(f_\lambda) df_\lambda, \quad \text{for occupied level } \lambda, \tag{2.84a}$$

$$E^{N+1}_{f_\lambda=1} - E^{N}_{f_\lambda=0} = \int_0^1 \varepsilon_\lambda(f_\lambda) df_\lambda, \quad \text{for empty level } \lambda. \tag{2.84b}$$

If $\lambda$ is the highest occupied molecular orbital (HOMO) level or the valence band maximum (VBM) in the case of crystals, (2.84) gives the **ionization potential** (IP). Similarly, if $\lambda$ is the lowest unoccupied molecular orbital (LUMO) level or the conduction band minimum (CBM) in the case of crystals, (2.84) gives the **electron affinity** (EA). From these equations, it is clear that, if $\varepsilon_\lambda(f_\lambda)$ does not depend on the occupation number $f_\lambda$, $\varepsilon_\lambda$ ($\lambda$ = HOMO or LUMO) is identical to the minus of the ionization potential (IP) or the minus of the electron affinity (EA). However, in DFT, the density $\rho(\boldsymbol{r})$ generally changes according to the value of $f_\lambda$, therefore $\varepsilon_\lambda(f_\lambda)$ generally depends on the occupation number $f_\lambda$. For this reason, the energy difference between the CBM and VBM, $\varepsilon_{\text{CBM}} - \varepsilon_{\text{VBM}}$, is not equal to the experimental band gap $E_g = E^{N+1} + E^{N-1} - 2E^N$ of semiconductors or insulators, where the host electrons cannot completely screen one additional electron even in the limit $N \to \infty$, because of the existence of the finite gap energy; this energy is required to change the electronic ground state [105]. The only exception is the ionization potential (IP) in the case of the "asymptotically exact" DFT, as shown by Almbladh and von Barth [106], who considered the correct asymptotic form of the exchange-correlation potential of DFT in the vacuum far outside solid surfaces.

The Kohn–Sham eigenvalue $\varepsilon_\lambda(f_\lambda)$ is obtained by solving the Kohn–Sham equation (2.66). Therefore, whether $\varepsilon_\lambda(f_\lambda)$ depends on $f_\lambda$ is directly related to whether the LDA or GGA effective potential depends on $f_\lambda$. Although the first and second terms in the curly brackets in the l.h.s. of (2.66), i.e., the kinetic energy operator $-\nabla^2$ and the external potential $v(\boldsymbol{r})$, are independent of $f_\lambda$, the third term, i.e., the Hartree potential, is the increasing function of $f_\lambda$. The fourth term, i.e., the exchange-correlation potential, has a minus sign but it does not cancel out the increment in the Hartree potential. Thus, the ordinary LDA and GGA cannot have the constant $\varepsilon_\lambda(f_\lambda)$ that is independent of $f_\lambda$. How about the case of the Hartree-Fock approximation? In the Hartree–Fock approximation, the increment in the Hartree potential exactly cancels out with the Fock term because the wavefunction $\psi_\lambda(\boldsymbol{r})$ does not feel any interaction caused by the same orbital. In this sense, the Hartree–Fock approximation is self-interaction free, and the resulting energy eigenvalue $\varepsilon_\lambda(f_\lambda)$ is independent of $f_\lambda$ for an occupied level $\lambda$. In contrast, the LDA and GGA have self-interaction, and the resulting energy eigenvalue $\varepsilon_\lambda(f_\lambda)$ depends on $f_\lambda$. In order to improve the energy gap in the LDA and GGA, it is necessary to remove the self-interaction from the theory.

### 2.2.8 Self-interaction Correction (SIC)

In the Hartree–Fock approximation, the Coulomb interaction energy of an electron in a given eigenstate with itself is exactly cancelled by its exchange interaction since the orbitals involved are the same in both cases. This term is called the self-interaction. Because of this exact cancellation, there was no problem in the Hartree–Fock approximation in the treatment of the Coulomb interaction and correlation of an electron with itself. However, in the LDA, the Coulomb term always contains a self-interaction term, since the total electronic charge density includes a contribution from the electronic level under consideration, and the cancellation between the self-interaction terms in the Coulomb term and the exchange–correlation term is generally incomplete when the electron charge density is not uniform. Thus the introduction of the self-interaction correction (SIC) is required in density functional formalisms such as the LDA, LSDA and GGA.

That is, we require that the equation

$$\Xi(\lambda\sigma;\lambda\sigma) + E_{\mathrm{xc}}[n_{\lambda\sigma}, 0] = 0 \tag{2.85}$$

holds, even for DFT. Here, we assumed the system is diagonal in spin index and wrote the level index $\lambda$ as $\lambda\sigma$; $\Xi(\lambda\sigma;\lambda\sigma)$ $(= U[n_{\lambda\sigma}])$ represents the self-interaction term of the direct Coulomb energy,

$$\Xi(\lambda\sigma;\nu\sigma') = \frac{1}{2}\int d\boldsymbol{r}\int d\boldsymbol{r}' \frac{\psi^*_{\lambda\sigma}(\boldsymbol{r})\psi^*_{\nu\sigma'}(\boldsymbol{r}')\psi_{\lambda\sigma}(\boldsymbol{r})\psi_{\nu\sigma'}(\boldsymbol{r}')}{|\boldsymbol{r}-\boldsymbol{r}'|}, \tag{2.86}$$

and $E_{\mathrm{xc}}[n_{\lambda\sigma}, 0]$ expresses symbolically the one-electron value of the exchange–correlation energy functional $E_{\mathrm{xc}}[n_\uparrow, n_\downarrow]$. If we introduce the Hartree–Fock-like exchange energy

$$E_{\mathrm{x}}(\lambda\sigma;\nu\sigma) = -\frac{1}{2}\int d\boldsymbol{r}\int d\boldsymbol{r}' \frac{\psi^*_{\lambda\sigma}(\boldsymbol{r})\psi^*_{\nu\sigma}(\boldsymbol{r}')\psi_{\nu\sigma}(\boldsymbol{r})\psi_{\lambda\sigma}(\boldsymbol{r}')}{|\boldsymbol{r}-\boldsymbol{r}'|}, \tag{2.87}$$

one may readily find that the equation

$$\Xi(\lambda\sigma;\lambda\sigma) + E_{\mathrm{x}}(\lambda\sigma;\lambda\sigma) = 0 \tag{2.88}$$

is fulfilled automatically. Therefore, defining the correlation energy as $E_{\mathrm{c}} = E_{\mathrm{xc}} - E_{\mathrm{x}}$, we find

$$E_{\mathrm{c}}(\lambda\sigma;\lambda\sigma) = 0. \tag{2.89}$$

Equations (2.85)–(2.89) form the basis for the self-interaction correction discussed by Perdew and Zunger [79].

Suppose now that we have $E^{\mathrm{approx}}_{\mathrm{xc}}[n_\uparrow, n_\downarrow]$ in some approximation such as the LDA, LSDA or GGA. Perdew and Zunger defined the self-interaction-corrected version of this approximation by

$$E_{\mathrm{xc}}^{\mathrm{SIC}}[n_\uparrow, n_\downarrow] = E_{\mathrm{xc}}^{\mathrm{approx}}[n_\uparrow, n_\downarrow] - \sum_{\lambda\sigma} \delta_{\lambda\sigma}, \tag{2.90}$$

where

$$\delta_{\lambda\sigma} = \Xi(\lambda\sigma; \lambda\sigma) + E_{\mathrm{xc}}^{\mathrm{approx}}[n_{\lambda\sigma}, 0] \tag{2.91}$$

is the self-interaction of the orbital $\lambda\sigma$.

A one-electron Schrödinger equation

$$\left[-\frac{1}{2}\nabla^2 + \bar{v}_{\lambda\sigma}(\boldsymbol{r})\right]\bar{\psi}_{\lambda\sigma}(\boldsymbol{r}) = \varepsilon_{\lambda\sigma}^{\mathrm{SIC}}\bar{\psi}_{\lambda\sigma}(\boldsymbol{r}) \tag{2.92}$$

with

$$\begin{aligned}\bar{v}_{\lambda\sigma}(\boldsymbol{r}) &= \left\{v(\boldsymbol{r}) - 2\mu_{\mathrm{B}}\boldsymbol{\sigma}\cdot\boldsymbol{B}(\boldsymbol{r}) + \Xi([\bar{n}];\boldsymbol{r}) + v_{\mathrm{xc}\,\sigma}^{\mathrm{approx}}([\bar{n}_\uparrow, \bar{n}_\downarrow];\boldsymbol{r})\right\} \\ &\quad - \left\{\Xi([\bar{n}_{\lambda\sigma}];\boldsymbol{r}) + v_{\mathrm{xc}\,\sigma}^{\mathrm{approx}}([\bar{n}_{\lambda\sigma}, 0];\boldsymbol{r})\right\}\end{aligned} \tag{2.93}$$

is derived by minimizing the total energy with $E_{\mathrm{xc}}^{\mathrm{approx}}$ replaced by $E_{\mathrm{xc}}^{\mathrm{SIC}}$. In (2.93), we have used a bar to denote quantities which are self-consistent with a self-interaction-corrected potential. The term, $-2\mu_{\mathrm{B}}\boldsymbol{\sigma}\cdot\boldsymbol{B}(\boldsymbol{r})$, represents the Zeeman energy of a spin $\boldsymbol{\sigma}$ in an external magnetic field $\boldsymbol{B}(\boldsymbol{r})$.

As is shown in the paper by Perdew and Zunger [79], the results for the correlation energy, the first ionization energy and the binding energy of negative ions (i.e. the electron affinity) obtained by the LSDA or the Hartree–Fock are greatly improved, in the case of isolated atoms. However, in the case of solids, if one regards the state $(\lambda\sigma)$ as a Bloch state, the magnitude of the SIC decreases, generally, towards zero as the system size goes to infinity. In order to solve this difficulty, $(\lambda\sigma)$ should be considered as a localized orbital. Usually, such a localized orbital $(\lambda\sigma)$ is chosen so as to minimize the self-interaction-corrected total energy.

The SIC is important in many insulators such as oxides, where the LDA might predict a metal, contrary to experiment. In insulators where the ionicity is already very strong, the LDA will lead to a gap, although smaller than the experimental value; but in systems where the energy levels of the constituents are similar, and the charge transfer small, the self-interaction in the LDA tends to delocalize the electrons and lead to a more covalent bond. As previously mentioned, the LDA is not always expected to give reasonable results in such cases, since it has been fitted to the results for the uniform electron gas. The SIC enhances the ionicity due to electron correlations and opens a gap in such systems. Recently, Vogel et al. [107] introduced self-interaction-corrected (SIC) and self-interaction- and relaxation-corrected (SIRC) pseudopotentials and discussed the effect of semicore d electrons, for example, in GaN crystal.

### 2.2.9 LDA + U, GGA + U

Despite the great success in the LDA and GGA in DFT, these methods can not treat e.g. 3d transition metal oxides, in particular the systems called Mott insulators, properly. For these systems, not only the band gap and other optical properties, but also the ground state properties such as the magnetic structures and other magnetic properties are not well described within the LDA and GGA. These materials are named as "strongly correlated systems," but we should study from the basics, not simply classify them as strongly correlated. What we should do is to compute numerically "how electrons are strongly correlated?" This is a difficult subject even by the present day supercomputer power, and therefore it is needed to introduce some phenomenological models, such as the Hubbard $U$ to express the strong Coulomb repulsion. The important physics in these systems is that $\uparrow$ and $\downarrow$ electrons cannot easily occupy the same site due to the strong Coulomb repulsive interaction between each other. This fact cannot be treated within the ordinary LDA and GGA. The simplest way to overcome this difficulty is to introduce the Hubbard $U$ parameter, defined as [108]

$$U = E(d^{n+1}) + E(d^{n-1}) - 2E(d^n), \tag{2.94}$$

i.e., the Coulomb repulsive energy cost to occupy two electrons at the same site. In order to restrict the interaction only to the case where the $\uparrow$ and $\downarrow$ electrons occupy the same site, we can introduce the Hubbard interaction

$$H_U = \sum_{\lambda\lambda'} U \hat{n}_{i\lambda\uparrow} \hat{n}_{i\lambda'\downarrow}, \tag{2.95}$$

where $\hat{n}_{i\lambda s}$ is the number operator of the $\lambda$ orbital at the site $i$ with spin $s$. Within the mean-field approximation, where the fluctuations around the average occupancies $\langle \hat{n}_{i\lambda s} \rangle$ are completely ignored, (2.95) becomes

$$\begin{aligned} H_U^{\mathrm{MF}} &= \sum_{\lambda,\lambda'} U \left( \hat{n}_{i\lambda\uparrow} \langle \hat{n}_{i\lambda'\downarrow} \rangle + \langle \hat{n}_{i\lambda\uparrow} \rangle \hat{n}_{i\lambda'\downarrow} - \langle \hat{n}_{i\lambda\uparrow} \rangle \langle \hat{n}_{i\lambda'\downarrow} \rangle \right) \\ &= \sum_{\lambda,\lambda'} U \left[ \frac{1}{2} n_{i\lambda} (\hat{n}_{i\lambda'\uparrow} + \hat{n}_{i\lambda'\downarrow}) + m_{i\lambda} (\hat{n}_{i\lambda'\uparrow} - \hat{n}_{i\lambda'\downarrow}) - \frac{1}{4} \{ n_{i\lambda} n_{i\lambda'} - m_{i\lambda} m_{i\lambda'} \} \right], \end{aligned} \tag{2.96}$$

where we put $n_{i\lambda} = \langle \hat{n}_{i\lambda\uparrow} \rangle + \langle \hat{n}_{i\lambda\downarrow} \rangle$ and $m_{i\lambda} = \langle \hat{n}_{i\lambda\uparrow} \rangle - \langle \hat{n}_{i\lambda\downarrow} \rangle$. When $U$ is much larger than the LDA or GGA band width $W$, the spin-unpolarized excited state with $n_{i\lambda} = 1$ and $m_{i\lambda} = 0$ has too high energy at $U/4$. However, for the spin-polarized ground state with $n_{i\lambda} = 1$ and $m_{i\lambda} = 1$, the energy drops to zero. The $d$ band is split into the occupied and unoccupied levels located, respectively, at $-U/2$ and $+U/2$, and the energy gap $U$ appears in this band.

According to Anisimov, Zaanen and Andersen [108], the total energy functional in the LDA $+U$ method or the GGA $+U$ method (DFT + U method) is defined as

$$E = E^{\mathrm{DFT}} + \frac{1}{2} \sum_{\lambda,\lambda',s} U(n_{i\lambda s} - n^0)(n_{i\lambda' -s} - n^0)$$

$$+\frac{1}{2} \sum_{\substack{\lambda,\lambda' s \\ (\lambda \neq \lambda')}} (U - J)(n_{i\lambda s} - n^0)(n_{i\lambda' -s} - n^0). \tag{2.97}$$

Using this formula, they identified all late-$3d$-transition-metal monoxides, as well as the parent compounds of the high-$T_c$ compounds, to be large-gap magnetic insulators, and predicted that $LiNiO_2$ is a low-spin ferromagnet and NiS a local-moment $p$-type metal. Using a basis-set independent generalization of the LDA $+U$, Liechtenstein, Anisimov and Zaanen [109] analyzed the Mott-Hubbard insulator $KCuF_3$. More recent review on the LDA $+U$ approach is, for example, given by Cococcioni [110].

### 2.2.10 Hybrid Functionals

Now the question is, what does the exact form of the density functional look like? Although we don't have an answer to this question, one might expect that a significant portion of electron correlations is included in the LDA or GGA. Then, what is missing in the LDA and GGA might be the true description of the Fock exchange functional, which cannot be simply written as a function of the density only. As will be seen in Sect. 2.7, the nonlocal description of the self-energy functional including the Fock exchange as a part is essentially important in the quasiparticle (QP) theory. DFT and QPT are similar to each other in a sense that both theories present one-particle pictures. However, DFT and QPT are philosophically quite different; in particular, the former is the ground-state theory and there is no meaning in the Kohn–Sham wavefunctions, while the latter is the excited-state theory and there is a definite physical meaning in the QP wavefunctions. Nevertheless, there should be some hint in the QPT to develop the exact DFT. That is, it would be possible to mimic the self-energy functional in the QP theory to construct more accurate density functional. In this sense, the first step towards the exact density functional might be to include the Fock exchange term in the density functional. However, if we use the Fock exchange term together with the correlation part of the LDA or GGA as a new DFT functional, the resulting thermochemical properties, in particular, intermolecular bonds becomes unsatisfactory as demonstrated by Clementi and Chakravorty [111]. This is due to the fact that the LDA and GGA exchange-correlation functionals are not so simply separable into the exchange and correlation parts. The Fock exchange is certainly exact in certain sense, but presents only a poor representation as the exchange-correlation functional.

According to the discussion given by Becke [112], let us suppose that we have the exact DFT exchange-correlation potential. Then, the noninteracting electron system in the nuclear Coulomb attractive potential and the exact exchange-correlation potential has the true electron density $\rho(\boldsymbol{r})$, i.e. that of the fully interacting electron system in the nuclear Coulomb attractive potential only. If the strength of the electron-electron Coulomb repulsive interaction, i.e., the coupling constant, is controlled by an arbitrary factor $g$ as $(g/2)\sum_{i\neq j}(1/|\boldsymbol{r}_i - \boldsymbol{r}_j|)$, $g = 0$ and $g = 1$ correspond, respectively, to the noninteracting Kohn–Sham system and to the fully interacting real system. It would be possible to interpolate these two extremes, i.e., to assume a partially interacting system with a fractal value of the coupling constant $g$, but still having the correct electron density $\rho(\boldsymbol{r})$. In such a system, we have to introduce the effective one-body potential $v_{\text{eff}}^{g}(\boldsymbol{r})$ apart from the external potential $v_{\text{ext}}(\boldsymbol{r})$ as a function of $g$ to maintain the correct electron density $\rho(\boldsymbol{r})$. Of course, $v_{\text{eff}}^{0}(\boldsymbol{r})$ should be equal to the "exact" Kohn–Sham potential at $g = 0$ and $v_{\text{eff}}^{1}(\boldsymbol{r}) = 0$ at $g = 1$. Then, the Hamiltonian and the total energy of this system are, respectively, written as

$$H^g = -\frac{1}{2}\sum_i^N \nabla_i^2 + \sum_i^N [v_{\text{ext}}(\boldsymbol{r}_i) + v_{\text{eff}}^g(\boldsymbol{r}_i) + \frac{g}{2}\sum_{i\neq j}\frac{1}{|\boldsymbol{r}_i - \boldsymbol{r}_j|}, \tag{2.98}$$

$$E^g = |\Psi_g\rangle H^g \langle\Psi_g| \tag{2.99}$$

$$= T_s[\rho] + E_{\text{int}}^g + \int [v_{\text{ext}}(\boldsymbol{r}) + v_{\text{eff}}^g(\boldsymbol{r})]\rho(\boldsymbol{r})d\boldsymbol{r} \tag{2.100}$$

with the definition of the interaction energy

$$E_{\text{int}}^g = \frac{g}{2}|\Psi_g\rangle \sum_{i\neq j}\frac{1}{|\boldsymbol{r}_i - \boldsymbol{r}_j|}\langle\Psi_g|. \tag{2.101}$$

Since the electron density $\rho(\boldsymbol{r})$ is independent of $g$, we can easily take the derivative of (2.100) to get

$$\frac{dE^g}{dg} = \frac{E_{\text{int}}^g}{g} + \int \frac{dv_{\text{eff}}^g(\boldsymbol{r})}{dg}\rho^g(\boldsymbol{r})d\boldsymbol{r}. \tag{2.102}$$

Then, by integrating this equation with respect to $g$ from zero to unity, we have

$$E^1 - E^0 = \int_0^1 \frac{dg}{g} E_{\text{int}}^g - \int v_{\text{eff}}^0(\boldsymbol{r})\rho(\boldsymbol{r})d\boldsymbol{r}. \tag{2.103}$$

On the other hand, from (2.100) with $g = 0$, $E^0$ is given by

$$E^0 = T_s[\rho] + \int [v_{\text{ext}}(\boldsymbol{r}) + v_{\text{eff}}^0(\boldsymbol{r})]\rho(\boldsymbol{r})d\boldsymbol{r} \tag{2.104}$$

for the noninteracting system. Combining these two equations, we have

$$E^1 = T_s[\rho] + \int v_{\text{ext}}(\boldsymbol{r})\rho(\boldsymbol{r})d\boldsymbol{r} + \int_0^1 \frac{dg}{g} E_{\text{int}}^g. \tag{2.105}$$

If the exact exchange-correlation functional $E_{\mathrm{xc}}[\rho]$ is known, $E^1$ can be written as

$$E^1 = T_s[\rho] + E_{\mathrm{xc}}[\rho] + \frac{1}{2}\int d\boldsymbol{r} \int d\boldsymbol{r}' \frac{\rho(\boldsymbol{r})\rho(\boldsymbol{r}')}{|\boldsymbol{r}-\boldsymbol{r}'|} + \int v_{\mathrm{ext}}(\boldsymbol{r})\rho(\boldsymbol{r})d\boldsymbol{r}. \quad (2.106)$$

Therefore, we finally obtain the following equation [73, 113, 114]

$$E_{\mathrm{xc}}[\rho] = \int_0^1 \frac{dg}{g} E_{\mathrm{xc}}^g, \quad (2.107)$$

where we put

$$E_{\mathrm{xc}}^g = E_{\mathrm{int}}^g - \frac{g}{2}\int d\boldsymbol{r} \int d\boldsymbol{r}' \frac{\rho(\boldsymbol{r})\rho(\boldsymbol{r}')}{|\boldsymbol{r}-\boldsymbol{r}'|}. \quad (2.108)$$

Equation (2.107) adiabatically connects the noninteracting system ($g = 0$) to the fully interacting system ($g = 1$) through a continuum of partially interacting systems ($0 \geq g \geq 1$). Although the integrand of the r.h.s. of (2.107) refers explicitly to potential energy only, the integration over $g$ effectively generates the the kinetic energy part of the exchange-correlation energy also; see [112] and Appendix of [115]. Equation (2.107) provides the foundation for constructing approximate exchange-correlation functionals. Let us consider the form of $E_{\mathrm{xc}}^g/g$ (i.e., (2.108) divided by $g$) in the integrand of the r.h.s. of (2.107). Particularly important is the noninteracting limit $g = 0$. In this limit, $\lim_{g\to 0} E_{\mathrm{xc}}^g/g$ becomes exactly equal to the pure Fock exchange energy of the Slater determinant made of the noninteracting Kohn–Sham orbitals [112]; Important fact here is that that the "exact" Kohn–Sham potential $v_{\mathrm{eff}}^0$ does not appear in (2.107); this term disappeared during the derivation of the final equation (2.107). Therefore, in this limit, we could identify the exact form of $\lim_{g\to 0} E_{\mathrm{xc}}^g/g$ to be $E_{\mathrm{x}}^{HF}$. In the other limit $g = 1$, we unfortunately don't know the exact form of $E_{\mathrm{xc}}^1$. If we could know this form, we can identify the exact exchange correlation functional $E_{\mathrm{xc}}[\rho]$. Becke [112] proposed to approximate $E_{\mathrm{int}}^1$ to be the LDA exchange-correlation energy. Then, linearly interpolating $E_{\mathrm{xc}}^g$ as

$$E_{\mathrm{xc}}^g \approx E_{\mathrm{xc}}^{LDA} + (E_{\mathrm{x}}^{HF} - E_{\mathrm{xc}}^{LDA})(1-g) \quad (2.109)$$

and integrating (2.107) with (2.112) from $g = 0$ to $g = 1$, he obtained

$$E_{\mathrm{xc}}[\rho] \approx \frac{1}{2}E_{\mathrm{x}}^{HF} + \frac{1}{2}E_{\mathrm{xc}}^{LDA}. \quad (2.110)$$

Using this hybrid functional, he showed that atomization energies, ionization potentials, proton affinities are well reproduced. Right after this work, Becke proposed the so-called Becke's three-parameter exchange functional [116].

Next year, Kim and Jordan [117] used a functional comprised of Becke's three-parameter exchange functional and Lee–Yang–Parr correlation functional [118] and demonstrated that it works very well for the geometrical and vibrational properties of water monomer and dimer. This Becke three-parameter Lee–Yang–Parr (B3LYP) functional is given by

$$E_{rmxc}^{\mathrm{B3LYP}} = E_{\mathrm{x}}^{\mathrm{LDA}} + a_0(E_{rmx}^{\mathrm{HF}} - E_{\mathrm{x}}^{\mathrm{LDA}}) + a_{\mathrm{x}}(E_{\mathrm{x}}^{\mathrm{GGA}} - E_{\mathrm{x}}^{\mathrm{LDA}}) + E_{\mathrm{c}}^{\mathrm{LDA}} + a_{\mathrm{c}}(E_{\mathrm{c}}^{\mathrm{GGA}} - E_{\mathrm{c}}^{\mathrm{LDA}}), \tag{2.111}$$

where $E_{\mathrm{x}}^{\mathrm{GGA}}$ and $E_{\mathrm{c}}^{\mathrm{GGA}}$ denote, respectively, Becke's exchange [92] and Lee–Yang–Parr's correlation [118] functionals in the GGA, while $E_{\mathrm{x}}^{\mathrm{LDA}}$ and $E_{\mathrm{c}}^{\mathrm{LDA}}$ denotes, respectively, the ordinary exchange (2.56) and Vosko–Wilk–Nusair's correlation [119] functionals in the LDA. The mixing parameters $a_0 = 0.20$, $a_{\mathrm{x}} = 0.72$, $a_{\mathrm{c}} = 0.81$ are used as the same as the original paper by Becke [116]. Although these parameters are purely empirical, the results for the ground-state properties are usually very good. However, the B3YLP and other functionals have been developed mainly for organics, especially with Gaussian software. Therefore in general, they should be carefully used for inorganic materials, etc.

Perdew, Ernzerhof and Burke [120] proposed to use the following interpolation formula in the adiabatic connection formula (2.107):

$$E_{\mathrm{xc}}^{g} \approx E_{\mathrm{x}}^{\mathrm{PBE}} + E_{\mathrm{c}}^{\mathrm{PBE}} + (E_{\mathrm{x}}^{\mathrm{HF}} - E_{\mathrm{x}}^{\mathrm{PBE}})(1-g)^n, \tag{2.112}$$

where $n$ is an integer, $E_{\mathrm{x}}^{\mathrm{PBE}}$ and $E_{\mathrm{c}}^{\mathrm{PBE}}$ are the exchange and correlation parts of the Perdew-Burke-Ernzerhof (PBE) functional [95]. Then, by integration, they obtained

$$E_{\mathrm{xc}} \approx \frac{1}{n} E_{\mathrm{x}}^{\mathrm{HF}} + \frac{4-n}{n} E_{\mathrm{x}}^{\mathrm{PBE}} + E_{\mathrm{c}}^{\mathrm{PBE}}. \tag{2.113}$$

The so-called PBE0 functional given by (2.112) with $n = 4$, in which the Fock and PBE exchange energies are mixed in a ratio 1:3, reproduces well the atomization energies.

In 2003, Heyd, Scuseria, and Ernzerhof [121] proposed to use a screened Coulomb potential only to the exchange interaction in order to screen the long-range part of the Fock exchange in particular for metallic systems. All other Coulomb interactions were assumed unscreened. The Coulomb potential is split into short-range (SR) and long-range (LR) parts:

$$\frac{1}{r} = \frac{\mathrm{erfc}(\omega r)}{r} + \frac{\mathrm{erf}(\omega r)}{r}, \tag{2.114}$$

where the first and second terms represent SR and LR, respectively; $\omega$ is an adjustable parameter to describe the crossover from SR to LR, $\mathrm{erf}(\omega r)$ is the error function, and $\mathrm{erfc}(\omega r) = 1 - \mathrm{erf}(\omega r)$. The so-called Heyd–Scuseria–Ernzerhof (HSE) functional is obtained by applying (2.112) to SR only:

$$E_{\mathrm{xc}}^{!\mathrm{PBEh}} = aE_{\mathrm{x}}^{\mathrm{HF,SR}}(\omega) + (1-a)E_{\mathrm{x}}^{\mathrm{PBE,SR}}(\omega) + E_{\mathrm{x}}^{\mathrm{PBE,LR}}(\omega) + E_{\mathrm{c}}^{\mathrm{PBE}}, \tag{2.115}$$

where $E_x^{HF,SR}(\omega)$ denotes the correct SR Fock functional, while $E_x^{PBE,SR}(\omega)$ and $E_x^{PBE,LR}(\omega)$ denote the SR and LR exchange parts of the PBE functional, respectively; the mixing parameter $a$ is fixed at $1/4$, and the range-separation parameter $\omega$ is set at 0.3 in HSE03 and at 0.2 in HSE06 ($\omega = 0$ corresponds to HSE0).

The so-called meta hybrid GGA or Minnesota density functionals have been proposed by Truhlar and co workers since 2005. They have published 14 highly parametrized local (M06-L [122], M11-L [123], MN12-L [124], and MN15-L [125]), global hybrid (M05 [126], M05-2X [127], M06, M06-2X [128], M06-HF [129], M08-HX, M08-SO [130], and MN15 [131]), and range-separated hybrid (M11 [132] and MN12-SX [133]) density functionals with kinetic energy density dependence. They are commonly called Minnesota density functionals. Of these 14 functionals, 10 have been of the metageneralized gradient approximation (meta-GGA) variety, while 4 (MN12-SX, MN15, MN12-L, and MN15-L) have been of the meta-nonseparable gradient approximation (meta-NGA) variety. Recently, Mardirossian and Head-Gordon [134] performed a thorough benchmark test of these functionals, and reported that M06-2X and MN15-L well reproduce thermodynamic properties, while M08-HX, M08-SO, MN15, and MN12-L are more suitable for the calculations of activation energies, and M06-2X, M08-HX, M08-SO, MN15, MN12-L, and MN15-L are more appropriate for the systems where self-interaction error is significant. However, none of them are not state-of-the-art for the full spectrum of noncovalent interactions and isomerization energies, although M06-2X is recommended among the ten hybrid Minnesota functionals and M06-L is the best among the four local Minnesota functionals. They are restricted to specific materials, not for general cases, especially to predict new materials, users should be very careful.

## 2.3 Explicit Methods to Calculate Electronic Structures

### *2.3.1 Periodic Systems*

The atomic positions of a pure bulk crystal are unchanged (invariant) when they are translated by any sum $\boldsymbol{R}$ of integer multiples of three "primitive translation vectors" $\boldsymbol{a}_1, \boldsymbol{a}_2, \boldsymbol{a}_3$:

$$\boldsymbol{R} = n_1\boldsymbol{a}_1 + n_2\boldsymbol{a}_2 + n_3\boldsymbol{a}_3. \tag{2.116}$$

This property is called "translational invariance", and the vector $\boldsymbol{R}$ is called a "lattice vector". The rhombohedral cell defined by the primitive translation vectors forms the basic unit of the lattice periodicity and is called the "unit cell". There is at least one atom in the unit cell. Figure 2.3 represents the unit cells of a BCC lattice and an FCC lattice [135].

In an effective one-electron problem, the eigenvalue equation reads

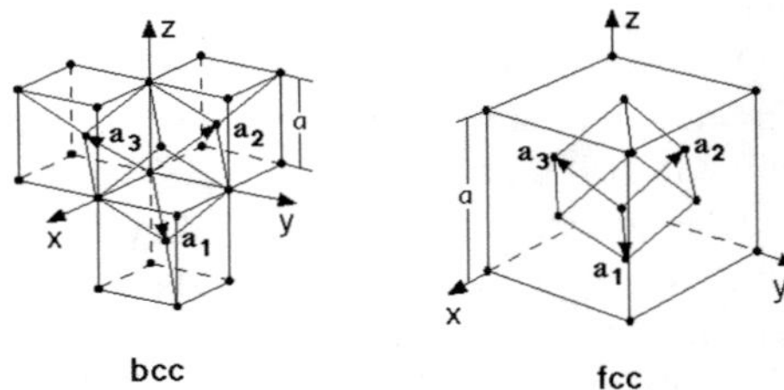


**Fig. 2.3** The unit cell of the BCC and FCC lattices

$$\mathcal{H}\psi_\lambda(\boldsymbol{r}) = \Big[-\frac{1}{2}\nabla^2 + V(\boldsymbol{r})\Big]\psi_\lambda(\boldsymbol{r}) = \varepsilon_\lambda\psi_\lambda(\boldsymbol{r}), \tag{2.117}$$

where the potential $V(\boldsymbol{r})$ felt by the electron is given by the sum of the three terms on the left-hand side of (2.54). In a crystal, $V(\boldsymbol{r})$ has a perfect periodicity and translational invariance. That is, it satisfies

$$V(\boldsymbol{r}+\boldsymbol{R}) = V(\boldsymbol{r}). \tag{2.118}$$

Then, owing to the translational invariance of the Hamiltonian, any effective one-electron wavefunction may be represented by the product of an exponential factor $\exp(i\boldsymbol{k}\cdot\boldsymbol{r})$ with the wave vector $\boldsymbol{k}$ and a periodic function $u_{\boldsymbol{k}\lambda}(\boldsymbol{r})$ (Fig. 2.4):

$$\psi_{\boldsymbol{k}\lambda}(\boldsymbol{r}) = \mathrm{e}^{i\boldsymbol{k}\cdot\boldsymbol{r}}u_{\boldsymbol{k}\lambda}(\boldsymbol{r}). \tag{2.119}$$

Equation (2.119) is referred to as the "Bloch theorem", and this form of the wavefunction is called the Bloch function or the Bloch state. The periodic function $u_{\boldsymbol{k}\lambda}(\boldsymbol{r})$ satisfies

$$u_{\boldsymbol{k}\lambda}(\boldsymbol{r}+\boldsymbol{R}) = u_{\boldsymbol{k}\lambda}(\boldsymbol{r}). \tag{2.120}$$

The wave vector $\boldsymbol{k}$ is a good quantum number. The Bloch function $\psi_{\boldsymbol{k}\lambda}(\boldsymbol{r})$ satisfies

$$\psi_{\boldsymbol{k}\lambda}(\boldsymbol{r}+\boldsymbol{R}) = \mathrm{e}^{i\boldsymbol{k}\cdot\boldsymbol{R}}\psi_{\boldsymbol{k}\lambda}(\boldsymbol{r}) \tag{2.121}$$

and the orthonormality condition

$$\int \psi^*_{\boldsymbol{k}\lambda}(\boldsymbol{r})\psi_{\boldsymbol{k}'\lambda'}(\boldsymbol{r})d\boldsymbol{r} = \delta_{\boldsymbol{k}\boldsymbol{k}'}\delta_{\lambda\lambda'}, \tag{2.122}$$

where the integral is over the whole system.

It is easy to verify from (2.117) that the Bloch function $\psi_{\boldsymbol{k}\lambda}(\boldsymbol{r})$ and the periodic function $u_{\boldsymbol{k}\lambda}(\boldsymbol{r})$ satisfy the eigenvalue equations

$$\mathcal{H}\psi_{\boldsymbol{k}\lambda}(\boldsymbol{r}) = \Big[-\frac{1}{2}\nabla^2 + V(\boldsymbol{r})\Big]\psi_{\boldsymbol{k}\lambda}(\boldsymbol{r}) = \varepsilon_{\boldsymbol{k}\lambda}\psi_{\boldsymbol{k}\lambda}(\boldsymbol{r}) \tag{2.123}$$

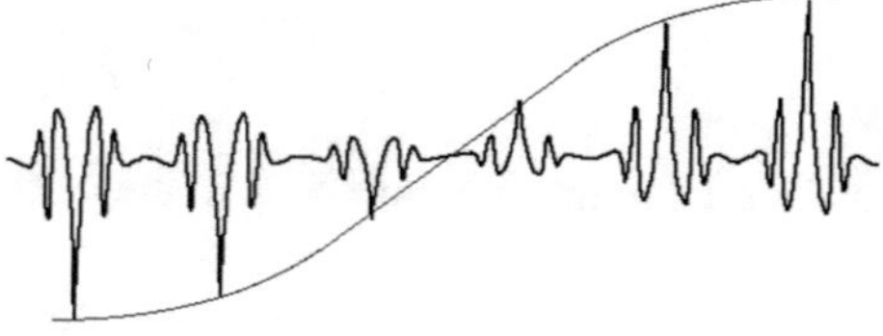

**Fig. 2.4** A wavefunction satisfying the Bloch theorem

and

$$\mathcal{H}_k u_{k\lambda}(\boldsymbol{r}) = \left[\frac{1}{2}(\boldsymbol{k} - i\nabla)^2 + V(\boldsymbol{r})\right] u_{k\lambda}(\boldsymbol{r}) = \varepsilon_{k\lambda} u_{k\lambda}(\boldsymbol{r}) \tag{2.124}$$

at the wave vector $\boldsymbol{k}$. The energy eigenvalue $\varepsilon_{k\lambda}$ of each level $\lambda$ changes smoothly as a function of $\boldsymbol{k}$ and forms a curve along one direction of $\boldsymbol{k}$ in the $\varepsilon$-$\boldsymbol{k}$ space. Several of these curves together look like a bundle, which is called an energy band or simply a band. The structure of the energy bands is referred to as the band structure.

It is possible to decompose the whole space into cells which surround each individual atom. To be concrete, one may introduce a polyhedron which is bounded by the planes perpendicularly bisecting the line segments from one atom under consideration (at the center of the cell) to its neighboring atoms. This polyhedral cell is called the "Wigner–Seitz cell". Alternatively, one may draw a sphere which has the same volume as the Wigner–Seitz cell around the central atom: this sphere is called the "Wigner–Seitz sphere".

Now we introduce three primitive *reciprocal*-lattice vectors $\boldsymbol{b}_1, \boldsymbol{b}_2, \boldsymbol{b}_3$ so as to satisfy

$$\boldsymbol{a}_i \cdot \boldsymbol{b}_j = 2\pi\delta_{ij}. \tag{2.125}$$

A vector made by adding arbitrary integer multiples of these primitive reciprocal-lattice vectors,

$$\boldsymbol{G} = l_1\boldsymbol{b}_1 + l_2\boldsymbol{b}_2 + l_3\boldsymbol{b}_3, \tag{2.126}$$

is called simply a "reciprocal-lattice vector". The space spanned by the reciprocal-lattice vectors is called the "reciprocal space" or "Fourier space" in contrast to the "real space". (Note that the reciprocal lattices for the simple cubic (SC), FCC and BCC lattices in real space are, respectively, the SC, BCC and FCC lattices in reciprocal space.) Then the above periodic function in real space can be expanded as a Fourier series:

$$u_{k\lambda}(\boldsymbol{r}) = \frac{1}{\sqrt{\Omega}} \sum_{\boldsymbol{G}} \widetilde{u}_{k\lambda}(\boldsymbol{G}) \mathrm{e}^{i\boldsymbol{G}\cdot\boldsymbol{r}}, \tag{2.127}$$

$$\widetilde{u}_{k\lambda}(\boldsymbol{G}) = \frac{1}{\sqrt{\Omega}} \int_{\text{cell}} \mathrm{e}^{-i\boldsymbol{G}\cdot\boldsymbol{r}} u_{k\lambda}(\boldsymbol{r}) d\boldsymbol{r}, \tag{2.128}$$

where $\Omega$ denotes the volume of the unit cell. Substituting (2.127) into the Bloch theorem (2.119), we have

$$\psi_{k\lambda}(\boldsymbol{r}) = \frac{1}{\sqrt{\Omega}} \sum_{\boldsymbol{G}} \widetilde{u}_{k\lambda}(\boldsymbol{G}) e^{i(\boldsymbol{k}+\boldsymbol{G})\cdot\boldsymbol{r}}. \tag{2.129}$$

If the total number of unit cells in the system is given by $\mathcal{N}$, the basis function

$$\langle \boldsymbol{r}|\boldsymbol{k}+\boldsymbol{G}\rangle = \frac{1}{\sqrt{\mathcal{N}\Omega}} e^{i(\boldsymbol{k}+\boldsymbol{G})\cdot\boldsymbol{r}} \tag{2.130}$$

is called a "plane wave" (PW), and the representation of an effective one-electron wavefunction in a periodic potential by (2.118) is referred to as the "plane wave expansion". In real applications, the maximum value of the kinetic energy $\frac{1}{2}G^2$ of the plane waves is limited to some value, and this maximum value is often referred to the "cutoff energy". Of course, the higher the cutoff energy the more accurate the result.

On the other hand, it is possible to represent the effective one-electron wavefunction in the periodic potential as a superposition of "localized orbitals", $\phi_\lambda(\boldsymbol{r})$, centered at each unit cell:

$$\psi_{k\lambda}(\boldsymbol{r}) = \frac{1}{\sqrt{\mathcal{N}}} \sum_{\boldsymbol{R}} e^{i\boldsymbol{k}\cdot\boldsymbol{R}} \phi_\lambda(\boldsymbol{r}-\boldsymbol{R}). \tag{2.131}$$

It is easy to verify that (2.131) satisfies (2.121) and therefore the Bloch theorem (2.119). In general, a sum of the form (2.131) is called a "Bloch sum". If the localized orbitals in the Bloch sum are the correct eigenfunctions of the effective one-electron Hamiltonian, they are different from the eigenstates of isolated atoms and are deformed specifically for the many-atom system.

Note that it is in principle possible to make the localized orbital $\phi_\lambda$ $\boldsymbol{k}$-independent. This can be done if we Fourier-transform (2.131) back into $\boldsymbol{R}$ space from $\boldsymbol{k}$ space:

$$\phi_\lambda(\boldsymbol{r}-\boldsymbol{R}) = \frac{1}{\sqrt{\mathcal{N}}} \sum_{\boldsymbol{k}} e^{-i\boldsymbol{k}\cdot\boldsymbol{R}} \psi_{k\lambda}(\boldsymbol{r}). \tag{2.132}$$

The localized orbitals $\phi_\lambda(\boldsymbol{r})$ introduced by (2.132) are called the "Wannier functions" or the "Wannier states". Since $\psi_{k\lambda}(\boldsymbol{r})$ and $\psi_{k'\lambda'}(\boldsymbol{r})$ are orthogonal (see (2.122)) if $\boldsymbol{k} \neq \boldsymbol{k}'$ or $\lambda \neq \lambda'$, one finds that the Wannier functions $\phi_\lambda(\boldsymbol{r}-\boldsymbol{R})$ and $\phi_{\lambda'}(\boldsymbol{r}-\boldsymbol{R}')$ are also orthogonal if $\boldsymbol{R} \neq \boldsymbol{R}'$ or $\lambda \neq \lambda'$:

$$\int \phi^*_\lambda(\boldsymbol{r}-\boldsymbol{R})\phi_{\lambda'}(\boldsymbol{r}-\boldsymbol{R}')d\boldsymbol{r} = \delta_{\lambda,\lambda'}\delta_{\boldsymbol{R}\boldsymbol{R}'}. \tag{2.133}$$

Because (2.131) has an ambiguity in the phase factor $\exp[i\theta(\boldsymbol{k})]$, it is in principle possible to enhance the spatial localization of the Wannier function $\phi_\lambda(\boldsymbol{r}-\boldsymbol{R})$ by choosing appropriately the phase factor. Indeed, Souza et al. showed how to construct the maximally localized Wannier functions [136], which have been used for the

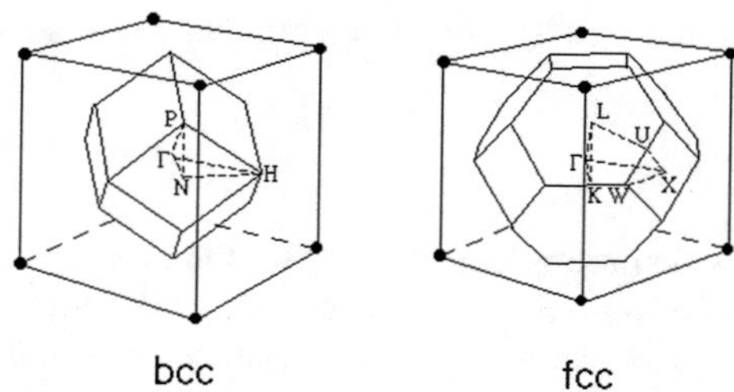


**Fig. 2.5** The first Brillouin zone for the BCC and FCC lattices

calculations of topological properties such as Berry phase, etc., electron-phonon interaction (see Sect. 2.4.5), and transport properties.

To see the relation between (2.131) and the Bloch theorem, it is good to note that the Fourier transform of the periodic function appearing in the Bloch theorem, $\widetilde{u}_{\boldsymbol{k}\lambda}(\boldsymbol{G})$, is related to this localized orbital via the relations

$$\widetilde{u}_{\boldsymbol{k}\lambda}(\boldsymbol{G}) = \frac{1}{\sqrt{N}}\widetilde{\phi}_\lambda(\boldsymbol{k}+\boldsymbol{G}) \tag{2.134}$$

and

$$\widetilde{\phi}_\lambda(\boldsymbol{k}+\boldsymbol{G}) = \frac{1}{\sqrt{\Omega}}\int \phi_\lambda(\boldsymbol{r})\mathrm{e}^{-i(\boldsymbol{k}+\boldsymbol{G})\cdot\boldsymbol{r}}d\boldsymbol{r}, \tag{2.135}$$

where the integral in (2.135) is taken over the whole space. That is, the Fourier transform $\widetilde{u}_{\boldsymbol{k}\lambda}(\boldsymbol{G})$ of the periodic function $u_{\boldsymbol{k}\lambda}(\boldsymbol{r})$ within the unit cell given by (2.128) is equal to the Fourier transform $\widetilde{\phi}_\lambda(\boldsymbol{k}+\boldsymbol{G})$ of the localized function $\phi_\lambda(\boldsymbol{r})$ in an infinite space.

In (2.131), one may see that the wavefunction $\psi_{\boldsymbol{k}\lambda}(\boldsymbol{r})$ does not change at all if one adds an arbitrary reciprocal-lattice vector $\boldsymbol{G}$ to the wave vector $\boldsymbol{k}$, i.e. $\psi_{\boldsymbol{k}\lambda}(\boldsymbol{r}) = \psi_{\boldsymbol{k}+\boldsymbol{G},\lambda}(\boldsymbol{r})$. From this fact, one may confine the wave vector $\boldsymbol{k}$ within a minimum region bounded by the planes bisecting perpendicularly the lines from the origin (0, 0, 0) to the neighboring reciprocal-lattice points. This region is called the "first Brillouin zone". From a simple analysis, it can be shown that some energy levels are split so to have an energy gap at the surfaces of the first Brillouin zone (see below). In Fig. 2.5, we show the first Brillouin zones for the FCC and the BCC structures. (Note that the first Brillouin zones for the SC, FCC and BCC structures have the same shapes as the Wigner–Seitz cells of, respectively, the SC, BCC and FCC structures.)

Here it is instructive to consider the free-electron limit in which the periodic potential $V(\boldsymbol{r})$ is infinitesimal. In this limit, Bloch wavefunctions become simply plane waves

$$\psi_{\boldsymbol{k}\lambda} = \frac{1}{N\Omega}\exp\left[i(\boldsymbol{k}+\boldsymbol{G}_\lambda)\cdot\boldsymbol{r}\right] \tag{2.136}$$

and the energy-band diagram becomes parabolic

$$\varepsilon_{\boldsymbol{k}\lambda} = \frac{1}{2}(\boldsymbol{k} + \boldsymbol{G}_\lambda)^2, \tag{2.137}$$

which are identical to those obtained by folding the free-electron dispersion $\varepsilon_{\boldsymbol{k}} = k^2/2$ onto the inside of the first Brillouin zone.

In order to see why an energy level splits at the zone boundaries, let us consider two wavefunctions $\psi_{-\boldsymbol{G}/2}(\boldsymbol{r})$ and $\psi_{+\boldsymbol{G}/2}(\boldsymbol{r})$ at the opposite zone boundaries $\boldsymbol{k} = \pm\boldsymbol{G}/2$. These two states are independent and degenerate (having the same energy $G^2/8$) in the free-electron limit. If we introduce a periodic potential $V(\boldsymbol{r})$ as a small perturbation (this treatment is called the nearly free-electron approximation), the doubly-degenerate energy level splits into two non-degenerate levels

$$\varepsilon_\pm = \frac{G^2/8}{8} \pm |\widetilde{V}(\boldsymbol{G})| \tag{2.138}$$

according to the first-order perturbation theory for degenerate states

$$\begin{vmatrix} G^2/8 - \varepsilon_\pm & \widetilde{V}^*(\boldsymbol{G}) \\ \widetilde{V}(\boldsymbol{G}) & G^2/8 - \varepsilon_\pm \end{vmatrix}, \tag{2.139}$$

where

$$\widetilde{V}(\boldsymbol{G}) = \langle \psi_{+\boldsymbol{G}/2} | V | \psi_{-\boldsymbol{G}/2} \rangle = \frac{1}{\Omega} \int_{\text{cell}} \mathrm{e}^{-i\boldsymbol{G}\cdot\boldsymbol{r}} V(\boldsymbol{r}) d\boldsymbol{r} \tag{2.140}$$

is the Fourier transform of the periodic potential.

This nearly free-electron approximation can be more rigorously formulated on the basis of perturbation theory. For example, using second-order perturbation theory, one may derive

$$\varepsilon_{\boldsymbol{k}\lambda} = \frac{1}{2}(\boldsymbol{k} + \boldsymbol{G}_\lambda)^2 + 2 \sum_{\nu(\neq\lambda)} \frac{|\widetilde{V}(\boldsymbol{G}_\nu - \boldsymbol{G}_\lambda)|^2}{(\boldsymbol{k} + \boldsymbol{G}_\lambda)^2 - (\boldsymbol{k} + \boldsymbol{G}_\nu)^2}, \tag{2.141}$$

where the sum runs over all reciprocal-lattice vectors $\boldsymbol{G}_\nu$ except for $\boldsymbol{G}_\lambda$. The formula (2.141) is known to be a good approximation in particular for simple metals, if one uses pseudopotentials (see Sect. 2.2.7) for $\widetilde{V}(\boldsymbol{G})$.

In summary, in a crystal, the effective one-electron wavefunctions are given by the Bloch functions, and the ground-state charge density is a sum over the occupied levels (bands) of Bloch functions (see also (2.52)):

$$\rho(\boldsymbol{r}) = 2 \sum_{\lambda}^{O} \sum_{\boldsymbol{k}\in\text{1stBZ}} |\psi_{\boldsymbol{k}\lambda}(\boldsymbol{r})|^2, \tag{2.142}$$

where the prefactor 2 comes from spin multiplicity and $\psi_{\boldsymbol{k}\lambda}(\boldsymbol{r})$ is an eigenstate of (2.123). Furthermore, because of the translational symmetry,

$$\langle \psi_{\boldsymbol{k}\lambda} | \mathcal{H} | \psi_{\boldsymbol{k}'\lambda'} \rangle = \varepsilon_{\boldsymbol{k}\lambda} \delta_{\boldsymbol{k}\boldsymbol{k}'} \delta_{\lambda\lambda'} \tag{2.143}$$

if both $\boldsymbol{k}$ and $\boldsymbol{k}'$ belong to the first Brillouin zone.

For semiconductors and insulators, where the valence band and the conduction band are well separated by an energy gap $E_g$, the special $k$-point method of Chadi and Cohen [137] or Monkhorst and Pack [138] can be used for sampling the $k$-points in the first Brillouin zone. On the other hand, in the case of metals, where no clear band gap exists, the charge density (2.142) should be constructed by summing all the wavefunctions at the levels below the Fermi level $E_F$. In order to determine the Fermi level, it is necessary to calculate the (electronic) density of states (DOS).

### 2.3.2 Group Theory

There are many symmetry operations which transform the crystal into itself. Any symmetry operation $\mathcal{T}$ transforms a lattice vector $\boldsymbol{R}$ into another lattice vector $\boldsymbol{R}'$. It does not change the effective one-electron Hamiltonian $\mathcal{H}$, which has the full crystal symmetry, i.e. $\mathcal{T}\mathcal{H}\mathcal{T}^{-1} = \mathcal{H}$. In other words, $\mathcal{T}$ commutes with $\mathcal{H}$:

$$[\mathcal{T}, \mathcal{H}] = \mathcal{T}\mathcal{H} - \mathcal{H}\mathcal{T} = 0. \tag{2.144}$$

Therefore, for a given symmetry operator $\mathcal{T}$, one may find wavefunctions $\psi_\lambda(\boldsymbol{r})$ satisfying the Schrödinger equation (2.117) which are also eigenstates of $\mathcal{T}$:

$$\mathcal{T}\psi_\lambda(\boldsymbol{r}) = \psi_\lambda(\mathcal{T}\boldsymbol{r}) = f\psi_\lambda(\boldsymbol{r}). \tag{2.145}$$

This is an important statement which prescribes the possible forms of wavefunctions in crystals. As the simplest symmetry operator, the identity operator $E$ maps $\boldsymbol{r}$ onto itself. If $\mathcal{T}$ and $\mathcal{T}'$ are symmetry operators, then their product $\mathcal{T}\mathcal{T}'$ is also a symmetry operator. They obviously satisfy the associativity $\mathcal{T}(\mathcal{T}'\mathcal{T}'') = (\mathcal{T}\mathcal{T}')\mathcal{T}''$. Moreover, one may see that $\mathcal{T}^{-1}$, which satisfies $\mathcal{T}^{-1}\mathcal{T} = \mathcal{T}\mathcal{T}^{-1} = E$, is also a symmetry operator of the crystal. A set of all symmetry operators $\{\mathcal{T}\}$ whose members satisfy all the above properties is called a "group" or "symmetry group". Clearly, for a given crystal, the set of all symmetry operators forms a group, which is generally called the "space group".

First, let us consider a translation symmetry. A corresponding symmetry operator $T$ transforms an arbitrary space point $\boldsymbol{r}$ to

$$T\boldsymbol{r} = \boldsymbol{r} + \boldsymbol{R}, \tag{2.146}$$

where $\boldsymbol{R}$ is a lattice vector. There are three primitive translations $T_1$, $T_2$, $T_3$ which satisfy

$$\begin{aligned} T_1\boldsymbol{r} &= \boldsymbol{r} + \boldsymbol{a}_1, \\ T_2\boldsymbol{r} &= \boldsymbol{r} + \boldsymbol{a}_2, \\ T_3\boldsymbol{r} &= \boldsymbol{r} + \boldsymbol{a}_3, \end{aligned} \tag{2.147}$$

where $\boldsymbol{a}_1$, $\boldsymbol{a}_2$, $\boldsymbol{a}_3$ denote the primitive translation vectors. The set of all translation operators $\{T\}$ forms a group called the translation group. The translation group is a very special group since all its elements commute with one another. Any element of the translation group can be represented as

$$T = T_1^{n_1} T_2^{n_2} T_3^{n_3}, \tag{2.148}$$

where $n_1, n_2, n_3$ are any integer numbers. If $n_j$ $(j = 1, 2, 3)$ is equal to the length of one edge of the crystal $L_j$ in units of the lattice constants, $T_j^{L_j}$ transforms $\boldsymbol{r}$ into itself, provided that we impose periodic boundary conditions on the whole crystal. (Note that the $L$ are generally large numbers and will be regarded as infinite.) Therefore, we have $T_j^{L_j} = E$. The eigenvalue which satisfies this relation and $T_j\psi(\boldsymbol{r}) = f_j\psi(\boldsymbol{r})$ is identified to be $f_j = \exp(2\pi i m_j/L_j)$ $(j = 1, 2, 3)$, where $m_j$ stands for an arbitrary integer. Then, for a general $T$ given by (2.148), the eigenvalue $f$ which satisfies (2.145) is found to be

$$f = \mathrm{e}^{i\boldsymbol{k}\cdot\boldsymbol{R}}, \tag{2.149}$$

with (2.116) and

$$\boldsymbol{k} = \frac{m_1}{L_1}\boldsymbol{b}_1 + \frac{m_2}{L_2}\boldsymbol{b}_2 + \frac{m_3}{L_3}\boldsymbol{b}_3 \tag{2.150}$$

($\boldsymbol{b}_1$, $\boldsymbol{b}_2$, $\boldsymbol{b}_3$ are the primitive reciprocal-lattice vectors). For every $j = 1, 2, 3$, there are only $L_j$ independent integer values for $m_j$, and one may confine the wave vector $\boldsymbol{k}$ to the inside of the first Brillouin zone. The already-stated Bloch theorem (2.119) (or (2.121)) can now be readily derived from (2.145) and (2.149).

Many crystals have not only the translation symmetry but also rotation, mirror and inversion symmetries. Such symmetry operators other than translations form also a group, called the point group. The point group also exists for molecules [139]. Unlike the translation group, the elements in the point group do not, in general, commute with each other, i.e. the group is non-Abelian. In the case of a non-Abelian group, all the group elements $P'$ which can be made from one group element $P$ by applying similarity transformations $P' = Q^{-1}PQ$ (where $Q$ is also a group element) form a subgroup called a "class" of the group. For example, let us consider the point group $C_{3v}$ characterizing a symmetry of a two-dimensional graphite (honeycomb) sheet or a planar benzene molecule (or more simply a regular triangle), which is illustrated in Fig. 2.6. The elements of this group are given as follows: $I$ is the inversion operation; $C_{3z}$ and $C_{3z}^{-1}$ are rotations through $2\pi/3$ about the $Oz$ axis perpendicular

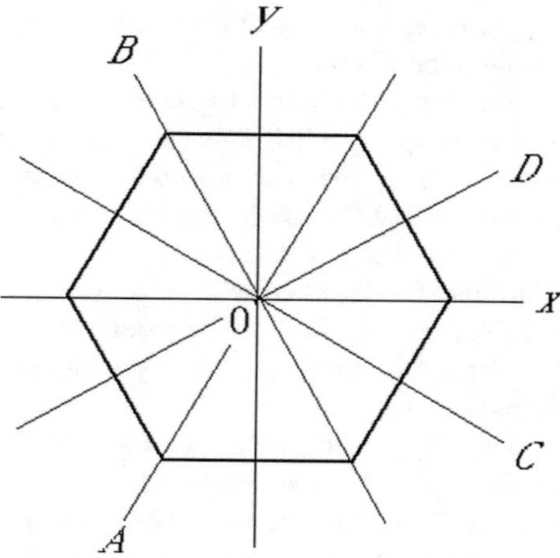


**Fig. 2.6** Axes defining symmetry operations in the point group $C_{3v}$

to the horizontal plane of this page; and $IC_{2y}$, $IC_{2C}$ and $IC_{2D}$ are reflections in the planes which are normal to $Oy$, $OC$ and $OD$ axes, respectively, and through the origin $O$. Then, applying the transformation $Q^{-1}PQ$ to the threefold rotation operator $R_{3z}$ around the $z$ axis gives $J^{-1}R_{3z}J = R_{3z}$ and $(IR_{2y})^{-1}R_{3z}(IR_{2y}) = R_{3z}^{-1}$. Therefore, $R_{3z}$ and $R_{3z}^{-1}$ belong to the same class in this group. Sometimes, point-group operations may be accompanied by a fractional translation. A space group in which this occurs is called a nonsymmorphic group and is distinguished from the usual symmorphic group. Hereafter, we restrict ourselves to the case of symmorphic space groups.

The translation group and the point group are both subgroups of the space group. In the case of isolated molecules (clusters), the point group defines the complete symmetry of the wavefunctions. However, in the case of crystals, the situation becomes slightly more complicated. Let us use the symbol $P$ for an element of the point group. Using the Bloch sum (2.131), we obtain

$$\begin{aligned} P\psi_{\boldsymbol{k}}(\boldsymbol{r}) &= \psi_{\boldsymbol{k}}(P\boldsymbol{r}) = \sum_{\boldsymbol{R}} \mathrm{e}^{i\boldsymbol{k}\cdot\boldsymbol{R}}\phi(P\boldsymbol{r}-\boldsymbol{R}) \\ &\times \sum_{\boldsymbol{R}'=P^{-1}\boldsymbol{R}} \mathrm{e}^{iP^{-1}\boldsymbol{k}\cdot\boldsymbol{R}'}\phi(P\{\boldsymbol{r}-\boldsymbol{R}'\}). \end{aligned} \tag{2.151}$$

This is none other than the Bloch-sum wavefunction at $\boldsymbol{k}^* = P^{-1}\boldsymbol{k}$ (or $\boldsymbol{k}^* = P^{-1}\boldsymbol{k}+\boldsymbol{G}$). The various points $\boldsymbol{k}^* = P^{-1}\boldsymbol{k}+\boldsymbol{G}$ appearing inside the first Brillouin zone are called the "stars" of $\boldsymbol{k}$. For example, if we write the inversion operator as $I$, obviously $I\boldsymbol{k} = -\boldsymbol{k}$ is a star of $\boldsymbol{k}$. Clearly, the corresponding electronic levels at these different stars of $\boldsymbol{k}$ are degenerate energetically. Owing to the commutation relation $[\mathcal{H}, P] = 0$, the eigenfunctions of the point group $\{P\}$ can be constructed as linear combinations of wavefunctions $\psi_{\boldsymbol{k}^*}(\boldsymbol{r})$ at the stars $\boldsymbol{k}^*$ of $\boldsymbol{k}$. Note that they are *not* the eigenfunctions of the translation group, except for $\boldsymbol{k} = 0$; this point ($\boldsymbol{k} = 0$) is called the $\Gamma$ point.

Because all the stars of $\boldsymbol{k}$ have the same energy, it is useful to confine ourselves to the irreducible first Brillouin zone, which is the minimum volume of the first Brillouin zone folded by the point-group operations. In the case of full cubic symmetry, this area is $1/48$ of the whole first Brillouin zone: see Fig. 2.5 for BCC and FCC. In this irreducible first Brillouin zone, all the $\boldsymbol{k}$ points are independent and the electronic states can be classified with respect to the wave vector $\boldsymbol{k}$. (For example, the charge density $\rho(\boldsymbol{r})$ in (2.142) can be produced by summing over $\boldsymbol{k}$ inside only the irreducible first Brillouin zone and then by symmetrizing $\rho(\boldsymbol{r})$ by means of the point group operations.)

Therefore, using information on the symmetries of $\mathcal{H}$ reduces greatly the task of computing for the ground state of the system. Good quantum numbers can be identified and many fewer states need to be computed. There are symmetry points, symmetry axes or symmetry planes in the Brillouin zone such that some of the stars of $\boldsymbol{k}$ coincide with $\boldsymbol{k}$ itself. The $\Gamma$ point is the highest symmetry point; all the stars of $\boldsymbol{k} = 0$ lie at $\boldsymbol{k} = 0$. Also, all the vertices of the polyhedron of the first Brillouin zone are symmetry points. If there is a twofold rotational operator $R_2$ around the (100) direction in the point group, the $X$ axis in the Brillouin zone is a symmetry axis, since $R_2\boldsymbol{k} = \boldsymbol{k}$ on this axis. Usually, capital Roman or Greek letters ($X$, $K$, $M$, ...) are used to indicate symmetry points and symmetry axes. If there is a mirror symmetry on the $XY$ plane, this plane in the Brillouin zone is also a symmetry plane: $M\boldsymbol{k} = \boldsymbol{k}$ on this plane ($M$ is the reflection operator).

The group theory is especially useful in understanding the degeneracy of electronic energy levels, as well as in constructing the symmetrized wavefunction, at certain $\boldsymbol{k}$ points. For these higher-symmetry $\boldsymbol{k}$ points, there are several point-group operations $P_{\boldsymbol{k}}$ which transform $\boldsymbol{k}$ into itself. The set $\{P_{\boldsymbol{k}}\}$ forms again a group (a subgroup of the point group), $\mathcal{G}(\boldsymbol{k})$, which we will call the *point group of the wave vector* $\boldsymbol{k}$. Suppose that a set of $n$-dimensional (i.e. $n \times n$) matrices $M(P_{\boldsymbol{k}})$ forms a representation of the point group at the $\boldsymbol{k}$ point, and that $\psi_{\boldsymbol{k}1}(\boldsymbol{r}), \psi_{\boldsymbol{k}2}(\boldsymbol{r}), \ldots, \psi_{\boldsymbol{k}n}(\boldsymbol{r})$ are a set of linearly independent wavefunctions of the $n$-fold degenerate level such that

$$P_{\boldsymbol{k}}\psi_{\boldsymbol{k}\lambda} = \sum_{\nu=1}^{n} M(P_{\boldsymbol{k}})_{\lambda\nu}\psi_{\boldsymbol{k}\nu}(\boldsymbol{r}), \qquad \lambda = 1, 2, \ldots, n. \tag{2.152}$$

If we adopt, in place of the $\psi$s, the functions $\phi_{\boldsymbol{k}1}(\boldsymbol{r}), \phi_{\boldsymbol{k}2}(\boldsymbol{r}), \ldots, \phi_{\boldsymbol{k}n}(\boldsymbol{r})$ which are related to the $\psi$s by the similarity transformation

$$\phi_{\boldsymbol{k}\lambda}(\boldsymbol{r}) = \sum_{\nu=1}^{n} A_{\lambda\nu}\psi_{\boldsymbol{k}\nu}(\boldsymbol{r}), \qquad \lambda = 1, 2, \ldots, n, \tag{2.153}$$

as basis functions, then the matrix representation $M(P_{\boldsymbol{k}})$ changes to $M'(P_{\boldsymbol{k}}) = A^{-1}M(P_{\boldsymbol{k}})A$. Therefore, each representation is associated with its own basis functions. By means of this similarity transformation, any basis function can be rearranged so as to satisfy the orthonormality condition:

$$\int \phi^*_{k\lambda}(r)\phi_{k\nu}(r)dr = \delta_{\lambda\nu}. \tag{2.154}$$

A representation $M(P_k)$ for basis functions which satisfy the orthonormality condition (2.154) is unitary, and is called a "unitary representation".

If a matrix representation of a group $\mathcal{G}(k)$ can be written in a block-diagonal form by applying a suitable similarity transformation, this representation is called a "reducible representation". If such a transformation is not possible by any similarity transformation, the representation is called an "irreducible representation".

*Schur's lemma* states that a matrix which commutes with all the matrices of an irreducible representation must be unit matrix. Thus, if a matrix which is different from unit matrix commutes with all matrices of a representation, the representation is reducible.

This lemma appears at an intermediate stage of the proof of the following orthogonality theorem for matrices, which is very important for the rest of the theory.

The orthogonality theorem for matrices states that the orthogonality equation

$$\sum_{P_k} M^i(P_k)^*_{\kappa\mu} M^j(P_k)_{\lambda\nu} = \frac{m}{n_i}\delta_{ij}\delta_{\kappa\lambda}\delta_{\mu\nu} \tag{2.155}$$

holds for two unitary, irreducible representations $M^i$ and $M^j$ of a group, which are not equivalent for $i \neq j$. In (2.155), the summation is over all the elements $P_k$ of the group (of the wave vector $k$), $m$ is the "order" (i.e. the number of independent elements) of the group; and $n_i$ is the dimension of $M^i$.

Now we define the "character" $\chi(P_k)$ as the trace (i.e. the sum of all the diagonal matrix elements) of the $n$-dimensional matrix representation of a group $\mathcal{G}(k)$:

$$\chi(P_k) = \sum_{i=i}^{n} M(P_k)_{ii}. \tag{2.156}$$

The set of characters corresponding to a representation is called the "character system" of the representation. Note that the characters are unaltered under similarity transformations, and a necessary and sufficient condition for the equivalence of two representations is the identity of their character systems. That is, equivalent representations have the same character system. Moreover, if $P_k$ and $P'_k$ belong to the same class of the group of the wave vector $k$, the corresponding characters $\chi(P_k)$ and $\chi(P'_k)$ are identical, since there is a group element $Q_k$ such that $P'_k = Q_k^{-1} P_k Q_k$ by definition, and $M(P'_k) = [M(Q_k)]^{-1} M(P_k) M(Q_k)$. One should also note that the character of the identity element is equal to the dimension of the representation $n$, because the identity element is represented by a unit matrix.

Evaluating the trace of $M^i(P_k)^*_{\kappa\mu}$ and $M^j(P_k)_{\lambda\nu}$ in (2.155), we find

$$\sum_{P_k} \chi^i(P_k)^* \chi^j(P_k) = m\delta_{ij} \tag{2.157}$$

for an irreducible representation. In particular, for $i = j$, we have

$$\sum_{P_k} |\chi(P_k)|^2 = m, \tag{2.158}$$

where $m$ is the number of elements of the group $\mathcal{G}(\boldsymbol{k})$. Since this equation does not hold for reducible representations, it can be used to check if the representation is irreducible or not. If we classify all the group elements $\{P_k\}$ into classes $\{C_\alpha\}$, (2.157) can be rewritten as

$$\sum_{\alpha} n_\alpha \chi^i(C_\alpha)^* \chi^j(C_\alpha) = m\delta_{ij}, \tag{2.159}$$

where the summation is over all the different classes, and $n_\alpha$ is the number of elements in the $\alpha$th class $C_\alpha$. Since the basis functions belonging to the different representations are orthogonal to each other, we can derive the orthogonality relation for the characters:

$$\sum_{i} n_\alpha \chi^i(C_\alpha)^* \chi^i(C_\beta) = m\delta_{\alpha\beta} \tag{2.160}$$

One may prove the following additional properties of the character systems.

(1) The number of inequivalent irreducible representations of a group $\mathcal{G}(\boldsymbol{k})$ is equal to the number of classes of $\mathcal{G}(\boldsymbol{k})$.
(2) The sum of the squares of the dimensions of the inequivalent irreducible representations is equal to the number $m$ of elements in the group.

Using these properties together with (2.159) and (2.160), one can construct character systems directly, without first obtaining explicit forms for the matrices. The character systems of the irreducible representations are conveniently displayed in the form of a character table. The classes of the group are usually listed along the top of the table, and the irreducible representations down the left-hand side. As a consequence of the property (1) mentioned above, this table is always square. For example, character tables for the point groups $C_{3v}$ (for a two-dimensional graphite sheet) and $O_h$ (for cubic symmetries including SC, FCC and BCC) are given in Tables 2.1 and 2.2, respectively. In the case of $C_{3v}$, we find that $\Gamma^1$ and $\Gamma^2$ are one-dimensional, and that $\Gamma^3$ is two-dimensional. One may easily check that all the properties of characters mentioned above are satisfied. For other symmetries, one may refer to [140, 141], for example.

There is a different kind of symmetry called the time–reversal symmetry. This symmetry is related to taking the complex conjugate of the Schrödinger (or Kohn–Sham) equation. The time–reversal operator $\mathcal{T}$, which changes $t$ to $-t$, transforms $\boldsymbol{r}$, $\boldsymbol{p}$ and $\boldsymbol{\sigma}$, respectively, onto $\boldsymbol{r}$, $-\boldsymbol{p}$ and $-\boldsymbol{\sigma}$. However, the Hamiltonian $\mathcal{H}$ does

**Table 2.1** Character table for the point group $C_{3v}$. $C$ and $D$ are the $xy$ in-plane axes, which are rotated $\pm 60°$ the $y$ axis

| | $E$ | $R_{3z}, R_{3z}^{-1}$ | $IR_{2y}, IR_{2C}, IR_{2D}$ |
|---|---|---|---|
| $\Gamma^1$ | 1 | 1 | 1 |
| $\Gamma^2$ | 1 | 1 | −1 |
| $\Gamma^3$ | 2 | −1 | 0 |

**Table 2.2** Character table for the $O_h$ group (for cubic symmetries) of the wave vector $\boldsymbol{k}$ corresponding to the $\Gamma$, $H$ and $R$ points. There are ten classes in the $O_h$ group. The number of the elements in each class is indicated under one typical operation at the top of this table. There are four one-dimensional, two two-dimensional and four three-dimensional representations

| | $E$ | $R_{3\alpha}$ | $R_{2x}$ | $R_{4x}$ | $R_{2a}$ | $I$ | $IR_{3\alpha}$ | $IR_{2x}$ | $IR_{4x}$ | $IR_{2a}$ |
|---|---|---|---|---|---|---|---|---|---|---|
| | 1 | 8 | 3 | 6 | 6 | 1 | 8 | 3 | 6 | 6 |
| $\Gamma_1$ | 1 | 1 | 1 | 1 | 1 | 1 | 1 | 1 | 1 | 1 |
| $\Gamma_2$ | 1 | 1 | 1 | −1 | −1 | 1 | 1 | 1 | −1 | −1 |
| $\Gamma_{12}$ | 2 | −1 | 2 | 0 | 0 | 2 | −1 | −1 | 0 | 0 |
| $\Gamma_{15'}$ | 3 | 0 | −1 | 1 | −1 | 3 | 0 | 0 | 1 | −1 |
| $\Gamma_{25'}$ | 3 | 0 | −1 | −1 | 1 | 3 | 0 | 0 | −1 | 1 |
| $\Gamma_{1'}$ | 1 | 1 | 1 | 1 | 1 | −1 | −1 | −1 | −1 | −1 |
| $\Gamma_{2'}$ | 1 | 1 | 1 | −1 | −1 | −1 | −1 | −1 | 1 | 1 |
| $\Gamma_{12'}$ | 2 | −1 | 2 | 0 | 0 | −2 | 1 | −2 | 0 | 0 |
| $\Gamma_{15}$ | 3 | 0 | −1 | 1 | −1 | −3 | 0 | 1 | −1 | 1 |
| $\Gamma_{25}$ | 3 | 0 | −1 | −1 | 1 | −3 | 0 | 1 | 1 | −1 |

not change, i.e. $\mathcal{H}$ and $\mathcal{T}$ commute. This statement holds also when the spin–orbit coupling exists in $\mathcal{H}$ (see (2.222) in Sect. 2.2.10), unless there is neither external magnetic field nor spontaneous magnetization. On the other hand, the operator $\mathcal{T}$ transforms $\psi_{\boldsymbol{k}\lambda\uparrow}(\boldsymbol{r})$ onto $\psi_{-\boldsymbol{k}\lambda\downarrow}(\boldsymbol{r})$, and $\psi_{\boldsymbol{k}\lambda\downarrow}(\boldsymbol{r})$ onto $\psi_{-\boldsymbol{k}\lambda\uparrow}(\boldsymbol{r})$. Therefore, in the absence of an external magnetic field or a spontaneous magnetization, we find

$$\varepsilon_{-\boldsymbol{k}\lambda\downarrow} = \varepsilon_{\boldsymbol{k}\lambda\uparrow}, \qquad \varepsilon_{-\boldsymbol{k}\lambda\uparrow} = \varepsilon_{\boldsymbol{k}\lambda\downarrow}. \tag{2.161}$$

In addition, if the crystal has also an inversion symmetry, i.e. if the Hamiltonian $\mathcal{H}$ commutes also with the inversion operator $I$, we have

$$\varepsilon_{\boldsymbol{k}\lambda\uparrow} = \varepsilon_{\boldsymbol{k}\lambda\downarrow}. \tag{2.162}$$

Another interesting symmetry argument is possible for bulk antiferromagnetic materials. The readers may refer to Ref. [726, 727] for details.

### 2.3.3 LCAO, OPW and Mixed-Basis Approaches

One of the standard approaches used in ab initio calculations is to expand the effective one-electron wavefunctions by means of basis functions, and to construct and diagonalize the Hamiltonian matrix $H(\boldsymbol{k})$ for each $\boldsymbol{k}$ in the first Brillouin zone. We call this approach the "spectral method". Note that, in calculations for isolated clusters, the wave vector $\boldsymbol{k}$ can be always set to zero.

For example, the localized orbitals $\phi_\lambda(\boldsymbol{r}-\boldsymbol{R})$ used in the Bloch sum (2.131) may be expressed by a superposition of several "atomic orbitals" (AOs):

$$\phi_\lambda(\boldsymbol{r}-\boldsymbol{R}) = \sum_i \sum_{n_i} C^{\lambda}_{i,n_i} \varphi_{n_i}(\boldsymbol{r}-\boldsymbol{R}-\boldsymbol{\tau}_i), \tag{2.163}$$

where $\varphi_{n_i}(\boldsymbol{r}-\boldsymbol{R}-\boldsymbol{\tau}_i)$ is the $n_i$th atomic orbital located at the $i$th atomic position $\boldsymbol{R}+\boldsymbol{\tau}_i$ inside the unit cell denoted by $\boldsymbol{R}$. Note that the Bloch sum (2.131) constructed from the localized orbitals given by (2.163) is equivalent to the linear combination

$$\psi_{\boldsymbol{k}\lambda}(\boldsymbol{r}) = \sum_i \sum_{n_i} C^{\boldsymbol{k}\lambda}_{i,n_i} \langle \boldsymbol{r}|\boldsymbol{k}, i, n_i\rangle \tag{2.164}$$

of "Bloch-sum basis functions" $\langle \boldsymbol{r}|\boldsymbol{k}, i, n_i\rangle$ defined as

$$\langle \boldsymbol{r}|\boldsymbol{k}, i, n_i\rangle = \frac{1}{\sqrt{N}} \sum_{\boldsymbol{R}} \mathrm{e}^{i\boldsymbol{k}\cdot\boldsymbol{R}} \varphi_{n_i}(\boldsymbol{r}-\boldsymbol{R}-\boldsymbol{\tau}_i), \tag{2.165}$$

where $N$ is the number of unit cells in the whole system. The method of constructing localized orbitals (2.163) first is often referred to as a localized-orbital approach. Such an approach is frequently used in $O(N)$ methods (see Sects. 2.3.8 and 3.3.1) or tight-binding methods (see Chap. 3). However, in the usual ab initio methods, a set of equations of the form (2.164) and (2.165) is more commonly used. This approach is called the "linear combination of atomic orbitals" (LCAO) and is similar to that used in the self-consistent-field approach to the Hartree–Fock method; see Sect. 2.2.2. In order to distinguish it from the Hartree–Fock approach, one may call it the LCAO density functional approach. When only one Bloch sum is introduced for each occupied electronic state, i.e. when the number of Bloch sums used in the calculation is the same as the number of electronic states in the unit cell, such a special choice of the Bloch sums (atomic orbitals) is called the minimal basis. However, strictly speaking, the minimal basis is not the LCAO in its original meaning. The use of the minimal basis and the semi-empirical two-center fitting of the matrix elements [142] yields the tight-binding method; see Chap. 3 for more details.

For ab initio calculations which use atomic orbitals as the basis, the Hamiltonian matrix elements are given by

$$
\begin{aligned}
H_{ij}(\boldsymbol{k}, n_i; \boldsymbol{k}', n_j) &= \langle \boldsymbol{k}, i, n_i | \mathcal{H} | \boldsymbol{k}', j, n_j \rangle \\
&= \frac{1}{N} \sum_{\boldsymbol{R},\boldsymbol{R}'} \mathrm{e}^{-i(\boldsymbol{k}\cdot\boldsymbol{R}-\boldsymbol{k}'\cdot\boldsymbol{R}')} \int \varphi^*_{n_i}(\boldsymbol{r}-\boldsymbol{R}-\boldsymbol{\tau}_i)\mathcal{H}\varphi_{n_j}(\boldsymbol{r}-\boldsymbol{R}'-\boldsymbol{\tau}_j)d\boldsymbol{r} \\
&= \delta_{\boldsymbol{k}\boldsymbol{k}'} \sum_{\boldsymbol{R}} \mathrm{e}^{-i\boldsymbol{k}\cdot\boldsymbol{R}} \int \varphi_{n_i}(\boldsymbol{r}-\boldsymbol{\tau}_i)\mathcal{H}\varphi_{n_j}(\boldsymbol{r}+\boldsymbol{R}-\boldsymbol{\tau}_j)d\boldsymbol{r}.
\end{aligned}
\tag{2.166}
$$

Similarly, the overlap matrix element is given by

$$
\begin{aligned}
S_{ij}(\boldsymbol{k}, n_i; \boldsymbol{k}', n_j) &= \langle \boldsymbol{k}, i, n_i | \boldsymbol{k}', j, n_j \rangle \\
&= \frac{1}{N} \sum_{\boldsymbol{R},\boldsymbol{R}'} \mathrm{e}^{-i(\boldsymbol{k}\cdot\boldsymbol{R}-\boldsymbol{k}'\cdot\boldsymbol{R}')} \int \varphi_{n_i}(\boldsymbol{r}-\boldsymbol{R}-\boldsymbol{\tau}_i)\varphi_{n_j}(\boldsymbol{r}-\boldsymbol{R}'-\boldsymbol{\tau}_j)d\boldsymbol{r} \\
&= \delta_{\boldsymbol{k}\boldsymbol{k}'} \sum_{\boldsymbol{R}} \mathrm{e}^{-i\boldsymbol{k}\cdot\boldsymbol{R}} \int \varphi_{n_i}(\boldsymbol{r}-\boldsymbol{\tau}_i)\varphi_{n_j}(\boldsymbol{r}+\boldsymbol{R}-\boldsymbol{\tau}_j)d\boldsymbol{r}.
\end{aligned}
\tag{2.167}
$$

Then, the Schrödinger equation (2.123) is written down in the form of a generalized eigenvalue problem

$$
\sum_j \sum_{n_j} H_{ij}(n_i, n_j; \boldsymbol{k}) C^{\boldsymbol{k}\lambda}_{j,n_j} = \varepsilon_{\boldsymbol{k}\lambda} \sum_j \sum_{n_j} S_{ij}(n_i, n_j; \boldsymbol{k}) C^{\boldsymbol{k}\lambda}_{j,n_j}
\tag{2.168}
$$

where $C^{\boldsymbol{k}\lambda}_{j,n_j}$ denotes the expansion coefficient of the $n_j$th AO basis function of the $j$th atom for a given electronic state $\boldsymbol{k}$ and $\lambda$. In order to simplify notations, one may write (2.168) as

$$
\sum_j H_{ij}(\boldsymbol{k}) C^{\boldsymbol{k}\lambda}_j = \varepsilon_{\boldsymbol{k}\lambda} \sum_j S_{ij}(\boldsymbol{k}) C^{\boldsymbol{k}\lambda}_j.
\tag{2.169}
$$

Here $i$ and $j$ are the abbreviations of $(i, n_i)$ and $(j, n_j)$, respectively.

The atomic orbitals used for Bloch sums are either Slater-type orbitals (STOs), Gaussian-type orbitals (GTOs), or numerical orbitals. To evaluate the matrix elements in (2.166) and (2.167), one may use either analytic formulae or numerical integration, using, for example, a Monte Carlo technique or a global mesh integration.

To treat finite systems such as clusters or molecules, one usually uses AOs as a basis; this is called the molecular-orbital (MO) approach. (The term, "LCAO–MO approach", is also used to distinguish this approach from the LCAO approach for crystals.) Such is the case for the discrete variational (DV) method (the DV–$X\alpha$ method) [143–148], where numerical orbitals and numerical integrations based on point sampling are used.

On the other hand, as mentioned in Sects. 2.2.4 and 2.2.7, one may alternatively introduce the plane wave (PW) expansion method, which uses plane waves given by (2.130) as basis functions. The PW basis set offers a complete orthonormal set, which is capable of describing spatially extended wavefunctions of valence orbitals with a rather limited number of PWs, and is convenient for expressing all the matrix elements analytically, provided that the Fourier components of the effective one-electron potential are known. Usually, the PW expansion method is combined with the pseudopotential approach (see Sect. 2.3.4), which enables one to omit core electrons, or the projector-augmented-wave (PAW) approach [149], which is basically same as the pseudopotential approach but establishes the relation to the core orbitals. In the PW expansion, the Hamiltonian matrix elements are given by

$$\begin{aligned} H_{\boldsymbol{G}\boldsymbol{G}'}(\boldsymbol{k}) &= \langle \boldsymbol{k}+\boldsymbol{G}|\mathcal{H}|\boldsymbol{k}+\boldsymbol{G}'\rangle = \frac{1}{\Omega}\int \mathrm{e}^{-i(\boldsymbol{k}+\boldsymbol{G})\cdot\boldsymbol{r}}\,\mathcal{H}\,\mathrm{e}^{i(\boldsymbol{k}+\boldsymbol{G}')\cdot\boldsymbol{r}}d\boldsymbol{r} \\ &= \frac{1}{2}|\boldsymbol{k}+\boldsymbol{G}|^2\delta_{\boldsymbol{G}\boldsymbol{G}'} + \widetilde{V}(\boldsymbol{G}-\boldsymbol{G}'), \end{aligned} \tag{2.170}$$

where $\widetilde{V}$ denotes a Fourier transform of the self-consistent potential $V(\boldsymbol{r})$, while the overlap matrix is the identity matrix:

$$S_{\boldsymbol{G}\boldsymbol{G}'}(\boldsymbol{k}) = \langle \boldsymbol{k}+\boldsymbol{G}|\boldsymbol{k}+\boldsymbol{G}'\rangle = \delta_{\boldsymbol{G}\boldsymbol{G}'}. \tag{2.171}$$

The naive PW expansion method has a problem of the convergence of the electronic states with respect to the number of the PW basis functions. In order to reduce the necessary number of PWs, the orthogonalized-plane-wave (OPW) method [150–153] was considered and used, historically, mainly before the pseudopotential approach has been established. The orthogonalized plane wave (OPW) denotes a PW orthogonalized to all the core orbitals. Since the wavefunctions of the valence states are all orthogonal to the core orbitals, they are conveniently represented by the OPWs. An OPW is constructed by superposing a PW and core orbitals. Writing the projection operator to the core orbitals $|i\rangle$ as

$$P = \sum_i^{\text{core}} |i\rangle\langle i|, \tag{2.172}$$

one may introduce an OPW as

$$|\text{OPW}\ \boldsymbol{k}+\boldsymbol{G}\rangle = (1-P)|\boldsymbol{k}+\boldsymbol{G}\rangle = |\boldsymbol{k}+\boldsymbol{G}\rangle - \sum_i^{\text{core}} |i\rangle\langle i|\boldsymbol{k}+\boldsymbol{G}\rangle, \tag{2.173}$$

where $|\boldsymbol{k}+\boldsymbol{G}\rangle$ is a plane wave with a wavevector $\boldsymbol{k}+\boldsymbol{G}$. Since true wavefunctions $|\boldsymbol{k}\lambda\rangle$ are given by linear combinations of OPWs, they are expressed by

$$|\boldsymbol{k}\lambda\rangle = (1-P)|\boldsymbol{k}\lambda), \tag{2.174}$$

where $|\boldsymbol{k}\lambda)$ takes a form of plane wave expansion

$$|\boldsymbol{k}\lambda) = \sum_{\boldsymbol{G}} C_{\boldsymbol{k}\boldsymbol{G}}^{\lambda}|\boldsymbol{k}+\boldsymbol{G}\rangle. \tag{2.175}$$

The OPW basis functions $|\text{OPW}\ \boldsymbol{k}+\boldsymbol{G}\rangle$ defined by (2.173) are made up of PWs orthogonalized to all core states. Therefore, the secular equation formed by the OPWs can give only valence and conduction energies and wavefunctions. In the OPW method, core states may be analytically represented by the STOs [150, 153, 154].

Deegan and Twose [155] used the muffin-tin potential (for the muffin-tin potential, see Sect. 2.2.8) to determine the core orbitals. Then they constructed the OPWs by making PWs orthogonal to these muffin-tin orbitals of core states. Euwema and Stukel [156] used Herman–Skillman's numerical core AOs defined inside non-overlapping atomic spheres. The evaluation of matrix elements required in the OPW method is essentially the same as that in the mixed-basis approach discussed below.

There is another approach, called the "mixed-basis approach" [157–161], which uses both plane waves (2.130) and Bloch sums made of atomic orbitals, (2.165), as the expansion basis (Fig. 2.7). In this mixed-basis approach also, the PW–PW matrix elements, i.e. $\langle \text{PW}|\mathcal{H}|\text{PW}\rangle$ and $\langle \text{PW}|\text{PW}\rangle$, are the same as those given by (2.170) and (2.171). The cross terms, i.e. the PW–AO elements, are given by

$$\langle \boldsymbol{k}+\boldsymbol{G}|\mathcal{H}|\boldsymbol{k}, i, n_i\rangle = \sum_{\boldsymbol{G}'}\left[\frac{1}{2}|\boldsymbol{k}+\boldsymbol{G}|^2\delta_{\boldsymbol{G}\boldsymbol{G}'} + \widetilde{V}(\boldsymbol{G}-\boldsymbol{G}')\right]\mathrm{e}^{-i(\boldsymbol{k}+\boldsymbol{G}')\cdot\boldsymbol{\tau}_i}\widetilde{\varphi}_{n_i}(\boldsymbol{k}+\boldsymbol{G}'), \quad (2.176)$$

$$\langle \boldsymbol{k}+\boldsymbol{G}|\boldsymbol{k}, i, n_i\rangle = \mathrm{e}^{-i(\boldsymbol{k}+\boldsymbol{G})\cdot\boldsymbol{\tau}_i}\widetilde{\varphi}_{n_i}(\boldsymbol{k}+\boldsymbol{G}), \quad (2.177)$$

where we put

$$\widetilde{\varphi}_{n_i}(\boldsymbol{k}+\boldsymbol{G}') = \frac{1}{\sqrt{\Omega}}\int \varphi_{n_i}(\boldsymbol{r})\mathrm{e}^{-i(\boldsymbol{k}+\boldsymbol{G})\cdot\boldsymbol{r}}d\boldsymbol{r} \quad (2.178)$$

(the integral is the whole space integral; see also (2.135)). Lastly, the AO–AO matrix elements are again the same as those given in the LCAO approach, i.e. (2.166) and (2.167).

In one kind of various mixed-basis formulations, one may introduce core AOs only in order to describe the core states accurately. Then, if the minimal set is used just for the core AOs, the method becomes closely related and very similar to the OPW method. However, unlike the OPW method, the secular equation includes also core states, so that, in this alternative method, not only the valence states but also the core states are obtained simultaneously by solving the secular equation. Such a mixed-basis method was developed first by Kunz [157] in 1968 and by Butler et al. [158] in 1969, and later by Bendt and Zunger [159] in 1982. In the same paper, Bendt and Zunger proposed also a new and efficient variational method with respect to the self-consistent potential (see also Sect. 2.2.3). All these authors introduced AOs only for core orbitals, which are determined in nonoverlapping atomic spheres. As an approximation, Bendt and Zunger replaced every Hamiltonian matrix element $\langle \text{AO}|\mathcal{H}|\text{PW}\rangle$, which is, in the usual programs, one of the most time-consuming parts, with the product of the corresponding overlap matrix element $\langle \text{AO}|\text{PW}\rangle$ and the diagonal AO–AO matrix element $\langle \text{AO}|\mathcal{H}|\text{AO}\rangle$. This replacement is justified only if these AOs are close to the exact eigenfunctions, and cannot be applied generally to the valence orbitals.

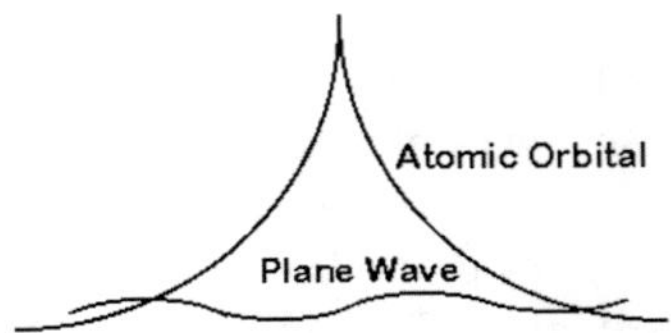


**Fig. 2.7** Representation of a wavefunction in the mixed-basis approach

To include valence AOs in the mixed-basis formulation, the use of GTOs is rather convenient, because the matrix elements between a PW and an AO can be evaluated conveniently by means of a summation with respect to the reciprocal-lattice vector $\boldsymbol{G}$. The Hamiltonian matrix elements between AOs are usually evaluated by means of real-space integrals. Euwema [160] proposed such an approach using the GTOs and confirmed that the wavefunctions are reasonably well described by the GTOs except for the deepest core states. (By this reason, Euwema's method has not been used widely.) This approach was conveniently and successfully combined with the pseudopotential approach, which will be described in the next subsection. In this case the method is called the pseudopotential mixed-basis approach. Such an approach was first developed by Louie et al. [161] in 1979 in order to treat the localized character of 3d electrons, and has also been used up to recently [162–164]. In the mixed-basis approach, if one introduces enough PWs, overcompleteness of the basis set makes it hard to determine uniquely the coefficients of the expansion of the eigenstates in the basis functions.

As has been stated so far, it is a natural extension of the mixed-basis method to incorporate AOs for both core and valence orbitals, in order to take all electrons into account. Then one may avoid the complexity of introducing pseudopotentials and may treat core levels also, although the calculation in core regions needs accurate treatments. We call such an approach, which represents both valence and core electrons in terms of AOs and PWs, the "all-electron mixed-basis approach". The use of STOs is one way to formulate this approach. In 1977, Friedli and Ashcroft [165] first applied such an approach to metallic hydrogen, although a hydrogen system is very special and simple. The extension of a similar all-electron mixed-basis approach using STOs to first-row elements has been performed by the author's group and applied mainly to molecular-dynamics simulations of carbon systems [166–170]. The STO damping parameters for valence orbitals may be chosen so as to confine the AOs inside nonoverlapping spheres. Then, it becomes unnecessary to evaluate any matrix elements between two adjacent AOs. This enables one to eliminate a complicated computation of two- and three-center integrals which is usually required for ab initio methods using the STO–LCAO basis. In the case of the mixed-basis approach, the interatomic region which is not described in the shrunken (or truncated) AOs can be supplemented by PWs. As another advantage of such choice of valence AOs, the overcompleteness problem can be avoided simultaneously for most purposes, also. Since the atomic orbitals of the core states, say 1s, are extremely localized, one has to evaluate the charge density, potential, matrix elements and forces analytically as far as possible. In order to deal with heavier elements accurately, representation by a single STO is not enough to describe each core orbital.

For heavier elements, it is desirable to use numerical atomic orbitals (NAOs), which are determined precisely by an atomic code. A complete all-electron mixed-basis program using not only core but also valence AOs has been developed in the author's group [171–173]. It uses NAOs, which are determined precisely (for core orbitals) by using (a logarithmic-mesh version of) the Herman–Skillman atomic code [174]. The charge density, potential, matrix elements and forces must also be computed numerically as accurately as possible, either in real space along the radial direction or in $\boldsymbol{G}$ space. The Coulomb potentials used for PWs and AOs are evaluated separately by solving the Poisson equation $\nabla^2 V(\boldsymbol{r}) = \rho(\boldsymbol{r})$ in reciprocal space and real space, respectively. The exchange–correlation potential is also evaluated separately. This method has been applied to static calculations of Ne–Kr, C, Si, GaP and GaAs [172] and molecular-dynamics calculations for a defect in Fe crystal [173, 175] and foreign atom insertions into $C_{60}$ [176–182]. A similar all-electron mixed-basis approach has been developed independently by Wijers et al. [183], although these authors used muffin-tin approximation (see Sect. 2.2.8) and presented only a band structure calculation of Cu and Si. There are some other reports using the all-electron mixed-basis approach for different purposes [184, 185].

In any spectral method, the greater the ability of the basis expansion, the better the effective one-electron wavefunction (and in turn the potential felt by one electron). In all these approaches, the Schrödinger equation is written down in the form of a generalized eigenvalue problem, as given by (2.168). In particular, in the simple plane wave expansion, the overlap matrix $S$ is not necessary and (2.168) reduces to the standard eigenvalue equation. On the other hand, in the approaches using atomic orbitals (LCAO or mixed-basis,), it is necessary to evaluate the potential-energy, kinetic-energy and overlap matrix elements between different atomic orbitals. The generalized eigenvalue problem can be transformed into the standard eigenvalue problem by using a technique called the "Choleski decomposition" [161, 186], which decomposes the overlap matrix $S$ into the product of a lower triangular matrix $T$ and its Hermitian conjugate:

$$S = TT^{\dagger}. \tag{2.179}$$

Then, defining the coefficient vector for the orthonormalized basis as

$$D_i^{\boldsymbol{k}\lambda} = \sum_j T_{ij}^{\dagger} C_j^{\boldsymbol{k}\lambda} \tag{2.180}$$

and the Hamiltonian matrix for the orthonormalized basis as

$$\mathcal{H}' = T^{-1} \mathcal{H} T^{\dagger -1}, \tag{2.181}$$

we obtain the transformed equation (with the same eigenvalues $\varepsilon_{\boldsymbol{k}\lambda}$)

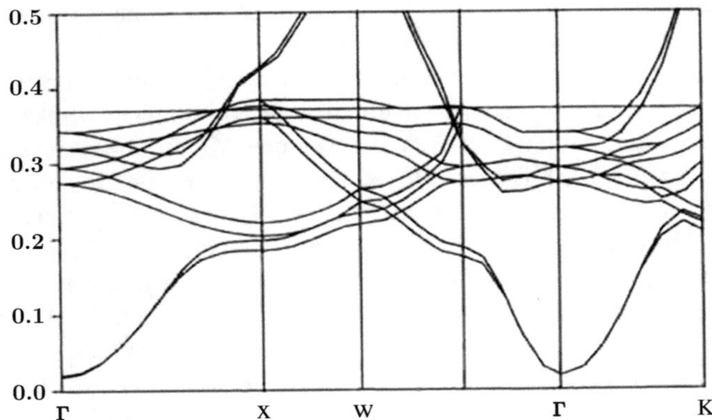


**Fig. 2.8** The band structure of FCC Ni calculated with the LSDA and the all-electron mixed-basis approach

$$\sum_j H'_{ij} D_j^{k\lambda} = \varepsilon_{k\lambda} D_i^{k\lambda} \tag{2.182}$$

in the space of the orthonormalized basis set.

The elements of the lower half-matrix $T^{-1}$ can be easily determined from the overlap matrix elements in a recursive way. After the problem for the $n \times n$ submatrix of $T^{-1}$ has been solved, the next problem for the $(n+1) \times (n+1)$ submatrix can be solved as follows. The additional elements of $T^{-1}$ are determined by

$$(T^\dagger)_{i,n+1} = \sum_{j=1}^{n} (T^{-1})_{i,j} S_{j,n+1} \tag{2.183}$$

$$(T^{-1})_{n+1,n+1} = \left( S_{n+1,n+1} - \sum_{i=1}^{n} (T^\dagger)^*_{i,n+1} (T^\dagger)_{i,n+1} \right)^{-1/2} \tag{2.184}$$

and

$$(T^{-1})_{n+1,i} = -(T^{-1})_{n+1,n+1} \sum_{j=i}^{n} (T^\dagger)^*_{j,n+1} (T^{-1})_{j,i}. \tag{2.185}$$

Figure 2.8 shows the energy band structure of FCC Ni calculated with the LSDA and the all-electron mixed-basis approach using numerical AOs. Each band is split into two subbands, corresponding to the up-spin state and the down-spin state. The horizontal straight line shows the position of the Fermi level. The band structures of many crystals of different elements can be found in compilations [78, 187, 188].

### 2.3.4 *Pseudopotential Approach*

Usually, the core electrons are not important in describing, for example, the nature of the bonding between atoms in a crystal; only the valence electrons which surround the core region contribute to it. A convenient technique which neglects the core electrons completely in the calculation scheme is the "pseudopotential" approach [58, 189] (For a comparison of the all-electron and pseudopotential approaches, see the discussion given later in Sect. 2.2.10.) In this scheme, the effective one-electron potential is given by the sum of this pseudopotential and the Coulomb potential due to the average valence electron density. The wavefunctions are normally described by a plane wave (PW) expansion, by the LCAO approach or by the mixed-basis approach.

The pseudopotential–PW expansion method was originally derived from the OPW method [58, 189] described in the previous section. Inserting (2.174) into the Kohn–Sham equation, we have

$$\left[-\frac{1}{2}\nabla^2 + V_{\rm pp}(\boldsymbol{r})\right]|\boldsymbol{k}\lambda) = \varepsilon_{\boldsymbol{k}\lambda}|\boldsymbol{k}\lambda), \tag{2.186}$$

$$V_{\rm pp}(\boldsymbol{r}) = V(\boldsymbol{r}) + \sum_i^{\rm core}(\varepsilon_{\boldsymbol{k}\lambda} - \varepsilon_i)|i\rangle\langle i|\,, \tag{2.187}$$

where $V(\boldsymbol{r})$ denotes the self-consistent potential. The nonlocal function $V_{\rm pp}(\boldsymbol{r})$ introduced in (2.187) can be regarded also as a potential seen by plane waves, and is an example of a pseudopotential. Since the nonlocal-operator form of these potentials is not convenient, they were sometimes replaced by local forms and determined empirically to fit experiments. Many such empirical pseudopotentials were constructed in the 1970s.

The "norm-conserving ab initio pseudopotential" [190, 191], which is used most commonly nowadays, is constructed in a way that the pseudowavefunction has the same value as the true wavefunction outside the cutoff radius $r_{\rm c}$, which indicates the radius of the core region. The integrated value of the absolute square of the wavefunction has the same norm as the true wavefunction inside $r_{\rm c}$. The pseudowavefunction is made as smooth as possible inside the sphere of radius $r_{\rm c}$ and is connected continuously to the true wavefunction, i.e. it has the same value and the same derivative as those of the true wavefunction (the matching condition), at the cutoff radius. The pseudopotentials are usually nonlocal, i.e. they depend on $\boldsymbol{G}$ and $\boldsymbol{G}'$ independently. They can be separated into a local part, which is a function depending only on the difference between $\boldsymbol{G}$ and $\boldsymbol{G}'$, and a nonlocal part:

$$V_{\rm pp}(\boldsymbol{k}+\boldsymbol{G}, \boldsymbol{k}+\boldsymbol{G}') = V_{\rm local}(\boldsymbol{G}-\boldsymbol{G}') + \sum_l V_{{\rm nonlocal},l}(\boldsymbol{k}+\boldsymbol{G}, \boldsymbol{k}+\boldsymbol{G}'). \tag{2.188}$$

Since the nonlocal part is dependent on the angular momentum ($l$) and spatially localized, its matrix elements should be calculated as

$$V_{\text{nonlocal},l}(\boldsymbol{k}+\boldsymbol{G},\boldsymbol{k}+\boldsymbol{G}') = \frac{2l+1}{4\pi\Omega} P_l(\cos\gamma) \times \int_0^\infty V_{\text{nonlocal},l}(r) j_l(|\boldsymbol{k}+\boldsymbol{G}|r) j_l(|\boldsymbol{k}+\boldsymbol{G}'|r) r^2 dr, \quad (2.189)$$

where $\Omega$ is the cell volume, $V_{\text{nonlocal},l}$ is the nonlocal potential, $j_l(x)$ is the spherical Bessel function and $P_l(\cos\gamma)$ is the Legendre polynomial, with

$$\cos\gamma = \frac{(\boldsymbol{k}+\boldsymbol{G})\cdot(\boldsymbol{k}+\boldsymbol{G}')}{|\boldsymbol{k}+\boldsymbol{G}||\boldsymbol{k}+\boldsymbol{G}'|}. \quad (2.190)$$

For this nonlocal part, one may use the separable form given by Kleinman and Bylander [192]:

$$V^{\text{KB}}_{\text{nonlocal},l}(\boldsymbol{q},\boldsymbol{q}') = \frac{\sum_{m=-l}^{l} Y_{lm}(\hat{\boldsymbol{q}}) Y^*_{lm}(\hat{\boldsymbol{q}}')}{\Omega \int \Phi_l^2(r) V_{\text{nonlocal},l}(r) r^2 dr} \times \left\langle \int \Phi_l(r) V_{\text{nonlocal},l}(r) j_l(qr) r^2 dr \right\rangle \left\langle \int \Phi_l(r) V_{\text{nonlocal},l}(r) j_l(qr) r^2 dr \right\rangle, \quad (2.191)$$

where $l$ and $m$ indicate the angular-momentum and magnetic quantum numbers, and $Y_{lm}(\hat{\boldsymbol{q}})$ is the spherical harmonics.

In the construction of the norm-conserving pseudopotential, there are some arbitrary parameters, such as the cutoff radius $r_c$, the connectivity conditions at $r_c$ and the choice of pseudowavefunction at $r < r_c$. Therefore, by choosing appropriate parameters, it is possible to reduce the required cutoff energy for PWs. To do it, Rappe et al. [193] considered the convergence of the Fourier series of the atomic pseudowavefunction, and Troullier and Martins [194] devised an efficient way to connect the pseudowavefunction with the true wavefunction at $r = r_c$.

Although the Kleinman–Bylander form of the nonlocal part is used very often, it sometimes brings in a "ghost state" which is unphysical and may destroy important chemical properties. The appearence of ghost states is due to the invalidity of the Wronskian theorem in the use of separable pseudopotentials. A way to avoid this problem is discussed by Gonze et al. [195].

Beyond the idea of this norm-conserving pseudopotential, a non-norm-conserving pseudopotential, called the "ultrasoft pseudopotential", was proposed in 1990 by Vanderbilt [196]. Using this, one can make the pseudowavefunction as smooth as possible in the core region and reduce the necessary number of plane waves; see Fig. 2.9. The norm conservation is taken into account afterwards by solving a generalized eigenvalue equation.

For example, for first-row elements such as B, C, N and O, the core electrons, which play a crucial role in screening the strong Coulomb potential caused by the nucleus, are only the two electrons in the 1s level. Moreover, the higher angular-momentum state that appears first in the valence state (1s, 2p, 3d, 4f, ...) is, in general, rather localized compared to the other valence states (2s, 3s, 3p, 4s, 4p, 4d, ...), since

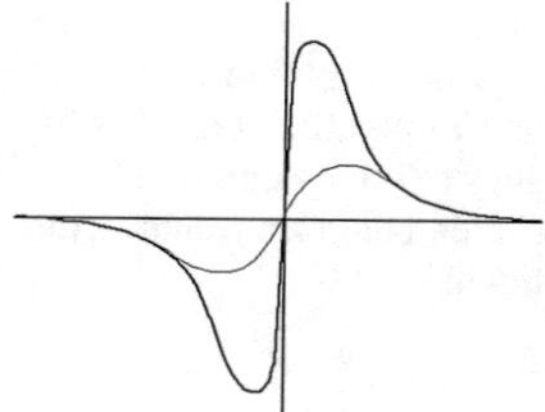

**Fig. 2.9** The pseudowavefunction for the ultrasoft pseudopotential (*thin line*) and the true wavefunction (*heavy line*)

it has no counterpart with the same angular momentum in the core state (0s, 1p, 2d, 3f, ...), to which they are orthogonal in the radial direction, and it may easily invade the core region to screen the strong Coulomb potential caused by the nucleus. Hence, for the first-row elements, it is not an easy task to produce an efficient pseudopotential which requires a relatively small number of plane waves to express the valence 2p orbitals well. For the same reason, the 3d orbitals in the transition-metal elements and the 4f orbitals in the rare-earth-metal elements also have a localized character, and devising an efficient pseudopotential is not easy. For these cases, the ultrasoft pseudopotential is rather effective, and the cutoff energy can be reduced up to half of that required with the norm-conserving pseudopotential. Recently, ab initio MD simulations have been performed with the ultrasoft-pseudopotential. However, if one wishes to treat a magnetically ordered structure accurately, the core contribution must generally be included in the calculation.

In the pseudopotential approach, the connection to the all-electron calculation becomes somewhat ambiguous. In order to solve this problem, Kresse and Joubert [149] reformulated the projector-augmented-wave (PAW) approach originally invented by Blöchl [197]. Nowadays, PAW usually means Kresse–Joubert's reformulation of the pseudopotential approach.

### 2.3.5 APW Method

There is an approach completely different from spectral methods, called the "cellular method", which assumes a simplified form of the potential. In the "muffin-tin approximation", the potential seen by an electron is decomposed into two regions. One is a spherically symmetric part around each atom and is called the "muffin-tin" (Fig. 2.10). The muffin-tin should be inside the Wigner–Seitz cell. The radius $r_{\mathrm{MT}}$ of the muffin-tin (the "muffin-tin radius") is an adjustable parameter. The potential in the outer region (this interstitial region is often called the "muffin-pan") is flat and constant, $V_{\mathrm{muffin-pan}}$; see Fig. 2.10. This shape of potential is called the muffin-tin potential. (The deviation of the real potential from the spherically symmetric potential can be corrected afterwards, and such a method is often called the "full-potential method".)

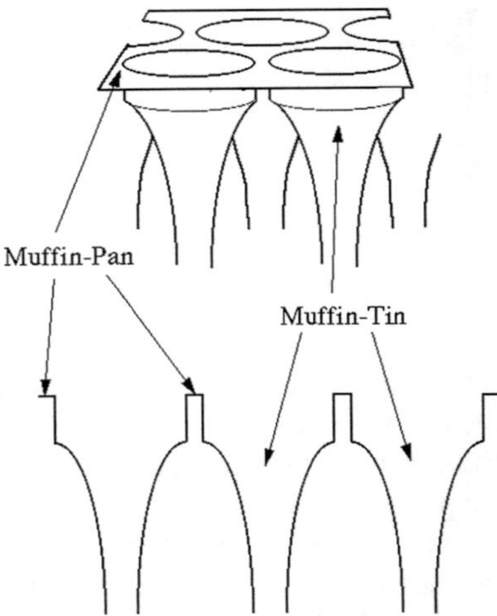


**Fig. 2.10** Schematic illustration of muffin-tin and muffin-pan. Each muffin-tin goes to minus infinity at the center, but is truncated in the figure

Inside the muffin-tin centered at the atomic position $Rs$, the effective one-electron wavefunction in the spherically symmetric potential $V(|\boldsymbol{r}-\boldsymbol{R}|)$ can be expressed in the form

$$\psi_{\boldsymbol{k}}^{\text{in}}(\boldsymbol{r}) = \mathrm{e}^{i\boldsymbol{k}\cdot\boldsymbol{R}} \sum_{lm} A_{lm} Y_{lm}(\theta_{\boldsymbol{r}}, \phi_{\boldsymbol{r}}) R_l(|\boldsymbol{r}-\boldsymbol{R}|), \tag{2.192}$$

where $l$ and $m$ are the quantum numbers related to the angular momentum, while $Y_{lm}(\theta_{\boldsymbol{r}}, \phi_{\boldsymbol{r}})$ represents the spherical harmonics (see [1, 2, 4]). The radial wavefunction $R_l(r)$ can be obtained by integrating the radial Schrödinger equation

$$-\frac{1}{2r^2}\frac{d}{dr}\left(r^2\frac{dR_l(r)}{dr}\right) + \left[\frac{l(l+1)}{2r^2} + V(r)\right] R_l(r) = \varepsilon R_l(r) \tag{2.193}$$

numerically with respect to $r$, provided that the eigenvalue energy $\varepsilon$ is given. This can be done by using the Herman–Skillman atomic code [174] which was devised originally for isolated atoms. Such wavefunctions determined inside the muffin-tin are called muffin-tin orbitals (MTOs).

On the other hand, in the muffin-pan region, the wavefunctions are identical to plane waves with the wave vector $\boldsymbol{k}$:

$$\psi_{\boldsymbol{k}}^{\text{out}}(\boldsymbol{r}) = \sum_{\boldsymbol{G}} C_{\boldsymbol{k}}(\boldsymbol{G}) \mathrm{e}^{i(\boldsymbol{k}+\boldsymbol{G})\cdot\boldsymbol{r}}. \tag{2.194}$$

Here we note that the exponential function appearing in (2.194) can be expanded [1] in terms of products of spherical Bessel functions and spherical harmonics as follows:

$$e^{i(\boldsymbol{k}+\boldsymbol{G})\cdot\boldsymbol{r}} = 4\pi \sum_{l,m} i^l j_l(|\boldsymbol{k}+\boldsymbol{G}|r) Y_{lm}(\theta_{\boldsymbol{r}}, \phi_{\boldsymbol{r}}) Y_{lm}(\theta_{\boldsymbol{k}+\boldsymbol{G}}, \phi_{\boldsymbol{k}+\boldsymbol{G}}). \tag{2.195}$$

Because the wavefunctions thus obtained inside and outside the muffin-tin sphere should be connected at the muffin-tin radius $r_{\mathrm{MT}}$, we have

$$A_{lm}(\boldsymbol{k}+\boldsymbol{G}) R_l(r_{\mathrm{MT}}) = 4\pi i^l C_{\boldsymbol{k}}(\boldsymbol{G}) j_l(|\boldsymbol{k}+\boldsymbol{G}|r_{\mathrm{MT}}) Y_{lm}(\theta_{\boldsymbol{k}+\boldsymbol{G}}, \phi_{\boldsymbol{k}+\boldsymbol{G}}), \tag{2.196}$$

where $\theta_{\boldsymbol{k}+\boldsymbol{G}}$ and $\phi_{\boldsymbol{k}+\boldsymbol{G}}$ define the solid angle of the vector $\boldsymbol{k}+\boldsymbol{G}$. Note that we put here an additional argument $\boldsymbol{k}+\boldsymbol{G}$ for $A_{lm}$, because we want to perform the "augmentation", i.e. the matching with the muffin-tin orbital, for each component of the plane waves. The wavefunction thus obtained for each $\boldsymbol{G}$ is called the "augmented plane wave" (APW). The APWs are not yet continuous in slope on the muffin-tin sphere at this stage, so that one has to solve the variational problem of the energy expectation value in a special manner.

In order to minimize the energy and to achieve matching inside and outside the muffin-tin radius, we first note that the Kohn–Sham equation $(\mathcal{H}-\varepsilon)\psi_{\boldsymbol{k}}(\boldsymbol{r}) = 0$ follows from the variational problem of minimizing the following integral over the unit cell:

$$\int_{\mathrm{cell}} \left\{ \frac{1}{2} [\nabla \psi_{\boldsymbol{k}}^*(\boldsymbol{r})][\nabla \psi_{\boldsymbol{k}}(\boldsymbol{r})] + \psi_{\boldsymbol{k}}^*(\boldsymbol{r}) \, [V(\boldsymbol{r}) - \varepsilon] \, \psi_{\boldsymbol{k}}(\boldsymbol{r}) \right\} d\boldsymbol{r}. \tag{2.197}$$

Performing the integral of the kinetic-energy term by parts and using Gauss's theorem, we find that (2.197) can be rewritten as

$$\int_{\mathrm{cell}} \psi_{\boldsymbol{k}}^*(\boldsymbol{r}) \, (\mathcal{H}-\varepsilon) \, \psi_{\boldsymbol{k}}(\boldsymbol{r}) d\boldsymbol{r} + \frac{1}{2} \sum_{i=1}^{n} \int_{r=r_{\mathrm{MT}_i}} \psi_{\boldsymbol{k}}^{\mathrm{in}*}(\boldsymbol{r}) \left( \frac{\partial \psi_{\boldsymbol{k}}^{\mathrm{in}}(\boldsymbol{r})}{\partial r} - \frac{\partial \psi_{\boldsymbol{k}}^{\mathrm{out}}(\boldsymbol{r})}{\partial r} \right) dS_i, \tag{2.198}$$

where the second term consists of a sum of surface integrals on the $i$th muffin-tin sphere ($i = 1, \ldots, n$); $n$ is the number of atoms inside the unit cell. Minimizing (2.198) with respect to $C_{\boldsymbol{k}}(\boldsymbol{G})$, we obtain

$$\sum_{s} \left[ \langle \boldsymbol{k}+\boldsymbol{G}'|\mathcal{H}-\varepsilon|\boldsymbol{k}+\boldsymbol{G}\rangle + \sum_{i=1}^{n} \langle \boldsymbol{k}+\boldsymbol{G}'|S_i|\boldsymbol{k}+\boldsymbol{G}\rangle \right] C_{\boldsymbol{k}}(\boldsymbol{G}) = 0, \tag{2.199}$$

where the notations $|\boldsymbol{k}+\boldsymbol{G}\rangle$ and $|\boldsymbol{k}+\boldsymbol{G}'\rangle$ are used to indicate the APWs associated with the different reciprocal-lattice vectors $\boldsymbol{G}$ and $\boldsymbol{G}'$, and $\langle \boldsymbol{k}+\boldsymbol{G}'|S_i|\boldsymbol{k}+\boldsymbol{G}\rangle$ indicates an APW matrix element of the surface integral appearing in the second term on the right-hand side of (2.198). Inside the muffin-tin, an APW is an eigenfunction of $\mathcal{H}$.

Therefore, the first term of (2.199) can be evaluated only in the muffin-pan region, to give

$$\langle \boldsymbol{k}+\boldsymbol{G}'|\mathcal{H}-\varepsilon|\boldsymbol{k}+\boldsymbol{G}\rangle = \left[\frac{1}{2}(\boldsymbol{k}+\boldsymbol{G})^2-\varepsilon\right]\int_{\text{muffin-pan}} \mathrm{e}^{i(\boldsymbol{G}-\boldsymbol{G}')\cdot\boldsymbol{r}}\,d\boldsymbol{r}$$

$$= \left[\frac{1}{2}(\boldsymbol{k}+\boldsymbol{G})^2-\varepsilon\right]\left[\Omega\delta_{\boldsymbol{G}\boldsymbol{G}'} - 4\pi\sum_{i=1}^{n}\mathrm{e}^{i(\boldsymbol{G}-\boldsymbol{G}')\cdot\boldsymbol{\tau}_i}r_{\mathrm{MT}_i}^2\frac{j_1(|\boldsymbol{G}-\boldsymbol{G}'|r_{\mathrm{MT}_i})}{|\boldsymbol{G}-\boldsymbol{G}'|}\right], \tag{2.200}$$

where $\Omega$ is the unit cell volume, $j_1(x)$ is the spherical Bessel function and $\boldsymbol{\tau}_i$ is the $i$th atomic position inside the unit cell. The calculation of the surface integral is straightforward. After some calculations, we have

$$\left[\frac{1}{2}(\boldsymbol{k}+\boldsymbol{G})^2-\varepsilon\right]C_{\boldsymbol{k}}(\boldsymbol{G}) + \sum_{\boldsymbol{G}'}\Gamma_{\boldsymbol{G}\boldsymbol{G}'}(\varepsilon)C_{\boldsymbol{k}}(\boldsymbol{G}') = 0 \tag{2.201}$$

for the coefficients $C_{\boldsymbol{k}}(\boldsymbol{G})$ of (2.194). Here, the kernel $\Gamma_{\boldsymbol{G}\boldsymbol{G}'}(\varepsilon)$ is given by

$$\begin{aligned}\Gamma_{\boldsymbol{G}\boldsymbol{G}'}(\varepsilon) &= \sum_{i=1}^{n}\mathrm{e}^{i(\boldsymbol{G}-\boldsymbol{G}')\cdot\boldsymbol{\tau}_i}\left\{\frac{4\pi r_{\mathrm{MT}_i}}{\Omega}\sum_{l}(2l+1)P_l(\cos\theta_{\boldsymbol{G}\boldsymbol{G}'})\right.\\ &\times j_l(|\boldsymbol{k}+\boldsymbol{G}|r_{\mathrm{MT}_i})j_l(|\boldsymbol{k}+\boldsymbol{G}'|r_{\mathrm{MT}_i})\left[\frac{d}{dr}\ln R_l(r) - \frac{d}{dr}\ln j_l(|\boldsymbol{k}+\boldsymbol{G}|r)\right]_{r=r_{\mathrm{MT}_i}}\\ &\left. - \left[\frac{1}{2}(\boldsymbol{k}+\boldsymbol{G})^2-\varepsilon\right]\frac{4\pi r_{\mathrm{MT}_i}}{\Omega}\frac{j_1(|\boldsymbol{G}-\boldsymbol{G}'|r_{\mathrm{MT}_i})}{|\boldsymbol{G}-\boldsymbol{G}'|}\right\}.\end{aligned} \tag{2.202}$$

Equation (2.201) is equivalent to the secular equation

$$\mathrm{Det}\left|\left\{\frac{1}{2}(\boldsymbol{k}+\boldsymbol{G})^2-\varepsilon\right\}\delta_{\boldsymbol{G}\boldsymbol{G}'} + \Gamma_{\boldsymbol{G}\boldsymbol{G}'}(\varepsilon)\right| = 0. \tag{2.203}$$

Note that this equation is different from the secular equation for standard eigenvalue equations, since the matrix elements $\Gamma_{\boldsymbol{G}\boldsymbol{G}'}(\varepsilon)$ depend also on $\varepsilon$ in a nonlinear way. Therefore, one has to determine the energy eigenvalue $\varepsilon$ self-consistently.

The method described here was formulated first by Slater in 1937 [198] and 1953 [199] and is called most commonly the "APW method". For more details of this method, readers may refer to the good monographs by Slater [200] and by Loucks [201]. As a simpler and more convenient method, Andersen [202] introduced the "linearized augmented-plane-wave" (LAPW) method in 1975. In the LAPW method, all the equations are linearized in the energy $\varepsilon$ around a special value $\varepsilon_0$. The energy derivative of a muffin-tin orbital $\dot{R}_l = \partial R_l/\partial\varepsilon|_{\varepsilon=\varepsilon_0}$ is obtained by solving the equation

$$-\frac{1}{2r^2}\frac{d}{dr}\left[r^2\frac{d\dot{R}_l(r)}{dr}\right] + \left[\frac{l(l+1)}{2r^2} + V(r) - \varepsilon\right]\dot{R}_l(r) = R_l(r). \tag{2.204}$$

If the muffin-tin orbital is normalized to unity in the muffin-tin sphere, i.e.,

$$4\pi \int_0^\infty R_l^2(r) r^2 dr = 1, \tag{2.205}$$

we find that $R_l$ and $\dot{R}_l$ are orthogonal:

$$\int_0^\infty R_l(r)\dot{R}_l(r) r^2 dr = 0. \tag{2.206}$$

Then, the kernel $\Gamma_{GG'}(\varepsilon)$ is linearized with respect to $\varepsilon$ around $\varepsilon_0$ as

$$\Gamma_{GG'}(\varepsilon) \sim \Gamma_{GG'}(\varepsilon_0) + (\varepsilon - \varepsilon_0)\frac{d\Gamma_{GG'}(\varepsilon_0)}{d\varepsilon_0}. \tag{2.207}$$

Equation (2.201) now becomes the generalized eigenvalue equation for $\varepsilon$, which can be solved, for example, by the Choleski decomposition (2.179)–(2.182), and the energy level is obtained in a small region around $\varepsilon_0$.

### 2.3.6 KKR, LMTO and ASW Methods

The Green's function method is also based on the muffin-tin approximation. It was first proposed in 1947 by Korringa [203], who used the theory of lattice interferences, and rederived later, in 1954, by Kohn and Rostoker [204], who used the Green's function defined in the free-electron limit by

$$\left(\varepsilon + \frac{1}{2}\nabla^2\right) G_{\boldsymbol{k}}^0(\boldsymbol{r}, \boldsymbol{r}'; \varepsilon) = \delta(\boldsymbol{r} - \boldsymbol{r}') \tag{2.208}$$

and

$$G_{\boldsymbol{k}}^0(\boldsymbol{r} + \boldsymbol{R}, \boldsymbol{r}'; \varepsilon) = \mathrm{e}^{i\boldsymbol{k}\cdot\boldsymbol{R}} G_{\boldsymbol{k}}^0(\boldsymbol{r}, \boldsymbol{r}'; \varepsilon) \tag{2.209}$$

the variational principle satisfied by this Green's function:

$$\delta\Lambda = 0, \tag{2.210}$$

where

$$\begin{aligned} \Lambda = & \int_{\text{cell}} \psi_{\boldsymbol{k}\lambda}^*(\boldsymbol{r}) V(\boldsymbol{r}) \psi_{\boldsymbol{k}\lambda}(\boldsymbol{r}) d\boldsymbol{r} \\ & - \int\int_{\text{cell}} \psi_{\boldsymbol{k}\lambda}^*(\boldsymbol{r}) V(\boldsymbol{r}) G_{\boldsymbol{k}}^0(\boldsymbol{r}, \boldsymbol{r}'; \varepsilon) V(\boldsymbol{r}') \psi_{\boldsymbol{k}\lambda}(\boldsymbol{r}') d\boldsymbol{r} d\boldsymbol{r}'. \end{aligned} \tag{2.211}$$

This variational principle is equivalent to the standard integral equation

$$\psi_{k\lambda}(\boldsymbol{r}) = \int_{\text{cell}} G_{\boldsymbol{k}}^{0}(\boldsymbol{r}, \boldsymbol{r}'; \varepsilon_{\boldsymbol{k}\lambda}) V(\boldsymbol{r}') \psi_{\boldsymbol{k}\lambda}(\boldsymbol{r}') d\boldsymbol{r}'. \tag{2.212}$$

(One can verify by operating $(\varepsilon_{\boldsymbol{k}\lambda} + \nabla^2/2)$ on both sides of (2.212) that this integral equation is identical to (2.123).)

This method is now more commonly called the Korringa–Kohn–Rostoker method, or simply the "KKR method". The secular equation derived in the KKR method is different from (2.203) and is given by [205]

$$\text{Det}\left| B_{lm;l'm'}(\boldsymbol{\tau}_i - \boldsymbol{\tau}_j) + \kappa \cos \eta_{l,i}(\varepsilon) \delta_{ll'} \delta_{mm'} \delta_{ij} \right| = 0, \tag{2.213}$$

where $\kappa = \sqrt{\varepsilon - V_{\text{muffin-pan}}}$, and $\eta_{l,i}(\varepsilon)$ denotes the phase shift for the spherical wave with an angular momentum $l$ at the $i$th atom:

$$\tan \eta_{l,i} = \frac{\left[\frac{d}{dr} j_l(\kappa r) - j_l(\kappa r) \frac{d}{dr} \ln R_l(r)\right]_{r=r_{\text{MT}_i}}}{\left[\frac{d}{dr} n_l(\kappa r) - n_l(\kappa r) \frac{d}{dr} \ln R_l(r)\right]_{r=r_{\text{MT}_i}}}, \tag{2.214}$$

where $j_l(x)$ and $n_l(x)$ are the spherical Bessel function and the spherical Neumann function, respectively. The structure constants $B_{L;L'}(\boldsymbol{\tau}_i - \boldsymbol{\tau}_j)$ in (2.213) (for simplicity we introduced the abbreviation $L$ in order to indicate both $l$ and $m$) are given by

$$B_{L;L'}(\boldsymbol{\tau}_i - \boldsymbol{\tau}_j) = 4\pi\kappa \sum_{L''} C_{LL'L''} D_{L''}(\boldsymbol{\tau}_i - \boldsymbol{\tau}_j) \mathrm{e}^{i\boldsymbol{k}\cdot(\boldsymbol{\tau}_i - \boldsymbol{\tau}_j)}. \tag{2.215}$$

In this equation, $C_{LL'L''}$ and $D_L(\boldsymbol{\tau}_i - \boldsymbol{\tau}_j)$ are given, respectively, by

$$C_{LL'L''} = \int d\hat{\boldsymbol{r}} Y_L^*(\hat{\boldsymbol{r}}) Y_{L'}(\hat{\boldsymbol{r}}) Y_L(\hat{\boldsymbol{r}}) \tag{2.216}$$

and

$$D_L(\boldsymbol{\tau}_i - \boldsymbol{\tau}_j) = D_L^{(1)}(\boldsymbol{\tau}_i - \boldsymbol{\tau}_j) + D_L^{(2)}(\boldsymbol{\tau}_i - \boldsymbol{\tau}_j) + \delta_{L,0} \delta_{\boldsymbol{\tau}_i, \boldsymbol{\tau}_j} D^{(3)}, \tag{2.217}$$

where

$$D_L^{(1)}(\boldsymbol{\tau}_i - \boldsymbol{\tau}_j) = \frac{4\pi}{\Omega\kappa} \mathrm{e}^{\varepsilon/q_0} \sum_{\boldsymbol{G}} \mathrm{e}^{i\boldsymbol{G}\cdot(\boldsymbol{\tau}_i - \boldsymbol{\tau}_j)} Y_L(\boldsymbol{k}+\boldsymbol{G}) |\boldsymbol{k}+\boldsymbol{G}|^l \frac{\mathrm{e}^{-(\boldsymbol{k}+\boldsymbol{G})^2/(2q_0)}}{\varepsilon - \frac{1}{2}(\boldsymbol{k}+\boldsymbol{G})^2}, \tag{2.218}$$

$$\begin{aligned} D_L^{(2)}(\boldsymbol{\tau}_i - \boldsymbol{\tau}_j) &= \frac{(-2)^{l+1} i^l}{\sqrt{\pi}\kappa} {\sum_{\boldsymbol{R}}}' \mathrm{e}^{i\boldsymbol{k}\cdot(\boldsymbol{R} - \boldsymbol{\tau}_i + \boldsymbol{\tau}_j)} \\ &\times Y_L(\boldsymbol{R} - \boldsymbol{\tau}_i + \boldsymbol{\tau}_j) |\boldsymbol{R} - \boldsymbol{\tau}_i + \boldsymbol{\tau}_j|^l \int_{q_0}^{\infty} z^{2l} \mathrm{e}^{-(\boldsymbol{R} - \boldsymbol{\tau}_i + \boldsymbol{\tau}_j)^2 z^2/4 + \varepsilon/z^2} dz, \end{aligned} \tag{2.219}$$

and

$$D^{(3)} = -\frac{\sqrt{q_0}}{2\pi}\sum_{n=0}^{\infty}\frac{(\varepsilon/q_0)^n}{n!(2n-1)}. \tag{2.220}$$

The summation $\sum'_{\boldsymbol{R}}$ does not contain the term $\boldsymbol{R} = \boldsymbol{\tau}_i - \boldsymbol{\tau}_j$, and $q_0$ is a parameter introduced for the Ewald sum. The structure constants $B_{lm;l'm'}(\boldsymbol{\tau}_i - \boldsymbol{\tau}_j)$ do not depend on the potential form inside the muffin-tin and contain only information about atomic positions. Therefore, they can be tabulated at the outset as functions of $\varepsilon$ and $\boldsymbol{k}$ for each type of lattice. The KKR method can give an answer which is quite similar to that of the APW method, although the sizes of the secular equation are different between the APW method and the KKR method. The secular equation in the former is labeled by the reciprocal-lattice vectors and that in the latter by the angular momenta. It is also possible to derive an equivalent secular equation from (2.213) labeled by reciprocal-lattice vectors instead of angular momenta [206–208]. One finds that the resulting secular equation is similar to but not exactly the same as (2.203) for the APW method.

The "full-potential LAPW" (FLAPW) method [209–216] and the "full-potential KKR" (FP-KKR) method [217] have appeared more recently; they expand the deviation from the spherically symmetric muffin-tin potential in terms of spherical harmonics.

The "muffin-tin orbital" (MTO) method proposed by Andersen and Wooly [218] in 1973 provides an alternative method. In this method, the wavefunction in the interstitial muffin-pan region is not expressed by the superposition of plane waves but approximated by the superposition of spherical waves centered at different atoms. These waves all have a definite kinetic energy $\varepsilon_{\text{kin}} = \varepsilon - V_{\text{muffin-pan}}$. Then the wavefunctions inside and outside the muffin-tin sphere are connected by a matching condition similar to that used for the APW and KKR methods. However, even if this is done rigorously, these spherical waves do not form a good basis outside the sphere. Therefore, in the MTO method, for simplicity, one should use overlapping spheres for the muffin-tins, and make the volume of the muffin-pan region as small as possible. If one decomposes the whole space into Wigner–Seitz cells and approximates them by Wigner–Seitz spheres, which may have small overlap regions, then one can neglect the muffin-pan completely. Such an approximation is called the atomic-sphere approximation (ASA), in which the sum of the volumes of the atomic spheres is just equal to the total volume of the unit cell.

Andersen [202] introduced in 1975 a convenient method called the linear muffin-tin orbital (LMTO) method, in which the wavefunctions outside the muffin-tin are approximated by a linear combination of two independent algebraic solutions of the Laplace equation $\nabla^2\chi_L = 0$, which is energy-independent (instead of using a solution of the Helmholtz equation $(\frac{1}{2}\nabla^2 + \varepsilon - V_{\text{muffin-pan}})\chi_L = 0$):

$$\chi_L^-(\boldsymbol{r}) = \left(\frac{r}{r_{\text{c}}}\right)^{-l-1} Y_L(\hat{\boldsymbol{r}}), \quad \chi_L^+(\boldsymbol{r}) = \left(\frac{r}{r_{\text{c}}}\right)^{l} Y_L(\hat{\boldsymbol{r}}). \tag{2.221}$$

Since the spherical waves centered at different atomic sites can be expanded by means of spherical harmonics centered at the origin, one can express the wavefunctions in the interstitial muffin-tin region in the form of a linear combination of such spherical harmonics. This function is then matched to the muffin-tin orbitals $R_l(r)$ and $\dot{R}_l(r)$ at the muffin-tin sphere. Finally, the Hamiltonian and overlap matrix elements are evaluated to give a secular equation for the spherical waves. This generalized eigenvalue problem can be solved in a similar way to the LAPW method. A good bibliography is also available for this method [219]. The LMTO method combined with the atomic-sphere approximation is called the LMTO–ASA method [202]. Sometimes, for example in the case of layered crystals, empty spheres should be introduced to describe the wavefunction in large interatomic regions, in the LMTO methods. The full-potential formalism can also be used for the LMTO method, as is described in [220–225]

In the usual LMTO methods, one uses spherical waves outside the muffin-tin spheres that have zero kinetic energy. In 1979, Williams et al. [226] proposed a slightly different method called the "augmented spherical wave" (ASW) method (see also [205]). This method is, so to say, a descendant of the LMTO method. They proposed to construct a spherical wave outside the muffin-tin sphere by solving the Helmholtz equation $(\frac{1}{2}\nabla^2 + \varepsilon_0)\chi_L = 0$, with $\varepsilon_0$ different from either $\varepsilon$ or $\varepsilon - V_{\text{muffin-pan}}$. For negative $\varepsilon_0$, the solution is given by a spherical Hankel function which is localized at the origin. (In all practical calculations, it is known that a fixed negative and small value ($\varepsilon_0 \sim 0.01\,\text{Ry}$) gives better results.) The spherical Hankel functions centered at different atomic sites can be expanded in terms of products of spherical Bessel functions and spherical harmonics centered at the origin. The wavefunction in the interstitial muffin-pan region is expressed by a linear combination of these spherical waves. It is matched with the muffin-tin orbitals defined inside the muffin-tin centered at the origin. Then, one may construct the Hamiltonian and overlap matrices between those orbitals and diagonalize it as if one was performing an LCAO calculation.

To express wavefunctions having localized character in the framework of an all-electron approach, the APW, KKR and LMTO methods are usually used as a powerful tool. However, there is a question as to whether the muffin-tin approximation is good enough except for simple metals. To improve it, the full-potential method is desirable.

### 2.3.7 Some General Remarks

In the two previous sections, we have seen that there are several ab initio methods based on the LDA or GGA of DFT. Each method named is unique but does not necessarily prescribe the details of all the algorithms used in the many specific codes treated independently by different groups. For example, the method of dividing the space into a mesh or of evaluating integrals may be different in different programs. As an another example, some parts of the calculation may be performed either in real space or by Fourier-space summation. Usually, such details are not mentioned in the literature. However, such differences may sometimes become important. This is the case when specific algorithms are chosen to achieve a decisive and important conceptual merit. This merit may sometimes be more accuracy and sometimes an acceleration of the computational speed. Typical examples related to the latter purpose are the various order-$N$ (or $O(N)$) techniques, which are the theme of the next section. In these techniques, many elaborate algorithms have been developed and tested to reduce the necessary computational cost from $O(N^3)$ to $O(N)$.

In previous sections, we learned that the effective one-electron wavefunctions determined by DFT are related to the true electron charge distribution. Experimentally also some information on the wavefunctions can be obtained, for example, by using scanning tunneling microscopy (STM) [227]. An explicit formulation including the dependence on the distance between the STM tip and the substrate was given in 1961 by Bardeen [228]. Detailed work on comparison between the calculated tunneling current and the experimentally observed STM images has been performed by Tersoff and Hamann [229] and by Tsukada et al. [230]. When the sample bias voltage is not too high, one may find that the STM images observed with positive and negative bias correspond roughly to the local density of states of the lowest unoccupied molecular orbital (LUMO) level and that of the highest occupied molecular orbital (HOMO) level, respectively, when the tip is grounded. STM images have been compared with the spatial distribution of the local density of states calculated using first-principles methods. Successful examples include the cases of $C_{60}$ adsorbed on substrates [231, 232] and $C_2H_4$ chemisorbed on the Si(001) surface [449].

Here, it is instructive to compare the advantages and shortcomings of all-electron approaches compared to pseudopotential approaches. So far, quite a large number of electronic-structure calculations and molecular-dynamics simulations of a wide variety of crystals, surfaces, clusters and so on have been very successfully performed by using pseudopotential approaches combined with the plane wave expansion. The method is simple and accurate, although its performance depends on the way the pseudopotential is constructed. For structural optimization and the determination of electronic structures near the Fermi level of nonmagnetic materials, the pseudopotential approach generally gives a good answer. The shortcomings of pseudopotential approaches are that they meet some obvious troubles when one wants to treat the effect of core electrons.

On the other hand, in all-electron approaches, it is possible to discuss the absolute values of the total energy and energy levels. An all-electron approach is also desirable for many other purposes; for example, to treat magnetically ordered structures [78, 233–237] in which the exchange–correlation potential acting on the localized orbitals is sensitive to the chemical environment; to calculate the exchange constant $J$ [238, 239]; to calculate the magnetic susceptibility in metals [240] and in semiconductors [172, 241]; to look at the effect of an external (electric or magnetic) field on systems in which a gradient of the core potential appears [242]; to discuss X-ray photoemission spectroscopy in which a core electron is excited [243–246]; to treat chemical shifts [247–250] or hyperfine interactions [251–253], in which the amplitude of the electron wavefunctions at the nucleus is important; and to discuss the electron momentum distribution in Compton profiles [254–257] where core states contribute very much.

The shortcomings of all-electron approaches are that they generally require much heavier computation than pseudopotential approaches, since they use a much larger number of electrons and therefore energy levels, in particular when one wants to treat the asphericity of the core region rigorously by using full-potential techniques.

So far, we have not discussed the relativistic terms in the Hamiltonian. The relativistic effect generally becomes more important as the atomic number increases. For elements heavier than the 3d transition-metal elements, usually only the terms up to $O(v^2/c^2)$ are treated. (Sometimes, a full relativistic calculation using the Dirac equation and spinor formalism is required for very heavy elements such as uranium. The full relativistic formulation of DFT is reviewed by Engel and Dreizler [258].) The relativistic Schrödinger equation in a centrifugal potential reads

$$\mathcal{H} = \mathcal{H}_{\mathrm{NR}} - \frac{p^4}{8m^3c^2} + \frac{\hbar^2}{4m^2c^2}\frac{1}{r}\frac{dV}{dr}(\boldsymbol{\sigma} \cdot \boldsymbol{L}) + \frac{\hbar^2}{8m^2c^2}\nabla^2 V, \tag{2.222}$$

where $\mathcal{H}_{\mathrm{NR}}$ denotes the nonrelativistic Hamiltonian. The first term in the correction, $-p^4/8m^3c^2$, is called the mass–velocity term and represents the relativistic correction to the kinetic energy $p^2/2m$. The second term, proportional to $\boldsymbol{\sigma} \cdot \boldsymbol{L}$, represents the spin–orbit coupling. (The spin–orbit coupling plays an important role, for example, in the discussion of the spin anisotropy in the exchange interactions between spins.) The last term is called the Darwin term and represents the correction to the centrifugal potential.

If the potential $V(r)$ is given by a pure Coulomb potential $-Ze^2/r$, then the Darwin term is equal to $\pi Z e^2 \hbar^2 \delta(\boldsymbol{r})/2m^2c^2$ and affects only the s-like wavefunction, which has a finite amplitude at the origin. However, this is not the case for higher s-states (3s, 4s, ...) of actual atoms, and the expectation value of the Darwin term for these states may become even negative. On the other hand, the mass-velocity term is always negative and its absolute (expectation) value is bigger in s-states than in p- and d-states. The mass-velocity and Darwin terms represent the semi-relativistic effect or scalar relativistic effect. When only these scalar relativistic terms are included in the calculation, for relatively heavy elements, we can see that only the energy levels with s character in valence and conduction bands are shifted down. For example, for germanium, the wavefunction at the conduction-band bottom is p-like and the wavefunction at the valence-band top is s-like. Therefore, if we include the scalar relativistic effect into the calculation, the valence band is shifted up, while the conduction band remains almost unchanged. As a result, in the usual type of LDA calculation for germanium, we have a negative band gap both at the $\Gamma$ and L points, which indicates a semimetal. To overcome this problem, one has to go beyond the LDA, by using, for example, the GW approximation (GWA), which will be described in Sect. 2.7.3.

### 2.3.8 Ab Initio $O(N)$ and Related Methods

In standard ab initio density functional methods, the effort required for the electronic-structure calculation is roughly proportional to $N^3$ for $N$-electron systems, but many order-$N$ methods, which become faster than the standard methods when $N$ exceeds a certain value, have been proposed. In the order-$N$ methods, the computational task scales linearly with $N$. The key point in these methods is to treat the Hamiltonian in real space and to neglect the interaction between regions separated by more than a finite distance $r_c$. Additional attention needs to be paid to the self-consistent minimization procedure of the total energy functional.

In 1991, Yang proposed the "divide-and-conquer" method, which can in principle be applicable to any ab initio approach, although Yang first discussed his idea in the context of the LCAO density functional approach [259]. In fact, he later extended it to the density matrix approach [260]. In the divide-and-conquer method, the total system of size $N$ is spatially divided into many subsystems (of a certain size, say $M$) which may have mutual overlap regions (of size $\alpha$). The number of subsystems is given by $N/(M-\alpha/2)$. Then, the Hamiltonian for each subsystem can be diagonalized independently, provided that the charge density distribution is given for each subsystem. This is possible even when a subsystem includes the potential (or other effect) of the adjacent subsystems. This procedure involves a numerical workload proportional to the number of subsystems $N/(M-\alpha/2)$ times $M^3$. It is repeated until the whole system achieves self-consistency. This method has been successfully applied to large molecules [261–264].

Although one can adopt the conventional approaches to solve directly the Schrödinger equation or the associated eigenvalue problem in order-$N$ methods, the use of the density matrix or the Green's function may also offer a convenient background for formulating the order-$N$ methods. The density matrix is defined by

$$\hat{\rho}(\boldsymbol{r},\boldsymbol{r}') = 2\sum_{\boldsymbol{k}\lambda}^{O} \psi_{\boldsymbol{k}\lambda}(\boldsymbol{r})\psi^*_{\boldsymbol{k}\lambda}(\boldsymbol{r}'), \tag{2.223}$$

(the factor of 2 is for the spin multiplicity and $O$ denotes the sum over only occupied states) and the Green's function is defined as

$$G(\boldsymbol{r},\boldsymbol{r}';\omega) = \sum_{\boldsymbol{k}}\sum_{\lambda} \frac{\psi_{\boldsymbol{k}\lambda}(\boldsymbol{r})\psi^*_{\boldsymbol{k}\lambda}(\boldsymbol{r}')}{\omega - \varepsilon_{\boldsymbol{k}\lambda} - i\delta_{\boldsymbol{k}\lambda}}, \tag{2.224}$$

where $\psi_{\boldsymbol{k}\lambda}(\boldsymbol{r})$ denotes an eigenfunction of the Kohn–Sham Hamiltonian $\mathcal{H}$, and $\delta_{\boldsymbol{k}\lambda}$ an infinitesimal quantity such that $\delta_{\boldsymbol{k}\lambda} = 0^+$ for $\varepsilon_{\boldsymbol{k}\lambda} < \varepsilon_F$ and $\delta_{\boldsymbol{k}\lambda} = 0^-$ for $\varepsilon_{\boldsymbol{k}\lambda} > \varepsilon_F$, provided that $\varepsilon_F$ is the Fermi energy. The Green's function is related to the density matrix via the contour integral

$$\hat{\rho}(\boldsymbol{r},\boldsymbol{r}') = 2\frac{1}{2\pi i}\oint_{C_F} G(\boldsymbol{r},\boldsymbol{r}';z)dz, \tag{2.225}$$

where $C_F$ is a contour in the complex energy plane enclosing all poles associated with occupied-state eigenvalues only. Moreover, the density matrix is related to the density through

$$\rho(\boldsymbol{r}) = \hat{\rho}(\boldsymbol{r}, \boldsymbol{r}). \tag{2.226}$$

The density matrix has two other important characteristics. One is the idempotency relation

$$\hat{\rho}(\boldsymbol{r}, \boldsymbol{r}') = \int \hat{\rho}(\boldsymbol{r}, \boldsymbol{r}'')\hat{\rho}(\boldsymbol{r}'', \boldsymbol{r}')d\boldsymbol{r}'' \tag{2.227}$$

and the other is its short-ranged nature at the ground state, i.e. one may assume that

$$\hat{\rho}(\boldsymbol{r}, \boldsymbol{r}') \sim 0 \ \text{ for } |\boldsymbol{r} - \boldsymbol{r}'| > r_c, \tag{2.228}$$

where $r_c$ is typically of the order of $1/\sqrt{E_g}$, where $E_g$ is the energy gap between the highest occupied state and the lowest unoccupied state. In the case of metals, the density matrix is known to have an algebraic decay. This short-ranged nature of the density matrix, which was first pointed out by Kohn [265] in 1959, is essential for the order-$N$ implementation.

In 1992, Baroni and Giannozzi [266] introduced a Green's function method by discretizing the real space into grid points. They evaluate the diagonal elements of the Green's function by truncating approximately the continued fraction in the recursion method due to Haydock [267]. Then the electron density is estimated from (2.225) and (2.226). The self-consistent potential function $V(\boldsymbol{r})$, the total self-consistent potential energy $E_{\text{pot}}$ and its contribution to the total energy $F_{\text{pot}}$ are calculated within the LDA. On the other hand, the kinetic-energy contribution to the total energy can also be evaluated from the diagonal elements of the Green's function through

$$E_{\text{kin}} = 2\sum_{\boldsymbol{k},\lambda} \varepsilon_{\boldsymbol{k}\lambda}\theta(\varepsilon_F - \varepsilon_{\boldsymbol{k}\lambda}) - E_{\text{pot}} \tag{2.229}$$

and

$$\sum_{\boldsymbol{k},\lambda} \varepsilon_{\boldsymbol{k}\lambda}\theta(\varepsilon_F - \varepsilon_{\boldsymbol{k}\lambda}) = \frac{1}{2\pi i}\int d\boldsymbol{r} \oint_{C_F} zG(\boldsymbol{r}, \boldsymbol{r}; z)dz. \tag{2.230}$$

A similar method was also discussed by Horsfield [268]. Instead of invoking the recursion method, one may use the locally self-consistent multiple-scattering method and truncate the multiple scattering processes at some distance $r_c$ [269, 270]. This cutoff radius $r_c$ must not be chosen to be very small. However, Abrikosov et al. [271, 272] suggested that the cutoff radius $r_c$ and hence the computational effort may be considerably reduced by embedding the truncated region in an effective medium.

The density matrix formulation also works very well as an order-$N$ method. As was seen in Sect. 2.2.3, the density matrix obeys a variational principle and plays a central role in DFT, which may also be applicable to finite temperatures. Since we know that the density matrix is related to the charge density $\rho(\boldsymbol{r})$ through (2.226), the electronic-structure calculation can be formulated very elegantly using the density matrix. The key point in the application of the density matrix to the order-$N$ theory is, as we have seen above, that the density matrix in the ground state approaches zero as $|\boldsymbol{r} - \boldsymbol{r}'|$ increases [265]. How to find the density matrix which satisfies the idempotency relation (2.227) is not easy and several methods have been proposed [273, 274]. The method of constructing the density matrix directly from the Kohn–Sham Hamiltonian was described by Daw [275]. The density matrix can be discussed either in real space or in the space of basis functions $(\varphi_i(\boldsymbol{r}))$ as

$$\hat{\rho}(\boldsymbol{r}, \boldsymbol{r}') = 2\sum_{\boldsymbol{k}\lambda}^{O} \psi_{\boldsymbol{k}\lambda}(\boldsymbol{r})\psi_{\boldsymbol{k}\lambda}^{*}(\boldsymbol{r}') = \sum_{ij} \rho_{ij}\varphi_i(\boldsymbol{r})\varphi_j^{*}(\boldsymbol{r}'), \tag{2.231}$$

$$\rho_{ij} = 2\sum_{\boldsymbol{k}\lambda}^{O} C_i^{\boldsymbol{k}\lambda} C_j^{\boldsymbol{k}\lambda^*}. \tag{2.232}$$

This latter formulation using $\rho_{ij}$ has been discussed mostly in the tight-binding approximation [273, 276–278]; see Sect. 3.3.1 for more details. The exception is the work of Yang and Lee [260] who used it together with the divide-and-conquer method to perform both density functional and Hartree–Fock calculations. If one considers the density matrix in real space, one encounters some trouble, since one cannot work directly with the six-dimensional function $\hat{\rho}(\boldsymbol{r}, \boldsymbol{r}')$. However, this problem has been overcome by introducing approximately some factorization or parameterizations [279–283].

In 1992, Singh, Krakauer, Haas and Liu [284] presented a new idea called the projector-basis technique, which is applicable to the nonlocal pseudopotential approach (accompanied by either a plane wave expansion or a mixed-basis approach) and also to the linearized augmented-plane-wave (LAPW) method. Their implementation of the projector-basis technique aims at reducing the computational workload by a factor $N$, i.e. to $O(N^2)$ from $O(N^3)$, in these standard methods. The basic idea is to perform multiplications of the kinetic-energy operator $-\nabla^2/2$ and the self-consistent potential $V(\boldsymbol{r})$ by the wavefunctions independently, in proper representations in which these operators become diagonal. That is, for plane waves, the kinetic and potential energies are evaluated by Fourier-space and real-space multiplications, respectively. These multiplications then cost $O(N)$ for each level. How about the use of a nonlocal pseudopotential which depends on $\boldsymbol{G}$ and $\boldsymbol{G}'$ separately? In this case, the usual multiplication costs $O(N^2)$ for each level. To get rid of this excessive work, Singh et al. [284] introduced a projector basis, which is defined as the plane wave counterpart defined in real space inside the atomic spheres. The nonlocal part of the pseudopotential is spatially localized around an atomic center, and its multiplication by wavefunctions is more easily performed in real space inside the atomic spheres. The multiplication by the projector basis costs $O(N)$. In order to transform the plane waves into real space, it is additionally necessary to use a fast

Fourier transformation, which costs $O(N \log N)$. Since the number of occupied levels is $O(N)$, the total amount of calculation involved in the projector-basis technique becomes $O(N^2 \log N)$.

Concerning the total-energy minimization, Galli and Parrinello [285] pointed out in 1992 that a considerable reduction of the computational task (at least, from $O(N^3)$ to $O(N^2)$) may be achieved in the standard density functional calculation if one constructs the localized orbitals $\phi_i(\boldsymbol{r})$, in a sense similar to the Wannier functions or the localized-orbital approach.

In order to make the orbitals $\phi_i(\boldsymbol{r})$ more localized when one uses the plane wave basis, Galli and Parrinello [285] proposed to add an artificial nonlocal potential

$$V = A \sum_{i,j}^{N} [1 - \theta(|\boldsymbol{R} + \boldsymbol{\tau}_i - \boldsymbol{r}| - r_c)] S_{ij}^{\dagger^{-1}} |i\rangle\langle j| \tag{2.233}$$

to the total-energy operator $\mathcal{F}$; for the definition of $\mathcal{F}$ see (2.70) in Sect. 2.2.3. They checked the validity of this scheme by investigating a system of 64 Si atoms.

To see the advantage of using these localized orbitals more generally, consider a system composed of $N$ spin-up and $N$ spin-down electrons without spin polarization. The ground state of this system can be represented by the $N$ proper orbitals $\phi_i(\boldsymbol{r})$ $(i = 1, \ldots, N)$, which are the $N$ independent linear combinations of the true eigenfunctions corresponding to the occupied states of the Kohn–Sham Hamiltonian $\mathcal{H}$. Introducing an $N \times N$ matrix $F_{ij}$ of the total-energy operator sandwiched by $\phi_i^*$ and $\phi_j$, one finds that the total energy of the electronic system is given by the trace of this matrix $F_{ij}$:

$$E = 2\,\mathrm{Tr}\,\mathcal{F} = 2 \sum_{i=1}^{N} F_{ii}, \tag{2.234}$$

where $F_{ij} = \langle i|\mathcal{F}|j\rangle$ and the factor of 2 accounts for the spin multiplicity. Note that when the total energy $E$ is minimized with respect to the orbitals $\phi_i(\boldsymbol{r})$ in (2.234), the $\phi$s become automatically the $N$ independent linear combinations of the eigenfunctions for the occupied states of the Hamiltonian. In a traditional minimization procedure, the total energy (2.234) is minimized subject to the orthonormality condition

$$S_{ij} = \langle i|j\rangle = \delta_{ij}. \tag{2.235}$$

It can be shown that this constrained minimization procedure is equivalent to the unconstrained minimization of the energy functional

$$E = 2\,\mathrm{Tr}\,S^{-1}\mathcal{F} = 2 \sum_{i,j=1}^{N} S_{ij}^{-1} F_{ji}. \tag{2.236}$$

In this formulation, one has to evaluate the inverse of the overlap matrix, $S^{-1}$, so that the computational task required for this minimization is $O(N^3)$ when $\phi$s are spatially extended, while the task becomes $O(N^2)$ when the $\phi$s are spatially localized because the matrix $S$ becomes sparse in this case. Here, it is interesting to note that the energy functional which appears in (2.236) is very similar to that which appears in the density matrix approach defined in the basis-function space:

$$E = 2\,\mathrm{Tr}\,\hat{\rho}\mathcal{F} = 2\sum_{i,j=1}^{N} \hat{\rho}_{ij} F_{ji}, \tag{2.237}$$

where the density matrix $\hat{\rho}_{ij}$ is as defined in (2.232).

Therefore, one may see that there is a close similarity between the formulations based on the overlap matrix $S$ and on the density matrix $\hat{\rho}$.

Another interesting idea applicable to the plane wave expansion scheme is to introduce adaptive curvilinear coordinates, as has been discussed by Gygi [286] and Hamann [287]. In this scheme, the curvilinear coordinate system itself should be optimized, as well as the wavefunctions when the total energy is minimized. Then, the localized structure of the wavefunctions close to the nuclei can be described well with a relatively small cutoff energy for the plane waves.

Instead of using plane waves, one may introduce the finite-difference method, which discretizes real space into a grid. Such a method was proposed first by Chelikowsky et al. [288] and then extended to multigrids by Briggs et al. [289].

Further progress towards order-$N$ methods was made by Mauri et al. [290], Mauri and Galli [291], and Ordejón et al. [292]. These authors proposed a convenient functional of the orbitals $\phi_i(\boldsymbol{r})$, which is similar to the original $\mathcal{F}$ but with Lagrange multipliers $\Lambda_{ij}$:

$$E' = 2\left(\sum_{i=1}^{N} F_{ii} - \sum_{i,j=1}^{N} \Lambda_{ij}(S_{ji} - \delta_{ji})\right). \tag{2.238}$$

This new functional guarantees the same ground state as the conventional scheme based on (2.234) and (2.235) when the total energy $E$ is minimized with respect to the orbitals $\phi_i(\boldsymbol{r})$, but with an additional and important advantage that the orthonormality (2.235) is automatically realized without imposing it explicitly. Since the Lagrange multipliers $\Lambda_{ji}$ can be considered as independent variables, the minimization of $E'$ with respect to $\Lambda_{ij}$ yields

$$\frac{\delta E'}{\delta \Lambda_{ij}} = 0, \tag{2.239}$$

and, in turn, $S_{ij} = \delta_{ij}$ and $\Lambda_{ij} = F_{ij}$. One may therefore replace $\Lambda_{ij}$ by $F_{ij} = \langle i|\mathcal{F}|j\rangle$ when the $\phi$s are not too far from the optimum. This replacement yields a convenient functional

$$E' = 2\left(\sum_{i=1}^{N} F_{ii} - \sum_{i,j=1}^{N} F_{ij}(S_{ji} - \delta_{ji})\right). \tag{2.240}$$

Here it should be stressed again that no explicit orthonormalization is required in this minimization, since the term proportional to $F_{ij}$ on the right-hand side of (2.240) constrains all the orbitals $\phi_i(\boldsymbol{r})$ towards the orthonormality (2.235). Moreover, in this formulation, the inverse of the overlap matrix $S_{ij}$ is not required, in contrast to (2.236). One can therefore solve the problem by using an unconstrained minimization of the functional $E'$, such as a standard conjugate-gradient (CG) procedure; see Appendix D.

It is interesting to note that (2.240) is the first term in the expansion of $S^{-1} = [I-(I-S)]^{-1} = I+(I-S)+(I-S)^2+\cdots$ (here $I$ is the unit matrix), and that in this way it is related to the original equation (2.236). Thus one may construct higher order generalizations of (2.240). In any case, one can stabilize the minimization procedure at any level of the approximation where the series is truncated at an odd power [290]. Because of the close similarity between $S_{ij}^{-1}$ and $\rho_{ij}$, this expansion is also related to the relations used to get higher idempotency in the density matrix scheme [273, 275, 277, 291]; see also Sect. 3.3.1.

The treatment in cases where the number of bases and the number of states are different is discussed in papers by Ordejón et al. [293] and by Kim et al. [294]. One may expand the orbitals $\phi_i(\boldsymbol{r})$ by using the LCAO basis according to (2.163) in Sect. 2.2.6, in which $\phi$s are expanded in terms of the atomic orbitals $\varphi_{n_i}(\boldsymbol{r} - \boldsymbol{\tau}_i)$ centered at $\boldsymbol{\tau}_i$; for some applications, see [294, 295]. Related work has been reported along with the tight-binding approach [296].

Order-$N$ methods using localized orbitals have also been formulated and discussed by some other authors [297–299]. There are also some similar methods (for example, QUEST [300]) which, although not $O(N)$, are slightly slower than $O(N)$ but still much faster than the standard ab initio methods.

In the usual minimization procedures, the electron total energy, which is just a single number, is minimized with respect to the wavefunctions. In place of this technique, one can instead minimize the functional difference between the input and output charge densities, $\Delta\rho(\boldsymbol{r}) = \rho_{\text{out}}(\boldsymbol{r}) - \rho_{\text{in}}(\boldsymbol{r})$, or the functional difference between the input and output Kohn–Sham potential energies, $\Delta V(\boldsymbol{r}) = V_{\text{out}}(\boldsymbol{r}) - V_{\text{in}}(\boldsymbol{r})$. The advantage of using these functional minimization techniques is that much more information for the selection of the next guess is available compared to the case of the single-number minimization. That is, fewer "wrong guesses" occur before finding the unique solution. Such techniques were first discussed by Bendt and Zunger [159] in 1982 in their implementation of Broyden's optimization method (see Appendix D) and the all-electron mixed-basis approach, which includes only core atomic orbitals (see Sect. 2.2.6). These techniques have been more recently recalled in the light of $O(N)$ methods by Annett [301] and Gonze [302].

If the minimal basis is used in the LCAO approach and if all the matrix elements are still evaluated self-consistently, this treatment becomes one which may be called the ab initio tight-binding method. As a simplified approach to the ab initio tight-binding method, Sankey and Niklewski [303] introduced an approximate and convenient formalism, in which the two- and three-center integrals are replaced by molecular coordinates depending only on the relative positions of the adjacent two or three atoms. The molecular coordinates can be calculated once and stored for later usage. Simultaneously, the charge density of the system is approximated as $\rho_0(\boldsymbol{r}) + \delta\rho(\boldsymbol{r})$, where $\rho_0(\boldsymbol{r})$ is the superposition of the spherical charge densities centered at each atom, and the deviation from this, $\delta\rho(\boldsymbol{r})$, is treated only to first order. The Sankey–Niklewski formalism has often been used together with the Harris functional method [304–314].

## 2.4 Perturbation and Linear Response

In this section, we introduce many important formulae which are relevant in the calculation of the response of a system to an external field. Some of them are very fundamental ones which have been known for a long time, while some have been those derived rather recently. All the theories described below are based on first- or second-order perturbation theory in quantum mechanics [1, 2, 4] and linear-response theory [208, 316–319]. Some important applications, such as for the Hall coefficients [320, 321] and the magnetic Kerr effect which is mainly due to the spin–orbit coupling and is important in optomagnetic applications (by the relativistic LMTO method [322–327] or by the ASW method [328–330]), and also some applications concerning the nonlinear susceptibility [208, 331–335] are not described here; readers who are interested in these should refer to the original papers.

### 2.4.1 *Effective-Mass Tensor*

In order to consider the response of crystals to a weak external field, or to derive the effective masses in crystals, it is necessary to use perturbation theory or linear-response theory.

The effective-mass tensor $(1/m_\lambda^*)_{\alpha\beta}$ is related to the energy level $\varepsilon_{\boldsymbol{k}\lambda}$ via

$$\varepsilon_{\boldsymbol{k}+\Delta\boldsymbol{k},\lambda} = \varepsilon_{\boldsymbol{k}\lambda} + \frac{1}{2}\sum_{\alpha=1}^{3}\sum_{\beta=1}^{3}\left(\frac{1}{m_\lambda^*}\right)_{\alpha\beta}\Delta k_\alpha \Delta k_\beta, \tag{2.241}$$

when $\varepsilon_{\boldsymbol{k}\lambda}$ has an extremum at $\boldsymbol{k}$. A free-electron-like picture using the effective mass in place of the bare electron mass is called the effective-mass approximation. This effective-mass tensor can be calculated by the so-called $\boldsymbol{k}\cdot\boldsymbol{p}$ method [57, 58, 208], which was first proposed by Shockley [336] to discuss the energy surface for degenerate band of semiconductors for $\boldsymbol{k}\sim 0$. This is a method to calculate the energy eigenvalue at $\boldsymbol{k} + \Delta\boldsymbol{k}$ perturbatively, provided that the electronic states at $\boldsymbol{k}$ is given.

Since Bloch functions belonging to different wave vectors are orthogonal, any wavefunction $\psi_{\boldsymbol{k}+\Delta\boldsymbol{k},\lambda}(\boldsymbol{r})$ must be expanded not in terms of $\psi_{\boldsymbol{k}\nu}(\boldsymbol{r})$ but in terms of $\exp[i\Delta\boldsymbol{k}\cdot\boldsymbol{r}]\psi_{\boldsymbol{k}\nu}(\boldsymbol{r})$. Therefore we seek a function $\chi_{\boldsymbol{k}\lambda}(\boldsymbol{r})$ introduced as

$$\psi_{\boldsymbol{k}+\Delta\boldsymbol{k},\lambda}(\boldsymbol{r}) = \mathrm{e}^{i\Delta\boldsymbol{k}\cdot\boldsymbol{r}}\chi_{\boldsymbol{k}\lambda}(\boldsymbol{r}). \tag{2.242}$$

Substituting (2.242) into the Kohn–Sham equation (2.123), we find the perturbative term of the Hamiltonian to be

$$\begin{aligned}\mathcal{H}' &= \frac{1}{2}(\Delta\boldsymbol{k} - i\nabla)^2 - \frac{1}{2}\nabla^2 \\ &= \frac{1}{2}(\Delta\boldsymbol{k})^2 - i\Delta\boldsymbol{k}\cdot\nabla \\ &= \sum_{\alpha=1}^{3}\frac{1}{2}\left[(\Delta k_\alpha)^2 + 2\Delta k_\alpha p_\alpha\right],\end{aligned} \tag{2.243}$$

where we put $p_\alpha = -i\nabla_\alpha$. If the level $(\boldsymbol{k}, \lambda)$ is not degenerate, we find from the second-order perturbation theory that

$$\left(\frac{1}{m^*_\lambda}\right)_{\alpha\beta} = \delta_{\alpha\beta} + 2\sum_{\nu}{}' \frac{\langle\boldsymbol{k},\lambda|p_\alpha|\boldsymbol{k},\nu\rangle\langle\boldsymbol{k},\nu|p_\beta|\boldsymbol{k},\lambda\rangle}{\varepsilon_{\boldsymbol{k}\lambda} - \varepsilon_{\boldsymbol{k}\nu}}, \tag{2.244}$$

where the summation with respect to the level index $\nu$ runs over all the occupied and unoccupied levels except for $\lambda$. The prefactor 2 in front of the second term on the right-hand side is needed to cancel out the factor $1/2$ in the final expression for the effective kinetic energy.

When the energy band is degenerate at wave vector $\boldsymbol{k}$, one should obey perturbation theory for degenerate states. If the level under consideration is $m$-fold degenerate $(\lambda_1, \lambda_2 \ldots \lambda_m)$ at $\boldsymbol{k} = \boldsymbol{0}$, one has to solve the secular equation of a $m \times m$ perturbation matrix of the form

$$\left|\sum_{\alpha\beta} A^{\alpha\beta}_{ij}\Delta k_\alpha \Delta k_\beta - \delta_{ij}\Delta\varepsilon\right| = 0, \tag{2.245}$$

where $i, j = 1, 2, \ldots, m$ stand for the row and column indices, and the coefficients $A^{\alpha\beta}_{ij}$ are given by

$$A^{\alpha\beta}_{ij} = 2\sum_{\nu}\frac{\langle\boldsymbol{0}\lambda_i|p_\alpha|\boldsymbol{0}\nu\rangle\langle\boldsymbol{0}\nu|p_\beta|\boldsymbol{0}\lambda_j\rangle}{\varepsilon_{\boldsymbol{0}\lambda} - \varepsilon_{\boldsymbol{0}\nu}}, \tag{2.246}$$

when one ignores the spin–orbit coupling. The $m$ energy–eigenvalues at $\boldsymbol{k} = \Delta\boldsymbol{k}$ are then given by

$$\varepsilon_{\boldsymbol{k}+\Delta\boldsymbol{k},\lambda} = \frac{\Delta k^2}{2} + \Delta\varepsilon, \tag{2.247}$$

where $\Delta\varepsilon$ are the $m$ solutions of (2.245).

This method was extended by Dresselhaus et al. [337] and Kane [338] to include the spin–orbit coupling. They compared their results for several typical semiconductors with experiments and obtained very good agreement.

### 2.4.2 Dielectric Response

In order to include an electromagnetic field, it is convenient to introduce the vector potential $\boldsymbol{A}$. In the presence of the vector potential, the one-electron Hamiltonian becomes

$$\left[\frac{1}{2}\left(\frac{1}{c}\boldsymbol{A} - i\nabla\right)^2 + V(\boldsymbol{r})\right]\psi_{\boldsymbol{k}\lambda}(\boldsymbol{r}) = \varepsilon_{\boldsymbol{k}\lambda}\psi_{\boldsymbol{k}\lambda}(\boldsymbol{r}), \tag{2.248}$$

where the speed of light, $c$, is 137.0 in a.u. We assume an electric field of the form $E = D\exp[-i(\omega t - \boldsymbol{q}\cdot\boldsymbol{r})] = -(1/c)\partial\boldsymbol{A}/\partial t = (i\omega/c)\boldsymbol{A}$, where $D$ denotes the maximum amplitude (vector) of the electric field. The electric susceptibility (or polarizability) $\chi(\boldsymbol{q},\omega)$ is defined as the proportionality constant in the relation between the electric polarization $\boldsymbol{P}$ per unit volume (or the electric dipole moment $\boldsymbol{P}$) and the electric field $E$: $\boldsymbol{P} = \chi(\boldsymbol{q},\omega)E$. The dielectric constant $\epsilon(\boldsymbol{q},\omega)$ is related to the electric susceptibility via

$$\epsilon(\boldsymbol{q},\omega) = 1 + 4\pi\chi(\boldsymbol{q},\omega). \tag{2.249}$$

If we write the complex susceptibility as

$$\chi(\boldsymbol{q},\omega) = \chi_1(\boldsymbol{q},\omega) + i\chi_2(\boldsymbol{q},\omega), \tag{2.250}$$

the imaginary part $\chi_2(\boldsymbol{q},\omega)$ represents the absorption spectrum which is important in optical measurements. The real part and the imaginary part of the susceptibility are related via the following identities:

$$\begin{aligned}\chi_1(\boldsymbol{q},\omega) &= \frac{2}{\pi}\mathrm{P}\int_0^\infty \frac{\omega'\chi_2(\boldsymbol{q},\omega')}{\omega'^2-\omega^2}d\omega',\\ \chi_2(\boldsymbol{q},\omega) &= -\frac{2\omega}{\pi}\mathrm{P}\int_0^\infty \frac{\chi_1(\boldsymbol{q},\omega')}{\omega'^2-\omega^2}d\omega',\end{aligned} \tag{2.251}$$

where "P" means the principal part. These relations are called the Kramers–Kronig relations.

Assuming that the unperturbed Hamiltonian is that of (2.248) with $E = D = \boldsymbol{A} = 0$ and using second-order perturbation theory, one finds a formula for the electric susceptibility [208, 315]

$$\chi(\boldsymbol{q},\omega) = -\frac{2}{\Omega q^2}\sum_{\boldsymbol{k}}\sum_{\lambda}\sum_{\nu}\frac{|\langle \boldsymbol{k}+\boldsymbol{q},\nu|\mathrm{e}^{i\boldsymbol{q}\cdot\boldsymbol{r}}|\boldsymbol{k},\lambda\rangle|^2[f_0(\varepsilon_{\boldsymbol{k}+\boldsymbol{q}\nu}) - f_0(\varepsilon_{\boldsymbol{k}\lambda})]}{\omega - \varepsilon_{\boldsymbol{k}\lambda}+\varepsilon_{\boldsymbol{k}+\boldsymbol{q}\nu}+i\delta}, \tag{2.252}$$

where $\delta$ is an infinitesimal positive number $0^+$, the summations with respect to the level indices $\lambda$ and $\nu$ run over all levels, $\boldsymbol{k}$ is inside the first Brillouin zone and $f_0(\varepsilon)$ denotes the Fermi–Dirac distribution function. The factor of 2 in the prefactor is due to the spin multiplicity. Although one may directly evaluate this formula, its imaginary part $\chi_2(\boldsymbol{q},\omega)$) can be evaluated more easily by introducing a delta function with respect to energy and converting it to the density of states. Then, by using the Kramers–Kronig relation, one may get the real part $\chi_1(\boldsymbol{q},\omega)$ as well. For crystals, the electric susceptibility in a length scale less than the lattice constant is nonlocal and cannot be simply Fourier transformed into the $(\boldsymbol{q},\omega)$ space. Instead, we have to introduce the nonlocal Fourier transformation

$$\chi(\boldsymbol{r},t;\boldsymbol{r}',t') = \sum_{\boldsymbol{q}}\sum_{\boldsymbol{G}}\sum_{\boldsymbol{G}'} e^{i(\boldsymbol{q}+\boldsymbol{G})\cdot\boldsymbol{r}}\int_{-\infty}^{\infty}\frac{d\omega}{2\pi}e^{-i\omega(t-t')}\chi_{\boldsymbol{G}\boldsymbol{G}'}(\boldsymbol{q},\omega)e^{-i(\boldsymbol{q}+\boldsymbol{G}')\cdot\boldsymbol{r}'}, \tag{2.253}$$

where $\boldsymbol{G}$ and $\boldsymbol{G}'$ are reciprocal lattice vectors and represent the short range part oscillating rapidly inside the unit cell with its periodic boundary condition. In this representation, (2.252) is replaced by

$$\begin{aligned}\chi_{\boldsymbol{G}\boldsymbol{G}'}(\boldsymbol{q},\omega) &= -\frac{2}{\Omega q^2}\sum_{\boldsymbol{k}}\sum_{\lambda}\sum_{\nu}\frac{\langle \boldsymbol{k},\lambda|\mathrm{e}^{-i(\boldsymbol{q}+\boldsymbol{G})\cdot\boldsymbol{r}}|\boldsymbol{k}+\boldsymbol{q},\nu\rangle\langle \boldsymbol{k}+\boldsymbol{q},\nu|\mathrm{e}^{i(\boldsymbol{q}+\boldsymbol{G}')\cdot\boldsymbol{r}'}|\boldsymbol{k},\lambda\rangle}{\omega - \varepsilon_{\boldsymbol{k}\lambda}+\varepsilon_{\boldsymbol{k}+\boldsymbol{q}\nu}+i\delta}\\ &\times [f_0(\varepsilon_{\boldsymbol{k}+\boldsymbol{q}\nu}) - f_0(\varepsilon_{\boldsymbol{k}\lambda})]. \end{aligned} \tag{2.254}$$

Either one of the summations with respect to $\lambda$ and $\nu$ runs over all empty states ($\mathcal{E}$), while the other runs over the occupied states ($\mathcal{O}$) only. Since there are huge number of empty states, it is sometimes computationally demanding to include enough number of empty states. In such cases, a matrix algebra to bypass the empty-state summation is often adopted. Using the completeness condition for the Kohn–Sham eigenstates

$$\sum_{\nu}|\boldsymbol{k},\nu\rangle\langle\boldsymbol{k},\nu| = \sum_{\nu}|\boldsymbol{k}+\boldsymbol{q},\nu\rangle\langle\boldsymbol{k}+\boldsymbol{q},\nu| = 1, \tag{2.255}$$

and the projection operator

$$\mathcal{P}_{\boldsymbol{k}} = \sum_{\nu}^{\mathcal{O}}|\boldsymbol{k},\nu\rangle\langle\boldsymbol{k},\nu|, \quad \mathcal{P}_{\boldsymbol{k}+\boldsymbol{q}} = \sum_{\nu}^{\mathcal{O}}|\boldsymbol{k}+\boldsymbol{q},\nu\rangle\langle\boldsymbol{k}+\boldsymbol{q},\nu|, \tag{2.256}$$

to replace the summation over the empty states ($\mathcal{E}$) to the summation over the occupied states ($\mathcal{O}$) as

$$\sum_{\nu}^{\mathcal{E}} |\boldsymbol{k},\nu\rangle\langle\boldsymbol{k},\nu| = 1 - \mathcal{P}_{\boldsymbol{k}}, \quad \sum_{\nu}^{\mathcal{E}} |\boldsymbol{k}+\boldsymbol{q},\nu\rangle\langle\boldsymbol{k}+\boldsymbol{q},\nu| = 1 - \mathcal{P}_{\boldsymbol{k}+\boldsymbol{q}}, \tag{2.257}$$

one may show that (2.254) is mathematically equivalent to

$$\begin{aligned}\chi_{\boldsymbol{G}\boldsymbol{G}'}(\boldsymbol{q},\omega) = & -\frac{2}{\Omega q^2}\sum_{\boldsymbol{k}}\sum_{\lambda}^{O}\Bigg[\langle\boldsymbol{k}+\boldsymbol{q},\lambda|\mathrm{e}^{i(\boldsymbol{q}+\boldsymbol{G})\cdot\boldsymbol{r}}\frac{1}{\omega - H + \varepsilon_{\boldsymbol{k}+\boldsymbol{q}\lambda} + i\delta}\mathrm{e}^{-i(\boldsymbol{q}+\boldsymbol{G}')\cdot\boldsymbol{r}'}|\boldsymbol{k}+\boldsymbol{q},\lambda\rangle \\ & + \langle\boldsymbol{k},\lambda|\mathrm{e}^{-i(\boldsymbol{q}+\boldsymbol{G})\cdot\boldsymbol{r}}\frac{1}{\omega - H + \varepsilon_{\boldsymbol{k}\lambda} + i\delta}\mathrm{e}^{i(\boldsymbol{q}+\boldsymbol{G}')\cdot\boldsymbol{r}'}|\boldsymbol{k},\lambda\rangle\Bigg],\end{aligned} \tag{2.258}$$

where $H$ is the Kohn–Sham Hamiltonian matrix in the basis function space. This algebra is useful not only in the calculation of the dielectric susceptibility but also in the calculations of other quantities using the perturbation theory. This technique, however, cannot be used in the LCAO methods without plane waves because the basis functions are incomplete.

We notice that there is an important relation which can be derived from the commutation relation between $\mathcal{H}$ and $\boldsymbol{r}$:

$$[\mathcal{H},\boldsymbol{r}] = \mathcal{H}\boldsymbol{r} - \boldsymbol{r}\mathcal{H} = -\nabla. \tag{2.259}$$

Multiplying this by $\langle\boldsymbol{k}',\nu|$ from the left-hand side and $|\boldsymbol{k},\lambda\rangle$ from the right-hand side, one obtains

$$(\varepsilon_{\boldsymbol{k}'\nu} - \varepsilon_{\boldsymbol{k}\lambda})\langle\boldsymbol{k}',\nu|\boldsymbol{r}|\boldsymbol{k},\lambda\rangle = -\langle\boldsymbol{k}',\nu|\nabla|\boldsymbol{k},\lambda\rangle. \tag{2.260}$$

Using this relation, one may rewrite (2.252) in the limit $\boldsymbol{q} \to 0$ as

$$\begin{aligned}\chi(0,\omega) = & -\frac{2}{3\Omega}\sum_{\boldsymbol{k}}\sum_{\lambda}\sum_{\nu}\frac{\langle\boldsymbol{k},\lambda|\nabla|\boldsymbol{k},\nu\rangle\cdot\langle\boldsymbol{k},\nu|\nabla|\boldsymbol{k},\lambda\rangle}{\omega - \varepsilon_{\boldsymbol{k}\lambda} + \varepsilon_{\boldsymbol{k}\nu} + i\delta} \\ & \times \frac{[f_0(\varepsilon_{\boldsymbol{k}\nu}) - f_0(\varepsilon_{\boldsymbol{k}\lambda})]}{(\varepsilon_{\boldsymbol{k}\nu} - \varepsilon_{\boldsymbol{k}\lambda})^2}.\end{aligned} \tag{2.261}$$

In practical calculations, core states contribute very little to the final result, and for most cases one may ignore the core levels in the summation in (2.252). For example, in the cases of silicon and diamond, both the pseudopotential approach and the all-electron mixed-basis approach give the same result for $\epsilon(\mathbf{0},0)$: 14.5 for silicon and 5.9 for diamond. The results for $\epsilon(\mathbf{0},\omega)$ of a silicon crystal evaluated with the latter approach [339] are shown in Fig. 2.11. The experimental value of $\epsilon(\mathbf{0},0)$ for Si is 12.0, and the difference between the theoretical and experimental values can be attributed to the LDA.

For isolated atoms or molecules, equations similar to (2.252)–(2.261) but without $\Omega$ in the denominator or the summation over the wave vector $\boldsymbol{k}$ are well known in the analysis of the second-order Stark effect [340]. There, a quantity called the oscillator

**Fig. 2.11** The dielectric constant $\epsilon(\mathbf{0}, \omega)$ of Si evaluated with the all-electron mixed-basis approach

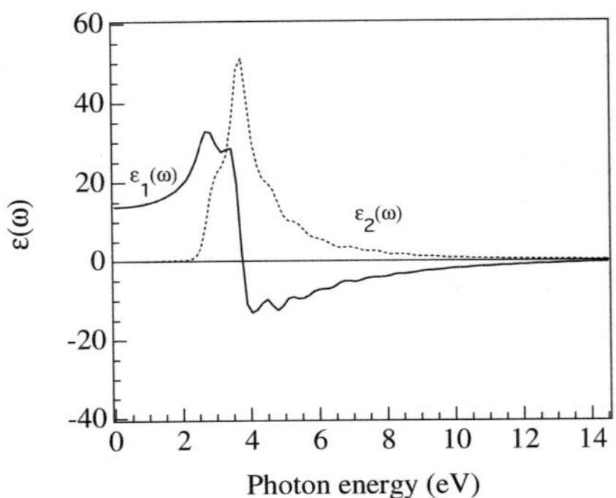


strength for the optical transition from the initial state $\boldsymbol{k}, \lambda$ to the final state $\boldsymbol{k}', \nu$,

$$\begin{aligned} f_{\boldsymbol{k},\lambda,\boldsymbol{k}',\nu} &= \frac{2}{3}|\langle \boldsymbol{k}, \lambda|\boldsymbol{r}|\boldsymbol{k}', \nu\rangle|^2(\varepsilon_{\boldsymbol{k}'\nu} - \varepsilon_{\boldsymbol{k}\lambda}) \\ &= \frac{2}{3}\frac{|\langle \boldsymbol{k}, \lambda|\nabla|\boldsymbol{k}', \nu\rangle|^2}{\varepsilon_{\boldsymbol{k}'\nu} - \varepsilon_{\boldsymbol{k}\lambda}}, \end{aligned} \tag{2.262}$$

has often been used, because it satisfies a famous sum rule called the "$f$-sum rule". Equation (2.260) was used to derive the second equality in (2.262). This oscillator strength gives the intensity of a spectral line corresponding to an optical transition of an electron from a state $\lambda$ to a state $\nu$, in the approximation $\exp(i\boldsymbol{q} \cdot \boldsymbol{r}) \approx 1$ over a region of the size of an atom (where $\boldsymbol{q}$ is the wave vector of the light); this approximation is referred to as the dipole approximation.

Now we derive the $f$-sum rule. We start from the commutation relation (2.259). Multiplying it by $\langle \boldsymbol{k}, \lambda|$ from the left-hand side and $|\boldsymbol{k}, \nu\rangle$ from the right-hand side, one obtains

$$(\varepsilon_{\boldsymbol{k}'\nu} - \varepsilon_{\boldsymbol{k}\lambda})\langle \boldsymbol{k}, \lambda|\boldsymbol{r}|\boldsymbol{k}', \nu\rangle = \langle \boldsymbol{k}, \lambda|\nabla|\boldsymbol{k}', \nu\rangle. \tag{2.263}$$

Further, we multiply (2.260) by $\langle \boldsymbol{k}, \lambda|\boldsymbol{r}|\boldsymbol{k}', \nu\rangle$, and (2.263) by $\langle \boldsymbol{k}', \nu|\boldsymbol{r}|\boldsymbol{k}, \lambda\rangle$, add these two and finally summing over all $\nu$, we find

$$\begin{aligned} &2\sum_{\nu}(\varepsilon_{\boldsymbol{k}'\nu} - \varepsilon_{\boldsymbol{k}\lambda})|\langle \boldsymbol{k}, \lambda|\boldsymbol{r}|\boldsymbol{k}', \nu\rangle|^2 \\ &= -\sum_{\nu}\langle \boldsymbol{k}, \lambda|\boldsymbol{r}|\boldsymbol{k}', \nu\rangle\langle \boldsymbol{k}', \nu|\nabla|\boldsymbol{k}, \lambda\rangle + \sum_{\nu}\langle \boldsymbol{k}, \lambda|\nabla|\boldsymbol{k}', \nu\rangle\langle \boldsymbol{k}', \nu|\boldsymbol{r}|\boldsymbol{k}, \lambda\rangle \\ &= -\langle \boldsymbol{k}, \lambda|(\nabla\boldsymbol{r} - \boldsymbol{r}\nabla)|\boldsymbol{k}, \lambda\rangle = -3. \end{aligned} \tag{2.264}$$

If we sum this equation with respect to the (initial) state $\lambda$ for all the occupied levels ($\mathcal{O}$), the summation on the left-hand side with respect to the (final) state $\nu$ can be reduced to one for the empty levels ($\mathcal{E}$), since the rest (i.e. the sum with respect to $\nu$ for all the occupied states) is exactly zero because of the antisymmetry of the equation. Then we have the $f$-sum rule:

$$\sum_{\lambda}^{O}\sum_{\nu}^{\mathcal{E}} f_{\boldsymbol{k}\lambda,\boldsymbol{k}'\nu} = \frac{2}{3}\sum_{\lambda}^{O}\sum_{\nu}^{\mathcal{E}} |\langle \boldsymbol{k},\lambda|\boldsymbol{r}|\boldsymbol{k}',\nu\rangle|^2(\varepsilon_{\boldsymbol{k}'\nu} - \varepsilon_{\boldsymbol{k}\lambda}) = N, \tag{2.265}$$

where $N$ is the total number of electrons in the unit cell.

The polarizability (2.252) (without $\Omega$ in the denominator, because of the difference between the electric susceptibility and the polarizability) for $\boldsymbol{q} = \omega = 0$ can be rewritten using the oscillator strength, as

$$\chi(\mathbf{0}, 0) = \sum_{\lambda}^{O}\sum_{\nu}^{\mathcal{E}} \frac{f_{\lambda,\nu}}{(\varepsilon_\nu - \varepsilon_\lambda)^2}. \tag{2.266}$$

This summation can be evaluated by using the $f$-sum rule if the energy denominator is approximated by the size of the energy gap, $E_g$, between the highest occupied molecular-orbital (HOMO) and the lowest unoccupied molecular-orbital (LUMO) levels:

$$\chi(\mathbf{0}, 0) \sim \frac{N e^4 a_\mathrm{B}}{E_\mathrm{g}^2}, \tag{2.267}$$

where $N$ is the number of electrons, $e$ the electron charge and $a_\mathrm{B}$ the Bohr radius (note that $e = a_\mathrm{B} = 1$ in a.u.). The polarizability has dimensions of $a_\mathrm{B}^3$. In molecules, the polarizability can be roughly estimated to be $\sim$ (radius of the molecule)$^3$. It will be discussed again in Sects. 4.2.1 and 4.2.2, in connection with interparticle potentials between polar or nonpolar atoms or molecules. For example, the electrostatic polarizability of He was first estimated by Pauling [40] to be 0.205 Å$^3$, which should be compared with the experimental values $0.2057 \pm 0.0004$ Å$^3$ [341] and $0.20505 \pm 0.00001$ Å$^3$ [342]. Polarizabilities of many different atoms and molecules can be found in a recent book by Bonin and Kresin [343].

### 2.4.3 Magnetic Susceptibility

In a similar way, one may consider the magnetic susceptibility of semiconductors or insulators [241]. Because there is no conducting electron, the magnetic susceptibility in these solids is negative, i.e. the diamagnetic susceptibility. (The magnetic susceptibility of metals can be calculated in a different way [240]). The magnetic susceptibility is defined as the second derivative of the total energy per unit volume with respect to the magnetic field $\boldsymbol{B}$. Similarly to the above dielectric problem, let us consider a magnetic field applied in the $z$ direction with a strength $b$, oscillating with a finite wave vector $\boldsymbol{q}$. Then $\boldsymbol{B}(\boldsymbol{r}) = b[0, 0, \sqrt{2}\cos(\boldsymbol{q}\cdot\boldsymbol{r})] = \nabla \times \boldsymbol{A}$, with the vector potential $A(\boldsymbol{r}) = b[0, \sqrt{2}\sin(\boldsymbol{q}\cdot\boldsymbol{r})/q, 0]$. ($\sqrt{2}$ is needed in these formulae to relate the root-mean-square value $b$ to the maximum value $\sqrt{2}b$.) Writing $\mathcal{H}^{(1)} = \boldsymbol{p}\cdot\boldsymbol{A}/c$ and $\mathcal{H}^{(2)} = A^2/2c^2$, we find the magnetic susceptibility at $T = 0$ to be

$$\begin{aligned}\chi_{zz}(\boldsymbol{q}) &= -\frac{d^2E}{db^2} \\ &= -\frac{4}{\Omega b^2}\sum_{\boldsymbol{k}'}\sum_{\lambda}^{\mathcal{O}}\left(\sum_{\boldsymbol{k}}\sum_{\nu}^{\mathcal{E}}\frac{|\langle\psi_{\boldsymbol{k}'\lambda}|\mathcal{H}^{(1)}|\psi_{\boldsymbol{k}\nu}\rangle|^2}{\varepsilon_{\boldsymbol{k}'\lambda}-\varepsilon_{\boldsymbol{k}\nu}} + \langle\psi_{\boldsymbol{k}'\lambda}|\mathcal{H}^{(2)}|\psi_{\boldsymbol{k}'\lambda}\rangle\right) \\ &= -\frac{2}{\Omega c^2 q^2}\sum_{\boldsymbol{k}}[g(\boldsymbol{k}+\boldsymbol{q},\boldsymbol{k}) + g(\boldsymbol{k}-\boldsymbol{q},\boldsymbol{k})] - \frac{\mathcal{N}}{\Omega c^2 q^2},\end{aligned} \tag{2.268}$$

where $\mathcal{N}$ is the number of electrons in the whole system, and

$$g(\boldsymbol{k}+\boldsymbol{q},\boldsymbol{k}) = \sum_{\lambda}^{\mathcal{O}}\sum_{\nu}^{\mathcal{E}}\frac{|\langle\boldsymbol{k}+\boldsymbol{q},\lambda|e^{i\boldsymbol{q}\cdot\boldsymbol{r}}\nabla_y|\boldsymbol{k},\nu\rangle|^2}{\varepsilon_{\boldsymbol{k}+\boldsymbol{q}\lambda}-\varepsilon_{\boldsymbol{k}\nu}}, \tag{2.269}$$

where the summations with respect to the level indices $\lambda$ and $\nu$ run over all occupied ($\mathcal{O}$) and all empty ($\mathcal{E}$) levels, respectively. In the limit $\boldsymbol{q} \to 0$, the two terms on the right-hand side of (2.268) individually diverge. However, because of the $f$-sum rule (2.265),

$$2\sum_{\boldsymbol{k}}\sum_{\lambda}^{\mathcal{O}}\sum_{\nu}^{\mathcal{E}} f_{\boldsymbol{k}\lambda,\boldsymbol{k}\nu} = \mathcal{N} = -4\int\sum_{\boldsymbol{k}} g(\boldsymbol{k},\boldsymbol{k}) \tag{2.270}$$

(here we are taking the spin multiplicity into account), one may see that $\chi_{zz}(\boldsymbol{q})$ remains finite. By inserting (2.270) into (2.268), we obtain the formula

$$\chi_{zz}(\boldsymbol{q}) = -\frac{2}{\Omega c^2}\sum_{\boldsymbol{k}}\frac{g(\boldsymbol{k}+\boldsymbol{q},\boldsymbol{k}) - 2g(\boldsymbol{k},\boldsymbol{k}) + g(\boldsymbol{k}-\boldsymbol{q},\boldsymbol{k})}{q^2}, \tag{2.271}$$

which is finite in the limit $\boldsymbol{q} \to 0$.

Unlike the electric susceptibility, in the present case of the magnetic susceptibility the core contribution may become as big as the other contributions. If one divides the occupied bands $\mathcal{O}$ into core bands $C$ and valence bands $\mathcal{V}$, one has

$$\chi = \chi_{C,\mathcal{E}} + \chi_{\mathcal{V},\mathcal{E}} = \chi_{\text{core}} - \chi_{C,\mathcal{V}} + \chi_{\mathcal{V},\mathcal{E}}. \tag{2.272}$$

Here $\chi_{C,\mathcal{E}}$ is given by (2.269) and (2.271) with the sum over $\lambda$ and $\nu$ in (2.269) running over the sets of core states $C$ and of empty states $\mathcal{E}$; $\chi_{\text{core}}$ is the susceptibility of the core electrons, which is not sensitive to the chemical environment and thus can be computed rather accurately by considering the isolated atoms, i.e.

$$\chi_{\text{core}} = \chi_{C,\mathcal{E}} + \chi_{C,\mathcal{V}} \simeq -\frac{1}{\Omega c^2}\sum_i\sum_{\lambda\in C}\langle\phi_\lambda(\boldsymbol{r}-\boldsymbol{R}_i)|x^2|\phi_\lambda(\boldsymbol{r}-\boldsymbol{R}_i)\rangle, \tag{2.273}$$

where $\phi_\lambda(\boldsymbol{r} - \boldsymbol{R}_i)$ denotes a core atomic wavefunction of an atom at $\boldsymbol{R}_i$. For estimating the contribution $\chi_{C,\mathcal{V}}$, it is necessary to use an all-electron approach. An all-electron mixed-basis calculation [172] shows, however, that this contribution is relatively small for typical semiconductors and inert-gas solids. Therefore, one can use a pseudopotential code to calculate the magnetic susceptibility using (2.272) and (2.273), and neglect $\chi_{C,\mathcal{V}}$. The resulting values for several inert-gas solids and several semiconductors are tabulated in Tables 2.3 and 2.4, respectively.

### 2.4.4 Chemical Shift

Chemical-shift ($\sigma$) spectra measured by nuclear magnetic resonance (NMR) are useful in the determination of the molecular geometry and the chemical structure of materials. A uniform external magnetic field $\boldsymbol{B}_{\text{ext}}$ applied to a sample induces an electronic current density $\boldsymbol{J}_{\text{in}}(\boldsymbol{r})$. This current produces an induced magnetic field $\boldsymbol{B}_{\text{in}}(\boldsymbol{r})$. When $\boldsymbol{B}_{\text{ext}}$ is small enough, which is a condition realized in NMR experiments, then

$$\boldsymbol{B}_{\text{in}}(\boldsymbol{r}) = -\overleftrightarrow{\sigma}(\boldsymbol{r})\boldsymbol{B}_{\text{ext}}. \tag{2.274}$$

Here $\overleftrightarrow{\sigma}(\boldsymbol{r})$ is the chemical-shift tensor. With NMR spectroscopy, it is possible to measure the symmetric part of $\overleftrightarrow{\sigma}$, i.e. $\sigma(\boldsymbol{r}) = (1/3)\,\mathrm{Tr}[\overleftrightarrow{\sigma}(\boldsymbol{r})]$, at the position of the nonzero-spin nuclei. Ab initio calculations of $\sigma$ have been performed in finite systems such as isolated molecules or clusters [247, 248], and recently extended to crystals [249].

**Table 2.3** LDA magnetic susceptibilities [172] for inert-gas solids in units of $10^{-6}$ cm$^3$/mole. The total LDA susceptibility is $\chi^{\text{LDA}} = \chi_{\text{core}} + \chi_{\mathcal{V},C} + \chi^{\text{LDA}}_{\mathcal{V},\mathcal{E}}$. The last two columns represent the LDA and experimental values for the isolated inert-gas atoms

| | $\chi_C$ | $\chi_{\mathcal{V},C}$ | $\chi_{\mathcal{V},\mathcal{E}}$ | $\chi^{\text{LDA}}$ | $\chi^{\text{LDA}}_{\text{atom}}$ | $\chi^{\text{exp}}_{\text{atom}}$ |
|---|---|---|---|---|---|---|
| Ne | −0.05 | 0.03 | −7.79 | −7.81 | −7.80 | −7.2 |
| Ar | −1.19 | 0.07 | −20.04 | −21.16 | −20.84 | −19.4 |
| Kr | −5.41 | 0.23 | −25.13 | −30.31 | −31.37 | −28 |

**Table 2.4** LDA magnetic susceptibilities [172] for semiconductors in units of $10^{-6}$cm$^3$/mole of $C_2$, $Si_2$, GaAs and GaP. The theoretical value inside parentheses for Si is the result obtained by expanding the LDA gap at the $\Gamma$ point so as to fit the experimental band gap

| | $\chi^{\text{LDA}}$ | $\chi^{\text{exp}}$ |
|---|---|---|
| C | −10.6 | −11.8 |
| Si | −1.2 (−7.6) | −6.4 |
| GaAs | −28.4 | −33 |
| GaP | −26.0 | −30 |

In a periodic system, $\overleftrightarrow{\sigma}(\boldsymbol{r})$ is also periodic. It may be obtained as a Fourier series,

$$\overleftrightarrow{\sigma}(\boldsymbol{r}) = \sum_{\boldsymbol{G}} \overleftrightarrow{\sigma}(\boldsymbol{G}) e^{i\boldsymbol{G}\cdot\boldsymbol{r}}, \tag{2.275}$$

where $\boldsymbol{G}$ runs over the reciprocal-lattice vectors. The value of $\overleftrightarrow{\sigma}(\boldsymbol{G}=\boldsymbol{0})$ depends on the shape of the sample, and is determined by the shape-dependent total energy. In the case of a spherical sample, we have

$$\overleftrightarrow{\sigma}(\boldsymbol{G}=\boldsymbol{0}) = -\frac{8\pi}{3}\overleftrightarrow{\chi}(0,0), \tag{2.276}$$

where $\overleftrightarrow{\chi}(0,0)$ is the macroscopic susceptibility given by (2.271) in the limit $\boldsymbol{q} \to 0$. On the other hand, the Fourier coefficients for $\boldsymbol{G} \neq 0$ are given by

$$\overleftrightarrow{\sigma}(\boldsymbol{G}) = -4\pi\overleftrightarrow{\chi}(\boldsymbol{G},0). \tag{2.277}$$

Here $\overleftrightarrow{\chi}(\boldsymbol{G},0)$ is the magnetic susceptibility, which is defined by the second derivative of the total energy $E(\boldsymbol{B})$ per unit volume in the external magnetic field $\boldsymbol{B}(\boldsymbol{r}) = B_0\hat{\boldsymbol{b}}_0 + B_{-\boldsymbol{G}}\hat{\boldsymbol{b}}_{-\boldsymbol{G}}\exp(-i\boldsymbol{G}\cdot\boldsymbol{r})$ ($\hat{\boldsymbol{b}}_0$ and $\hat{\boldsymbol{b}}_{-\boldsymbol{G}}$ being unit vectors) with respect to $B_0$ and $B_{-\boldsymbol{G}}$:

$$\hat{\boldsymbol{b}}_{-\boldsymbol{G}} \cdot \overleftrightarrow{\chi}(\boldsymbol{G},0)\hat{\boldsymbol{b}}_0 = -\left.\frac{\partial^2 E(\boldsymbol{B})}{\partial B_0 \partial B_{-\boldsymbol{G}}}\right|_{\boldsymbol{B}=0}. \tag{2.278}$$

Using second-order perturbation theory, one obtains [249]

$$\begin{aligned}\hat{\boldsymbol{b}}_{-\boldsymbol{G}} \cdot \overleftrightarrow{\chi}(\boldsymbol{G},0)\hat{\boldsymbol{b}}_0 = -\lim_{q\to 0}\frac{1}{2qc^2G^2}&[f(\hat{\boldsymbol{q}}\times\hat{\boldsymbol{b}}_0, \boldsymbol{G}\times\hat{\boldsymbol{b}}_{-\boldsymbol{G}}, \boldsymbol{G}, \boldsymbol{q}) \\ &- f(\hat{\boldsymbol{q}}\times\hat{\boldsymbol{b}}_0, \boldsymbol{G}\times\hat{\boldsymbol{b}}_{-\boldsymbol{G}}, \boldsymbol{G}, -\boldsymbol{q})],\end{aligned} \tag{2.279}$$

where $\hat{\boldsymbol{q}} = \boldsymbol{q}/q$, and the function $f(\boldsymbol{a}_0, \boldsymbol{a}_1, \boldsymbol{G}, \boldsymbol{q})$ is given by

$$\begin{aligned}&f(\boldsymbol{a}_0, \boldsymbol{a}_1, \boldsymbol{G}, \boldsymbol{q}) \\ &= -\frac{1}{2}\sum_{\boldsymbol{k}}\sum_{\lambda}^{O}\sum_{\nu}^{\mathcal{E}}\left[\frac{\langle \boldsymbol{k},\lambda|\boldsymbol{a}_1\cdot\mathcal{D}_{-\boldsymbol{G}-\boldsymbol{q}}|\boldsymbol{k}+\boldsymbol{q},\nu\rangle\langle\boldsymbol{k}+\boldsymbol{q},\nu|\boldsymbol{a}_0\cdot\mathcal{D}_{\boldsymbol{q}}|\boldsymbol{k},\lambda\rangle}{\varepsilon_{\boldsymbol{k}\lambda}-\varepsilon_{\boldsymbol{k}+\boldsymbol{q}\nu}}\right. \\ &\qquad \left.+\frac{\langle \boldsymbol{k},\lambda|\boldsymbol{a}_0\cdot\mathcal{D}_{\boldsymbol{q}}|\boldsymbol{k}-\boldsymbol{q},\nu\rangle\langle\boldsymbol{k}-\boldsymbol{q}\,\nu|\boldsymbol{a}_1\cdot\mathcal{D}_{-\boldsymbol{G}-\boldsymbol{q}}|\boldsymbol{k},\lambda\rangle}{\varepsilon_{\boldsymbol{k}\lambda}-\varepsilon_{\boldsymbol{k}-\boldsymbol{q},\nu}}\right],\end{aligned} \tag{2.280}$$

with $\mathcal{D}_{\boldsymbol{q}} = e^{i\boldsymbol{q}\cdot\boldsymbol{r}}\nabla + \nabla e^{i\boldsymbol{q}\cdot\boldsymbol{r}}$. The chemical shift $\sigma$ computed for the HF molecule from (2.279) with (2.280) is 28.4 ppm, whereas its experimental value is 28.5 ppm [249]. The value for an HF crystal at 4.2 K is given by the same equation as 20.71 ppm, whereas the value for HF liquid at 214 K is measured experimentally as 21.45 ppm.

Also the chemical shift of $NH_3$, $N_2H_3$, $CH_{4-n}(NH_2)_n$ ($n = 0 - 3$) and a few other C-H-N molecules have been determined by the same method [250].

### 2.4.5 *Phonon Spectrum*

By using similar perturbation and linear-response theories, one may also compute the phonon spectra of insulators, semiconductors and metals. Within the adiabatic approximation, the lattice-distortion-associated phonons can be seen as a static perturbation acting on the electrons. Here we describe one way of calculating the force constants, which was presented by Giannozzi et al. [344]. The matrix of the force constants is defined as

$$W_{i\alpha,j\beta}(\boldsymbol{R}-\boldsymbol{R}') = \frac{\partial^2 E}{\partial u_{i\alpha}(\boldsymbol{R})\partial u_{j\beta}(\boldsymbol{R}')}, \tag{2.281}$$

where $E$ is the total energy of the system and $u_{n\alpha}(\boldsymbol{R})$ the atomic displacements. Here $\boldsymbol{R}$, $\boldsymbol{R}'$ denote the lattice vectors indicating unit cells, $i$, $j$ indicate the atomic positions inside the unit cells, and $\alpha, \beta\,(=x, y, z)$ are the polarization directions. Because of its translational invariance, it is convenient to transform the force constants into Fourier space via the relation

$$W_{i\alpha,j\beta}(\boldsymbol{R}) = \sum_{\boldsymbol{q}} \mathrm{e}^{i\boldsymbol{q}\cdot\boldsymbol{R}} \widetilde{W}_{i\alpha,j\beta}(\boldsymbol{q}). \tag{2.282}$$

In this wave vector representation, the lattice distortion is written as

$$u_{i\alpha}(\boldsymbol{R}) = \sum_{\boldsymbol{q}} \mathrm{e}^{i\boldsymbol{q}\cdot\boldsymbol{R}} \widetilde{u}_{i\alpha}(\boldsymbol{q}). \tag{2.283}$$

For the way to calculate the phonon dispersion from the force constant matrix $\widetilde{W}_{i\alpha,j\beta}(\boldsymbol{q})$, see Appendix C.

The force constants can be divided into the ionic and electronic contributions:

$$\widetilde{W}_{i\alpha,j\beta}(\boldsymbol{q}) = \widetilde{W}^{\text{ion}}_{i\alpha,j\beta}(\boldsymbol{q}) + \widetilde{W}^{\text{elec}}_{i\alpha,j\beta}(\boldsymbol{q}). \tag{2.284}$$

The ionic contribution $\widetilde{W}^{\text{ion}}$ is essentially the second derivative of an Ewald sum and can be obtained from (C.9) in Appendix C:

$$\begin{aligned}\widetilde{W}^{\text{ion}}_{i\alpha,j\beta}(\boldsymbol{q}) &= \frac{\partial^2 V_{\text{ion-ion}}(\boldsymbol{q})}{\partial\widetilde{u}_{i\alpha}(\boldsymbol{q})\partial\widetilde{u}_{j\beta}(\boldsymbol{q})} \\ &= \frac{4\pi Z_i Z_j \mathrm{e}^2}{\Omega}\sum_{\boldsymbol{G}\neq-\boldsymbol{q}}\frac{\mathrm{e}^{-(\boldsymbol{q}+\boldsymbol{G})^2/4\varepsilon}}{(\boldsymbol{q}+\boldsymbol{G})^2}\mathrm{e}^{i(\boldsymbol{q}+\boldsymbol{G})\cdot(\boldsymbol{\tau}_i-\boldsymbol{\tau}_j)}(q_\alpha+G_\alpha)(q_\beta+G_\beta) \\ &- \delta_{ij}\frac{4\pi Z_i \mathrm{e}^2}{\Omega}\sum_{\boldsymbol{G}\neq 0}\frac{\mathrm{e}^{-G^2/4\varepsilon}}{G^2}\sum_l Z_l\cos[\boldsymbol{G}\cdot(\boldsymbol{\tau}_i-\boldsymbol{\tau}_l)]G_\alpha G_\beta, \qquad (2.285)\end{aligned}$$

where $\varepsilon$ is chosen large enough to neglect the term representing real-space summation. Here, $\boldsymbol{\tau}_i$ denotes the atomic position inside the unit cell. The electronic contribution $\widetilde{W}^{\text{elec}}$ in (2.283) is identified as

$$\widetilde{W}^{\text{elec}}_{i\alpha,j\beta}(\boldsymbol{q}) = \int\left[\frac{\partial\rho(\boldsymbol{r})}{\partial\widetilde{u}_{i\alpha}(\boldsymbol{q})}\right]^*\frac{\partial v_{\text{ion}}(\boldsymbol{r})}{\partial\widetilde{u}_{j\beta}(\boldsymbol{q})}d\boldsymbol{r} + \delta_{ij}\int\rho_0(\boldsymbol{r})\frac{\partial^2 v_{\text{ion}}(\boldsymbol{r})}{\partial\widetilde{u}_{i\alpha}(\boldsymbol{q}=0)\partial\widetilde{u}_{j\beta}(\boldsymbol{q}=0)}d\boldsymbol{r}, \qquad (2.286)$$

if the Hellmann–Feynman theorem [17, 345] is used for the force (this force is called the Hellmann–Feynman force; see Sect. 2.4). Here, $\rho(\boldsymbol{r})$ and $\rho_0(\boldsymbol{r})$ are the perturbed (by ion displacements) and nonperturbed electron charge densities; $v_{\text{ion}}(\boldsymbol{r})$ represents the Coulomb potential for electrons caused by the nuclear charges, when an all-electron approach is used; alternatively, it should be replaced by the ionic pseudopotential when the pseudopotential method is used. All-electron formalisms such as the LAPW method [346] and the LMTO method [222] have also been applied to linear-response calculations of phonon spectra.

The evaluation of the derivatives of $v_{\text{ion}}(\boldsymbol{r})$ with respect to the atomic coordinates is straightforward, since its explicit dependence on the atomic positions is given at the outset. Therefore, the second term on the right-hand side of (2.286) can be evaluated without perturbation theory. The first term can be expressed as

$$\sum_{\boldsymbol{G}}\left[\frac{\partial\widetilde{\rho}(\boldsymbol{q}+\boldsymbol{G})}{\partial\widetilde{u}_{i\alpha}(\boldsymbol{q})}\right]^*\frac{\partial\widetilde{v}_{\text{ion}}(\boldsymbol{q}+\boldsymbol{G})}{\partial\widetilde{u}_{j\beta}(\boldsymbol{q})} \qquad (2.287)$$

when $v_{\text{ion}}(\boldsymbol{r})$ is local. (The expression for a nonlocal pseudopotential is given in [344].) Here, we notice that

$$\frac{\partial\widetilde{v}_{\text{ion}}(\boldsymbol{q}+\boldsymbol{G})}{\partial\widetilde{u}_{j\beta}(\boldsymbol{q})} = -i(q_\beta+G_\beta)\mathrm{e}^{-i(\boldsymbol{q}+\boldsymbol{G})\cdot\boldsymbol{\tau}_j}\widetilde{v}(\boldsymbol{q}+\boldsymbol{G}), \qquad (2.288)$$

where $\widetilde{v}(\boldsymbol{q}+\boldsymbol{G})$ is the Fourier transform $\boldsymbol{r}+\boldsymbol{R}\to\boldsymbol{q}+\boldsymbol{G}$ of the local ion potential $v_{\text{ion}}(\boldsymbol{r}+\boldsymbol{\tau}_j+\boldsymbol{R})$; for all-electron approaches, it is given by $-Z_j/[\Omega(\boldsymbol{q}+\boldsymbol{G})^2]$. By means of first-order perturbation theory, we have

$$\Delta\widetilde{\rho}(\boldsymbol{q}+\boldsymbol{G}) = \frac{4}{M\Omega}\sum_{\boldsymbol{k}}\sum_{\lambda\in O}\sum_{\nu\in\mathcal{E}}\frac{\langle\boldsymbol{k},\lambda|\mathrm{e}^{-i(\boldsymbol{q}+\boldsymbol{G})\cdot\boldsymbol{r}}|\boldsymbol{k}+\boldsymbol{q},\nu\rangle\langle\boldsymbol{k}+\boldsymbol{q},\nu|\,\Delta V_{\mathrm{SCF}}\,|\boldsymbol{k},\lambda\rangle}{\varepsilon_{\boldsymbol{k}\lambda}-\varepsilon_{\boldsymbol{k}+\boldsymbol{q}\nu}}. \tag{2.289}$$

$\Delta V_{\mathrm{SCF}}$ denotes the deviation of the self-consistent potential in the Kohn–Sham equation (2.54) due to the lattice distortion. For convenience, an expression using a finite difference has been used here, but it means a derivative with respect to an atomic displacement. Up to the first order in perturbation, this deviation is given by

$$\Delta V_{\mathrm{SCF}}(\boldsymbol{r}) = \Delta v_{\mathrm{ion}}(\boldsymbol{r}) + \int\frac{\Delta\rho(\boldsymbol{r}')}{|\boldsymbol{r}-\boldsymbol{r}'|}d\boldsymbol{r}' + \Delta\rho(\boldsymbol{r})\left[\frac{\delta\mu_{\mathrm{xc}}}{\delta\rho}\right]_{\rho=\rho_0(\boldsymbol{r})}. \tag{2.290}$$

Equations (2.289) and (2.290) form a system that can be solved iteratively. The matrix element $\langle\boldsymbol{k}+\boldsymbol{q},\nu|\Delta V_{\mathrm{SCF}}(\boldsymbol{r}')|\boldsymbol{k},\lambda\rangle$ appearing in (2.289) can be rewritten as

$$\langle\boldsymbol{k}+\boldsymbol{q},\nu|\,\Delta V_{\mathrm{SCF}}\,|\boldsymbol{k},\lambda\rangle = \sum_{\boldsymbol{G}'}\langle\boldsymbol{k}+\boldsymbol{q},\nu|\mathrm{e}^{i(\boldsymbol{q}+\boldsymbol{G}')\cdot\boldsymbol{r}'}|\boldsymbol{k},\lambda\rangle\Delta\widetilde{V}_{\mathrm{SCF}}(\boldsymbol{q}+\boldsymbol{G}'). \tag{2.291}$$

In Fourier space, the three terms in the r.h.s. of (2.290) are given, respectively, by $\Delta\widetilde{v}_{\mathrm{ion}}(\boldsymbol{q}+\boldsymbol{G})$ that is identical to (2.288),

$$\Delta\widetilde{V}_{\mathrm{H}}(\boldsymbol{q}+\boldsymbol{G}) = \frac{\partial\widetilde{V}_{\mathrm{H}}(\boldsymbol{q}+\boldsymbol{G})}{\partial\widetilde{u}_{j\beta}(\boldsymbol{q})} = \Omega\widetilde{v}(\boldsymbol{q}+\boldsymbol{G})\Delta\widetilde{\rho}(\boldsymbol{q}+\boldsymbol{G}), \tag{2.292}$$

and

$$\Delta\widetilde{V}_{\mathrm{xc}}(\boldsymbol{q}+\boldsymbol{G}) = \frac{\partial\widetilde{V}_{\mathrm{xc}}(\boldsymbol{q}+\boldsymbol{G})}{\partial\widetilde{u}_{j\beta}(\boldsymbol{q})} = \Omega\sum_{\boldsymbol{G}'}\widetilde{f}_{\mathrm{xc}}(\boldsymbol{q}+\boldsymbol{G},\boldsymbol{q}+\boldsymbol{G}')\Delta\widetilde{\rho}(\boldsymbol{q}+\boldsymbol{G}'), \tag{2.293}$$

where $\widetilde{f}_{\mathrm{xc}}(\boldsymbol{q}+\boldsymbol{G},\boldsymbol{q}+\boldsymbol{G}')$ is the nonlocal Fourier transformation of the second derivative of the exchange-correlation energy $E_{\mathrm{xc}}$ with respect to the electron density

$$f_{\mathrm{xc}}(\boldsymbol{r},\boldsymbol{r}') = \left.\frac{\delta^2E_{\mathrm{xc}}[\rho]}{\delta\rho(\boldsymbol{r})\delta\rho(\boldsymbol{r}')}\right|_{\rho=\rho_0(\boldsymbol{r})}. \tag{2.294}$$

In the case of LDA, $f_{\mathrm{xc}}(\boldsymbol{r},\boldsymbol{r}')$ and $\widetilde{f}_{\mathrm{xc}}(\boldsymbol{q}+\boldsymbol{G},\boldsymbol{q}+\boldsymbol{G}')$ become proportional to $\delta(\boldsymbol{r}-\boldsymbol{r}')$ and $\delta_{\boldsymbol{G}\boldsymbol{G}'}$, respectively.

However, (2.289) together with (2.291) is not easy to calculate, because one has to sum over all $\boldsymbol{G}$, $\boldsymbol{G}'$, $\nu\in\mathcal{E}$, and $\lambda\in O$ for each $\boldsymbol{q}$. More convenient way is to use the matrix algebra. Using density functional perturbation theory (DFPT), Baroni et al. [347, 348] and de Gironcoli [349] formulated the way to do it. Linearization of (2.142) with respect to a small deviation from the unperturbed system leads to

$$\Delta\rho(\boldsymbol{r}) = 4\mathrm{Re}\sum_{\lambda}^{O}\sum_{\boldsymbol{k}\in 1\mathrm{stBZ}} \psi^*_{\boldsymbol{k}\lambda}(\boldsymbol{r})\Delta\psi_{\boldsymbol{k}+\boldsymbol{q}\lambda}(\boldsymbol{r}), \tag{2.295}$$

where Re means to take the real part. Using standard first-order perturbation theory, one may derive an equation for the variation of a Kohn–Sham orbital, $\Delta\psi_{\boldsymbol{k}+\boldsymbol{q}\lambda}(\boldsymbol{r})$,

$$(H - \varepsilon_{\boldsymbol{k}\lambda})\Delta\psi_{\boldsymbol{k}+\boldsymbol{q}\lambda}(\boldsymbol{r}) = -\left[\Delta V_{\mathrm{SCF}}(\boldsymbol{r}) - \Delta\varepsilon_{\boldsymbol{k}\lambda}\right]\psi_{\boldsymbol{k}\lambda}(\boldsymbol{r}), \tag{2.296}$$

where $\Delta\varepsilon_{\boldsymbol{k}\lambda} = \langle\boldsymbol{k}+\boldsymbol{q}, \lambda|\,\Delta V_{\mathrm{SCF}}\,|\boldsymbol{k}, \lambda\rangle$ is the first-order variation of the Kohn–Sham eigenvalue $\varepsilon_{\boldsymbol{k}\lambda}$. Note that the perturbation $\Delta\widetilde{V}_{\mathrm{SCF}}(\boldsymbol{q}+\boldsymbol{G})$ carries the momentum transfer $\boldsymbol{q}$. Comparing (2.296) and (2.289), we notice that the perturbation should couple the occupied states $\psi_{\boldsymbol{k}\lambda}$ to the empty states $\psi_{\boldsymbol{k}+\boldsymbol{q}\nu}$ only, and should not couple the occupied states $\psi_{\boldsymbol{k}\lambda}$ to the occupied states $\psi_{\boldsymbol{k}+\boldsymbol{q}\lambda'}$. That is, the result of the perturbation acting onto an occupied state $\psi_{\boldsymbol{k}\lambda}$ must be expanded by a linear combination of the empty states $\psi_{\boldsymbol{k}+\boldsymbol{q}\nu}$ only. In this sense, we used the notation $\Delta\psi_{\boldsymbol{k}+\boldsymbol{q}\lambda}$ instead of $\Delta\psi_{\boldsymbol{k}\lambda}$ in (2.295) and (2.296). This can be also seen from the fact that the l.h.s. of (2.296) is singular because the linear operator $\mathcal{L} = H - \varepsilon_{\boldsymbol{k}\lambda}$ has zero (in the case $\varepsilon_{\boldsymbol{k}\lambda} = \varepsilon_{\boldsymbol{k}+\boldsymbol{q}\lambda'}$) or very small (in the case $\varepsilon_{\boldsymbol{k}\lambda} \sim \varepsilon_{\boldsymbol{k}+\boldsymbol{q}\lambda'}$) eigenvalues for the occupied state $\psi_{\boldsymbol{k}+\boldsymbol{q}\lambda'}$ and its simple inversion $\mathcal{L}^{-1}$ diverges. Therefore, we have to remove the occupied-state components from the r.h.s. of (2.296) by using the projection operator $\mathcal{P}_{\boldsymbol{k}+\boldsymbol{q}}$ defined in (2.256) and add a term $\alpha\mathcal{P}_{\boldsymbol{k}+\boldsymbol{q}}$ to the Hamiltonian in order to remove all small or null eigenvalues of the operator $\mathcal{L}$; this term has no effect on $\Delta\psi_{\boldsymbol{k}+\boldsymbol{q}\lambda}$, since $\mathcal{P}_{\boldsymbol{k}+\boldsymbol{q}}\Delta\psi_{\boldsymbol{k}+\boldsymbol{q}\lambda} = 0$ by construction. Thus, (2.296) can be rewritten as

$$(H + \alpha\mathcal{P}_{\boldsymbol{k}+\boldsymbol{q}} - \varepsilon_{\boldsymbol{k}\lambda})\Delta\psi_{\boldsymbol{k}+\boldsymbol{q}\lambda}(\boldsymbol{r}) = -(1 - \mathcal{P}_{\boldsymbol{k}+\boldsymbol{q}})\Delta V_{\mathrm{SCF}}\psi_{\boldsymbol{k}\lambda}(\boldsymbol{r}). \tag{2.297}$$

By solving (2.297), one may obtain the variation of the Kohn–Sham orbitals, $\Delta\psi_{\boldsymbol{k}+\boldsymbol{q}\lambda}(\boldsymbol{r})$, to be used in (2.295). The operator $H + \alpha\mathcal{P}_{\boldsymbol{k}+\boldsymbol{q}} - \varepsilon_{\boldsymbol{k}\lambda}$ is made non-singular by choosing the parameter $\alpha$ somewhat larger than $\varepsilon_{\mathrm{F}} - \varepsilon_{\boldsymbol{k}\lambda}$ [348, 349]. This algorithm works well not only for semiconductors and insulators but also for metals, although some cares are necessary as explained by de Gironcoli [349].

To solve (2.289) or (2.295) and (2.297) together with (2.290), the starting perturbation $\Delta V_{\mathrm{SCF}}$ is set equal to the electron-nuclei potential (2.288). Then $\Delta\widetilde{\rho}(\boldsymbol{q}+\boldsymbol{G})$ is calculated by (2.289) or (2.295) and (2.297), and $\Delta V_{\mathrm{SCF}}$ is updated by (2.290) in Fourier space. The cycle is repeated until the change in $\Delta V_{\mathrm{SCF}}$ between two successive steps becomes smaller than a set tolerance.

The matrix elements $\langle\boldsymbol{k}+\boldsymbol{q}, \nu|\,\Delta V_{\mathrm{SCF}}\,|\boldsymbol{k}, \lambda\rangle$ in (2.291) can be used in the calculation of the electron-phonon interaction as described in detail in a recent review by Giustino [350]. The atomic displacement in the $\alpha$ $(= x, y, z)$ component of the $i$th atom in the $p$th unit cell can be expressed as

$$\Delta u_{i\alpha p} = \sqrt{\frac{M_0}{N_p M_i}}\sum_{\boldsymbol{q},n} \mathrm{e}^{i\boldsymbol{q}\cdot\boldsymbol{R}_p} e_{i\alpha n}(\boldsymbol{q}) l_{\boldsymbol{q}n}(a_{\boldsymbol{q}n} + a^\dagger_{-\boldsymbol{q}n}), \tag{2.298}$$

where $l_{qn}$ is the zero-point displacement amplitude,

$$l_{qn} = \sqrt{\frac{\hbar}{2M_0\omega_{qn}}}, \tag{2.299}$$

$M_0$ is an arbitrary reference mass introduced to give the dimension of a length to $l_{qn}$, $e_{i\alpha\nu}$ is the eigenvector of the dynamical matrix associated with the normal mode $n$ (see Appendix C), $N_p$ is the number of unit cells, and $a_{qn}$ and $a^{\dagger}_{-qn}$ are the phonon annihilation and creation operators of a phonon of the $n$th branch with a momentum $\boldsymbol{q}$. Then, the electron-phonon coupling constant can be calculated as

$$g_{\nu\lambda n}(\boldsymbol{k}, \boldsymbol{q}) = \langle \boldsymbol{k}+\boldsymbol{q}, \nu|\, \Delta_{qn} V_{\rm SCF}\, |\boldsymbol{k}, \lambda\rangle, \tag{2.300}$$

where $\Delta_{qn} V_{\rm SCF}$ is expressed by $\Delta V_{\rm SCF}$ as

$$\Delta_{qn} V_{\rm SCF} = l_{qn} \sum_{i,\alpha} \sqrt{\frac{M_0}{M_i}} e_{i\alpha n}(\boldsymbol{q}) \Delta V_{\rm SCF}. \tag{2.301}$$

In the case of polar materials, where Born effective charges $Z_i^*$ are nonzero, the dipole contribution to the potential $\Delta V_{\rm SCF}$ becomes dominant. This contribution is long range and decays as $|\boldsymbol{R}_p|^{-2}$, which diverges as $|\boldsymbol{q}|^{-1}$ in the limit $\boldsymbol{q}$ going to zero in Fourier space. Therefore fine $\boldsymbol{q}$ grid is required. Physically, this singularity corresponds to the so-called "Fröhlich electron-phonon coupling" [351], this long range behavior can be expressed as

$$\begin{aligned} g^{\mathcal{L}}_{\nu\lambda n}(\boldsymbol{k}, \boldsymbol{q}) &= i\frac{4\pi}{\Omega}\frac{e^2}{4\pi\varepsilon_0} \sum_i \sqrt{\frac{\hbar}{2N_p M_i \omega_{qn}}} \sum_{\boldsymbol{G}\neq -\boldsymbol{q}} \frac{(\boldsymbol{q}+\boldsymbol{G})\cdot \boldsymbol{Z}_i^* \cdot \boldsymbol{e}_{in}(\boldsymbol{q})}{(\boldsymbol{q}+\boldsymbol{G})\cdot \boldsymbol{\epsilon}^{\infty} \cdot (\boldsymbol{q}+\boldsymbol{G})} \\ &\times \langle \boldsymbol{k}+\boldsymbol{q}, \nu| \mathrm{e}^{i(\boldsymbol{q}+\boldsymbol{G})\cdot(\boldsymbol{r}-\boldsymbol{\tau}_i)} |\boldsymbol{k}, \lambda\rangle, \end{aligned} \tag{2.302}$$

where $\boldsymbol{Z}_i^*$ and $\boldsymbol{\epsilon}^{\infty}$ denote the Born effective charge tensor and the electronic permittivity tensor in the limit $\omega = \infty$. Although $g^{\mathcal{L}}_{\nu\lambda n}(\boldsymbol{k}, \boldsymbol{q})$ dominates for very small $\boldsymbol{q}$, it is necessary to smoothly interpolate the DFPT results by including this singular contribution over wider region. In this course of calculations, Wannier interpolation is often used because the maximally localized Wannier function rapidly damps at long distances and only a limited number of unit cells can be used in the Wannier representation and the calculation of the huge number of $\boldsymbol{k}$ and $\boldsymbol{q}$ points can be avoided [350, 352, 353].

The relaxation time $\tau_{\rm el-ph}$ related to the electron-phonon interaction has been recently calculated [350, 352, 354, 355] from the imaginary part of the electron-phonon self-energy,

$$\text{Im}\,\Sigma_{k\nu}^{\text{el-ph}} = \sum_{\lambda,n,q} |g_{\nu\lambda n}(k,q)|^2 \text{Im}\left[\frac{N_{qn}+1-f(\varepsilon_{k\lambda})}{\varepsilon_{k\nu}-\varepsilon_{\lambda k+q}-\omega_{qn}-i\delta} + \frac{N_{qn}+f(\varepsilon_{k\lambda})}{\varepsilon_{k\nu}-\varepsilon_{\lambda k+q}+\omega_{qn}-i\delta}\right], \tag{2.303}$$

through the relation $1/\tau_{\text{el-ph}} = \text{Im}\,\Sigma_{k\nu}^{\text{el-ph}}$, where Im means to take the imaginary part, $N_{qn}$ and $f(\varepsilon_{k\lambda})$ denote the Bose (phonon) and Fermi (electron) distribution functions, respectively.

The linear-response technique for calculating the force constants can be compared with the so-called "direct" method [223, 224, 356–362], which calculates the force constants directly by changing slightly one atomic position at a time in a large unit cell. The direct method offers also an accurate estimate of the dynamical matrix, in particular at high-symmetry points in the Brillouin zone. see also Appendix C for phonons.

Ab initio calculational methods for phonon spectra have also been applied to the investigation of ferroelectricity [363] and piezoelectricity [364, 365].

### 2.4.6 Electrical Conductivity

In the presence of a weak electric field $E(\boldsymbol{r},t)$, the perturbative Hamiltonian describing an interaction between electrons and a external field is given by

$$\mathcal{H}'(t) = \frac{1}{c}\int d\boldsymbol{r}\,\boldsymbol{A}(\boldsymbol{r},t)\cdot\boldsymbol{j}(\boldsymbol{r}), \tag{2.304}$$

where $\boldsymbol{r}_i$ and $\boldsymbol{p}_i$ are, respectively, the position and momentum of the $i$th atom, $\boldsymbol{A}(\boldsymbol{r},t)$ is the vector potential associated with the electric field $E(\boldsymbol{r},t)$, and $\boldsymbol{j}(\boldsymbol{r},t)$ is the paramagnetic part (2.45) of the total current operator

$$\boldsymbol{J}(\boldsymbol{r},t) = \boldsymbol{j}(\boldsymbol{r},t) - \frac{N}{c}\boldsymbol{A}(\boldsymbol{r},t) \tag{2.305}$$

($N$ is the total number of electrons) defined by

$$\boldsymbol{J}(\boldsymbol{r},t) = \frac{1}{2}\sum_i \left\{\left[\frac{1}{\mathrm{i}}\nabla - \frac{1}{c}\boldsymbol{A}(\boldsymbol{r},t)\right]\delta(\boldsymbol{r}-\boldsymbol{r}_i) + \delta(\boldsymbol{r}-\boldsymbol{r}_i)\left[\frac{1}{\mathrm{i}}\nabla - \frac{1}{c}\boldsymbol{A}(\boldsymbol{r},t)\right]\right\} \tag{2.306}$$

Using an interaction representation [1, 2, 4], one may write the temporal evolution of $\boldsymbol{j}(\boldsymbol{r},t)$ according to the unperturbed Hamiltonian $\mathcal{H}_0$ as follows:

$$\boldsymbol{j}_{\text{I}}(\boldsymbol{r},t) = \mathrm{e}^{i\mathcal{H}_0 t}\boldsymbol{j}(\boldsymbol{r},t)\mathrm{e}^{-i\mathcal{H}_0 t} \tag{2.307}$$

Then, the expectation value for the total current is given to the first-order in $\mathcal{H}'$ by

$$\langle \boldsymbol{J}(\boldsymbol{r},t)\rangle = \langle \boldsymbol{j}(\boldsymbol{r},t)\rangle + i\int_{-\infty}^{t} dt' \langle [\mathcal{H}'(t'), \boldsymbol{j}_{\mathrm{I}}(\boldsymbol{r},t)]\rangle_0 - \frac{N}{c}\boldsymbol{A}(\boldsymbol{r},t), \quad (2.308)$$

where $\langle ... \rangle_0$ stands for the expectation value evaluated for the unperturbed system. Thus, we are led to the Kubo formula [316],

$$\langle J_\alpha(\boldsymbol{r},t)\rangle = \langle j_\alpha(\boldsymbol{r},t)\rangle + \frac{1}{c}\int_{-\infty}^{\infty} dt' \int d\boldsymbol{r}' \sum_{\beta=1}^{3} S_{\alpha\beta}(\boldsymbol{r}-\boldsymbol{r}', t-t') A_\beta(\boldsymbol{r}', t'), \quad (\alpha = 1,2,3) \quad (2.309)$$

with

$$S_{\alpha\beta}(\boldsymbol{r}-\boldsymbol{r}', t-t') = \mathcal{S}_{\alpha\beta}(\boldsymbol{r}-\boldsymbol{r}', t-t') - N\delta_{\alpha\beta}\delta(\boldsymbol{r}-\boldsymbol{r}')\delta(t-t') \quad (2.310)$$

and

$$\mathcal{S}_{\alpha\beta}(\boldsymbol{r}-\boldsymbol{r}', t-t') = -i\theta(t-t')\langle [j_\alpha(\boldsymbol{r},t), j_\beta(\boldsymbol{r}',t')]\rangle. \quad (2.311)$$

The second term in (2.310) is called the diamagnetic term, and $\mathcal{S}_{\alpha\beta}(\boldsymbol{r}-\boldsymbol{r}', t-t')$ is called the response function.

We know that the retarded form of the response function in (2.311) can be calculated from the time-ordered correlation function defined as

$$\mathcal{S}^{\mathrm{T}}_{\alpha\beta}(\boldsymbol{r}-\boldsymbol{r}', t-t') = -i\langle [\mathrm{T} j_\alpha(\boldsymbol{r},t), j_\beta(\boldsymbol{r}',t')]\rangle, \quad (2.312)$$

where $T$ is the time-ordering operator. The relation is called the fluctuation–dissipation theorem and is given in Fourier-transformed $\omega$ space by

$$\mathrm{Re}\mathcal{S}_{\alpha\beta}(\boldsymbol{r}-\boldsymbol{r}', \omega) = \mathrm{Re}\mathcal{S}^{\mathrm{T}}_{\alpha\beta}(\boldsymbol{r}-\boldsymbol{r}', \omega) \quad (2.313)$$

and

$$\mathrm{Im}\mathcal{S}_{\alpha\beta}(\boldsymbol{r}-\boldsymbol{r}', \omega) = \frac{\omega}{|\omega|}\mathrm{Im}\mathcal{S}^{\mathrm{T}}_{\alpha\beta}(\boldsymbol{r}-\boldsymbol{r}', \omega). \quad (2.314)$$

The time-ordered correlation function can more conveniently be calculated perturbatively by using Wick's theorem [318, 319].

The conductivity tensor is defined by Ohm's law, which can be written in $\omega$ space as

$$J_\alpha = \sum_{\beta=1}^{3} \sigma_{\alpha\beta}(\omega) E_\beta = \frac{i\omega}{c}\sum_{\beta=1}^{3} \sigma_{\alpha\beta}(\omega) A_\beta. \quad (2.315)$$

Therefore, conductivity is obtained from the response function as

$$\sigma_{\alpha\beta}(\omega) = \frac{S_{\alpha\beta}(\omega)}{i\omega}, \tag{2.316}$$

and the d.c. conductivity is given from this equation in the limit $\omega \to 0$. Since we expect a finite d.c. conductivity in any case, we notice that $S_{\alpha\beta}(\omega)$ cancels the diamagnetic term precisely in the limit $\omega \to 0$. As was pointed out first by London [366, 367] in 1935, the superconductivity can be considered as a result of the absence of this cancellation between the paramagnetic and diamagnetic parts. Also, for bulk metals which have perfect crystal forms, the d.c. conductivity becomes infinity at the absolute-zero temperature, as seen from (2.316). The finite d.c. conductivity is obtained from the effect of imperfections and lattice vibrations.

A more convenient and equivalent expression was derived by Greenwood [368] in 1958 as

$$\begin{aligned} \sigma_{\alpha\beta} &= \frac{2\pi}{\Omega\omega} \sum_{\boldsymbol{k}} \sum_{\lambda} \sum_{\nu} \langle \boldsymbol{k}, \lambda | \nabla_\alpha | \boldsymbol{k}, \nu \rangle \langle \boldsymbol{k}, \nu | \nabla_\beta | \boldsymbol{k}, \lambda \rangle \\ &\quad \times [f_0(\varepsilon_{\boldsymbol{k}\lambda}) - f_0(\varepsilon_{\boldsymbol{k}\nu})] \delta(\varepsilon_{\boldsymbol{k}\lambda} - \varepsilon_{\boldsymbol{k}\nu} - \omega) \end{aligned} \tag{2.317}$$

where $\Omega$ is the volume of the unit cell, and the spin multiplicity is taken into account. This non-interacting formula is commonly called the Kubo–Greenwood formula. Using this formula, we find that the imaginary part of the electric susceptibility in the limit $\boldsymbol{q} \to 0$ given by (2.261) can be related to the electriccolorRedal conductivity $\sigma(\omega)$ by

$$\mathrm{Im}\chi_{\alpha\beta}(0, \omega) = \frac{\sigma_{\alpha\beta}(\omega)}{\omega}. \tag{2.318}$$

The calculation of the conductivity has been performed by several authors [369–372] for disordered systems by using the Kubo–Greenwood formula and averaging it along the trajectories obtained from ab initio molecular-dynamics simulations. The resulting $\sigma(\omega)$ can be fitted to the Drude function

$$\sigma(\omega) = \frac{\rho_e e^2 \tau}{\Omega m} \frac{1}{1 + \omega^2 \tau} \tag{2.319}$$

to derive the relaxation time $\tau$ as discussed by Silvestrelli et al. [372].

A completely different approach to consider the electric conductivity is to use the Landauer formula [317, 373]. In this approach, one calculates the transmission coefficient $T$ instead of the response function. It is mainly used for 1D or pseudo-1D model systems. Here we do not go into the details of this approach, although it has often been used in the treatments of mesoscopic systems such as quantum dots and wires. We only note that this approach was used to calculate a single wire under a magnetic field [374] and for a sandwiched carbon nanotube by using the tight-binding approach [375, 377, 379].

### 2.4.7 Thermal Conductivity

Similar to the electrical conductivity, the thermal conductivity can also be obtained from a linear-response formulation. The driving field is in this case a temperature gradient. By assimilating the thermal gradients to an external mechanical field Luttinger provided a formal derivation now known as the Green–Kubo (GK) formula [316, 376, 378]. In its classical form, the GK formula states that the thermal conductivity tensor is the time integral of heat current autocorrelation evaluated *at equilibrium*:

$$\kappa^{\alpha\beta} = \frac{1}{\Omega k_B T^2} \int_0^\infty dt < J^\alpha(t) J^\beta(0) > \qquad (2.320)$$

where the Greek letters represent the cartesian components, $\Omega$ is the system volume, $< \cdots >$ represents the equilibrium average, $T$ is the temperature, and $\boldsymbol{J}$ is the heat current defined by the volume integral of the heat current density $\mathcal{J}$ vector, itself defined by $\partial\mathcal{H}(\boldsymbol{r},t)/\partial t + \nabla \cdot \mathcal{J}(\boldsymbol{r},t) = 0$. Note that this equation defines only the longitudinal part of the current. We will later show that the requirement of locality imposes a condition on the transverse component, which uniquely defines the heat current density.

The quantum version of this relation (less often used) is

$$\kappa^{\alpha\beta} = \frac{1}{\Omega T} \lim_{\eta\to 0} \int_0^\beta d\lambda \int_0^\infty dt\, e^{-\eta t} \;< J^\alpha(t+i\lambda) J^\beta(0) >, \qquad (2.321)$$

where we put $\beta = 1/k_\mathrm{B}T$.

The GK is the fundamental definition of the thermal conductivity which works in all cases. The thermal average is an equilibrium average, implying that without perturbing the system, one can compute its *linear* response to external perturbations. There are electron and phonon contributions to $\kappa$. Usually in metals the electronic contribution dominates the phonons perhaps by an order of magnitude, but in thermoelectric materials and in doped semiconductors, the two contributions might be of the same order. In such a case, one may use the Wiedemann–Franz law [380]

$$\kappa_{el} = LT\sigma \qquad (2.322)$$

for the electronic contribution to $\kappa$, although it does not exactly hold even for metals. Here $L = (\pi k_\mathrm{B}/e)^2/3$ is the Lorentz number. Therefore, the accurate calculation of the relaxation time $\tau$ involving various electron-phonon and impurity scattering processes is required. The relaxation time associated with electron-phonon scatterings $\tau_{\mathrm{el-ph}}$ may be obtained by (2.303) using the calculation of the electron-phonon interaction, while the relaxation time associated with impurity scatterings $\tau_{\mathrm{imp}}$ may be obtained by the fitting of the electric conductivity to (2.319). Then the total relaxation time is obtained using Mathiessen's rule

$$\frac{1}{\tau} = \frac{1}{\tau_{\text{el-ph}}} + \frac{1}{\tau_{\text{imp}}} \tag{2.323}$$

to obtain the electronic contribution to $\kappa$. Another way to determine $\tau_{\text{el-ph}}$ in somewhat more empirical fashion is discussed recently by Juneja et al. [381].

For the phonon contribution, *equilibrium* molecular dynamics method, which will be described later, can be used to estimate $\kappa$. The evaluation of autocorrelations is discussed in the appendix on statistical data analysis.

An equivalent formulation is that of "Einstein", which is based on the time integral of $\boldsymbol{J}$, the *energy-displacement* vector $\boldsymbol{G}$ such that $d\boldsymbol{G}/dt = \boldsymbol{J}$. This formulation may be more efficient to calculate $\kappa$. In terms of $\boldsymbol{G}$, thermal conductivity can be shown to become:

$$\kappa^{\alpha\beta} = \frac{1}{\Omega k_B T^2} \lim_{t\to\infty} \frac{< [G^\alpha(t) - G^\alpha(0)][G^\beta(t) - G^\beta(0)] >}{2t} \tag{2.324}$$

or in case of isotropic materials, in 3D, this simplifies to:

$$\kappa = \frac{1}{\Omega k_B T^2} \lim_{t\to\infty} \frac{< | \boldsymbol{G}(t) - \boldsymbol{G}(0) |^2 >}{6t} \tag{2.325}$$

Again, this is an equilibrium average, and in the case where there is no macroscopic work done on the system, we can identify heat with energy and use these two terms equivalently.

The question then becomes, what is the microscopic definition of $\boldsymbol{J}$ or $\boldsymbol{G}$ to use in the above formulas? Let us start with the Hamiltonian $\mathcal{H}$ and the continuity equation which yields the heat current density. Integrating the continuity equation with respect to time, we find that $\mathcal{H} + \nabla \cdot \mathcal{G} = 0$, from which one can extract the longitudinal expression for $\mathcal{G}$.[1] Extraction of $\mathcal{G}$ becomes easier using the Fourier transform defined by:

$$\mathcal{H}(\boldsymbol{q}, t) = \int \frac{d\boldsymbol{r}}{\Omega} \mathcal{H}(\boldsymbol{r}, t)\, e^{i\boldsymbol{q}\cdot\boldsymbol{r}} \; ; \quad \mathcal{H}(\boldsymbol{r}, t) = \int \frac{d\boldsymbol{q}}{(2\pi)^3} \mathcal{H}(\boldsymbol{q}, t)\, e^{-i\boldsymbol{q}\cdot\boldsymbol{r}}$$

$\mathcal{H}(\boldsymbol{q}) - i\boldsymbol{q} \cdot \mathcal{G}(\boldsymbol{q}) = 0$, implying $\mathcal{G}_l(\boldsymbol{q}) = -i\boldsymbol{q}\mathcal{H}(\boldsymbol{q})/q^2$. The space integration leading to the energy displacement $\boldsymbol{G}$, corresponds to taking the $\boldsymbol{q} \to 0$ limit of $\mathcal{G}(\boldsymbol{q})$. So, now we are left with the calculation of the total energy of the system, or in other words, defining the total Hamiltonian. Typically for a set of particles interacting with a pair potential $V(\boldsymbol{r})$, we can define the *local energy* $e_i$ corresponding to atom $i$ as follows[2]:

[1] Recall $\partial\mathcal{G}/\partial t = \mathcal{J}$. The constant of integration may be absorbed in the definition of the total energy

[2] For more complicated potentials involving many-body forces the definition of the local energy is not unique, but one can show that it does not affect the total thermal conductivity [382].

$$H = \sum_i e_i \text{ where } e_i = \frac{p_i^2}{2m_i} + \frac{1}{2}\sum_{j \neq i} V(\boldsymbol{r}_{ij}) \tag{2.326}$$

The energy density $\mathcal{H}$ is then defined as

$$\mathcal{H}(\boldsymbol{r}, t) =< \sum_i e_i \, \delta(\boldsymbol{r} - \boldsymbol{r}_i(t)) > \tag{2.327}$$

where $\delta$ is the Dirac delta function. The effect of ensemble averaging $< \cdots >$ is then to broaden the delta function according to the distribution of particle positions $\boldsymbol{r}_i(t)$. In the canonical ensemble, for a classical system, the ensemble average is defined to be:

$$< A(t) >_{\text{ens}} = \frac{\int d\boldsymbol{r}_1 ... d\boldsymbol{r}_N d\boldsymbol{p}_1 ... d\boldsymbol{p}_N \, A(\boldsymbol{r}_1, \ldots, \boldsymbol{r}_N, \boldsymbol{p}_1, \ldots, \boldsymbol{p}_N, t) \mathrm{e}^{-\beta H(\boldsymbol{r}, \boldsymbol{p})}}{\int d\boldsymbol{r}_1 ... d\boldsymbol{r}_N d\boldsymbol{p}_1 ... d\boldsymbol{p}_N \, \mathrm{e}^{-\beta H(\boldsymbol{r}, \boldsymbol{p})}} \tag{2.328}$$

while in the microcanonical case, the energy conservation must be imposed:

$$< A(t) >_{\text{ens}} = \frac{\int d\boldsymbol{r}_1 ... d\boldsymbol{r}_N d\boldsymbol{p}_1 ... d\boldsymbol{p}_N \, A(\boldsymbol{r}_1, \ldots, \boldsymbol{r}_N, \boldsymbol{p}_1, \ldots, \boldsymbol{p}_N, t) \delta(E - H(\boldsymbol{r}, \boldsymbol{p}))}{\int d\boldsymbol{r}_1 ... d\boldsymbol{r}_N d\boldsymbol{p}_1 ... d\boldsymbol{p}_N \, \delta(E - H(\boldsymbol{r}, \boldsymbol{p}))} \tag{2.329}$$

There is no dynamics in the ensemble averaging process. One has to evaluate an expression at all points in the phase space. Practically, the ensemble average is replaced by a dynamical simulation of the system, and is obtained by repeating the simulation with a different set of initial particle coordinates and momenta corresponding to the same temperature, volume, and number of particles, in order to scan all possible regions of the phase space. The integral over $\boldsymbol{r}_i$ and $\boldsymbol{p}_i$ is really over all possible initial conditions, since their later time evolution is uniquely determined from the equations of motion.

In a computer simulation, assuming the system is ergodic,[3] the equilibrium ensemble average is usually replaced by the time average

$$< A >_{\text{time}} = \frac{1}{N_t} \sum_{j=1}^{N_t} A(t_j) \tag{2.330}$$

where $N_t$ is the number of time steps and $A(t_j)$ are the successive values of the observable $A$ measured at those times according to the values of the dynamical variables $(\boldsymbol{r}_i(t_j), \boldsymbol{p}_i(t_j))$. For a better numerical averaging, since in practice the system becomes ergodic after a very long time, it is better to also perform ensemble averaging by repeating the same computer simulation with several different initial conditions, and average over all these runs as well. The meaning of equilibrium average in (2.328) is indeed averaging over all possible initial conditions $(\boldsymbol{r}_i(0), \boldsymbol{p}_i(0))$

[3]Ergodic means the whole volume of the phase space will eventually be "visited" provided the simulation is continued for a long enough time.

weighted by their thermodynamic ensemble weight: $e^{-\beta H(\boldsymbol{r},\boldsymbol{p})}$ in the canonical case and $\delta(E - H(\boldsymbol{r}, \boldsymbol{p}))$ in the microcanonical case.

$$< A >_{\text{time,ens}} = \frac{1}{N_t N_{ens}} \sum_{s=1}^{N_{ens}} \sum_{j=1}^{N_t} A^{[s]}(t_j) \tag{2.331}$$

The choice of the initial conditions depends on the ensemble: In the microcanonical case all initial conditions must have the same total energy while in the canonical case, they can have any energy chosen from the probability distribution given by $e^{-\beta H(\boldsymbol{r},\boldsymbol{p})}$. It is preferable that the calculations of thermal conductivity using GK formulas be done in the microcanonical ensemble as in other ensembles the dynamics are modified due to the addition of thermostat forces (to keep the kinetic energy per particle around $3k_B T/2$).

Back to the (2.327), we can easily evaluate it in the $\boldsymbol{q}$ space after a Fourier transformation:

$$\mathcal{H}(\boldsymbol{q}, t) = \int \frac{d\boldsymbol{r}}{\Omega} \mathcal{H}(\boldsymbol{r}, t)\, e^{i\boldsymbol{q}\cdot\boldsymbol{r}} = \frac{1}{\Omega} < \sum_i e_i\, e^{i\boldsymbol{q}\cdot\boldsymbol{r}_i} >$$

The $\boldsymbol{q} \to 0$ limit maybe obtained from a Taylor expansion:

$$\mathcal{H}(\boldsymbol{q} \to 0, t) \approx \frac{1}{\Omega} < \sum_i e_i (1 + i\boldsymbol{q} \cdot \boldsymbol{r}_i) > = \frac{1}{\Omega} \left( E + i\boldsymbol{q} \cdot < \sum_i e_i \boldsymbol{r}_i > \right)$$

Choosing the energy reference so that $E = 0$ allows us to identify $\boldsymbol{G}$ with $< \sum_i e_i \boldsymbol{r}_i >$, hence the name energy displacement vector. It is the sum of the particle positions weighted by their local energy, very similar to the dipole moment. Only here the atomic charges have been replaced by atomic energies. With the choice $E = 0$ this vector will be independent of the choice of coordinates origin. With the energy displacement vector now defined as the energy weighted position of the atomic degrees of freedom, we are now able to evaluate the thermal conductivity from Eq. (2.320) or (2.325) using the following definitions:

$$\boldsymbol{G}(t) = \sum_i e_i \boldsymbol{r}_i; \quad \boldsymbol{J}(t) = \sum_i \frac{de_i}{dt} \boldsymbol{r}_i + e_i \boldsymbol{v}_i \tag{2.332}$$

Note that in the 'Einstein' formulation only the difference $\boldsymbol{G}(t) - \boldsymbol{G}(0)$ appears. In a microcanonical simulation where the total energy $E = \sum_i e_i$ is conserved, the choice of the origin of coordinates will not affect the result. In practice, this constraint might not be exactly satisfied, so one might instead define the energy displacement by: $\boldsymbol{G}(t) = \sum_i (e_i(t) - \frac{E(t)}{N}) \boldsymbol{r}_i$ so that at all times $\sum_i (e_i(t) - \bar{e}(t)) = 0$.[4]

[4] Note that the breakdown of the total energy between the local energies of the particles is not unique. Here we divided it into $N$ equal parts, but for convenience one may adopt other choices.

#### 2.4.7.1 Case of Crystalline Solids

In the case of solids, crystalline or not, the second term in $\boldsymbol{J}$, the advective term, will not contribute to the thermal conductivity, and the first term of $\boldsymbol{J}(t)$ in (2.332) may be approximated by:

$$\boldsymbol{J}(t) = \sum_i \frac{de_i}{dt}\boldsymbol{R}_i = \sum_i \boldsymbol{R}_i \left( \boldsymbol{v}_i \cdot \boldsymbol{F}_i + \frac{1}{2}\sum_{j\neq i} \boldsymbol{F}_{ij}\cdot \boldsymbol{v}_{ij} \right) = \sum_{i,j\neq i} \boldsymbol{R}_{ij}\left( \boldsymbol{F}_{ij} \cdot \frac{\boldsymbol{v}_i + \boldsymbol{v}_j}{4} \right) \tag{2.333}$$

where $\boldsymbol{F}_{ij} = -\boldsymbol{F}_{ji}$ is the pair force on atom $j$ from atom $i$, and we replaced the instantaneous position $\boldsymbol{r}_i(t)$ by the equilibrium position $\boldsymbol{R}_i$. Note that the sum in $\boldsymbol{J}$ is restricted to pairs of interacting atoms only and thus for short-range forces its computation is of order $N$ just like the total energy. This expression of the heat current might be thought of as the volume integral of the stress tensor ($\sigma(\boldsymbol{r}) = \sum_{i,j\neq i} \boldsymbol{R}_{ij} \cdot \boldsymbol{F}_{ij}\delta(\boldsymbol{r} - \frac{\boldsymbol{r}_i+\boldsymbol{r}_j}{2})$) times the velocity field.

— $\kappa$ from Lattice Dynamics (LD) —

The GK formula, provides only a single number at a given temperature for a given superlattice size. Size scaling analysis needs furthermore to be implemented to extrapolate this quantity to systems of infinite size. MD combined with the lattice dynamics (LD) theory however provides information on the contribution of each vibrational mode to the thermal conductivity, and is therefore a powerful tool to analyze individual mode contributions to the heat current. In a lattice dynamics approach, the force on each atom is expanded in powers of atomic displacements. In this case the heat current is a product of a force constant by a velocity by integer powers of atomic displacements. Peirels [383] has independently shown that this reduces to the now well-known formula:

$$\kappa^{\alpha\beta} = \frac{1}{\Omega}\sum_{k\lambda} C_{k\lambda}\, v^{\alpha}_{k\lambda}\, v^{\beta}_{k\lambda}\, \tau_{k\lambda} \tag{2.334}$$

In this formula, phonons have been used to describe lattice excitations. The Boltzmann theory, within the relaxation time approximation (RTA) has been used to express $\kappa$ in this form. In the above equation, as usual, $\boldsymbol{k}\lambda$ indicates the phonon state, and $\tau_{k\lambda}$ represents the relaxation time which can be thought as the average lifetime of a phonon in the state $\boldsymbol{k}\lambda$. Its inverse, the scattering rate $1/\tau_{k\lambda}$ is the combination of all scattering processes which are assumed to be additive for independent processes, which are usually among one of the following:

(a) *3-phonon scattering* also obtained from Fermi's golden rule in which 3 phonons are involved. Energy conservation must be satisfied in this process. If momentum is conserved, we are dealing with *Normal* processes, while if the momentum change is a reciprocal lattice vector, we are dealing with *Umklapp* processes. The latter are responsible for thermal resistance. These intrinsic processes are always present, but dominate at higher temperatures. Indeed as T is increased the number of phonons and the likelyhood of them scattering is increased. Four (and higher number of) phonon processes are less likely because they require larger phonon populations. They lead to temperature dependences proportional to $1/T^2$ and higher powers, while the most dominant multiphonon process at low or moderate temperatures is the 3-phonon one, proportional to $n_{k\lambda}$, itself linear in $T$ at high temperatures.[5] At low frequencies it is usually (it really depends on the crystalline structure [384, 385]) quadratic in frequency, so that it can be approximated as $1/\tau_{k\lambda}^{3-phonon} \propto \omega_{k\lambda}^2 T$. The exact formula for the phonon lifetime due to 3-phonon processes comes from perturbation theory. The corresponding phonon self-energy is given by [386]:

$$\Sigma(q, \omega) = -\frac{1}{2\hbar^2} \sum_{1,2,\epsilon=\pm 1} |V_{12q}|^2 \left( \frac{1 + n_1 + n_2}{\omega_1 + \omega_2 + \epsilon\omega_c} + \frac{n_2 - n_1}{\omega_1 - \omega_2 + \epsilon\omega_c} \right) \tag{2.335}$$

where $1, 2, q$ are the three phonon modes involved in the process, $\omega_c = \omega + i0^+$, $n_i$ is the Bose-Einstein distribution function of state $i$, and the 3-phonon matrix element $V$ is defined in terms of the cubic force constants $\Psi_{ijk}$ as:

$$V_{q_1,q_2,q_3} = \left(\frac{\hbar}{2}\right)^{3/2} \sum_{ijk} \frac{\Psi_{ijk}\, e_i^{q_1} e_j^{q_2} e_k^{q_3}}{\sqrt{m_i m_j m_k\, \omega_{q_1}\, \omega_{q_2}\, \omega_{q_3}}}\, e^{i(q_1.R_i + q_2.R_j + q_3.R_k)} \tag{2.336}$$

Translational invariance of $\Psi$ implies that the sum of three momenta has to be either 0 (Normal process) or a reciprocal lattice vector $\boldsymbol{G}$ (Umklapp process). Note that depending on how the phonon wave numbers $q_i$ are defined the classifications of Normal versus Umklapp processes will be different. Also note that this expression is invariant under translations by a reciprocal lattice vector $: V_{q_1,q_2,q_3} = V_{q_1+G,q_2,q_3}$ so that it is enough to keep all 3 vectors within the first Brillouin zone. Finally the phonon lifetime is given by $1/2\tau_{k\lambda} = \mathrm{Im}(\Sigma(\boldsymbol{k}, \omega_{k\lambda}))$.

(b) *Boundary scattering*, for which one may adopt the very approximate formula $1/\tau_{k\lambda}^{boundary} \approx v_{k\lambda}/L$ where $L$ is the characteristic distance between "walls". This term being frequency independent is dominant at low frequencies and therefore at low temperatures where only low frequency phonons are populated. More accurate expressions have been developed by Fuchs [387] and Sondheimer [388] in the context of electrons, where the boundary scattering was introduced as a boundary condition on the solution to the Boltzmann equation. In later theories [389, 390], this was extended as an effective rate correcting the 3-phonon mean free paths due to boundary scatterings.

(c) *Impurity/defect scattering*: in the case of localized defects one usually uses the Raleigh formula $1/\tau_{k\lambda}^{impurity} \propto \omega_{k\lambda}^4$ or the more accurate formula (still from perturbation theory) due to Tamura [391]:

---

[5]At temperatures below the Debye Temperature, the number of Normal processes involve low frequency phonons for which the phonon number can be approximated by $k_B T/\hbar\omega_{k\lambda}$, while in an Umklapp process, which involves large wavenumbers, or modes near the Brillouin zone boundary, the phonon frequency is on the order or larger than the Debye frequency and thus is proportional to $e^{-\omega_D/T}$.

$$1/\tau_{k\lambda}^{impurity} = \frac{\pi n_{\text{atom}}}{6} \omega_{k\lambda}^2 \, DOS(\omega_{k\lambda}) \sum_i f_i \left( \frac{\delta m_i}{\bar{m}} \right)^2 \tag{2.337}$$

where $n_{\text{atom}}$ is the atomic density and $f_i$ is the volume fraction of atom of type $i$. Note that the rate does not vary with temperature, and behaves as $\omega^4$ at low frequencies. Point defects therefore mostly scatter short wavelength high frequency phonons, but do not affect much the contribution of long wavelength phonons to the thermal conductivity.

— $\kappa$ from Equilibrium Molecular Dynamics (EMD) —

Direct application of the Green–Kubo (GK) formula (Eq. (2.320)) to trajectories generated from a MD simulation performed in a supercell leads to an estimation of the thermal conductivity. As previously mentioned care must be taken to perform proper ensemble averaging to reduce the numerical noise, and size scaling to extrapolate to infinite size. A simple use of GK leads to a single number and might not be rewarding. Combining the results of a MD simulation with the LD information on individual phonon modes allows one to study the contribution of each mode to the total thermal conductivity. In an insightful publication [392], this approach was introduced and validated by A. Henry who called it Green–Kubo Modal Analysis (GKMA). In this approach, the heat current is projected onto each individual mode $\lambda$ by substituting in (2.333) atomic velocities by mode velocities, and the correlation of this contribution with the total initial heat current is interpreted as the thermal conductivity due to the mode $\lambda$:

$$\kappa_\lambda^{\alpha\beta} = \frac{1}{\Omega k_B T^2} \int_0^\infty dt < J_\lambda^\alpha(t) J^\beta(0) > \tag{2.338}$$

This formulation is particularly useful in the context of amorphous structures where phonon modes are replaced by normal vibrational modes of the amorphous system, which could be localized (locons) or extended wave-like (propagons) or not having any spatial periodicity (diffusons). The latter terminology was first introduced by Allen and Feldman [393] to classify vibrational modes of an amorphous material. It is then possible to separate the contribution of propagons, diffusons and locons (which is non-zero for anharmonic structures) in the total thermal conductivity.

In the case of crystalline solids, it is likewise possible to obtain from the autocorrelation, the lifetime of a given mode:

$$\tau_\lambda = \int_0^\infty dt \, \frac{< \boldsymbol{J}_\lambda(t) \cdot \boldsymbol{J}_\lambda(0) >}{< \boldsymbol{J}_\lambda(0) \cdot \boldsymbol{J}_\lambda(0) >} \tag{2.339}$$

An alternative way to find the phonon lifetimes is to compute the broadening of the $\boldsymbol{k}$-projected DOS obtained from the $\boldsymbol{k}$-projected velocity autocorrelation:

$$DOS(\boldsymbol{k}, \omega) = \int_0^\infty dt \, e^{i\omega t} \sum_\tau m_\tau < \boldsymbol{v}_{\boldsymbol{k}\tau}(t) \cdot \boldsymbol{v}_{-\boldsymbol{k}\tau}(0) > \tag{2.340}$$

For each $\boldsymbol{k}$ in the first Brillouin zone, this function plotted as a function of the frequency will have $3N$ peaks, the center of which is the anharmonic phonon frequency and the broadening of which is the corresponding phonon lifetime.[6] This is to be compared with the inverse imaginary part of the self-energy expressed in (2.335). This gives strictly the phonon lifetime but not necessarily the relaxation time in (2.339), which appears in the GK expression for the thermal conductivity.

— Non-Equilibrium Molecular Dynamics (NEMD) —

In this approach a long supercell of length $L$ is made with specific non-equilibrium boundary conditions on the temperature of specific layers of atoms. In order to have periodicity, the temperature difference can be applied at $L/4$ ($T = T_{\text{Hot}}$) and $3L/4$ ($T = T_{\text{Cold}}$). The resulting heat flux can be measured at each layer from the net work done from one layer $L$ to its neighboring layer $L+1$: $\Sigma_{i\in L, j\in L+1} \boldsymbol{F}_{ij} \cdot \boldsymbol{v}_{ij}$. This quantity should remain constant between $L/4$ and $3L/4$ and will have the opposite sign outside of this region. The ratio between the heat flux and the applied temperature difference by length $L/2$, will give an estimate of the thermal conductivity. Clearly, this number will depend on the choice of $L$, and therefore several runs are needed with different $L$ values. Using an assumption on the length-dependence one may then extrapolate the resutls to infinite length. Sellan et al. [394] have proposed an analytical length dependence based on the solution of the Boltzmann equation.

#### 2.4.7.2 Case of Amorphous Materials

Except for the lattice dynamics, most of the approaches mentioned above such as GK or NEMD can also be applied to amorphous materials. In the case of disordered materials, since translational symmetry is broken phonons with a wavenumber $\boldsymbol{k}$ are not a good description of the vibrational modes. At low temperatures, when the system is harmonic to a good degree, one can still diagonalize the force constants matrix, and use its eigenvalues, i.e. the normal modes to describe vibrations in an amorphous material. At higher temperatures, anharmonicity can be treated perturbatively, as structural disorder is mainly responsible for thermal resistance. The most popular harmonic quantum theory of thermal resistance was put forth by Allen and Feldman [393] where they classified heat carriers according to their spatial extent and polarization. At high frequencies near band edges vibrational modes are localized, and are thus called "locons". Evidence for these modes is provided by the inverse participation ratio (IPR):

$$IPR(\lambda) = \frac{\sum_i |e_i^\lambda|^4}{\sum_i |e_i^\lambda|^2}$$

For an extended mode the eigenvector $e_i^\lambda \propto \frac{1}{\sqrt{N}}$, so that $IPR(\lambda) \propto \frac{1}{N} \to 0$. For a localized mode however $e_i^\lambda$ is mostly zero except on a few sites where the vibrations are largest. In this case $IPR(\lambda) = O(1)$ is a finite number. Low frequency

[6] The derivation is discussed in the section on phonons.

modes, similar to sound waves, propagate coherently throughout the structure since due to their long wavelength nature, they cannot see the small inhomogeneities in the system. These modes are called "propagons" and behave like plane waves in terms of their polarization amplitudes. Intermediate modes between the two, which are extended but not planewave-like were called "diffusons". Allen and Feldman (AF) prescribed to compute the thermal conductivity from the quantum harmonic version of the Green–Kubo formula, also called the Kubo–Greenwood formula. This formula would be correct at low temperatures where quantum effects are important and anharmonicity negligible. At higher temperatures, typically on the order of the Debye temperature, it ceases to be accurate, but still provides reasonable numbers. AF define a diffusivity per normal mode $\lambda$, which is zero for localized modes in the harmonic approximation, and otherwise given by:

$$D_\lambda = \frac{\pi \Omega}{3\hbar^2 \omega_\lambda^2} \sum_{\mu \neq \lambda} |J_{\lambda\mu}|^2 \, \delta(\omega_\lambda - \omega_\mu) \tag{2.341}$$

where $J_{\lambda\mu}$ is the matrix element of the heat current operator evaluated in the harmonic approximation. Finally, in the spirit of the Peierls-Boltzmann theory they write the thermal conductivity as

$$\kappa = \frac{1}{\Omega} \sum_\lambda D_\lambda C_\lambda \tag{2.342}$$

where $C_\lambda$ is the heat capacity associated with the mode $\lambda$:

$$C_\lambda = k_B \left( \frac{\frac{\hbar\omega_\lambda}{2k_BT}}{\sinh \frac{\hbar\omega_\lambda}{2k_BT}} \right)^2$$

Using this methodology, it is therefore possible to compute, for a given interatomic potential, the force-constants matrix, the normal modes, the mode diffusivity, and finally the thermal conductivity of an amorphous material within the harmonic approximation. It is worth mentioning that the thermal diffusivity is usually different from the square of the speed of sound times the phonon lifetime coming from the broadening of the density of states (time-Fourier transform of the velocity autocorrelation).

#### 2.4.7.3 Discussion

Several methods to compute the thermal conductivity of ordered and disordered solids were introduced. For crystalline materials, the LD approach combined with the solution of the Boltzmann equation is a very rapid and accurate way to calculate the thermal conductivity. It is mostly valid at low temperatures where quantum effects are important and anharmonicity can be treated as a perturbation. At higher temperatures, typically on the order of the Debye temperature, as anharmonicity becomes more important, one might need to include higher-order terms or go beyond perturbative treatments. The GK and the NEMD approaches are however more reliable at high temperatures since they treat anharmonicity exactly. The catch in this case is that interparticle interactions are usually not exactly known. This is a challenge, since interactions can accurately be computed using ab-initio methods such as DFT. Performing MD simulations in a large supercell on time scales of a fraction of a nanosecond, however, are prohibitive, and not rewarding. One usually resorts to semi-empirical potentials fitted to ab initio calculations, or force fields which are Taylor expansion of the DFT forces [395]. The approximation in the latter case is the truncation of the expansion to, typically, fourth order and the truncation of the range of interactions, which is still quite larger than semi-empirical potentials. The GK method, being classical is not valid at low temperatures unless combined with a LD calculation where one can replace the heat capacity by its quantum version. The combination with LD data which provides information on normal modes (GKMA) [392] is a powerful tool to analyze contribution of different modes to the thermal conductivity. On the non-crystalline side, GKMA is still the method of choice. The second alternative which is valid at low-temperatures and includes quantum effects is the so-called Allen–Feldman method [393] which uses the Kubo–Greenwood harmonic formula, and computes in addition to the total thermal conductivity, the mode diffusivity.

### 2.4.8 *Time-Dependent Density Functional Theory*

The time-dependent density functional theory is based on the pioneering work by Runge and Gross [63] as was explained already in Sect. 2.2.3. To extract an excitation energy from the time-dependent density functional theory (TDDFT), one may exploit the fact that the frequency-dependent linear-response function has poles at the exact excitation energies, as was suggested by Petersilka, Gossmann and Gross [396]. If one writes the density response to an external perturbation $\Delta v_{\boldsymbol{q}}(\boldsymbol{r}, \omega)$ as $\Delta \rho_{\boldsymbol{q}}(\boldsymbol{r}, \omega)$, it satisfies

$$\Delta\rho_{\boldsymbol{q}}(\boldsymbol{r},\omega) = \int d\boldsymbol{r}' P_{\boldsymbol{q}}(\boldsymbol{r},\boldsymbol{r}';\omega)\Delta v_{\boldsymbol{q}}(\boldsymbol{r},\omega) + \int d\boldsymbol{r}'d\boldsymbol{r}'' P_{\boldsymbol{q}}(\boldsymbol{r},\boldsymbol{r}';\omega)$$
$$\times \sum_{\boldsymbol{R}} \mathrm{e}^{-i\boldsymbol{R}\cdot\boldsymbol{q}} \left\{ \frac{1}{|\boldsymbol{r}'-\boldsymbol{r}''-\boldsymbol{R}|} + f_{\mathrm{xc}}[\rho_0](\boldsymbol{r}',\boldsymbol{r}''+\boldsymbol{R};\omega) \right\} \Delta\rho_{\boldsymbol{q}}(\boldsymbol{r}'',\omega). \tag{2.343}$$

In (2.343), the sum runs over all the lattice vectors $\boldsymbol{R}$, the polarization part $P_{\boldsymbol{q}}(\boldsymbol{r}, \boldsymbol{r}'; \omega)$ is given by (2.479), and the function $f_{\mathrm{xc}}[\rho_0]$ is defined by

$$f_{\mathrm{xc}}[\rho_0](\boldsymbol{r}) = \frac{\delta \mu_{\mathrm{xc}}[\rho_0]}{\delta \rho_0(\boldsymbol{r})}, \tag{2.344}$$

where $\mu_{\mathrm{xc}}[\rho_0]$ is the nonlocal exchange correlation potential of the unperturbed system with the charge density $\rho_0$. The first term on the right-hand side of (2.343) is always finite, while the density response $\Delta\rho_{\boldsymbol{q}}(\boldsymbol{r}, \omega)$ has poles and diverges at the true excitation energies $\omega = \varepsilon$. Therefore, the excitation energies can be determined from the condition that the eigenvalue $\lambda(\omega)$ of

$$\lambda(\omega)\zeta_{\boldsymbol{q}}(\boldsymbol{r}, \omega) = \int d\boldsymbol{r}' d\boldsymbol{r}'' P_{\boldsymbol{q}}(\boldsymbol{r}, \boldsymbol{r}'; \omega)$$
$$\times \sum_{\boldsymbol{R}} \mathrm{e}^{-i\boldsymbol{R}\cdot\boldsymbol{q}} \left\{ \frac{1}{|\boldsymbol{r}' - \boldsymbol{r}'' - \boldsymbol{R}|} + f_{\mathrm{xc}}[\rho_0](\boldsymbol{r}', \boldsymbol{r}'' + \boldsymbol{R}; \omega) \right\} \zeta_{\boldsymbol{q}}(\boldsymbol{r}'', \omega) \tag{2.345}$$

satisfy

$$\lambda(\varepsilon) = 1. \tag{2.346}$$

This formulation is rigorous within the linear response theory. An application of this theory to the calculation of the X-ray absorption spectra of 3d transition metals was given by Schwitalla and Ebert [245] and their result is fairly in good agreement with experiments. More recently, using a similar method, Vasiliev et al. [397] calculated the optical absorption spectrum of Be, Mg, Ca, Sr, Zn and Cd atoms and small Na clusters.

### 2.4.9 *Dynamical Mean Field Theory, Density Matrix Renormalization Theory, and Tensor Network*

In this section, two new modern theories, i.e., dynamical mean field theory and density matrix renormalization theory both of which have been intended to treat strongly correlated materials, are briefly explained.

#### 2.4.9.1 Dynamical Mean Field Theory (DMFT)

Solving the problem of correlated electrons in a lattice is a very challenging one due to its many-body nature, and the variety of energy scales involved. The mean-field theory assumes each electron is in some average or mean-field static potential. Examples are the Hartree-Fock approximation and the local density approximation in DFT. In these theories, the rest of the system is replaced by a *static* field, which is the sum of the electrostatic and exchange(-correlation) potentials reflecting quantum effects such as the Pauli exclusion principle in addition to the Coulomb interaction.

Mean-field theory becomes exact in the infinite dimensions limit. This is because the mean-field, which is approximated by the sum of the fields created by neighboring electrons on that given site, is a fluctuating quantity, which we can represent by a random number. In the infinite-dimension limit where each site will have infinitely large number of neighbors, this sum of the fields due to neighboring electrons will have a Gaussian distribution of width tending to zero[7]. Therefore the effect of fluctuations becomes less and less important as the dimensionality is increased towards infinity, and mean-field theories become exact in this limit.

The DMFT [398, 399] goes beyond the mean-field approach as it assumes the local mean-field is dynamical, better reflecting the effect of quantum fluctuations due to other electrons on that given site. This mean-field is represented by a frequency-dependent function. In DMFT, the interacting problem is mapped onto a self-consistent quantum impurity problem in contact with a "bath" of non-interacting fermions [400]. The latter can be solved, and its effect is a self-energy of the form $\Delta(\omega) = \Sigma_l \frac{|V_l|^2}{\omega - \epsilon_l}$, where $V_l$ is the strength of the coupling to state $l$ of the bath of energy $\epsilon_l$. This is very similar to the second-order perturbation theory formula if the frequency $\omega$ is replaced by the on-site energy of the "impurity". The frequency-dependence of this self-energy, also called *the hybridization function* or *the dynamical mean-field*, which is effectively added to the impurity site Hamiltonian, makes this theory *dynamical*, thereby incorporating the effect of local quantum fluctuations.

The coupling and the bath spectrum have been arbitrary so far, but later we will require $\sigma(i\omega)$ to satisfy a self-consistency condition.

For the real lattice of interacting electrons, let us write the Green's function as a function of the true self-energy (Dyson's equation) as:

$$G_{ij}(i\omega) = \frac{\Omega}{(2\pi)^3} \int d^3k \, \frac{1}{i\omega + \mu - \omega_k - \Sigma(i\omega)} \tag{2.347}$$

where the integral is performed over the first Brillouin zone volume, $\Omega$ is the volume of the unit cell, and $\omega_k$ is the non-interacting spectrum of the lattice. Note that the self-energy is *local* as it does not depend on $k$. According to this exact Green's function, with a still unknown self-energy, the onsite or *local* Green's function (GF), which the same for all sites, is $G_{ii}(\omega) = G_o(\omega)$. Using the density of non-interacting states as $D(\omega) = \frac{\Omega}{(2\pi)^3} \int d^3k \, \delta(\omega - \omega_k)$, the local GF can be written as:

$$G_o(i\omega) = \int dE \, \frac{D(E)}{i\omega + \mu - E - \Sigma(i\omega)} \tag{2.348}$$

Now we have the relationship between the exact GF projected on an atomic site $G_o$ and the self-energy due to electron-electron interactions $\Sigma$.

[7]This is due to the central-limit theorem, which states that in the limit where N goes to infinity, the average of N random numbers $x_i$ has a Gaussian distribution of mean $\bar{x}$ and of variance $\sigma_x/\sqrt{N}$, where $\bar{x}$ is the true mean of the distribution and $\sigma_x$ its variance.

The self-consistency equation expresses the fact that the bath hybridization function $\Delta(\omega)$ or $G_o^{-1}$ must be such that the local self-energy $\Sigma$ can be deduced from the local GF and the dynamical mean-field $\Delta$ as:

$$G_o^{-1}(i\omega) = g^{-1}(i\omega) - \Sigma(i\omega) = i\omega + \mu - \Delta(i\omega) - \Sigma(i\omega) \tag{2.349}$$

In practice the above set of equations which form a closed set are solved iteratively until they become self-consistent. The algorithm could be described as follows:

- For each frequency $\omega$,

1. start with a trial local self-energy $\Sigma$ (it could initially be taken =0),
2. use it in equation 2.348 to get the local GF,
3. use the self-consistency equation 2.349 to get the dynamical field $\Delta$,
4. solve the interacting impurity problem to get the local self-energy
5. iterate until convergent ( go to step 2).

The most time-consuming part of the calculations is the "impurity solver" which solves the single impurity problem including electron interactions, as exactly as possible, in order to produce a new self energy $\Sigma$.

The impurity solver calculates the local action:

$$S_{loc} = -\int_0^\beta d\tau \int_0^\beta d\tau' c^+(\tau)\, g^{-1}(\tau - \tau')\, c(\tau') + \int_0^\beta d\tau\, U\, n_\uparrow(\tau)\, n_\downarrow(\tau) \tag{2.350}$$

where $g^{-1}(i\omega) = i\omega + \mu - \epsilon_0 - \Delta(i\omega)$. This local action represents the effective dynamics of the local site. The interacting impurity problem is usually solved either using the quantum Monte Carlo (QMC) method, followed by an analytic continuation to real frequencies [401], exact diagonalization, or using diagrammatic techniques and iterated perturbation theory (IPT). The output is a frequency-dependent self-energy $\Sigma$ which is reinjected into equation 2.348 to iterate the computation of the local GF.

DMFT becomes exact in the following 3 limits:

1. The atomic limit where the bandwidth goes to zero and the density of states becomes a Dirac delta function.
2. The non-interacting $U \to 0$ limit where $G^{-1}(i\omega) = g^{-1}(i\omega)$.
3. The infinite dimension or infinite coordination limit [402].

This theory can be systematically be improved and converged to the exact result if the single-site impurity is replaced by larger clusters and therefore the self-energy is made momentum-dependent. Examples are the dynamical cluster approximation (DCA) [403], and cellular DMFT [404]. This comes however at the cost of solving the many-body interacting impurity problem of a cluster of few sites instead of a single site, wherein depending on the method used, the computational time may exponentially increase with the number of electrons.

#### 2.4.9.2 Density Matrix Renormalization Group (DMRG) Theory

The density matrix renormalization group (DMRG) provides a method to truncate the Hilbert space and extract the major part of strongly correlated systems [25, 405–407]. The basic idea of the DMRG is to divide the system into two parts; one is the system (S) and the other is the environment (E) and take the trace of E first in order to evaluate the expectation value of a physical quantity.

To do this, we define a reduced density matrix $\hat{\rho}$,

$$\hat{\rho} = \mathrm{Tr_E}\, |\psi\rangle\langle\psi|, \tag{2.351}$$

where the states belonging to E have been traced out. Using the Wannier representation, we write the state $|\psi\rangle$ in a direct product of the S and E states as

$$|\psi\rangle = \sum_{i \in \mathrm{S}} \sum_{j \in \mathrm{E}} \psi_{ij} |i\rangle |j\rangle. \tag{2.352}$$

Then, it is easy to see

$$\langle i|\hat{\rho}|i'\rangle = \sum_{j \in \mathrm{E}} \psi_{ij}\psi^*_{i'j}. \tag{2.353}$$

The expectation value of the operator $\hat{A}$ is given by

$$\langle \hat{A} \rangle = \mathrm{Tr_S}\, \hat{\rho}\hat{A}. \tag{2.354}$$

Here we consider the eigenvalue problem for the density matrix $\hat{\rho}$

$$\hat{\rho}|w_\alpha\rangle = w_\alpha |w_\alpha\rangle. \tag{2.355}$$

There are $N^S$ eigenstates $|w_\alpha\rangle$ associated with eigenvalue $w_\alpha$, which satisfy

$$\sum_\alpha w_\alpha = 1 \tag{2.356}$$

and $w_\alpha \geq 0$. The ground state of the system is best described by the $M^S$ states $|w_\alpha\rangle$ with largest weight $w_\alpha$. If we project the original Fock space down to the reduced space spanned by the $M^S$ dominant eigenvectors $|w_\alpha\rangle$, the expectation value of the operator $\hat{A}$ is approximated by

$$\langle \hat{A} \rangle_{\mathrm{approx}} = \sum_{\alpha=1}^{M^S} w_\alpha \langle w_\alpha|\hat{A}|w_\alpha\rangle, \tag{2.357}$$

and the error for $\langle \hat{A} \rangle$ is bounded by

$$| \langle \hat{A} \rangle_{\text{approx}} - \langle \hat{A} \rangle | \leq \left( \sum_{\alpha > M^S}^{N^S} w_\alpha \right) c_A = \epsilon_\rho c_A, \tag{2.358}$$

where $c_A$ is the largest eigenvalue of the operator $\hat{A}$. Therefore the error is the order of the truncated weight

$$\epsilon_\rho = 1 - \sum_{\alpha=1}^{M^S} w_\alpha. \tag{2.359}$$

Now, we introduce an approximate wavefunction $|\widetilde{\psi}\rangle$ truncated to the Fock space spanned by only $M_S$ orthonormal states $|u^\alpha\rangle = \sum_{i \in \mathrm{S}} u_{\alpha i} |i\rangle$ as

$$|\widetilde{\psi}\rangle = \sum_{\alpha \in \mathrm{S}} \sum_{j \in \mathrm{E}} a_{\alpha j} |u^\alpha\rangle |j\rangle. \tag{2.360}$$

It is necessary to minimize the norm of the residue or its square,

$$S = || \, |\psi\rangle - |\widetilde{\psi}\rangle \, ||^2, \tag{2.361}$$

which is equal to

$$S = 1 - 2 \sum_{\alpha} \sum_{i \in \mathrm{S}} \sum_{j \in \mathrm{E}} \psi_{ij} a_{\alpha j} u_{\alpha i} + \sum_{\alpha} \sum_{j \in \mathrm{E}} a_{\alpha j}^2, \tag{2.362}$$

with respect to $a_{\alpha j}$ and $u_{\alpha i}$. Here we assumed real coefficients for simplicity and used $\langle \psi | \psi \rangle = 1$. From $\partial S / \partial a_{\alpha j} = 0$, we have

$$\sum_{i \in \mathrm{S}} \psi_{ij} u_{\alpha i} = a_{\alpha j}. \tag{2.363}$$

Using this equation in (2.364), we find that

$$S = 1 - \sum_{\alpha} \sum_{i \in \mathrm{S}} \sum_{i' \in \mathrm{S}} u_{\alpha i} \rho_{ii'} u_{\alpha i'} \tag{2.364}$$

should be stationary for the global minimum of the residue (2.361). Here we introduced the density matrix coefficients

$$\rho_{ii'} = \sum_{j \in \mathrm{E}} \psi_{ij} \psi_{i' j}. \tag{2.365}$$

Equation (2.364) is stationary according to the Rayleigh–Ritz principle, if $|u^\alpha\rangle$ is the eigenvectors of the density matrix. Choosing $|u^\alpha\rangle$ to be the eigenstates $|w_\alpha\rangle$ of the density matrix $\hat{\rho}$, we have

$$S = 1 - \sum_{\alpha=1}^{M^S} w_\alpha = \epsilon_\rho. \tag{2.366}$$

According to singular-value decomposition theory after a Schmidt decomposition, it is possible to write

$$|\psi\rangle = \sum_{i=1}^{N^S} \sum_{\alpha=1}^{N_{\text{Schmidt}}} \sum_{i=1}^{N^E} U_{i\alpha} \sqrt{w_\alpha} V_{\alpha j}^T |i\rangle |j\rangle = \sum_{\alpha=1}^{N_{\text{Schmidt}}} \sqrt{w_\alpha} |w_\alpha^{\text{S}}\rangle |w_\alpha^{\text{E}}\rangle \tag{2.367}$$

with $N_{\text{Schmidt}} \leq \min(N^S, N^E)$ and

$$|w_\alpha^{\text{S}} = \sum_{i=1}^{N^S} U_{i\alpha} |i\rangle; \quad |w_\alpha^{\text{E}} = \sum_{i=1}^{N^E} V_{j\alpha} |j\rangle. \tag{2.368}$$

Then we can construct

$$\hat{\rho}_{\text{S}} = \sum_{\alpha=1}^{N_{\text{Schmidt}}} w_\alpha |w_\alpha^{\text{S}}\rangle \langle w_\alpha^{\text{S}}|; \quad \hat{\rho}_{\text{E}} = \sum_{\alpha=1}^{N_{\text{Schmidt}}} w_\alpha |w_\alpha^{\text{E}}\rangle \langle w_\alpha^{\text{E}}|. \tag{2.369}$$

Although the system and the environment are different, both density matrices have similar structures. Quantum information theory may provide a measure of entanglement via the von Neumann entropy,

$$S_{\text{vN}} = -\operatorname{Tr} \hat{\rho} \log_2 \hat{\rho} = -\sum_{\alpha=1}^{N_{\text{Schmidt}}} w_\alpha \log_2 w_\alpha, \tag{2.370}$$

which is also called the entanglement entropy. Therefore, we know that the truncation to the $M^S$ states is performed with keeping the entanglement between the system (S) and the environment (E) as large as possible.

For the infinite system, the DMRG algorithm is described as follows (see Fig. 2.12):

(1) Prepare two lattices of some small size $l$; one consists of $M_1^S$ bases $|i_1\rangle$ for S and the other consists of $M_1^E$ bases $|j_1\rangle$ for E.
(2) Add one site (consisting of $M_2^S$ bases) to S to make new S'. Similarly add one site (consisting of $M_2^E$ bases) to E to make new E'.
(3) Build the Hamiltonian $\hat{H}_{2l+2}$ of the *superblock* of length $2l + 2$ composed of S' and E'.
(4) Find the ground state $|\psi\rangle$ by diagonalizing the Hamiltonian $\hat{H}_{2l+2}$ using a sparse matrix algorithm.
(5) Construct the reduced density matrix $\hat{\rho} = \operatorname{Tr}_{\text{E}}' |\psi\rangle\langle\psi|$ by (2.353). Diagonalize $\hat{\rho}$ to determine its eigenvalues $w_\alpha$ and eigenvectors $u_{i_1,i_2}^\alpha$. Form a new (reduced)

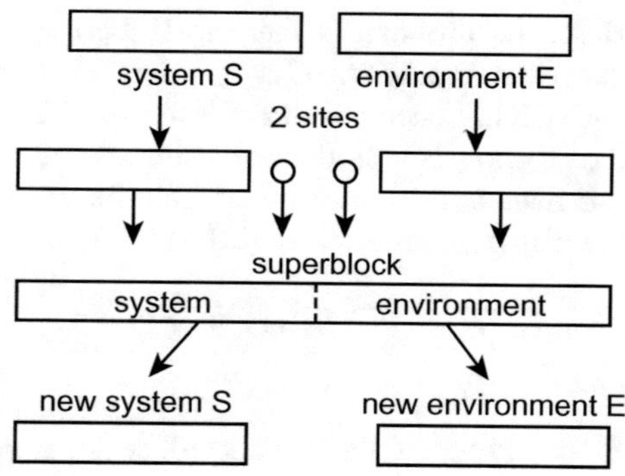


**Fig. 2.12** Schematic illustration of the DMRG process for an infinite system [25, 405–407]; system S (environment E) plus one site forms S' (E'), and the junction of S' and E' forms the superblock, from which new system S and new environment E are constructed

basis for S' by taking the $M_1^S$ eigenstates with the largest eigenvalues (weights) $w_\alpha$. Taking them as column vectors, form an $M_1^S M_2^S \times M_1^S$ rectangular matrix $T$. Do the reverse procedure for the environment E'.

(7) Transform the Hamiltonian onto the new $M_1^S$-state (or $M_1^E$-state) basis by $\hat{H}_{l+1}^{\text{tr}} = T^\dagger \hat{H}_{l+1} T$. Go to Step (2).

White and Feiguin [409] extended the DMRG method to incorporate real-time evolution by using the Suzuki–Trotter formula.

#### 2.4.9.3 Tensor Network

Another recent topic is the relationship to the tensor network; see also Sect. 5.4.5a. The DMRG can be regarded as a matrix (i.e., a second-order tensor) version of the tensor network and the renormalization group (TRG) [410] based on the higher-order singular value decomposition (HOTRG) [411].

While a matrix $M_{ij}$ can be graphically represented by a circle with two external lines $i$ and $j$, a tensor $T_{ijk...}$ may be represented by a circle with more than two external lines $i, j, k, ...$ If we connect two matrices $M$ and $M'$ by a line $k$ as $\sum_k M_{ik} M'_{kj} = M''_{ij}$, the result is a matrix $M''$. Similarly, if we connect two tensors $T$ and $T'$ by a line $k$ as $T_{ij...k...} T'_{i'j'...k...} = T''_{ij...i'j'...}$, the result becomes a tensor $T''$. This is called a "contraction." The single site wavefunction is represented as

$$|\psi\rangle = \sum_{\alpha_1} C_{\alpha_1} |\alpha_1\rangle. \tag{2.371}$$

As well, the multi-site wavefunction can be generally represented as [412]

$$|\psi\rangle = \sum_{\alpha_1} \sum_{\alpha_2} \cdots \sum_{\alpha_L} C_{\alpha_1 \alpha_2 ... \alpha_L} |\alpha_1\rangle |\alpha_2\rangle \cdots |\alpha_L\rangle. \tag{2.372}$$

Here, the problem is that the Hilbert space grows exponentially large in $L$ and that each coefficient $C_{\alpha_1\alpha_2\ldots\alpha_L}$ has $d^L$ elements ($d$ is the spatial dimensionality) and quickly becomes difficult to treat. The easiest approximation is to assume that $C\alpha_1\alpha_2\ldots\alpha_L$ is a product of a single sites as $C_{\alpha_1\alpha_2\ldots\alpha_L} = C_{\alpha_1}C_{\alpha_2}\cdots C_{\alpha_L}$. However, this decomposition is not suitable for entangled systems. Instead, we can consider the matrix product states (MPS) [413]:

$$|\psi\rangle = \sum_{\alpha_1}\cdots\sum_{\alpha_L}\sum_{\beta_1}\cdots\sum_{\beta_{L-1}} C^{\alpha_1}_{\beta_1}C^{\alpha_2}_{\beta_1\beta_2}\ldots C^{\alpha_{L-1}}_{\beta_{L-2}\beta_{L-1}}C^{\alpha_L}_{\beta_{L-1}}|\alpha_1\rangle|\alpha_2\rangle\cdots|\alpha_L\rangle. \quad (2.373)$$

Since the size of the tensors still scales exponentially with system size, we truncate the Hilbert space by capping the size of the $\beta$ indices at some point, say $\chi$ and choose components that have major contributions by keeping the entanglement information. That is, we have to truncate the Hilbert space to the $M^S$ states with keeping the entanglement between the system (S) and the environment (E) as large as possible. This is again measured by the entanglement entropy (2.370) for the reduced density matrix $\hat{\rho}$. General states satisfy $S_{\rm vN} \propto M^S$, while the ground state obeys an area law $S_{\rm vN} \propto \partial M^S$, where $\partial M^S$ means the minimal surface separating region S and E. Therefore, if one wants to find the ground state only, one can maintain the area-law states only. The correlation function scales with the length of the geodesic connecting the sites [414]:

$$C(x_1, x_2) = \langle\hat{O}(x_1)\hat{O}(x_2)\rangle \propto \exp\left[-\alpha D(x_1, x_2)\right]. \quad (2.374)$$

For the MPS, $S_{\rm vN} \propto$ const. and we have

$$C(x_1, x_2) \propto \exp\left[-|x_1 - x_2|/\xi\right], \quad (2.375)$$

while for the PEPS, $S_{\rm vN} \propto L$. Using the multi-scale entanglement renormalization ansatz (MERA), which is also called the holographic tensor network, we have $S_{\rm vN} \propto \log L$, $D_{\rm hol}(x_1, x_2) \propto \log|x_1 - x_2|$, and

$$C(x_1, x_2) \propto \exp\left[-\alpha D_{\rm hol}(x_1, x_2)\right] \propto \exp\left[-a\log|x_1 - x_2|\right] \propto |x_1 - x_2|^a. \quad (2.376)$$

Using the higher-order singular value decomposition (HOTRG) [408, 411], one may write

$$|\psi\rangle = \sum_{\alpha_1}\sum_{\chi}\cdots\sum_{\alpha_2}\cdots\sum_{\alpha_L} U_{\alpha_1,\chi}D_{\chi,\chi}V^{\dagger}_{\chi,\alpha_2\alpha_3\ldots\alpha_L}, \quad (2.377)$$

where $D_{\chi,\chi}$ is the diagonal matrix. Therefore, the coefficients $C_{\alpha_1\alpha_2\ldots\alpha_L}$ can be reshaped as the contraction of $U_{\alpha_1,\chi}$ and $V^{\dagger}_{\chi,\alpha_2\alpha_3\ldots\alpha_L}$, which can be regarded as the $C_{\alpha_2\alpha_3\ldots\alpha_L}$. As we have seen in the last section, this can be done with keeping the entanglement entropy maximum. Repeating this procedure step by step, we can finally obtain the MPS. After getting the MPS, we can do the procedure similar to the DFRG to evaluate the expectation value of physical quantities.

Operators can also be represented as a product of tensors, which is called the matrix product operator (MPO).

$$\hat{O} = \sum_{\alpha_1} \cdots \sum_{\alpha_L} \sum_{\alpha_1'} \cdots \sum_{\alpha_L'} \sum_{\beta_1} \cdots \sum_{\beta_{L-1}} W_{\beta_1}^{\alpha_1 \alpha_1'} W_{\beta_1 \beta_2}^{\alpha_2 \alpha_2'} \cdot W_{\beta_{L-2}\beta_{L-1}}^{\alpha_{L-1}\alpha_{L-1}'} W_{\beta_{L-1}}^{\alpha_L \alpha_L'} |\alpha_1 ... \alpha_L\rangle\langle\alpha_1' ... \alpha_L'|. \tag{2.378}$$

Then, its expectation value $\langle\psi|\hat{O}|\psi\rangle$ can be simply evaluated by the full contraction of the tensor network. For more details, see the paper by Goldsborough and Römer [415].

## 2.5 Ab Initio Molecular Dynamics

Since an electron is much lighter than a nucleus, one may usually assume the Born–Oppenheimer (BO) adiabatic approximation [416]. That is, one ignores the coupling between the motions of electrons and of nuclei and performs the integrations of these two motions separately at each time step. When the motion of nuclei is not very fast, and when the temperature is not very high, one assumes also that the electronic states are at the ground state. The term, "adiabatic potential surface"or "Born–Oppenheimer surface" (BO surface) is often used to indicate energy surface, in the $3N_{\mathrm{atom}}$-dimensional (atomic) configuration space, calculated using the BO approximation. Here we should make some remarks on the use of the BO approximation. First, for systems composed of light atoms such as hydrogen, there are cases in which, even in the adiabatic approximation, the zero-point vibration energy of nuclei may become important in a discussion of the subtle energy difference between stable atomic configurations [417]. In such a case, one has to include the contribution from the zero-point energy. Second, the BO approximation itself may break down if, for example, the electrons are not in the ground state or the atoms move at very high speeds. For more details, see Sect. 2.6.

If one makes the BO approximation and treats all nuclei as classical charged particles, one may perform ab initio molecular-dynamics simulations in the ways described in this section.

### 2.5.1 *Car–Parrinello Method*

Since the paper by Car and Parrinello [418] first appeared in 1985, ab initio MD techniques have developed very rapidly [419, 420]. In their paper, Car and Parrinello proposed to integrate a Newtonian equation of motion *for both the wavefunctions and the atomic coordinates*. Such an approach was a completely new idea, beyond the previous common-sense idea that the electronic structure must be calculated by means of matrix diagonalization and self-consistent iterations. Since the basic equation of motion in the Car–Parrinello method is derived from a Lagrangian, total energy conservation is automatically guaranteed in the formulation (the total energy in this method can be considered as the true total energy of the system if $\mu$, defined below, is sufficiently small), and their approach is especially suitable for microcanonical simulations (with constant total energy). Their subsequent very active studies with various applications demonstrated the validity and effectiveness of their approach. With this technique, it is no longer necessary to treat huge eigenvalue problems and it becomes possible to reduce significantly both the computational memory and the computational time. The reason for the recent very rapid development of ab initio MD techniques is mainly to this point.

In this section, we introduce first the Car–Parrinello method, and then several alternative methods. The basic equation of the Car–Parrinello method is given by the Lagrangian $\mathcal{L}$ as a functional of the wavefunction $\psi_\lambda$ ($\lambda$ denotes the level number of the electronic states) and the atomic position $\boldsymbol{R}_i$ ($i$ stands for the atom index):

$$\mathcal{L} = \sum_\lambda \frac{\mu_\lambda}{2} \int d^3r |\dot{\psi}_\lambda|^2 + \frac{1}{2} \sum_i M_i |\dot{\boldsymbol{R}}_i|^2 - E[\{\psi_\lambda\}, \{\boldsymbol{R}_i\}]. \tag{2.379}$$

Here, $\mu$ is a fictitious electron mass which governs the motion of the wavefunctions. Note that this Lagrangian coincides with the true Lagrangian of the system only in the limit $\mu \to 0$, although we set $\mu$ finite. The important meaning of the existence of a Lagrangian is that the system has no energy dissipation and the total energy, including the fictitious kinetic energy (KE) $\mu_\lambda |\dot{\psi}_\lambda|^2/2$ of the electrons, is conserved. Therefore, the atomic oscillation of this system is guaranteed to be preserved permanently without damping. From this Lagrangian, one may derive the equation of motion for the wavefunctions

$$\mu \frac{d^2}{dt^2} \psi_\lambda = -\mathcal{H}\psi_\lambda + \sum_\nu \Lambda_{\lambda\nu} \psi_\nu, \tag{2.380}$$

where the $\Lambda_{\lambda\nu}$ are Lagrange multipliers to ensure the orthogonality of the wavefunctions $\psi_\lambda$. This multiplier is determined so as to orthonormalize the wavefunctions at every time step. For this purpose, we first replace the differential equation by the difference equation

$$\psi_\lambda(t + \Delta t) = \phi_\lambda(t + \Delta t) + \frac{(\Delta t)^2}{\mu} \sum_\nu \Lambda_{\lambda\nu} \psi_\nu(t), \tag{2.381}$$

where

$$\phi_\lambda(t + \Delta t) = 2\psi_\lambda(t) - \psi_\lambda(t - \Delta t) + \frac{(\Delta t)^2}{\mu} \mathcal{H}\psi_\lambda(t) \tag{2.382}$$

means the wavefunction at time $t + \Delta t$ obtained from the equation with $\Lambda_{\lambda\nu} = 0$. From the orthonormality condition for the wavefunction $\psi_\lambda(t + \Delta t)$, we obtain the nonlinear matrix equation

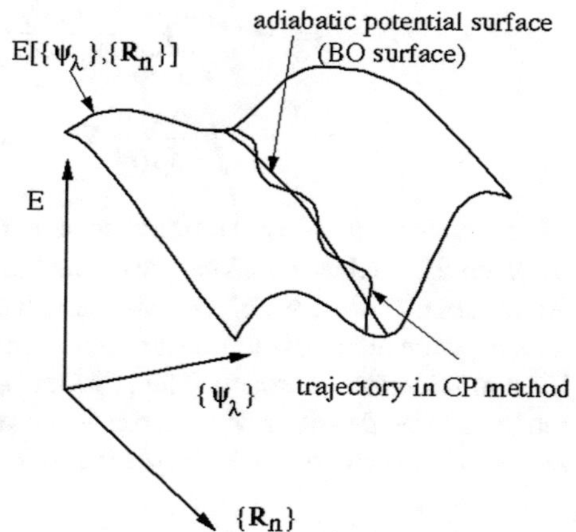


**Fig. 2.13** The oscillation around the adiabatic potential surface following the nuclear movement in the Car–Parrinello method

$$\widetilde{\Lambda}^{\mathrm{t}} = \frac{1}{2} s^{-1}(1 - p - \widetilde{\Lambda}^* \widetilde{\Lambda}^{\mathrm{t}}), \tag{2.383}$$

where $^{\mathrm{t}}$ and $^*$ stand for the transpose and the complex conjugate, respectively; we have used $\widetilde{\Lambda}_{\lambda\nu} = [(\Delta t)^2/\mu]\Lambda_{\lambda\nu}$, $s_{\lambda\nu} = \langle \phi_\lambda(t + \Delta t)|\psi_\nu(t)\rangle$ and $p_{\lambda\nu} = \langle \phi_\lambda(t + \Delta t)|\phi_\nu(t + \Delta t)\rangle$. To determine the unknown multiplier $\widetilde{\Lambda}_{\lambda\nu}$, one solves (2.383) recursively. This method is called the "Ryckaert method" [421].

To ensure the Born–Oppenheimer adiabatic approximation in the Car–Parrinello formalism, one has to choose the left-hand side of (2.380) to be small enough. That is, the fictitious electron mass $\mu$ should be chosen small enough. Here, one should comment on the fact that, even with this choice of $\mu$, the electrons' wavefunctions oscillate very rapidly around the BO surface, following the nuclear movement; see Fig. 2.13. This rapid oscillation of the electronic states is of course not realistic, but just due to the second-order differentiation with respect to time in (2.380) like a usual wave equation. However, for sufficiently small $\mu$, the amplitude of this oscillation is usually very small.

On the other hand, the nuclear movement obeys the Newtonian equation

$$M_i \frac{d^2}{dt^2} \boldsymbol{R}_i = -\nabla_i E \tag{2.384}$$

with $E$ being the sum of the Coulomb potential between the nuclei and the total energy of the electron system for the given nuclear positions, i.e. the physical total energy + $\sum_\lambda (\mu/2)|\dot{\psi}_\lambda|^2$, which is conserved. If the force on the right-hand side of (2.384) is obtained, this equation can be integrated by means of, for example, the Verlet method described in Appendix A.

On the basis of DFT, the left-hand side of this equation can be written as

$$-\nabla_i E = -\nabla_i \sum_{j(\neq i)} \frac{Z_i Z_j}{|\boldsymbol{R}_i - \boldsymbol{R}_j|} - \int \rho(\boldsymbol{r}) \nabla_i v_i(|\boldsymbol{r} - \boldsymbol{R}_i|) d\boldsymbol{r} - \int \frac{\delta E\{\rho\}}{\delta \rho(\boldsymbol{r})} \nabla_i \rho(\boldsymbol{r}) d\boldsymbol{r}, \tag{2.385}$$

where the first and second terms represent, respectively, the Coulomb repulsive force between the nuclei (which can be evaluated by means of the Ewald sum described in Appendix C in the case of a crystal) and the Coulomb attractive force exerted on the nuclei by the electron cloud. Here, $v_i(r)$ denotes either a pseudopotential, or $-Z_i/r$ in an all-electron approach. The third term represents the force acting on the nuclei via the charge density $\rho(\boldsymbol{r})$, but this force vanishes if the electronic state is precisely at the BO surface, because $\delta E/\delta\rho(\boldsymbol{r}) = \mathcal{H}$ and

$$-\int \sum_{\lambda} [(\nabla_i \psi_\lambda^*) \mathcal{H} \psi_\lambda + \psi_\lambda^* \mathcal{H} (\nabla_i \psi_\lambda)] d\boldsymbol{r} = 0 \tag{2.386}$$

(the Hellmann–Feynman theorem). So, this last force is called the "variational force" [422]. In contrast, the first and second terms of (2.385) represent the purely electrostatic force due to the explicit dependence on the nuclear positions, and this is called the "Hellmann–Feynman force" [17, 345].

### 2.5.2 Steepest Descent, Conjugate Gradient, and Block Davidson Methods

In place of (2.380) which uses a second-order differential equation and orthonormalization by the Ryckaert method, one may use a first-order differential equation together with the standard Gram–Schmidt orthogonalization technique. This alternative approach is called the "steepest-descent" (SD) method and its basic equation is given by

$$\mu \frac{d}{dt} \psi_\lambda(\boldsymbol{r}) = -(\mathcal{H} - \varepsilon_\lambda) \psi_\lambda(\boldsymbol{r}), \quad \varepsilon_\lambda = \int \psi_\lambda^*(\boldsymbol{r}) \mathcal{H} \psi_\lambda(\boldsymbol{r}) d\boldsymbol{r}. \tag{2.387}$$

In the numerical integration of this equation of motion by discretizing the time, one has to orthogonalize the wavefunctions at each time step from the bottom level to the top, one by one, by means of the Gram–Schmidt orthogonalization technique. Note that, in (2.387), the Lagrange multiplier is replaced by its expectation value [423]. The SD method is essentially an energetically dissipative algorithm, and is suitable for the determination of the most stable structure of clusters or crystals by ab initio MD. In such a case, first we fix the atomic positions to get a good convergence of the electronic states and then we continue to iterate the convergence procedure for the electronic states while gradually moving the atomic positions as in the usual ab initio MD method.

During this procedure, the kinetic energy of the atomic motions decreases gradually, a damped oscillation around the most stable configuration appears and finally the atoms stop moving at the most stable configuration. With the same technique, one may also approach a constant-energy simulation, by iterating the convergence procedure of the electronic states enough times between the successive update of the atomic positions in order to suppress the energy dissipation as far as possible. More recently, as a much faster convergence technique in place of the SD method, the "conjugate-gradient" (CG) method (or "preconditioned conjugate-gradient method" [419]) or the block Davidson (BD) method [424–426] has also been a popular method. All these methods are called iterative diagonalization methods. Those methods are described in Appendix D. The computational flow of this scheme is shown in Fig. 2.14. One starts by assuming the initial atomic positions and the initial charge-density distribution of the electrons. First, the main loop is iterated a sufficient number of times while fixing the atomic coordinates to achieve self-consistency of the electronic states by means of the steepest-descent methodcombined with the Gram–Schmidt orthogonalization technique or the CG (or BD) method. If necessary, the conventional matrix diagonalization technique is used for the first one or several iterations and then one switches to the SD, CG or BD technique. Then, one evaluates the forces acting on every atom and updates the atomic coordinates for a small time step $\Delta t$. Next one iterates the electronic structure updates several times by using the iterative diagonalization method, evaluates the forces and updates the atomic coordinates. This process is repeated many times.

The computational flow in the original Car–Parrinello method is similar but the iterative diagonalization method is replaced by a one-step update of the discretized version of the second-order differential equation (2.380) combined with the Ryckaert method.

### 2.5.3 Formulation with Plane Wave Basis

According to the original paper by Car and Parrinello [418], the basis functions used to represent the electron wavefunctions may be chosen arbitrarily. After this paper was published, Car and Parrinello applied their method to a variety of systems, using the pseudopotential approach combined with the plane wave expansion. For every $k$ in the first Brillouin zone, if one expands the wavefunction as:

$$\psi_\lambda(\boldsymbol{r}) = \sum_{\boldsymbol{G}} c_\lambda(\boldsymbol{G}) \frac{1}{\Omega} e^{i\boldsymbol{G}+\boldsymbol{k}\cdot\boldsymbol{r}}, \tag{2.388}$$

(2.380) becomes the equation for the expansion coefficients $c_\lambda(\boldsymbol{G})$:

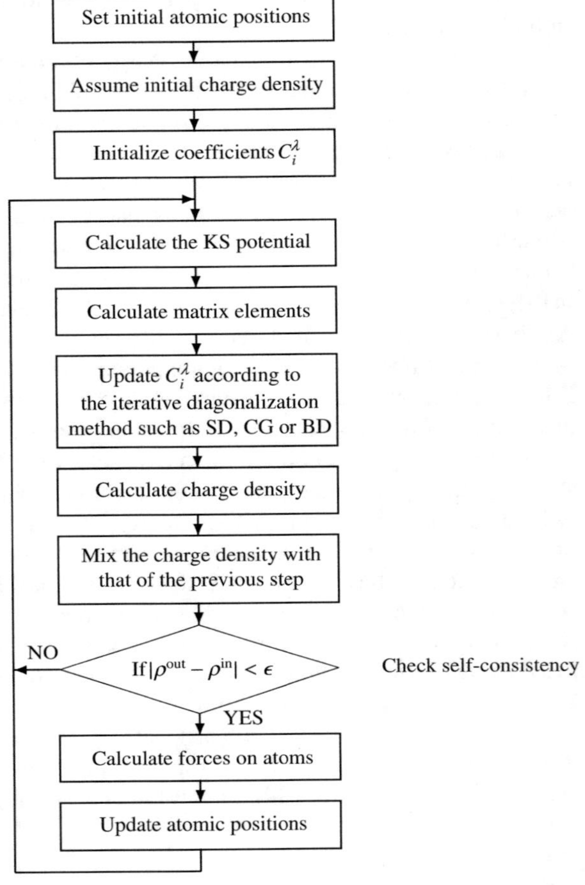


**Fig. 2.14** Flow diagram for ab initio molecular-dynamics using the conjugate-gradient (CG) or steepest-descent (SD) method

$$\mu \ddot{c}_\lambda(\boldsymbol{G}) = -\sum_{\boldsymbol{G}'} H(G, G'; k) c_\lambda(\boldsymbol{G}') + \sum_\nu \Lambda_{\lambda\nu} c_\nu(\boldsymbol{G}). \tag{2.389}$$

In the case of (2.387), we have

$$\begin{aligned} \mu \frac{d}{dt} c_\lambda(\boldsymbol{G}) &= -\sum_{\boldsymbol{G}'} H(\boldsymbol{G}, \boldsymbol{G}'; k) c_\lambda(\boldsymbol{G}') + \varepsilon_\lambda c_\lambda(\boldsymbol{G}), \\ \varepsilon_\lambda &= \sum_{\boldsymbol{G}', \boldsymbol{G}''} c_\lambda^*(\boldsymbol{G}') H(\boldsymbol{G}', \boldsymbol{G}'') c_\lambda(\boldsymbol{G}''). \end{aligned} \tag{2.390}$$

In this case, the Hamiltonian matrix $H(\boldsymbol{G}, \boldsymbol{G}'; k)$ becomes a large matrix with the numbers of rows and columns both equal to the number of plane waves, but when the effective one-electron potential is a local function, and $H(\boldsymbol{G}, \boldsymbol{G}'; k)$ is written as

$$H(\boldsymbol{G}, \boldsymbol{G}'; k) = \frac{1}{2}(\boldsymbol{G} + k)^2 + V(\boldsymbol{G} - \boldsymbol{G}'), \tag{2.391}$$

it is not necessary to store all the matrix elements in the computer memory. As previously mentioned, one can also calculate $T\psi$ in reciprocal space and $V\psi$ in real space to save computational time. (When one uses pseudopotentials, even within the LDA, the potential is usually nonlocal. However, for its nonlocal part, it is possible to use a separable form [192].) Since the ground state properties of semiconductors can be described by the pseudopotential formalism with high accuracy, they have also been a good target for the Car and Parrinello method. On this subject, Car, Parrinello and their collaborators have applied their method very actively to a wide variety of systems. One merit of using the plane wave expansion is that one needs to evaluate neither the variational force (the third term on the right-hand side of (2.385)) nor the force arising from the derivative of $(\mu/2)|\psi_\lambda|^2$ with respect to $\boldsymbol{R}_i$; these forces are always zero even if the system is off the BO surface, since neither the plane waves nor the coefficients $C_\lambda(\boldsymbol{G})$ depend explicitly on the atomic coordinates. Nevertheless, there still remains a deviation of the calculated force from the true force due to the finiteness of the number of plane waves, i.e. there is a so-called "incomplete basis-set force" [427]. When one relaxes the size of the unit cell by means of an ab initio MD method, another force called the "Pulay force" appears because the plane waves depend explicitly on the lattice constant through the reciprocal-lattice vectors. This force also becomes zero exactly at the BO surface [428].

There is a limit to the applicability of the pseudopotential–PW expansion method to large systems of the first-row elements or the 3d transition-metal elements. Recently, the implementation of force calculations has been reported in various alternative approaches and applied to ab initio MD simulations or structural optimizations.

### 2.5.4 Formulation with Other Bases

In ab initio MD methods based on the LCAO (or mixed-basis) approach, since the basis functions are not mutually orthogonal, it is necessary to introduce the overlap matrix $S$. When one uses the steepest-descent method together with the Gram–Schmidt orthogonalization technique, the basic evolution equation for the convergence of the wavefunctions reads

$$\mu \sum_j S_{ij} \frac{d}{dt} C_j^\lambda = -\sum_j (H_{ij} - \varepsilon_\lambda S_{ij}) C_j^\lambda, \quad \varepsilon_\lambda = \sum_{ij} C_i^{\lambda *} H_{ij} C_j^\lambda, \tag{2.392}$$

where $H_{ij}$ and $S_{ij}$ denote, respectively, the Hamiltonian- and overlap-matrix elements, and $C_i^\lambda$ is a coefficient of the $i$th AO. Although this time-evolution equation contains the overlap matrix $S$, one may rewrite it in a form without $S$ by means of the Choleski decomposition (2.179), (2.183)–(2.185) [161]. That is, we introduce the lower-triangular matrix $T$ satisfying (2.179). Then, defining $D_i^\lambda = \sum_j T_{ij}^\dagger C_j^\lambda$ and an $\mathcal{H}'$ which satisfies (2.181), one can derive the standard time-evolution equation, which has a form similar to (2.387):

$$\mu \frac{d}{dt} D_i^\lambda = -\sum_j (H'_{ij} - \varepsilon_\lambda \delta_{ij}) D_j^\lambda, \quad \varepsilon_\lambda = \sum_{ij} D_i^{\lambda *} H'_{ij} D_j^\lambda. \tag{2.393}$$

This equation is as tractable as the more common approach of the plane wave expansion. However, at each time step, one has to evaluate the charge density from the coefficients $C_i^\lambda = \sum_j T_{ij}^{\dagger -1} D_j^\lambda$ and calculate the Hamiltonian matrix elements $H_{ij}$ from the charge density, and then transform them to $H'_{ij}$.

If one uses a basis set which is different from plane waves, it becomes necessary to evaluate the variational force in the force calculation (in a simulation based on either the Car–Parrinello method or alternative methods), since the basis functions themselves depend explicitly on the atomic positions. For example, the calculation of the forces has been formulated and applied to various structural optimizations for various $O(N)$ LCAO approaches (see Sect. 2.3.8): for the pseudopotential mixed-basis approach, by Ho et al. in 1983 [429] and 1992 [162] (the same formulation, apart from the part associated with the pseudopotential, works well also for the all-electron mixed-basis approach and makes it possible to perform realistic large-scale molecular-dynamics simulations [173, 175–182]), and, with a slightly different formulation, by Kishi and Itoh in 1995 [430] and 1996 [431]; for the APW method, by Soler and Williams in 1989 [432], 1990 [433] and 1992 [434], by Yu, Singh and Krakauer in 1991 [435] and 1992 [436], by Goedecker and Mashke in 1992 [437], by Oguchi in 1993 [438, 439], and by Krimmel in 1994 [440] (some recent applications can be seen in [442, 443]); and for the FLAPW method, by Kohler et al. in 1996 [444], by Batilev et al. in 1996 [445] and by Wu et al. in 1996 [446]. A method of calculating the total energy and pressure was formulated by Janak [441] in the muffin-tin approximation in 1974.

A formulation and implementation of the force calculation were presented by Blöchl [197] in 1994 in his projector-augmented-wave (PAW) approach. This approach is somewhat related to the projector-basis technique which was introduced by Singh et al. [284] (see Sect. 2.3.8) but is not the same; in fact, Blöchl's work in 1994 did not aim at the reduction of the computational task to $O(N^2 \log N)$ as in Singh's projector-basis technique. There has been some confusion in the usage of the term "projector", as is mentioned in the paper by Blöchl [197]. Blöchl's projector-augmented-wave approach has been successfully applied [447–450] to structural optimizations and molecular-dynamics simulations, and compared [451] with the results of the pseudopotential approach and the LAPW method. The projector-augmented-wave (PAW) approach was later reformulated in connection to the psuedopotential approach by Kresse and Joubert [149], and now very widely used in place of the psuedopotential approach. Now, PAW usually means Kresse–Joubert's reformulation of the pseudopotential approach.

By the recent progress in ab initio molecular-dynamics methods, it has become possible to perform constant-energy simulations without relying on the original Car–Parrinello method. If one iterates the CG steps of the self-consistent electronic-states calculation long enough before updating the atomic positions, one can neglect the variational force because the electronic states have converged to the BO surface. Note, however, that, even in this case, there may still remain a deviation of the calculated force from the true one because of the incomplete basis set. In a case where the number of basis functions has not been enough chosen, even correcting by the variational force would not yield the correct physical force, since one has not reached the true ground state. Therefore, it is always desirable to adopt an efficient basis set (or pseudopotential) in carrying out a reliable ab initio molecular-dynamics simulation.

### *2.5.5 Application to Fullerene Systems*

The MD technique is suitable for studies of behavior on timescales from femtoseconds (fs) to nanoseconds (ns) and on length scales from angstroms ($1\,\text{Å} = 10^{-10}\,\text{m}$) to nanometers (nm). In this region, the laws of statistical mechanics do not apply. For example, the law of equipartition of energy is obviously violated in elementary processes in chemical reactions, and such systems do not obey the Boltzmann distribution at all. MD simulations at the atomic or electronic level play a powerful role in understanding nanoscale phenomena.

As an example of the application of ab initio MD methods to nanometerscale materials, we choose the $C_{60}$ molecule. Other clusters, such as smaller carbon clusters [452], silicon clusters [453, 454] aluminum clusters [455, 456] and copper clusters [457, 458] have also been studied intensively by the Car–Parrinello and other methods, but we shall not go into details of these clusters. For $C_{60}$ molecules, there are good references [459, 460]. The stable structures of carbon had been considered until recently to be of only two kinds, i.e. diamond and graphite. However, Kroto et al. discovered the $C_{60}$ molecule, which has a cage structure like a soccer ball (Fig. 2.15), by laser ablation or arc discharge experiments using a graphite rod in a helium gas atmosphere [461]. Moreover, the success in the mass production of $C_{60}$ molecules by Krätschmer et al. [462] has advanced research on $C_{60}$. This molecule has the structure of a truncated icosahedron, and its point group has the icosahedral symmetry $I_h$. Now, $C_{60}$ is known to exist abundantly. Besides $C_{60}$, similar cage-structure molecules such as $C_{70}$, $C_{82}$ and $C_{84}$ have also been found. Writing the number of facets, edges and vertices as $S$, $B$ and $V$, respectively, the Euler theorem on polyhedrons gives

$$S + V = B + 2. \tag{2.394}$$

Assuming the $sp^2$ bonding condition $B = 3V/2$, i.e. three edges always start from one vertex, one may easily derive $S = V/2+2 = B/3+2$. Moreover, assuming that

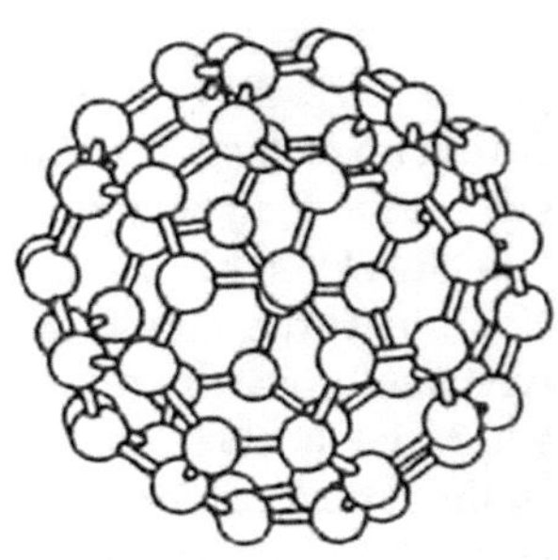

**Fig. 2.15** $C_{60}$ molecule with a soccer-ball structure

all the facets are either hexagons or pentagons, and denoting the numbers of these as $N_6$ and $N_5$, respectively, one obtains $N_5 = 12$. That is, a closed cage composed only of six-membered rings and five-membered rings always has 12 five-membered rings. Carbon clusters having this structure are called "fullerenes", after Buckminster Fuller, who is famous for constructing buildings with the same architecture. On the other hand, Iijima et al. [463] found the "carbon nanotube (CNT)", which has a cylindrical structure obtained by rolling a graphite sheet. Similarly, a carbon nanotube with one end closed by a half sphere of a fullerene, and the "carbon onion", which is essentially a fullerene with a multi-layer structure, have also been found. Furthermore, a nanotube which contains a seven-membered ring and spreads out like a funnel has also been observed. All of these structures may be promising as components of materials.

The further important fact about fullerenes and carbon nanotubes is that no structure with fused pentagons has been found: this rule is called the isolated-pentagon rule(IPR). If we assume the IPR, it is easily understood that the smallest cage is $C_{60}$, with $I_h$ symmetry. Figure 2.15 shows an isomer of $C_{60}$ with $C_{2v}$ symmetry. It is also be possible to consider $C_{60}$ isomers with other structures. However, they do not obey the IPR. In fact, the experimentally obtained $C_{60}$ consists only of the molecule with $I_h$ symmetry. The next higher fullerene is $C_{70}$. Crystals of $C_{60}$ and $C_{70}$ have been grown experimentally, and their electronic states have been compared with ab initio calculations [464–469, 567].

Here, we choose $C_{60}$ as an example for applying an ab initio molecular-dynamics simulation. An ab initio pseudopotential molecular-dynamics simulation and a related simulated-annealing study (Appendix D) on an isolated $C_{60}$ molecule were first performed by Zhang et al. [711] and by Feuston et al. [471] These authors determined the vibration spectra of $C_{60}$.

Concerning longer-time simulations involving formation, dissociation and other chemical reactions of $C_{60}$, all the preexisting molecular-dynamics studies had been restricted to the use of classical potentials [472–474] and tight-binding methods [475, 476] before 1996. The ab initio molecular-dynamics simulation shown below is one based on the all-electron mixed-basis approach, which was explained in Sects. 2.2.6 and 2.4.4. The structural stability and a simulation of the encapsulation process of a polonium atom into $C_{60}$ are the topics discussed here.

In the ab initio MD method based on the all-electron mixed-basis approach, one may use either STOs colorRed [166–170] or numerical AOs (NAOs) [171, 173, 175–182]. For each carbon atom, we use five AOs, i.e. the 1s, 2s, $2p_x$, $2p_y$ and $2p_z$ atomic orbitals.

The "supercell", which is the periodic unit cell chosen to be large enough for the cluster calculation (the treatment of a cluster or a slab using such a supercell is called the "supercell approximation"), is divided into a mesh of $64 \times 64 \times 64$ units, each unit corresponding to 0.196 Å. The number of PWs used in the calculation is either 2969 (corresponding to a 5.7 Ry cutoff energy)when STOs are used for AOs [168, 169], or 4169 (7.0 Ry) when numerical AOs are used [176, 177]. After the electronic states has been converged with fixed atomic positions, the dynamics simulation starts. The basic time step is chosen to be $\Delta t = 4\,\text{a.u.} = 0.97 \times 10^{-16}\,\text{s} \sim 0.1\,\text{fs}$. The nuclear motions are treated by solving the Newtonian equation (2.384) together with a velocity-scaling technique every 10 time steps to keep the temperature constant. The electron occupation is determined by the Fermi–Dirac distribution function

$$f_0(\varepsilon) = \frac{1}{1 + \exp[(\varepsilon - \mu)/k_B T]} \tag{2.395}$$

at the same temperature as the atomic dynamics. (There were several studies using the Car–Parrinello approach for the excited electrons [477] and for the finite-temperature dynamics [478].) The electronic states were tested and found to converge to the BO surface with five to six iterations of the SD step between successive updates of the atomic positions. To get rapid convergence, a new electron charge density is created by mixing 50–70% of the charge density at the previous step. In a case where the atomic orbitals are incorporated as basis functions, it is basically necessary to evaluate the variational force, since the AOs depend explicitly on the atomic coordinates. However, the variational force has the property of becoming zero whenever the electronic states are converged well. Therefore, the assumption that one can omit the variational force is valid in this case.

After the initial relaxation process of 500 time steps, the structure of $C_{60}$ is stabilized and the lengths of the single bond and the double bond are found, respectively, to be $1.46 \pm 0.02$ Å and $1.43 \pm 0.02$ Å in the case of a 5.7 Ry cutoff energy, and $1.460 \pm 0.005$ Å and $1.400 \pm 0.005$ Å in the case of a 7.0 Ry cutoff energy. (the experimental values are 1.46 Å and 1.40 Å). The reason why these values have an error of width 0.005–0.02 Å is the use of the supercell approximation and the insufficient number of plane waves.

A further important aspect of the results obtained by ab initio MD methods is the stability of $C_{60}$ at finite temperatures [168]. The ideal $C_{60}$ with $I_h$ symmetry is found to be stable up to 4000 K and unstable at 5000 K, if one includes the effect of electron excitations using the Fermi–Dirac distribution. While the $C_{60}$ isomer with $C_{2v}$ symmetry (Fig. 2.16) is found to be stable up to 1000 K, it becomes unstable at 2000 K. These results are nontrivial, because, if we assume the ground state for the electronic states, we observe that these molecules are stable up to at least 10 000 K! Thus we

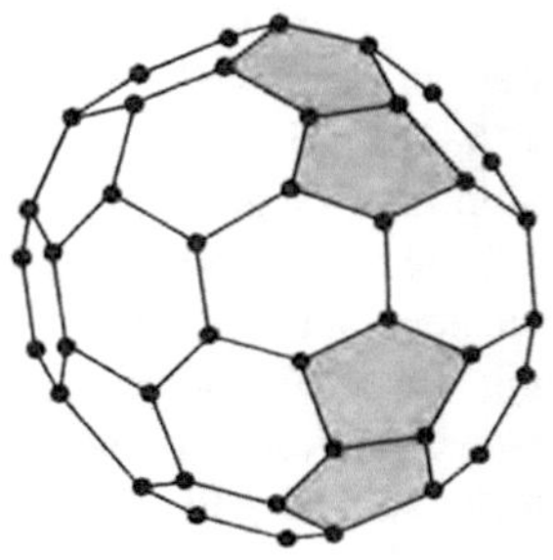

**Fig. 2.16** $C_{60}$ isomer with $C_{2v}$ symmetry

find that the finite-temperature effect on the electronic states is very important in discussing the stability of fullerenes at finite temperatures.

From numerical simulations indicating the finite-temperature effect on the electronic states, we find that an ideal $C_{60}$ molecule is more stable than its $C_{2v}$ isomer at high temperatures and that the reactivity of the $C_{2v}$ isomer with $C_2$ is stronger than that of ideal $C_{60}$. Moreover, it is found that $C_{60}$ with almost $I_h$ symmetry is selectively produced from the $C_{58} + C_2$ reaction [168]. All these results confirm and support the IPR. In real production process, there is an example, $C_{78}$, where one particular isomer with lower symmetry is much more abundant than the other isomers of $C_{78}$, and we should say that nature is very fascinating. In the near future, the reason for this may be identified by using ab initio MD simulations.

The next topic we discuss is a simulation concerning the encapsulation of a foreign atom inside the cage of $C_{60}$. The "endohedral fullerene", which has at least one atom inside the cage of the fullerene, has attracted much current interest. Experimentally, it has been confirmed that at least one lanthanum, yttrium, or scandium atom can be encapsulated inside $C_{82}$or $C_{84}$ using arc-discharge vaporization of composite rods made of graphite and the metal oxide [479, 480]. Also, laser vaporization techniques have revealed the production of endohedral $C_{60}$ in which an alkali-metal atom such as potassium or cesium is trapped. On the other hand, the authors have investigated foreign atom insertions into $C_{60}$ by using the all-electron mixed basis approach [169, 170, 176–182]. The collision of an alkali-metal cation and $C_{60}$ anion has been extensively studied by ab initio MD simulations to see whether the alkali-atom endohedral $C_{60}$, e.g. Li@$C_{60}$, can be produced, and which kind of process must occur if the insertion is to be possible [169, 484]. (Here, the notation Li@$C_{60}$, for example, means a $LiC_{60}$ molecule with the Li atom encapsulated inside the $C_{60}$ cage.) It has been also experimentally reported that endohedral $[Li@C_{60}]^+$ and $[Na@C_{60}]^+$ species are formed during collisions of alkali-metal ions with $C_{60}$ vapor molecules [481, 482]. More recently, global synthesis of $[Li@C_{60}]^+$ became possible with the plasma shower technique [483, 484].

The insertion of Be [170], heavier noble-gas elements (Kr and Xe) [176], As [177], N [179], Sb, Te [180], Se [181], and Po [182] into $C_{60}$ is presented by Ohtsuki et al. Figure 2.17 shows the snapshots of the simulation, where a Po atom

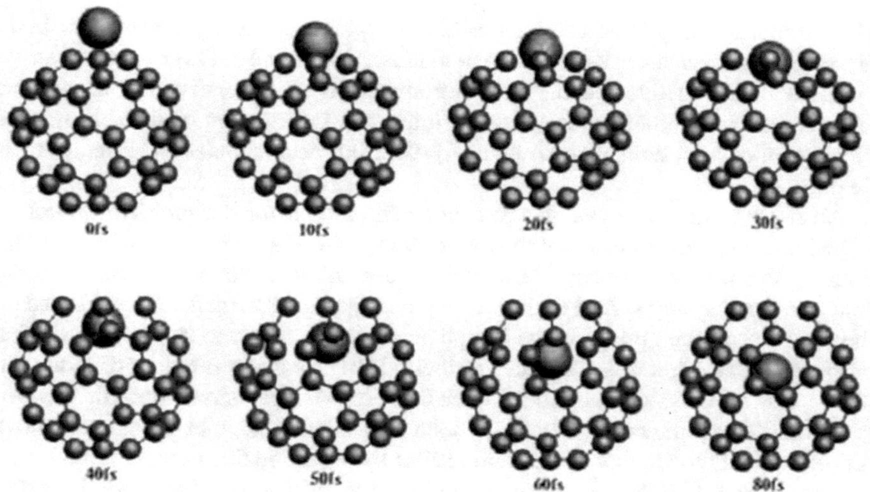


**Fig. 2.17** Insertion process where Po hits the center of a six-membered ring of $C_{60}$ with 40 eV kinetic energy [182]. The figure shows only the positions of the carbon atoms in $C_{60}$ (*small circles*) and Po atom (*large circle*). In this case, after 80 fs, Po@$C_{60}$ is created

with the kintetic energy 40 eV is vertically inserted into $C_{60}$ through the center of a six-membered ring [182].

### 2.5.6 Application to Point Defects in Crystals

Crystals have wide variety of defects (or imperfections) [485–487]. One type of imperfection is a point defect such as a single vacancy, an interstitial atom or a substitutional atom. If the interstitial or substitutional atom is different from the host atoms, it is called an impurity. Sometimes, not only the structural properties but also the electronic properties change significantly because of impurities. First of all, it is well known that transition-metal impurities in metals cause significant phenomena, such as the Kondo effect [319, 488–491] and the spin glass caused by the oscillatory behavior of the long-range RKKY (Ruderman–Kittel–Kasuya–Yoshida) interaction. Impurity centers such as the DX center has been investigated in connection with deep impurity levels [492–494]. Also transition metal impurities in semiconductors play an important role in changing the electronic structure [495]; a dilute magnetic semiconductor is one of such examples [496].

Other types of imperfections in bulk crystals include stacking faults, dislocations and grain boundaries. Defects are not always stationary but often migrate or diffuse. Atomic diffusion is the central problem of the study of defects [497, 498]. Point defects sometimes form clusters by spontaneous aggregation under an external stress.

Many impurity atoms have a strong tendency to spontaneously aggregate in the bulk, or to segregate near surfaces or grain boundaries. An external stress or damage caused by X-ray or radioactive sources often introduces divacancies, trivacancies, or, as a larger vacancy complex, cavities. Sometimes, crack formation or the catastrophic breakdown of a crystal called "fracture" [499, 500] occurs under an extreme stress field.

Ab initio molecular-dynamics methods can be used for the problem of atomic diffusion in crystals. Because of the loss of translational symmetry, reliable calculations for this problem are considerably more difficult than standard calculations for small-unit-cell systems. As a typical example of the Car–Parrinello method based on the pseudopotential approach and the plane wave expansion to study the diffusion problem, Buda, Chiarotti, Car and Parrinello [501] treated a problem of hydrogen diffusion in a Si crystal. They found from the results of their simulations that hydrogen atoms stay at the center of a Si–Si bond for a small time interval and then move on very quickly to another center, and diffuse by repeating this process [501].

The energetics of defect formation or migration are also very important in understanding the diffusion problem in crystals, and are challenging topics in the application of ab initio techniques. On the basis of the pseudopotential–PW expansion method using a supercell, the vacancy formation energy has been calculated, for example, in aluminum [502], lithium and sodium [503]. The pseudopotential mixed-basis approach has also been applied to the formation energy and the formation volume of a monovacancy in Mo [504] and the intermetallic compound $Fe_3Al$ [505]. The projector-augmented-wave implementation of Car–Parrinello molecular dynamics was used to investigate defects and diffusion in solid $Li_3N$ [447]. The full-potential KKR method enables one to estimate the vacancy formation energy without invoking the supercell approximation. By this method, Dederichs et al. [506] calculated the vacancy formation energy of various FCC crystals. The similar investigation using the GGA was performed by Asato et al. [507, 508]. Next, as an example of an ab initio MD study, we will describe a simulation for a BCC iron crystal [173].

One may estimate the vacancy formation energy of a crystal as the difference between the total energies of a perfect crystal and a system with a single vacancy. Here, 16 iron atoms are put inside a cubic unit cell of size $5.73 \times 5.73 \times 5.73$ Å$^3$ to form a perfect BCC lattice. Comparing the total energies of this perfect system (16-atom system) and a system with a single vacancy (15-atom system), we obtain a vacancy formation energy which is comparable to the experimental value of 1.8 eV. The method used in this study is the all-electron mixed-basis approach using numerical atomic orbitals (NAOs). The calculation was performed with the local-spin-density approximation (LSDA). The number of $k$-points in the irreducible Brillouin zone (1/48 of the first Brillouin zone) is 4 and the PW cutoff energy is 16.5 Ry (the number of PW's is 1419, and the unit cell is divided into a mesh of $64 \times 64 \times 64$ units). For atomic orbitals, we use the 1s, 2s, 2p, 3s, 3p, 3d and 4s AOs with a cutoff radius, $r_c = 2.0$ a.u. (1 a.u. $= 0.52918$ Å).

Next in order to simulate the diffusion of a self-interstitial atom, we put 55 iron atoms in a larger cubic cell of size $8.60 \times 8.60 \times 8.60$ Å$^3$ to form an imperfect BCC lattice. Here also the unit cell is divided into a mesh of $64 \times 64 \times 64$ units. The initial

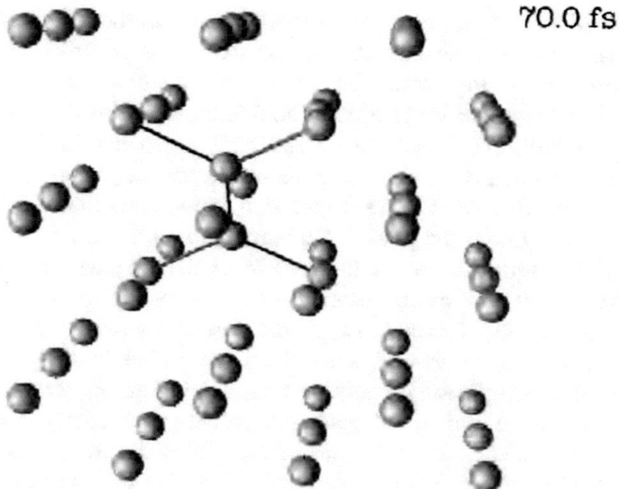


**Fig. 2.18** Optimized structure of the <110> dumbbell in an iron crystal

speed of the atoms is set to be 0 for every atom. For this system, we use the LDA instead of the LSDA, and the 1s, 2s, 2p, 3s, 3p and 3d atomic orbitals together with 3071 PWs corresponding to a cutoff energy 12 Ry. After the initial 50 time iterations, we perform a first-principles MD simulation with $\Delta t = 4\,\text{a.u.} = 0.1\,\text{fs}$ as the MD time step. During the MD steps, we iterate typically 6 SD steps between adjacent updates of the atomic positions. Starting from four different initial configurations, we optimized the local configurations. Figure 2.18 represents the optimized structure of the <110> dumbbell. The five bonds drawn in the figure are shorter than the other bonds. These bond lengths are as follows: the center bond is 1.86 Å long, the lower two 1.87 Å and the upper two 1.89 Å.

## 2.5.7 *Application to Other Systems*

An ab initio total-energy calculation or an ab initio MD technique can also be applied to perform "simulated annealing" (Appendix D) which determines the most stable structure. In this case, it sometimes becomes necessary to evaluate not only the force acting on each atom but also the force related to a change of the lattice constant.

The Car–Parrinello method is applicable also to problems of surface reconstruction [509–511] or surface adsorption [512–514]. In this case, to represent the crystal surface, one usually uses the "repeated slab approximation", which uses a periodic boundary condition in the $XY$ plane and assumes several layers with finite thickness in the $Z$ direction (this slab is repeated with some vacuum region in the $Z$ direction), i.e. it approximates the crystal surface by the surface of a slab which is repeated in the $Z$ direction. Another atom (or molecule) is put in the intermediate vacuum region. The simulation is performed until the particle is adsorbed on one side of the slab. In this method, the interaction between the adsorbate and the other side of the slab should be negligibly small, and the thickness of the slab should be large enough so that a bulk-like density of states (DOS) is reached in the middle of the slab. These are the conditions to get a good result, and the calculation necessarily becomes huge. That Štich et al. [511] succeeded in realizing the $7\times7$ reconstructed surface of Si(111) showed the power of this method. Diffusion of silicon adatoms on a stepped Si(001) surface was studied by Kim et al. [515], who discussed also the Schwoebel barriers as well as an earlier work by Oshiyama [516]. (see Sect. 5.4.7 for the Schwoebel effect). The microscopic mechanism for the mechanical polishing of the diamond (110) surface was discussed by Jarvis et al. [517].

The dissociation adsorption process on a surface is a strongly dynamic problem [417, 518] which is similar to the oxidation process on a catalytic surface [519] or, more generally, to chemical reactions of molecules. To understand fully such a process is not easy and it is very challenging. To bridge the gap between ab initio molecular-dynamics simulations of such catalytic behavior at surfaces and the basic theory such as that given by Newns [520] and by Suhl [521] may also be an important problem in the future. Also, how to make quantum dots or a self-organized surface would be an interesting issue in material science on a larger scale.

Ab initio molecular-dynamics simulations have been applied to determine the phase coexistence curve of the first-order phase transition of real materials. In order to determine the phase boundary, it is necessary to evaluate not only the pressure but also the chemical potential, or the free energy. Thepressure can be determined in the general case without a problem, by differentiating the total energy with respect to the lattice constant. However, the free energy must be determined, for example, by introducing a reference system and integrating the internal energy with respect to the switching constant $\lambda$ (Kirkwood's method [522–524]):

$$\mathcal{H}(\lambda) = \lambda\mathcal{H}_1 + (1-\lambda)\mathcal{H}_0, \tag{2.396}$$

$$F_1 - F_0 = \int_0^1 d\lambda \langle \mathcal{H}_1 - \mathcal{H}_0 \rangle_\lambda, \tag{2.397}$$

where $\mathcal{H}_i$ and $F_i$ are, respectively, the Hamiltonian and the Helmholtz free energy of the reference system ($i = 0$) and the system under consideration ($i = 1$). In the integrand, the average should be taken with respect to the ensemble of the system described by the Hamiltonian (2.396). This means that one should perform many molecular-dynamics simulations of systems with various values of $\lambda$. Such a procedure is quite time-consuming, especially for ab initio methods. One convenient method was proposed by Watanabe and Reinhardt for classical cases [525] and is applicable directly to ab initio cases also. Their method is to perform the $\lambda$ integral dynamically during the simulation, i.e. to assume that $\lambda$ (and therefore the Hamiltonian) is time-dependent such that $\lambda(t = t_0) = 0$ and $\lambda(t = t_1) = 1$ and to perform a simulation from $t = t_0$ to $t = t_1$. Then the free energy difference is given by

$$F_1 - F_0 = \int_{t_1}^{t_0} \frac{d\lambda(t)}{dt}[E_1(t) - E_0(t)]dt, \tag{2.398}$$

where $E_i(t)$ stands for the total energy of the system at time $t$ calculated with the Hamiltonian $\mathcal{H}_i$ ($i = 0, 1$). For the Hamiltonian of the reference system, one can use a very simple classical model which represents well the overall nature of the system. For the solid state, the reference system can be chosen to be a simple Einstein solid in which all the atoms move independently around lattice points in a parabolic potential. Two reports using ab initio methods came out in late 90's. One deals with the melting transition of Si [526] and the other with the melting transition of Al [527].

An alternative way to determine the free energy is to consider the difference in the Gibbs free energy between two phases by integrating a thermodynamic quantity from $(p_0, T_0)$ to $(p_1, T_1)$. For example, by integrating the pressure along an isotherm, one finds the following formula;

$$F(V_1, T) - F(V_0, T) = -\int_{V_0}^{V_1} p(V)dV. \tag{2.399}$$

However, such a method has not been extensively used even in classical simulations, because of some inconvenience [528].

The problem of structural phase transitions in the bulk has been so far restricted mainly to the problem at absolute-zero temperature, although some applications of the ab initio molecular-dynamics method have appeared which have dealt with high pressure. In order to simulate constant pressure, a variable-cell-shape algorithm together with, for example, Andersen's or Parrinello and Rahman's method (see also Appendix A) is needed. (One may use the constant-volume method also, but in this case one has to perform many simulations with cells of different volumes [529].) Such simulations are very involved and have been performed by Wentzcovitch's group [530–533] and Parrinello and Tosatti's group [534–539]. If it is possible to apply the ab initio method to the problem of structural phase transitions at finite temperatures, for example the ferroelectric phase transitions of perovskite materials such as $SrTiO_3$ or $BaTiO_3$, there is no doubt that broad applications of ab initio molecular-dynamics simulations at finite temperatures will open a new paradigm in computational materials science.

All the ab initio molecular-dynamics methods described so far are based on the Born–Oppenheimer adiabatic approximation. However, as was discussed in the introduction of this chapter, nonadiabatic effects may become essential in some important problems related to light atoms such as hydrogen, where quantum fluctuation of the nuclei is important. The path-integral representation of nuclei has been combined with the ab initio MD method to perform fully quantum mechanical simulations [540, 541]. Further, even for heavier elements, many important phenomena such as plasma oscillations, ionization–neutralization processes, electronic transition processes with photoemission or photoabsorption, electron tunneling and electric currents are essentially nonadiabatic, although quantum fluctuations of nuclei can be ignored in these cases. One possibility for treating this kind of nonadiabatic phenomenon is to use the time-dependent Schrödinger equation, and such an ab initio molecular-dynamics method will be described briefly in Sect. 2.6.

Many ab initio techniques have been described in this chapter from their fundamental basis to several applications. Nevertheless, there are many topics which have been skipped completely in this book. For example, the BCS theory [542] of superconductivity, the quantum Hall effect [543–545], the giant magnetoresistance (GMR) [546–548] and so on have not been described at all, but they can be found easily in many standard textbooks of statistical quantum theory. Not only these, but also many important recent examples of progress in ab initio theory could not be covered completely (for example, the study on giant magnetic moments [549–554], topological insulators [555], hyperfine interactions [556–558], photovoltaic solar cells [559–562], thin-film organic solar cells [563, 565], thermoelectric properties [566–568], and so on.). Some of the topics can be also found in other books, for example, [569–573].

### 2.5.8 Coherent Potential Approximation (CPA)

A very commonly encountered problem is the problem of alloys and disorder. One way to obtain a reasonable density of states is to use the CPA [574]. In this approximation, one assumes an average coherent crystal, i.e. a perfectly periodic crystal, with an average alloy potential, such that, on average, no scattering takes place. This is a much better and more consistent approximation that the virtual crystal approximation (VCA), where from the beginning, one takes the average of the Hamiltonian over the impurity distribution. In both cases, if one adopts a tight-binding picture, it is assumed that the hopping integrals do not change, but only the onsite energies take different values according to the occupation of that site. VCA would yield a band centered around the averaged onsite energy, and would therefore be unacceptable if the dispersion of the onsite energies is larger than the bandwidth. CPA is the best single-site approximation within the mean-field theory.

Since the DOS is given by the trace of the imaginary part of the Green's function (GF), one needs first to compute the diagonal elements of the GF of the crystal. In the case of disordered systems, one may use the T-matrix approach, where the multiple interactions with a single site are summed-up to renormalize the single impurity potential.

Assume the total Hamiltonian of the disordered system to be the sum of the averaged Hamiltonian (the VCA Hamiltonian) and the fluctuations from this term which are basically a sum of onsite energy terms:

$$H = H_0 + V = H_0 + \sum_i \varepsilon_i c_i^\dagger c_i, \tag{2.400}$$

where the impurity average of $V$ is zero: $\langle V \rangle = 0$.

The Green's functionis given by the Dyson equation in operator form:

$$G(z) = [z - H]^{-1} = G_0(z) + G_0(z)VG(z). \tag{2.401}$$

Here $z$ is the complex energy $z = E + i\eta$, where $E$ is the energy of the propagating electron and $\eta$ is a small positive real number. This equation can be formally solved for $G$ to give $G(z) = [1 - G_0(z)V]^{-1}G_0(z)$. Alternatively, one may define the T matrix by $G(z) = G_0(z) + G_0(z)T(z)G_0(z)$, in which case, we get

$$T(z) = V + VG_0(z)V + VG_0(z)VG_0(z)V + \cdots \tag{2.402}$$

The series can be summed to give $T(z) = V[1 - G_0(z)V]^{-1}$. As we can see, the above series correspond to the multiple scatterings of the electron wave against the impurity potential $V$. Since it is impossible to compute the impurity average of this T matrix exactly, one assumes in the CPA that there is only one impurity in the average medium. Its scattering properties should, however, average out to zero. In this case, where the potential $V$ becomes localized on one site only, the T matrix can be computed very easily: $t_i(z) = \varepsilon_i/[1 - G_{0ii}(z)\varepsilon_i] = \varepsilon_i/[1 - g_0(z)\varepsilon_i]$. $G_0$ being the GF of a translationally invariant system, its diagonal elements are all equal to each other, and we denote them here by the complex function $g_0(z)$.

As the eigenstates of $H_0$ are Bloch wavefunctions, we can write

$$H_0 = \sum_{k\lambda} \varepsilon_{k\lambda} c^{\dagger}_{k\lambda} c_{k\lambda}, \tag{2.403}$$

where $\boldsymbol{k}$ is a vector in the first Brillouin zone, and $\lambda$ is the band index (see also Sect. 2.3.1 on periodic systems and Bloch theorem). This way, one has all matrix elements of the GF in the site representation:

$$G_{0ij}(z) = \frac{1}{N} \sum_{k\lambda} \mathrm{e}^{i\boldsymbol{k}\cdot(\boldsymbol{R}_i - \boldsymbol{R}_j)} \frac{c^{i}_{k\lambda} c^{j*}_{k\lambda}}{z - \varepsilon_{k\lambda}}, \tag{2.404}$$

where $c^i_{k\lambda}$ and $c^j_{k\lambda}$ are the coefficients for the expansion of the eigenstates on the atomic basis centered at $i$ and $j$, respectively.

Following Soven [574] and Taylor [575], we assume that within the CPA, the GF is written as

$$G_{\mathrm{CPA}}(z) = [z - \Sigma(z) - H_0]^{-1}; \tag{2.405}$$

i.e. it still applies to a periodic system, but depending on the energy of the electron, the self-energy term $\Sigma(z)$ shifts its energy, hence the name of "coherent potential". The random potential at the impurity site is therefore replaced by $\varepsilon_i - \Sigma(z)$. The scattering is, however, still site dependent, but the effect of the fluctuations in the site potential are assumed to be already included in $G_{\mathrm{CPA}}$. In order to have self-consistency, therefore, the impurity average of the *single-site* T-matrix $t_i(z)$ must vanish.

$$0 = \langle t_i(z) \rangle = \langle [\varepsilon_i - \Sigma(z)][1 - g_{CPA}(z)(\varepsilon_i - \Sigma(z))]^{-1} \rangle, \tag{2.406}$$

where the ensemble average $\langle ... \rangle$ is taken with respect to the distribution function of $\varepsilon_i$ having zero mean and given higher moments. The solution of (2.406) gives the self energy $\Sigma(z)$. The above is for the situation of a single orbital per cell. In general, this equation has to be seen as a matrix equation where all the above quantities are matrices of the size of the number of orbitals per unit cell.

This approximation has been applied to a simple model of binary alloy with a single band. The comparison with the exact solution where the ensemble average was taken numerically was astonishingly good [575]. In this case, where there are two elements $A$ and $B$ with onsite energies (we assume one level per atom) $\varepsilon_A$ and $\varepsilon_B$ and occupation probabilities $p_A$ and $p_B$ ($p_A + p_B = 1$), the onsite term is given by $\varepsilon_i = \varepsilon_A$ ($\bar{\varepsilon} = p_A \varepsilon_A + p_B \varepsilon_B = 0$), if the site $i$ is occupied by atom $A$ for example. In this case, the distribution function of $\varepsilon_i$ is a sum of two delta functions centered at $\varepsilon_A$ and $\varepsilon_B$. (2.406) is reduced to

$$\frac{p_A(\varepsilon_A - \Sigma)}{1 - (\varepsilon_A - \Sigma) g_0(z - \Sigma)} + \frac{p_B(\varepsilon_B - \Sigma)}{1 - (\varepsilon_B - \Sigma) g_0(z - \Sigma)} = 0 \tag{2.407}$$

In the limits of weak disorder, where the difference between the onsite energies is much smaller than the band width, or in the opposite limit, the above equation can easily be solved [576]. If either of the probabilities $p_A$ or $p_B$ are low, the agreement with the exact numerical result becomes worse due to the appearence of localized impurity modes. Other good references on CPA are [577–580].

## 2.6 Beyond the Born

Despite recent progress in the fields of DFT and ab initio molecular-dynamics simulation, many simple effects cannot be described simply by these methods. For example, the Car-Parrinello MD method (and other ab initio MD methods) are based on the Born–Oppenheimer (BO) approximation. Thus, they cannot simulate the electron-phonon interaction, which is the basis of the resistance of metals at finite temperatures. Furthermore, the present treatment of DF theory cannot properly describe the excited states of a system. Therefore the interactions of radiation with matter cannot be treated by the usual DF methods either. Other examples are complex systems where many-body effects can become important, such as magnetic impurities and the Kondo effect, the (fractional) quantum Hall effect, High-$T_c$ superconductors, etc. The present functionals to treat correlations and exchange are still very primitive, and cannot take into account subtle many-body effects. For such phenomena, one still uses simple models and constructs model Hamiltonians with the necessary ingredients to take into account some specific behavior. At present, simulation of such phenomena is beyond the scope of first-principles methods.

We see then that there are basically two limitations in treating or simulating more complex systems. One is the inappropriateness of the theory (lack of better functionals), and the other the lack of more sophisticated methods to perform such simulations. Our interest at present is the second point, namely to improve the simulations methods in order to describe effects such as the electron-phonon interaction by computer simulation.

Most textbooks on solid-state physics start their developments by stating and, later on, using the Born-Oppenheimer approximation. In this approximation, the motion of atoms is separated from the electronic degrees of freedom. The effective interaction between the atoms is provided by the electrons; so one usually calculates, for a fixed atomic configuration, the electronic ground state and the resulting forces or interactions between the atoms, and then moves the atoms *classically* according to these forces, and Newton's equations of motion. This is called the adiabatic approximation, since in this treatment, electrons follow the atomic trajectories adiabatically. The assumption behind this is that atomic motions are usually much slower than those of electrons, which have a response time of the order of femtoseconds, and move much faster than atoms. So the electrons can adjust themselves almost instantaneously to the new atomic configuration. This assumption clearly breaks down when atoms move at very high speeds. Such is the case for high-energy collisions of molecules with a surface, for example. The other assumption usually made in the BO approximation is that the electrons always stay in their ground state. Besides the previous example, where electrons can obviously go to their excited states if the energies involved are high enough, there can also be electronic transitions to higher states if the gap is too small or zero as in open-shell clusters or in metals. In this case, even for low-energy atomic motions, electrons can gain energy from phonons to make transitions to excited states.

If we denote by $\theta_n(\boldsymbol{R}, \boldsymbol{r})$ the eigenstates of the Hamiltonian (for atoms fixed at $\boldsymbol{R}$ treated as parameters),

$$\mathcal{H} = \frac{p^2}{2m} + \frac{P^2}{2M} + V_{ei}(\boldsymbol{r}, \boldsymbol{R}) + V_{ii}(\boldsymbol{R}) + V_{ee}(\boldsymbol{r}) \tag{2.408}$$

(the capital letters and "$i$" refer to ions, $\boldsymbol{R}$ is the 3N-dimensional vector representing the atomic coordinates, and $r$ represents the coordinated of the electrons) then one can write the matrix elements of the full Hamiltonian in this complete basis as

$$\langle\theta_n|\mathcal{H}|\theta_l\rangle = \delta_{nl}[E_l(\boldsymbol{R}) + V_{ii}(\boldsymbol{R})] + \langle\theta_n|\frac{P^2}{2M}|\theta_l\rangle. \tag{2.409}$$

Here the integration is over the electronic degrees of freedom $\boldsymbol{r}$, and the electronic eigenstate $E_n(\boldsymbol{R})$ provides the effective interaction between the ions. The last term is clearly nonzero since the operator $P^2$ acting on $\theta_l$ yields a priori a linear combination of all states. This last term is precisely responsible for the complications in the treatment of atomic systems. The adiabatic approximation consists of neglecting the off-diagonal terms of the ionic kinetic-energy operator; the rationale being that it differs from the electronic kinetic energy by a factor of $m/M$, and is therefore small. The complete neglect of this term, even the diagonal contributions, yields the Born–Oppenheimer approximation.

There have been many attempts in the past to overcome these approximations by either perturbation theory or considering a nonlinear time-dependent phase factor in the wavefunction [581]. But none of these can be directly applied to computer simulations. One paper, however, attempts to treat nonadiabatic effects in computer simulations [582] by adding a random "frictional" force to Newton's equations of motion, similarly to Langevin's dynamics. This Brownian force would simulate the effect of the drag caused by electrons lagging behind the ions in the case of high-speed motion, for example. Other approaches are used by quantum chemists who use the configuration interaction theory, but these are limited to processes in very small molecules such as reactions of diatomic molecules.

Our approach is to directly integrate the time-dependent Schrödinger equation. Within the DFT, this reduces to the time-dependent Kohn–Sham equations. The latter are integrated by the finite-difference method, where the atomic positions vary with time according to their past trajectory and Newtonian equations of motion. The electronic time step must obviously be very small, maybe about 0.01 times the atomic time step, which is itself of the order of a femtosecond. Forces are calculated each femtosecond and atoms are moved accordingly, so that during the electronic integration steps the potential is changed. This way, the motion of atoms is very smooth and does not involve step functions at every atomic update. In this sense, the ionic potential is still treated adiabatically, but one allows electronic transitions to excited states. This is very different from the usual ab initio MD techniques, where electrons are kept at all times on the BO surface (ground state). No diagonalization is performed during the simulation, and the electronic state is characterized by its density matrix (DM) in the original electronic basis states. The advantage of using the DM formalism over the standard one is that the eigenstate indices, which lose their meaning during the time evolution, disappear and are replaced by basis indices, telling us where the electrons are. It has, furthermore, the advantage of being easily generalizable to finite temperatures. Within the BO approximation, the usual equations of motion for the electrons and ions read, respectively,

$$i\frac{d\varrho}{dt} = [\mathcal{H}\{\boldsymbol{R}(t)\}, \varrho] \tag{2.410}$$

$$M\frac{d^2\boldsymbol{R}}{dt^2} = -\nabla_{\boldsymbol{R}} E[\boldsymbol{R}(t), \varrho(t)] = -\nabla_{\boldsymbol{R}}(V_{\text{ii}} + V_{\text{ei}}). \tag{2.411}$$

The first equation can easily be derived from the time-dependent (TD) Schrödinger equation, itself obtained from a variational principle [583]. The second equation is the classical equation of motion for ions where forces are obtained from the Hellmann–Feynman (HF) theorem.

Within TD-LDA, these equations are both exact if the basis set is complete, orthonormal and independent of the atomic positions. Such is the case for plane waves. In practice, however, the basis set is always finite, and therefore the above equations are not exact. Correction terms must be added to them, in order to satisfy energy conservation.

Let us first consider the evolution of the wavefunction,

$$\mathrm{i}\frac{d\psi^{\lambda}(t)}{dt} = H(t)\psi^{\lambda}(t), \tag{2.412}$$

where $\lambda$ is the eigenstate index. After expansion in the basis functions $\psi_{\lambda} = \sum_i c_i^{\lambda}\,\phi_i(r)$ and projecting on the basis states, one obtains:

$$\mathrm{i}\dot{c}_i^{\lambda} = S^{-1}(H + \dot{\boldsymbol{R}}_J P_J)_{ij} c_j^{\lambda} = \mathcal{H}_{ij} c_j^{\lambda}, \tag{2.413}$$

where

$$\mathcal{H} = S^{-1}(H + \sum_J \dot{\boldsymbol{R}}_J P_J),$$

$$P_{Jij} = \langle \phi_i \,| - \mathrm{i}\nabla_J\ \phi_j \rangle$$

$$S_{ij} = \langle \phi_i \,|\, \phi_j \rangle$$

and

$$H_{ij} = \langle \phi_i \,| H_{KS} |\, \phi_j \rangle.$$

Multiplying (2.413) by $c_j^{\lambda *}$ and adding this equation to its complex conjugate, one can obtain the evolution equation for the DM $\varrho$:

$$i\dot{\varrho} = \mathcal{H}\varrho - \varrho\mathcal{H}. \tag{2.414}$$

We see that the original Hamiltonian $H$ must be replaced by a modified one $\mathcal{H}$ in order to take into account the overlap and the dependence of the basis set on the atomic coordinates. Similarly, the corrections for the HF forces may be obtained by requiring the conservation of the total energy [584] or writing down the Euler–Lagrange equations of motion derived from the following Lagrangian [583]:

$$\mathcal{L}(c, \dot{c}, \boldsymbol{R}, \dot{\boldsymbol{R}}) = \sum_J \frac{1}{2} M_J \dot{\boldsymbol{R}}_J{}^2 + \sum_{\lambda}^{occ} \sum_{ij} \Big[ c_j^{\lambda *} \langle \phi_j | (i\frac{\partial}{\partial t} - T) | \phi_i \rangle c_i^{\lambda} \Big]$$

$$- V_{\mathrm{ii}} - V_{\mathrm{ee}}^{C} - E_{\mathrm{xc}} - V_{\mathrm{ei}}.$$

As in most MD simulations, the ions are treated classically because of their large mass. The total energy of the system ions + electrons is written as

$$E = \frac{1}{2}\sum_{JL}' \frac{Z_J Z_L}{|\boldsymbol{R}_J - \boldsymbol{R}_L|} + \frac{1}{2}\sum_J M_J \dot{\boldsymbol{R}}_J^2 + \langle \Psi_{\mathrm{el}}|T_{\mathrm{e}} + V_{\mathrm{ei}}(\boldsymbol{R}_J) + V_{\mathrm{ee}}|\Psi_{\mathrm{el}}\rangle$$

We also have the constraint of conservation of the number of electrons: $\int \rho(r)dr = \mathrm{Tr}\varrho S = N$. This can be included if one introduces a Lagrange multiplier $\mu$ and minimizes the potential $\Omega = E - \mu\,(\mathrm{Tr}\,\varrho S - N)$. Energy conservation $(dE/dt = 0)$ leads to the following:

$$\sum_J \dot{\boldsymbol{R}}_J \cdot \left( M_J \ddot{\boldsymbol{R}}_J + \nabla_J \sum_L \frac{Z_J Z_L}{|\boldsymbol{R}_J - \boldsymbol{R}_L|} - \int \rho(r)\nabla_J \frac{Z_J}{|r - \boldsymbol{R}_J|} dr \right)$$

$$-\mu\,(\,\mathrm{Tr}\,\varrho\dot{S} + \dot{\boldsymbol{R}}_J\,\mathrm{Tr}\,\varrho\nabla_J S\,)$$

$$+\langle\frac{d\Psi_{\mathrm{el}}}{dt}|T_{\mathrm{e}} + V_{\mathrm{ei}} + V_{\mathrm{ee}}|\Psi_{\mathrm{el}}\rangle + \langle\Psi_{\mathrm{el}}|T_{\mathrm{e}} + V_{\mathrm{ei}} + V_{\mathrm{ee}}|\frac{d\Psi_{\mathrm{el}}}{dt}\rangle = 0$$

The sum of the last two terms is zero if $\Psi$ is exact or if the basis is formed of plane waves only. For a finite localized basis $\phi$, however, we end up with

$$\dot{\boldsymbol{R}}_J \left( \sum_{ijk} \varrho_{ij} \langle \frac{\partial \phi_j}{\partial \boldsymbol{R}_J} | \phi_k \rangle \langle \phi_k | H_{\mathrm{el}} | \phi_i \rangle \right.$$

$$+ \sum_{ijk} \varrho_{ij} \langle \phi_k | \frac{\partial \phi_i}{\partial \boldsymbol{R}_J} \rangle \langle \phi_j | H_{\mathrm{el}} | \phi_k \rangle$$

$$\left. - \sum_{ij} \varrho_{ij} \langle \frac{\partial \phi_j}{\partial \boldsymbol{R}_J} | H_{\mathrm{el}} | \phi_i \rangle - \sum_{ij} \varrho_{ij} \langle \phi_j | H_{\mathrm{el}} | \frac{\partial \phi_i}{\partial \boldsymbol{R}_J} \rangle \right).$$

This correction in $\dot{\boldsymbol{R}}_J$ must be added to the Hellmann–Feynman force acting on particle $J$. It appears due to the incompleteness of the basis. There is furthermore another correction due to the added chemical-potential term. The correction terms for the HF force become

$$F_J^{\mathrm{corr}} = \mathrm{Tr}\,\varrho(\mu\nabla_J S - \nabla_J H) + \mathrm{i}\mathrm{Tr}\,(H - \mu S)(S^{-1} D_J \varrho - \varrho S^{-1} D_J).$$

The evolution equations for $\{\varrho, \boldsymbol{R}_J\}$ can now be written as:

$$\mathrm{i}\dot{\varrho} = \mathcal{H}\varrho - \varrho\mathcal{H} \tag{2.415}$$

$$M_J \ddot{\boldsymbol{R}}_J = F_J^{\mathrm{HF}} + F_J^{\mathrm{corr}}. \tag{2.416}$$

Integration of these equations may be done by the predictor–corrector method. For a pure state, such as in a zero-temperature calculation, there is a further constraint of idempotency on $\varrho$:

$$\varrho S \varrho = \varrho.$$

One must therefore make sure after each update, that $\varrho$ stays idempotent. This can be done by using McWeeny's *purification transformation* [585] after each update: $\varrho \rightarrow 3\varrho^2 - 2\varrho^3$ is more idempotent than $\varrho$.

These equations are exact within the TD-LDA formalism and a finite, nonorthogonal basis set; see also the pioneering paper written by Saalman and Schmidt [584]. We see that corrections to the force due to overlap, incompleteness and electron-number-conservation constraint appear. Such a formalism can, to some extent, treat electronic excitations to higher excited states and the electron–phonon coupling in metals by ab-initio methods.

## 2.7 Quasiparticle Theory and Green's Function Methods

### *2.7.1 Quasiparticle (QP) Theory*

In the interacting electron systems, it is impossible to describe each electron independently. However, using the quasiparticle picture, one can treat the interacting electron systems as if each quasiparticle (QP) do not interact each other. A quite important hint lies in the experimental photoelectron spectroscopy measuring the photoemission or inverse photoemission (PE/IPE) energy spectra, which correspond to minus of the energies required by removing or produced by adding one electron from/to the system. What are drawn in the energy band diagrams are nothing more than these PE/IEP energy spectra. Therefore, it is quite important to calculate these spectra as accurate as possible from first principles. The Green's function method on the basis of the many-body perturbation theory (MBPT) can directly describe the PE/IPE energy spectra, which are usually called the quasiparticle (QP) energies. The number of the papers using this method such as the GW approximation has increased recent years. One difficulty of using this method is that the MBPT and the Green's function are not easy to understand for the researchers who are not familiar to the field theory. Therefore, the Green's function is not at all popular in the fields of quantum chemistry and materials science. Moreover, there is a serious problem related to the energy-dependence of the self-energy, which has been a long standing issue and the QP picture has sometimes been considered as an approximation. And, the main issue here is that the Green's function has been always defined on the neutral ground state only, while the initial electronic state in the PE/IPE spectroscopy can be any of the neutral or charged excited states.

The purpose of this section is to present a simple derivation of the QP theory, which is applicable to any of the initial electronic excited eigenstates $| \Psi_\gamma^M \rangle$ of the

$M$-electron system ($M$ can be different from the total number of protons $N$ in the system) [13] without invoking the complicated field theoretical MBPT. In the course of this derivation, we will show that the energy-dependence of the self-energy is not an essential difficulty [13], and the QP picture holds rigorously if there is no relaxation mechanism such as electron-photon and electron-phonon interactions or cascade damping in the continuum energy bands. (Even when there is such a relaxation mechanism in the system, the QP picture should still hold within the lifetime [586, 587], or equivalently within the mean free path.)

We start from the definition of the QPs. The initial state can be a ground or any excited state of the $N$-electron neutral system or any excited state of the $M(\neq N)$-electron charged system. We write this state as $|\,\Psi_\gamma^M\,\rangle$ and its total energy as $E_\gamma^M$. They satisfy the eigenvalue equation $H\,|\,\Psi_\gamma^M\,\rangle = E_\gamma^M\,|\,\Psi_\gamma^M\,\rangle$. The electrons are interacting each other by the Coulomb interaction, and therefore cannot be treated independently. However, if we introduce the quasiparticle (QP) picture, we have a simple way to treat $|\,\Psi_\gamma^M\,\rangle$ in terms of the QPs. The idea is based on Einstein's photoelectronic effect, which is called in modern terminology the photoelectron or photoemission (PE)/inverse photoemission (IPE) spectroscopy. After the photoemission process, the system becomes a $(M-1)$-electron excited state $|\,\Psi_\mu^{M-1}\,\rangle$, which satisfies the eigenvalue equation $H\,|\,\Psi_\mu^{M-1}\,\rangle = E_\mu^{M-1}\,|\,\Psi_\mu^{M-1}\,\rangle$. Alternatively, after the inverse photoemission process, the system becomes a $(M+1)$-electron excited state, which satisfy $H\,|\,\Psi_\nu^{M+1}\,\rangle = E_\nu^{M+1}\,|\,\Psi_\nu^{M+1}\,\rangle$. The corresponding total energies satisfy

$$E_\mu^{M-1} - E_\gamma^M = h\nu - K, \tag{2.417a}$$
$$E_\gamma^M - E_\nu^{M+1} = h\nu - K, \tag{2.417b}$$

according to the energy conservation law, where $h\nu$ is the energy of the incident photon and $K$ is the kinetic energy of the emitted (or incident) electron. Therefore, these total energy differences are the measurable quantities. We call these energies the QP energies and define them as

$$\varepsilon_\mu = E_\gamma^M - E_\mu^{M-1}, \tag{2.418a}$$
$$\varepsilon_\nu = E_\nu^{M+1} - E_\gamma^M. \tag{2.418b}$$

The QP energy $\varepsilon_\mu$corresponds to the minus of the energy required to remove one electron from the $|\,\Psi_\gamma^M\,\rangle$ state and add it at at the vacuum level; see Fig. 2.19. Similarly, the QP energy $\varepsilon_\nu$ corresponds to the minus of the energy gain when one electron is added to the $|\,\Psi_\gamma^M\,\rangle$ state from the vacuum level.

If $\gamma$ is the ground state, and $\mu$ is the highest occupied molecular orbital (HOMO) level or the valence band maximum (VBM) in the case of crystals, (2.418a) (see also (2.84)) gives the **ionization potential** (IP). Similarly, if $\gamma$ is the ground state, and $\nu$ is the lowest unoccupied molecular orbital (LUMO) levelor the conduction band minimum (CBM) in the case of crystals, (2.418) (see also (2.84)) gives the **electron affinity** (EA).

**Fig. 2.19** Meaning of the quasiparticle (QP) energies

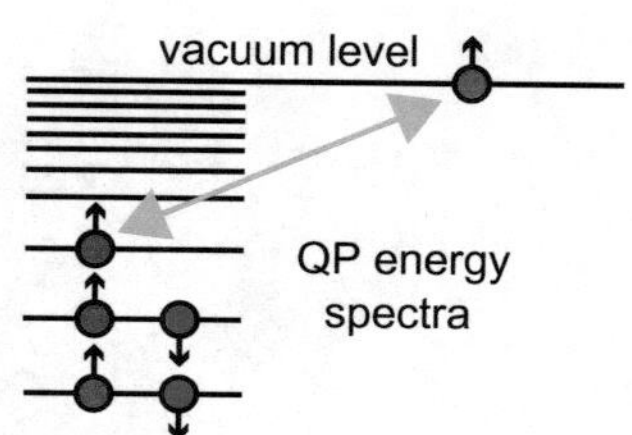


Note that the final states includes many ($M \pm 1$)-electron states, which can be created by adding or removing one electron from the initial state. This does not necessarily mean that the final states include all possible ($M \pm 1$)-electron states. For example, multiple scattered eigenstates where many electron configurations are different from the initial eigenstate are excluded or their amplitudes are quite small even if they are included.

The quasiparticle (QP) wavefunctions can be defined as an overlap between the initial state $|\Psi_\gamma^M\rangle$ with one-electron removed or added at a point $(\boldsymbol{r}, s)$ ( $\boldsymbol{r}$ and $s$ are spacial and spin coordinates of the electron) and the final state $|\Psi_\mu^{M-1}\rangle$ or $|\Psi_\nu^{M+1}\rangle$ (therefore, they are sometimes called the overlap amplitudes):

$$\phi_\mu(\boldsymbol{r}, s) = \langle \Psi_\mu^{M-1} | \hat{\psi}_s(\boldsymbol{r}) | \Psi_\gamma^M \rangle, \tag{2.419a}$$

$$\phi_\nu^*(\boldsymbol{r}, s) = \langle \Psi_\nu^{M+1} | \hat{\psi}_s^\dagger(\boldsymbol{r}) | \Psi_\gamma^M \rangle. \tag{2.419b}$$

Here $\hat{\psi}_s^\dagger(\boldsymbol{r})$ and $\hat{\psi}_s(\boldsymbol{r})$ are the (creation and annihilation) field operators. These QP wavefunctions represent the amplitude of the one-electron difference between the initial and the final states (see Fig. 2.20). The total electron spin density of the $|\Psi_\gamma^M\rangle$ state is given by

$$\begin{aligned}\rho_s(\boldsymbol{r}) &= \langle \Psi_\gamma^M | \hat{\psi}_s^\dagger(\boldsymbol{r}) \hat{\psi}_s(\boldsymbol{r}) | \Psi_\gamma^M \rangle \\ &= \sum_\mu^{\text{occ}} \langle \Psi_\gamma^M | \hat{\psi}_s^\dagger(\boldsymbol{r}) | \Psi_\mu^{M-1} \rangle \langle \Psi_\mu^{M-1} | \hat{\psi}_s(\boldsymbol{r}) | \Psi_\gamma^M \rangle \\ &= \sum_\mu^{\text{occ}} |\phi_\mu(\boldsymbol{r}, s)|^2,\end{aligned} \tag{2.420}$$

where the completeness condition for the eigenstates $|\Psi_\mu^{M-1}\rangle$ of the ($M-1$)-electron Hamiltonian

$$\sum_\mu^{\text{occ}} |\Psi_\mu^{M-1}\rangle\langle \Psi_\mu^{M-1}| = 1 \tag{2.421}$$

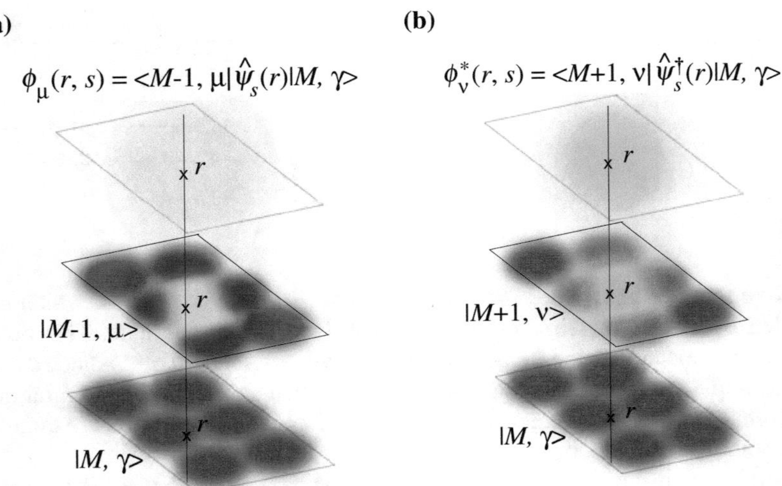


**Fig. 2.20** Meaning of the QP wavefunctions (overlap amplitudes). **a** represents the hole distribution as the difference between the $M$ particle state and the $(M-1)$-particle state, and **b** represents the electron distribution as the difference between the $M$ particle state and the $(M+1)$-particle state

was used. Similarly, the expectation value of the one-body part of the Hamiltonian

$$H^{(1)} = \hat{T} + \hat{v} = \sum_s \int \hat{\psi}_s^\dagger(\boldsymbol{r}) h_s^{(1)}(\boldsymbol{r}) \hat{\psi}_s(\boldsymbol{r}) d\boldsymbol{r},$$

$$h_s^{(1)}(\boldsymbol{r}) = -\frac{\hbar^2}{2m}\nabla^2 + v(\boldsymbol{r}) \tag{2.422}$$

($v(\boldsymbol{r})$ is the local potential such as the Coulomb potential caused by nuclei) are obtained by the QP wavefunctions as follows:

$$\begin{aligned}\langle \hat{T} \rangle &= -\frac{\hbar^2}{2m} \langle \Psi_\gamma^M \mid \sum_s \int \hat{\psi}_s^\dagger(\boldsymbol{r}) \nabla^2 \hat{\psi}_s(\boldsymbol{r}) d\boldsymbol{r} \mid \Psi_\gamma^M \rangle \\ &= -\frac{\hbar^2}{2m} \sum_\mu^{\mathrm{occ}} \sum_s \int \phi_\mu^*(\boldsymbol{r}, s) \nabla^2 \phi_\mu(\boldsymbol{r}, s) d\boldsymbol{r},\end{aligned} \tag{2.423}$$

$$\begin{aligned}\langle \hat{v} \rangle &= \langle \Psi_\gamma^M \mid \sum_s \int \hat{\psi}_s^\dagger(\boldsymbol{r}) v(\boldsymbol{r}) \hat{\psi}_s(\boldsymbol{r}) d\boldsymbol{r} \mid \Psi_\gamma^M \rangle \\ &= \sum_\mu^{\mathrm{occ}} \sum_s \int v(\boldsymbol{r}) \left| \phi_\mu(\boldsymbol{r}, s) \right|^2 d\boldsymbol{r}.\end{aligned} \tag{2.424}$$

In fact, the QP wavefunctions diagonalize the density matrix

$$\rho_s(\boldsymbol{r}', \boldsymbol{r}) = \langle \Psi_\gamma^M \mid \hat{\psi}_s^\dagger(\boldsymbol{r}')\hat{\psi}_s(\boldsymbol{r}) \mid \Psi_\gamma^M \rangle = \sum_\mu^{\text{occ}} \phi_\mu^*(\boldsymbol{r}', s)\phi_\mu(\boldsymbol{r}, s). \tag{2.425}$$

The spin density $\rho_s(\boldsymbol{r})$ is given by $\rho_s(\boldsymbol{r}, \boldsymbol{r})$, and the expectation values of the kinetic and local potential energies in the $\mid \Psi_\gamma^M \rangle$ state are given by

$$\langle \hat{T} \rangle = -\frac{\hbar^2}{2m} \sum_s \int \lim_{\boldsymbol{r}' \to \boldsymbol{r}} \nabla^2 \rho_s(\boldsymbol{r}', \boldsymbol{r}) d\boldsymbol{r}, \tag{2.426}$$

$$\langle \hat{v} \rangle = \sum_s \int v(\boldsymbol{r}) \rho_s(\boldsymbol{r}, \boldsymbol{r}) d\boldsymbol{r}. \tag{2.427}$$

In this sense, the QP wavefunctions $\phi_\mu(\boldsymbol{r}, s)$ have a property similar to the natural spin orbitals introduced by Löwdin more than 60 years ago [43, 588] in connection with the CI theory [27].

Here, it is beneficial to note that the completeness condition

$$\sum_\mu^{\text{occ}} \phi_\mu(\boldsymbol{r}, s)\phi_\mu^*(\boldsymbol{r}', s') + \sum_\nu^{\text{emp}} \phi_\nu(\boldsymbol{r}, s)\phi_\nu^*(\boldsymbol{r}', s')$$
$$= \sum_\lambda^{\text{all}} \phi_\lambda(\boldsymbol{r}, s)\phi_\lambda^*(\boldsymbol{r}', s') = \delta(\boldsymbol{r} - \boldsymbol{r}')\delta_{ss'}. \tag{2.428}$$

holds for the QP states. This can be easily seen by sandwiching the anticommutation relation of the field operators

$$\left\{ \hat{\psi}_s(\boldsymbol{r}), \hat{\psi}_{s'}^\dagger(\boldsymbol{r}') \right\} = \hat{\psi}_s(\boldsymbol{r})\hat{\psi}_{s'}^\dagger(\boldsymbol{r}') + \hat{\psi}_{s'}^\dagger(\boldsymbol{r}')\hat{\psi}_s(\boldsymbol{r}) = \delta(\boldsymbol{r} - \boldsymbol{r}')\delta_{ss'}, \tag{2.429}$$

with $\langle \Psi_\gamma^M \mid$ and $\mid \Psi_\gamma^M \rangle$ and using the other completeness condition for the eigenstates $\mid \Psi_\nu^{M+1} \rangle$ of the $(M+1)$-electron Hamiltonian

$$\sum_\nu^{\text{emp}} \mid \Psi_\nu^{M+1} \rangle \langle \Psi_\nu^{M+1} \mid = 1 \tag{2.430}$$

together with (2.421). On the other hand, the QP wavefunctions are not orthonormal, i.e., not normalized to unity, and even not linearly independent [589, 607], and therefore they are different from the natural spin orbitals in an exact sense.

What should be emphasize here is that there are vast amount of irrelevant $(M \pm 1)$-electron states, which can be ignored from the summation over all empty states or all occupied states in (2.420)–(2.425) and (2.428). For example, one may consider a lot of independent excitations that are not directly associated with the photoemission process under consideration. Even such excited states are the $(M-1)$-electron states, which can be included in the summation over all occupied states. Of course, such states do not contribute to the intermediate states in (2.420)–(2.425) and (2.428), because such multiple excitations cannot occur with the single creation or annihilation operator. This is a clear example of the non-orthonormality: The amplitude for such QP states is zero or almost zero. The simplest way is to simply remove such $(M \pm 1)$-electron states from the QP states. Then, the QP wave functions may satisfy orthonormalty and become quite similar to the natural spin orbitals that diagonalize the density matrix in the CI method.

Now, let us derive the equation satisfied by the QP wavefunctions and the QP energies. Since the Hamiltonian $H$ is given by $H = H^{(1)} + H^{(2)}$ with the one-body part (2.422) and the two-body part, i.e., the electron-electron Coulomb interaction,

$$H^{(2)} = \frac{1}{2}\sum_{ss'}\int \hat{\psi}_s^\dagger(\boldsymbol{r})\hat{\psi}_{s'}^\dagger(\boldsymbol{r}')h_{ss'}^{(2)}(\boldsymbol{r},\boldsymbol{r}')\hat{\psi}_{s'}(\boldsymbol{r}')\hat{\psi}_s(\boldsymbol{r})d\boldsymbol{r}d\boldsymbol{r}',$$

$$h_{ss'}^{(2)}(\boldsymbol{r},\boldsymbol{r}') = \frac{e^2}{4\pi\varepsilon_0}\frac{1}{|\boldsymbol{r}-\boldsymbol{r}'|}, \tag{2.431}$$

the commutation relation between $\hat{\psi}_s(\boldsymbol{r})$ and $H$ is given by

$$[\hat{\psi}_s(\boldsymbol{r}), H] = h_s^{(1)}\hat{\psi}_s(\boldsymbol{r}) + \sum_{s'}\int \hat{\psi}_{s'}^\dagger(\boldsymbol{r}')h_{s's}^{(2)}(\boldsymbol{r}',\boldsymbol{r})\hat{\psi}_{s'}(\boldsymbol{r}')\hat{\psi}_s(\boldsymbol{r})d\boldsymbol{r}'. \tag{2.432}$$

Then, by sandwiching this with $\langle \Psi_\mu^{M-1} |$ and $| \Psi_\gamma^M \rangle$ or with $\langle \Psi_\gamma^M |$ and $| \Psi_\nu^{M+1} \rangle$, and using (2.418) and (2.419), we derive

$$\varepsilon_\mu\phi_\mu(\boldsymbol{r},s) = h_s^{(1)}(\boldsymbol{r})\phi_\mu(\boldsymbol{r},s) + \sum_{s'}\int h_{s's}^{(2)}(\boldsymbol{r}',\boldsymbol{r}) \times \langle \Psi_\mu^{M-1} | \hat{\psi}_{s'}^\dagger(\boldsymbol{r}')\hat{\psi}_{s'}(\boldsymbol{r}')\hat{\psi}_s(\boldsymbol{r}) | \Psi_\gamma^M \rangle d\boldsymbol{r}', \tag{2.433a}$$

$$\varepsilon_\nu\phi_\nu(\boldsymbol{r},s) = h_s^{(1)}(\boldsymbol{r})\phi_\nu(\boldsymbol{r},s) + \sum_{s'}\int h_{s's}^{(2)}(\boldsymbol{r}',\boldsymbol{r}) \times \langle \Psi_\gamma^{M} | \hat{\psi}_{s'}^\dagger(\boldsymbol{r}')\hat{\psi}_{s'}(\boldsymbol{r}')\hat{\psi}_s(\boldsymbol{r}) | \Psi_\nu^{M+1} \rangle d\boldsymbol{r}', \tag{2.433b}$$

Here, the second term in the r.h.s. of these equations are rewritten as

$$\sum_{s'}\int h_{s's}^{(2)}(\boldsymbol{r}',\boldsymbol{r})\langle \Psi_\mu^{M-1} | \hat{\psi}_{s'}^\dagger(\boldsymbol{r}')\hat{\psi}_{s'}(\boldsymbol{r}')\hat{\psi}_s(\boldsymbol{r}) | \Psi_\gamma^M \rangle d\boldsymbol{r}'$$

$$= \int \Sigma_s(\boldsymbol{r},\boldsymbol{r}',E_\gamma^M - E_\mu^{M-1})\langle \Psi_\mu^{M-1} | \hat{\psi}_s(\boldsymbol{r}') | \Psi_\gamma^M \rangle d\boldsymbol{r}', \tag{2.434a}$$

$$\sum_{s'}\int h_{s's}^{(2)}(\boldsymbol{r}',\boldsymbol{r})\langle \Psi_\gamma^{M} | \hat{\psi}_{s'}^\dagger(\boldsymbol{r}')\hat{\psi}_{s'}(\boldsymbol{r}')\hat{\psi}_s(\boldsymbol{r}) | \Psi_\nu^{M+1} \rangle d\boldsymbol{r}'$$

$$= \int \Sigma_s(\boldsymbol{r}, \boldsymbol{r}', E_\nu^{M+1} - E_\gamma^M) \langle \Psi_\gamma^M \mid \hat{\psi}_s(\boldsymbol{r}') \mid \Psi_\nu^{M+1} \rangle d\boldsymbol{r}', \tag{2.434b}$$

by introducing the energy-dependent self-energy $\Sigma_s(\boldsymbol{r}, \boldsymbol{r}'; \varepsilon_\lambda)$ with $\varepsilon_\lambda = \varepsilon_\mu$ or $\varepsilon_\nu$ given by (2.418). (Here and hereafter, the self-energy includes the Hartree term, which describes the classical Coulomb interaction.) It is clear that both identities (2.434a) and (2.434b), which define the self-energy, are consistent with each other, in particular, in the excited-state formulation. In fact, since $\langle \Psi_\mu^{M-1} \mid$ and $\mid \Psi_\gamma^M \rangle$ are arbitrary excited eigenstates in (2.434a), they can be changed to $\langle \Psi_\gamma^M \mid$ and $\mid \Psi_\nu^{M+1} \rangle$, and then (2.434b) is readily obtained. In other words, in order to have the two identities (2.434a) and (2.434b) simultaneously, one must recognize that the ordinary formulation for the self-energy for the ground state should hold for arbitrary excited eigenstates also.

From (2.434a) and (2.434b), the quasiparticle (QP) equations

$$h_s^{(1)}(\boldsymbol{r})\phi_\mu(\boldsymbol{r}, s) + \int \Sigma_s(\boldsymbol{r}, \boldsymbol{r}', \varepsilon_\mu)\phi_\mu(\boldsymbol{r}', s) d\boldsymbol{r}' = \varepsilon_\mu \phi_\mu(\boldsymbol{r}, s), \tag{2.435a}$$

$$h_s^{(1)}(\boldsymbol{r})\phi_\nu(\boldsymbol{r}, s) + \int \Sigma_s(\boldsymbol{r}, \boldsymbol{r}', \varepsilon_\nu)\phi_\nu(\boldsymbol{r}', s) d\boldsymbol{r}' = \varepsilon_\nu \phi_\nu(\boldsymbol{r}, s) \tag{2.435b}$$

are derived. Since there is the energy-dependence in the self-energy, we have to solve these equations level to level. Note that the self-energy has a state dependence only through the dependence on a particular QP energy. In the case of the ground state, it has been demonstrated that the energy dependence can be approximately treated by linearization [586, 587, 592]. However, the energy-dependence is very important in the QP theory. Indeed, multiplying both sides of (2.434a) by $\frac{1}{2}\langle \Psi_\gamma^M \mid \hat{\psi}_s^\dagger(\boldsymbol{r}) \mid \Psi_\mu^{M-1} \rangle = \frac{1}{2}\phi_\mu^*(\boldsymbol{r}, s)$, summing up with respect to $\mu$ and $s$, integrating with respect to $\boldsymbol{r}$, and using the completeness condition (2.421) again, we obtain

$$\begin{aligned} E_{\text{int}} &= \frac{1}{2}\sum_{ss'} \int h_{ss'}^{(2)}(\boldsymbol{r}, \boldsymbol{r}') \langle \Psi_\gamma^M \mid \hat{\psi}_s^\dagger(\boldsymbol{r}) \hat{\psi}_{s'}^\dagger(\boldsymbol{r}') \hat{\psi}_{s'}(\boldsymbol{r}') \hat{\psi}_s(\boldsymbol{r}) \mid \Psi_\gamma^M \rangle d\boldsymbol{r} d\boldsymbol{r}' \\ &= \frac{1}{2}\sum_\mu^{\text{occ}} \sum_s \int \phi_\mu^*(\boldsymbol{r}, s) \Sigma_s(\boldsymbol{r}, \boldsymbol{r}'; \varepsilon_\mu) \phi_\mu(\boldsymbol{r}', s) d\boldsymbol{r} d\boldsymbol{r}' \end{aligned} \tag{2.436}$$

for the expectation value of the electron-electron Coulomb interaction for the $\mid \Psi_\gamma^M \rangle$ state, which is known as the Galitskii and Migdal formula [593],

$$E_{\text{int}} = -\frac{1}{2} \operatorname{Tr} \int \Sigma_s(\boldsymbol{r}, \boldsymbol{r}'; \omega) G_s(\boldsymbol{r}', \boldsymbol{r}; \omega) d\boldsymbol{r}', \tag{2.437}$$

$$\operatorname{Tr} = i \sum_s \int d\boldsymbol{r} \int_{-\infty}^{\infty} e^{i\eta\omega} \frac{d\omega}{2\pi}, \quad \text{with } \eta = 0^+, \tag{2.438}$$

in the QP representation [594] for the case of the ground state. The crucial points here are that the energy dependence of the self-energy appears in the same way in both (2.435a) and (2.436) in a general scheme of the non-orthonormal QP wavefunctions and that (2.436) holds not only for the ground state but also for arbitrary excited eigenstates. Equation (2.436) implies that the information of the self-energy is enough to determine the total energy $E_\gamma^M$ of the $|\,\Psi_\gamma^M\,\rangle$ state of the system, because the QP wavefunctions are uniquely obtained by solving the QP equation (2.435). In other words, if the self-energy is known, the QP energies and QP wavefunctions can be calculated by solving the QP equation (2.435) one by one, and then the electron spin density $\rho_s(\boldsymbol{r})$, the kinetic energy $\langle\hat{T}\rangle$, the local potential energy $\langle\hat{v}\rangle$, and the interaction energy $E_{\rm int}$ are obtained via (2.420), (2.423), (2.424), and (2.436), respectively. Consequently the total energy $E_\gamma^M$ is obtained from the knowledge of the self-energy only. How to determine the self-energy will be described in Sects. 2.7.3 and 2.7.5. In next section (Sect. 2.7.2), we will show that the equations derived in Sect. 2.7.1 can be rederived by using the Green's function.

### 2.7.2 Green's Function Formalism

The equations (2.434) and (2.435) can be also derived from the Dyson equation for the Green's function as follows. The Green's function $G$ on the eigenstate $|\,\Psi_\gamma^M\,\rangle$ is defined as

$$G_s(x,x') = -i\,\langle\,\Psi_\gamma^M\,|\,T[\hat{\psi}_s(x)\hat{\psi}_s^\dagger(x')]\,|\,\Psi_\gamma^M\,\rangle, \tag{2.439}$$

where the convention to write $x=(\boldsymbol{r},t)$ is used; $T$ denotes the time-ordered product, and $\hat{\psi}_s^\dagger(x) = e^{iHt}\hat{\psi}_s^\dagger(\boldsymbol{r})e^{-iHt}$ and $\hat{\psi}_s(x) = e^{iHt}\hat{\psi}_s(\boldsymbol{r})e^{-iHt}$ are, respectively, the creation and annihilation operators in the Heisenberg representation. By means of the completeness conditions (2.421) and (2.430),

$$\begin{aligned} G_s(x,x') &= i\sum_\mu^{\rm occ} \phi_\mu(x,s)\,\phi_\mu^*(x',s)\,\theta(t'-t) \\ &\quad - i\sum_\nu^{\rm emp} \phi_\nu(x,s)\,\phi_\nu^*(x',s)\,\theta(t-t') \end{aligned} \tag{2.440}$$

is readily derived. Here we put

$$\phi_\mu(x,s) = \langle\,\Psi_\mu^{M-1}\,|\,\hat{\psi}_s(x)\,|\,\Psi_\gamma^M\,\rangle = \phi_\mu(\boldsymbol{r},s)e^{-i\varepsilon_\mu t}, \tag{2.441a}$$

$$\phi_\nu(x,s) = \langle\,\Psi_\gamma^M\,|\,\hat{\psi}_s(x)\,|\,\Psi_\nu^{M+1}\,\rangle = \phi_\nu(\boldsymbol{r},s)e^{-i\varepsilon_\nu t}, \tag{2.441b}$$

and $\phi_\lambda(\boldsymbol{r},s)$ and $\varepsilon_\lambda$ ($\lambda=\mu$ or $\nu$) are the QP wavefunctions and the QP energies defined in (2.418) and (2.419).

The Heisenberg equation of motion for the annihilation operator is given by

$$i\frac{\partial}{\partial t}\hat{\psi}_s(x) = [\hat{\psi}_s(x), H] = h^{(1)}(\boldsymbol{r})\,\hat{\psi}_s(x)$$
$$+\sum_{s''}\int h^{(2)}_{s''s}(\boldsymbol{r}'', \boldsymbol{r})\left[\hat{\psi}^{\dagger}_{s''}(x'')\hat{\psi}_{s'}(x'')\right]_{t''=t} d\boldsymbol{r}''\,\hat{\psi}_s(x). \tag{2.442}$$

Operating both sides $\hat{\psi}^{\dagger}_s(x')$ from the right and the time-ordered operator $T$ from the left, and sandwiching them with $\langle\,\Psi^M_\gamma\mid$ and $\mid\Psi^M_\gamma\,\rangle$, we have

$$\left[i\frac{\partial}{\partial t} - h^{(1)}_s(\boldsymbol{r})\right] G_s(x, x') = \delta(x - x') + i\sum_{s''}\int h^{(2)}_{s''s}(\boldsymbol{r}'', \boldsymbol{r})$$
$$\times\langle\,\Psi^M_\gamma\mid T\big[\hat{\psi}_s(x)\hat{\psi}_{s''}(x'')\hat{\psi}^{\dagger}_{s''}(x''^{+})\hat{\psi}^{\dagger}_s(x')\big]\mid\Psi^M_\gamma\,\rangle\big|_{t''=t}\,d\boldsymbol{r}'' \tag{2.443}$$

where $x''^{+}$ with $t'' = t$ means $(\boldsymbol{r}'', t^{+})$ with $t^{+} = t+0^{+}$. The self-energy is introduced as

$$\int \Sigma_s(x, x'')G_s(x'', x')dx'' = i\sum_{s''}\int h^{(2)}_{s''s}(\boldsymbol{r}'', \boldsymbol{r})$$
$$\times\langle\,\Psi^M_\gamma\mid T\big[\hat{\psi}_s(x)\hat{\psi}_{s''}(x'')\hat{\psi}^{\dagger}_{s''}(x''^{+})\hat{\psi}^{\dagger}_s(x')\big]\mid\Psi^M_\gamma\,\rangle\big|_{t''=t}\,d\boldsymbol{r}'', \tag{2.444}$$

and then the Dyson equation

$$\left[i\frac{\partial}{\partial t} - h^{(1)}_s(\boldsymbol{r})\right] G_s(x, x') = \delta(x - x') + \int \Sigma_s(x, x'')G_s(x'', x')dx'' \tag{2.445}$$

holds for any eigenstate. Equation (2.445) is rewritten in the Fourier space as

$$\left[\omega - h^{(1)}_s(\boldsymbol{r})\right] G_s(\boldsymbol{r}, \boldsymbol{r}'; \omega) = \delta(\boldsymbol{r} - \boldsymbol{r}'') + \int \Sigma_s(\boldsymbol{r}, \boldsymbol{r}''; \omega)G_s(\boldsymbol{r}'', \boldsymbol{r}'; \omega)d\boldsymbol{r}''. \tag{2.446}$$

The Fourier transform of the Green's function (2.440) with (2.441) is given by

$$G_s(\boldsymbol{r}, \boldsymbol{r}'; \omega) = \sum_{\lambda}^{\text{all}} \frac{\phi_\lambda(\boldsymbol{r}, s)\,\phi^*_\lambda(\boldsymbol{r}', s)}{\omega - \varepsilon_\lambda - i\delta_\lambda}, \tag{2.447}$$

where $\delta_\lambda = 0^{+}$ for occupied states and $\delta_\lambda = -0^{+}$ for empty states, i.e., $\delta_\mu = 0^{+}$ and $\delta_\nu = -0^{+}$. Inserting this into (2.446) and approaching $\omega$ to one of the $\varepsilon_\lambda$'s ($\lambda$ can be either $\nu$ (emp) or $\mu$ (occ)), we obtain the QP equation

$$\left[\varepsilon_\lambda - h^{(1)}_s(\boldsymbol{r})\right]\phi_\lambda(\boldsymbol{r}, s) - \int \Sigma_s(\boldsymbol{r}, \boldsymbol{r}'; \varepsilon_\lambda)\phi_\lambda(\boldsymbol{r}', s)d\boldsymbol{r}' = 0, \tag{2.448}$$

which is equivalent to (2.435a) and (2.435b). If the eigenvalues are $m$-fold degenerate, one can choose the corresponding $m$ QP wavefunctions to be orthogonal to each other in this degenerate subspace. Then, substituting (2.447) with $\omega = \varepsilon_\lambda$ into (2.446), multiplying the resulting equation by the complex conjugate of one of the $m$ QP wavefunctions $\phi_\lambda^*(\mathbf{r}', s)$, and integrating with respect $\mathbf{r}'$, we derive (2.448).

Next, we derive a concrete method to calculate the self-energy $\Sigma_s$. Because here we discuss the energies at an instant time only, even when the system is slowly varying in time, we simply drop all the time dependence in the following expressions.

First, by the equations

$$\hat{\psi}_s(\mathbf{r}) \mid \Psi_\gamma^M \rangle = \sum_\mu^{\text{occ}} \mid \Psi_\mu^{M-1} \rangle \langle \Psi_\mu^{M-1} \mid \hat{\psi}_s(\mathbf{r}) \mid \Psi_\gamma^M \rangle$$

$$= \sum_\mu^{\text{occ}} \mid \Psi_\mu^{M-1} \rangle \phi_\mu(\mathbf{r}, s) = \sum_\mu^{\text{occ}} a_\mu \mid \Psi_\gamma^M \rangle \phi_\mu(\mathbf{r}, s), \tag{2.449a}$$

$$\hat{\psi}_s^\dagger(\mathbf{r}) \mid \Psi_\gamma^M \rangle = \sum_\nu^{\text{emp}} \mid \Psi_\nu^{M+1} \rangle \langle \Psi_\nu^{M+1} \mid \hat{\psi}_s^\dagger(\mathbf{r}) \mid \Psi_\gamma^M \rangle$$

$$= \sum_\nu^{\text{emp}} \mid \Psi_\nu^{M+1} \rangle \phi_\nu^*(\mathbf{r}, s) = \sum_\nu^{\text{emp}} a_\nu^\dagger \mid \Psi_\gamma^M \rangle \phi_\nu^*(\mathbf{r}, s), \tag{2.449b}$$

and $a_\nu \mid \Psi_\gamma^M \rangle = a_\mu^\dagger \mid \Psi_\gamma^M \rangle = 0$ for an empty QP state $\nu$ and an occupied QP state $\mu$, we introduce the annihilation and creation operators, $a_\lambda$ and $a_\lambda^\dagger$, as

$$\hat{\psi}_s(\mathbf{r}) = \sum_\lambda \phi_\lambda(\mathbf{r}, s) a_\lambda, \tag{2.450a}$$

$$\hat{\psi}_{s'}^\dagger(\mathbf{r}') = \sum_{\lambda'} \phi_{\lambda'}^*(\mathbf{r}', s') a_{\lambda'}^\dagger, \tag{2.450b}$$

where $\lambda$, $\lambda'$ includes both $\mu$ and $\nu$. Then, the anticommutation relations of the field operators are written as

$$\left\{\hat{\psi}_s(\mathbf{r}), \hat{\psi}_{s'}^\dagger(\mathbf{r}')\right\} = \sum_{\lambda\lambda'} \phi_\lambda(\mathbf{r}, s)\phi_{\lambda'}^*(\mathbf{r}', s') \left\{a_\lambda, a_{\lambda'}^\dagger\right\}$$

$$= \delta(\mathbf{r} - \mathbf{r}')\delta_{ss'}, \tag{2.451a}$$

$$\left\{\hat{\psi}_s(\mathbf{r}), \hat{\psi}_{s'}(\mathbf{r}')\right\} = \sum_{\lambda\lambda'} \phi_\lambda(\mathbf{r}, s)\phi_{\lambda'}(\mathbf{r}', s') \{a_\lambda, a_{\lambda'}\} = 0, \tag{2.451b}$$

$$\left\{\hat{\psi}_s^\dagger(\mathbf{r}), \hat{\psi}_{s'}^\dagger(\mathbf{r}')\right\} = \sum_{\lambda\lambda'} \phi_\lambda^*(\mathbf{r}, s)\phi_{\lambda'}^*(\mathbf{r}', s') \left\{a_\lambda^\dagger, a_{\lambda'}^\dagger\right\} = 0. \tag{2.451c}$$

From these equations, we can assume that the QP annihilation and creation operators satisfy the anticommutation relations

$$\left\{a_\lambda, a_{\lambda'}^\dagger\right\} = \delta_{\lambda\lambda'}, \quad \{a_\lambda, a_{\lambda'}\} = 0, \quad \left\{a_\lambda^\dagger, a_{\lambda'}^\dagger\right\} = 0. \tag{2.452}$$

Note that these relations hold even when the QP states $\lambda$, $\lambda'$ are not orthogonal to each other, because (2.451c) with (2.452) satisfies the completeness condition (2.428) only.

Second, we introduce a hypothetical non-interacting system, where the QPs are not interacting each other, although in real systems the QPs generally interact with each other at least at short distances. This hypothetical system is described by the Hamiltonian

$$H' = H^{(1)} + H^{(2)}_{\text{self}} \tag{2.453}$$

where $H^{(2)}_{\text{self}}$ is given by (2.431) whose operation is restricted to construct the self-energy $\Sigma_s$ only and ineffective for any other interaction. Its eigenvalue equation is given by

$$H' \mid \Phi_\alpha^{M+n} \rangle = E'^{M+n}_\alpha \mid \Phi_\alpha^{M+n} \rangle \tag{2.454}$$

where $M+n$ indicates the total number of electrons in this system. When $n = 0$ and $\alpha = \gamma$, we assume that $\mid \Phi_\gamma^M \rangle$ has the same energy eigenvalue $E'^M_\gamma = E^M_\gamma$ as the true system. All the other states of the hypothetical system are constructed by operating several $a_\lambda$'s and $a_\lambda^\dagger$'s on $\mid \Phi_\gamma^M \rangle$. We assume $H'$ as the unperturbed Hamiltonian and

$$H'' = H^{(2)} - H^{(2)}_{\text{self}} \tag{2.455}$$

as a perturbation, and adopt Brillouin–Wigner's perturbation theory, which is well known to be applicable not only to the ground state but also to any excited eigenstate of the full Hamiltonian. Multiplying both sides of

$$\left(E_\gamma^M - H'\right) \mid \Psi_\gamma^M \rangle = H'' \mid \Psi_\gamma^M \rangle, \tag{2.456}$$

by the projection operator

$$P = 1 - \mid \Phi_\gamma^M \rangle\langle \Phi_\gamma^M \mid, \tag{2.457}$$

and noting that $H'$ commutes with $P$, we have

$$\left(E_\gamma^M - H'\right) P \mid \Psi_\gamma^M \rangle = P H'' \mid \Psi_\gamma^M \rangle. \tag{2.458}$$

Equation (2.458) can be rewritten as

$$\mid \Psi_\gamma^M \rangle = \mid \Phi_\gamma^M \rangle\langle \Phi_\gamma^M \mid \Psi_\gamma^M \rangle + \frac{1}{E_\gamma^M - H'} P H'' \mid \Psi_\gamma^M \rangle. \tag{2.459}$$

Then, by iteration, we have

$$\frac{|\,\Psi_\gamma^M\,\rangle}{\langle\,\Phi_\gamma^M\,|\,\Psi_\gamma^M\,\rangle} = |\,\Phi_\gamma^M\,\rangle + \frac{1}{E_\gamma^M - H'}PH''\,|\,\Phi_\gamma^M\,\rangle + \frac{1}{E_\gamma^M - H'}PH''\frac{1}{E_\gamma^M - H'}PH''\,|\,\Phi_\gamma^M\,\rangle + \cdots . \tag{2.460}$$

So, the expectation value of $A(\boldsymbol{r},\boldsymbol{r}') = \hat{\psi}_s^\dagger(\boldsymbol{r})\hat{\psi}_{s'}^\dagger(\boldsymbol{r}')\hat{\psi}_{s'}(\boldsymbol{r}')\hat{\psi}_s(\boldsymbol{r})$ can be evaluated as

$$\langle A(\boldsymbol{r},\boldsymbol{r}')\rangle = \langle\,\Phi_\gamma^M\,|\,A(\boldsymbol{r},\boldsymbol{r}')\,|\,\Phi_\gamma^M\,\rangle + \langle\,\Phi_\gamma^M\,|\,A(\boldsymbol{r},\boldsymbol{r}')\frac{1}{E_\gamma^M - H'}PH^{(2)}\,|\,\Phi_\gamma^M\,\rangle + \langle\,\Phi_\gamma^M\,|\,A(\boldsymbol{r},\boldsymbol{r}')\frac{1}{E_\gamma^M - H'}PH^{(2)}\frac{1}{E_\gamma^M - H'}PH^{(2)}\,|\,\Phi_\gamma^M\,\rangle + \cdots , \tag{2.461}$$

where we replaced $H''$ with $H^{(2)}$ without restriction because it is obvious that these interactions are not related to the construction of the self-energy but simply related to the construction of the expectation value of $A(\boldsymbol{r},\boldsymbol{r}')$. Then, multiplying the both sides of (2.461) by $\frac{1}{2}h_{ss'}^{(2)}(\boldsymbol{r},\boldsymbol{r}')$, summing over $s, s'$, and integrating with respect to $\boldsymbol{r}$ and $\boldsymbol{r}'$, we have

$$E_{\text{int}} = \langle\,\Phi_\gamma^M\,|\,H^{(2)}\,|\,\Phi_\gamma^M\,\rangle + \langle\,\Phi_\gamma^M\,|\,H^{(2)}\frac{1}{E_\gamma^M - H'}PH^{(2)}\,|\,\Phi_\gamma^M\,\rangle + \langle\,\Phi_\gamma^M\,|\,H^{(2)}\frac{1}{E_\gamma^M - H'}PH^{(2)}\frac{1}{E_\gamma^M - H'}PH^{(2)}\,|\,\Phi_\gamma^M\,\rangle + \cdots . \tag{2.462}$$

In the $k$th order term in this perturbation series, there are $4(k+1)$ field operators sandwiched by $\langle\,\Phi_\gamma^M\,|$ and $|\,\Phi_\gamma^M\,\rangle$, which can be expanded as (2.450a) and (2.450b). For example, the first term in the r.h.s. of (2.462) is expressed as

$$\frac{1}{2}\sum_{ss'}\sum_{\mu_1\mu_2\mu_3\mu_4}^{\text{occ}}\int h_{ss'}^{(2)}(\boldsymbol{r},\boldsymbol{r}')\phi_{\mu_1}^*(\boldsymbol{r},s)\phi_{\mu_2}^*(\boldsymbol{r}',s')\phi_{\mu_3}(\boldsymbol{r}',s')\phi_{\mu_4}(\boldsymbol{r},s)d\boldsymbol{r}d\boldsymbol{r}' \times \langle\,\Phi_\gamma^M\,|\,a_{\mu_1}^\dagger a_{\mu_2}^\dagger a_{\mu_3}a_{\mu_4}\,|\,\Phi_\gamma^M\,\rangle. \tag{2.463}$$

In this matrix element, the QP states $\mu_4$ and $\mu_3$ are removed from and $\mu_2$ and $\mu_1$ are added to the $M$-particle state $|\,\Phi_\gamma^M\,\rangle$, and after these operations, the $M$-particle state must be restored to the original state $|\,\Phi_\gamma^M\,\rangle$; otherwise this matrix element becomes zero. That is, either ($\mu_1 = \mu_4$ and $\mu_2 = \mu_3$) or ($\mu_1 = \mu_3$ and $\mu_2 = \mu_4$) should be satisfied. Of course, this first order approximation rigorously corresponds to the Hartree–Fock energy in the ordinary or extended [595] Hartree–Fock approximation. In fact, (2.463) becomes

**Fig. 2.21** Diagrammatic representation of the interaction energy $E_{\mathrm{int}}$

$$E_{\mathrm{int}}^{\mathrm{HF}} = \frac{1}{2}\sum_{ss'}\sum_{\mu\mu'}^{\mathrm{occ}}\int h_{ss'}^{(2)}(\boldsymbol{r},\boldsymbol{r}')\left|\phi_\mu(\boldsymbol{r},s)\right|^2\left|\phi_{\mu'}(\boldsymbol{r}',s')\right|^2 d\boldsymbol{r}d\boldsymbol{r}'$$

$$-\frac{1}{2}\sum_{s}\sum_{\mu\mu'}^{\mathrm{occ}}\int h_{ss}^{(2)}(\boldsymbol{r},\boldsymbol{r}')\phi_\mu^*(\boldsymbol{r},s)\phi_{\mu'}^*(\boldsymbol{r}',s)\phi_\mu(\boldsymbol{r}',s)\phi_{\mu'}(\boldsymbol{r},s)d\boldsymbol{r}d\boldsymbol{r}'. \quad (2.464)$$

Its analogy holds for all orders. That is, all the intermediate QP states $\lambda_1, \lambda_2, \lambda_3, \lambda_4, \cdots$ must be pairwise ($\lambda_i$ can be either occupied states $\mu_i$ or empty states $\nu_i$); each pair consists of one annihilation operator $a_{\lambda_i}$ and one creation operator $a_{\lambda_i}^\dagger$ of the same QP state $\lambda_i$. All possible combinations of the product of the pairs have to be summed up to evaluate the total contribution at this order. Because of the existence of the projection operator $P$ in (2.462), the intermediate states cannot be the initial state $\mid \Phi_\gamma^M \rangle$. Moreover, the energy denominator in (2.462) is just a simple addition or subtraction of the QP energies $\varepsilon_\lambda$'s. If we draw a pair of the QP state by a thick sold line with an arrow and $H^{(2)}$ by a dotted line, (2.462) is expressed by the diagrams shown in Fig. 2.21. This expression is exactly the same as the interaction energy expressed by the skeleton diagrams, i.e., the diagrams expressed by the clothed Green's function (a thick solid line in Fig. 2.21), in the usual MBPT [17, 20, 596, 597]. They also correspond to the same order contribution in the Møller–Plesset theory if the initial excited-state configuration and the full self-consistency are imposed in the latter. Thus we derived exactly the same formula only by using the true eigenstates of the full Hamiltonian and their derivatives such as the QP energies and QP wavefunctions. An important point here is that we have never assumed the orthogonality of the QP wavefunctions in these equations.

Third, in order to obtain the expression for the self-energy $\Sigma_s$, we compare the expression (2.462) with the Galitskii and Migdal formula of the interaction energy (2.436). Looking at (2.436), we find that if we drop $\phi_\mu(\boldsymbol{r},s)\phi_\mu^*(\boldsymbol{r}',s)$ from (2.462), i.e., a thick solid line from Fig. 2.21, we should have the self-energy $\Sigma_s(\boldsymbol{r},\boldsymbol{r}';\varepsilon_\mu)$ equivalent to Fig. 2.22. In Fig. 2.22, we drew two external thin solid lines to indicate the dependence of the self-energy $\Sigma_s$ on the two positions $(\boldsymbol{r},s)$ and $(\boldsymbol{r}',s)$. This is again exactly the same skeleton diagrams for the self-energy in the MBPT [20, 597]. When we remove a pair of $\phi_\mu(\boldsymbol{r},s)\phi_\mu^*(\boldsymbol{r}',s)$ from the product of $H^{(2)}$'s in the expression of $E_{\mathrm{int}}$ (2.462), we also have to remove the summation with respect to this special QP state, $\mu$. Therefore, the corresponding QP energy $\varepsilon_\mu$ appearing in some parts of the energy denominators $(E_\gamma^M - H')^{-1}$ remains unsummed with the special index $\mu$ after taking the summations with respect to all the intermediate QP states, resulting in the $\varepsilon_\mu$ dependence in the final expression. Consequently, the resulting

**Fig. 2.22** Diagrammatic representation of the self-energy $\Sigma_s$. Two external thin solid lines are additionally drawn to indicate the dependence of the self-energy on the two positions $(\boldsymbol{r}, s)$ and $(\boldsymbol{r}', s)$.

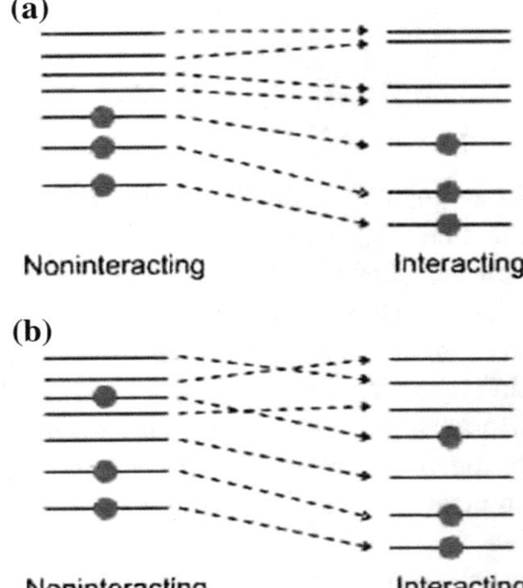


**Fig. 2.23** **a** Adiabatic switch on process for the ground state. Level crossing is not allowed between the occupied and empty QP levels. **b** Gradual switch on process for an excited state. Occupied and empty QP levels can intersect each other and change the order

self-energy must have the explicit $\varepsilon_\mu$ dependence, where $\mu$ corresponds to the QP state of the external thin line in Fig. 2.22.

Before ending this section, several important remarks on the excited eigenstates are given below [13].

(1) First of all, it is easy to see that the virial theorem holds for arbitrary excited eigenstates if the nuclear coordinates are fully relaxed as explained in Sect. 2.2.1.

(2) In the usual MBPT, the adiabatic switch on process in the MBPT can be extended to excited eigenstates as written in Sect. 5–4 of [597]; see also Sect. VI of [13]. In fact, Wick's theorem holds also for an excited eigenstate if we transform creation and annihilation operators from the coordinate representation to the one-particle state representation of the corresponding noninteracting ($H^{(2)} = 0$) system, exchange the role of creation and annihilation operators for occupied and empty states of the excited eigenstate, and introduce the normal product as usual. When the interaction $H^{(2)}$ is gradually switched on, the QP levels may apparently change the order. However, this does not matter at all, because, in the case of an excited state, it is not necessary to require that the noninteracting ground state moves onto the interacting ground state without level crossing. Even if the order of occupied and empty levels changes during the switch on process, we can keep track of each

QP state; see Fig. 2.23. In the noninteracting system, the QP states refer to the one-particle states. There is a one-to-one correspondence between the one-particle states of the noninteracting system and the QP states of the interacting systems, although the orders of these states may be different. There is no constraint on the orders of the one-particle states and the QP states, and they can be different to each other. Therefore, we should not use the standard words "the adiabatic switch on process" for this process because the word "adiabatic" involves the meaning of no level crossing. Instead, it would be better to use the words "the gradual switch on process" to avoid confusion. The gradual switch on of the interaction from time $t = -\infty$ to $t = 0$ transforms the noninteracting excited state (at $t = -\infty$) into the interacting excited state (at $t = 0$).

(3) The total energy of the system at a ground state or an excited eigenstate can be expressed according to Klein's derivation (see, e.g., Sect. 5–6 in [597]) as follows:

$$E_\gamma^M[G] - E_\gamma^M[G^0] = \Phi[G] + \mathrm{Tr}\left[\frac{G_s}{G_s^0} - 1 - \ln\frac{G_s}{G_s^0}\right], \tag{2.465}$$

where $G_s^0$ is the Green's function of the non-interacting system, Tr is defined in (2.438), and $\Phi[G]$ is the Luttinger–Ward functional [587, 597, 598] defined as

$$\Phi[G] = -\sum_{k=1}^{\infty}\frac{1}{2k}\,\mathrm{Tr}\int d\boldsymbol{r}'\,\Sigma_s^{(k)}(\boldsymbol{r},\boldsymbol{r}';\omega)G_s(\boldsymbol{r}',\boldsymbol{r};\omega), \tag{2.466}$$

where $\Sigma_s^{(k)}(\boldsymbol{r},\boldsymbol{r}';\omega)$ represents the $k$th order skeleton diagram of the self-energy. In (2.465), $\mathrm{Tr}[G_s/G_s^0 - 1]$ is equal to $\mathrm{Tr}[\Sigma_s G_s] = -2E_{\mathrm{int}}$ (see the Galitsukii–Migdal formula (2.438) and the Dyson equation (2.446) with $G_s^{0\,-1} = \omega - h_s^{(1)}$), and $-\mathrm{Tr}\ln G_s/G_s^0$ can be evaluated as $\sum_\mu^{\mathrm{occ}}\varepsilon_\mu - E_\gamma^M[G^0]$ and $E_\gamma^M[G^0] = \sum_\mu^{\mathrm{occ}}\varepsilon_\mu^0$ with the eigenvalue of the non-interacting system $\varepsilon_\mu^0$. We have also the variational principles $\delta\Phi[G]/\delta G_s = -\Sigma_s$ and $\delta E_\gamma^M[G]/\delta G_s = 0$. [597, 598]

(4) The Ward–Takahashi identity [599–601], i.e. gauge invariance,

$$\delta(x-z)\,G_s^{-1}(z,y) - G_s^{-1}(x,z)\,\delta(z-y) = i\sum_{\alpha=0}^{4}\frac{\partial}{\partial z_\alpha}\Gamma_s^{(\alpha)}(x,y,z), \tag{2.467}$$

which is equivalent to the continuity equation for the electron density, is guaranteed also for the excited eigenstate. Here $\Gamma_s^{(\alpha)}(x,y,z)$ ($\alpha = 1-4$) are the scalar ($\alpha = 0$) and vector ($\alpha = 1, 2, 3$) vertex functions defined as

$$\langle\,\Phi_\gamma^M\,|\,T[j_\alpha(z)\psi_s(x)\psi_s^\dagger(y)]\,|\,\Phi_\gamma^M\,\rangle = -\int G_s(x,x')\Gamma_s^{(\alpha)}(x',y',z)G_s(y',y)dx'dy', \tag{2.468}$$

with the four-current density $j_\alpha(x)$ defined as

$$j_0(x) = \rho(x) = \sum_s \psi_s^\dagger(x)\psi_s(x), \tag{2.469}$$

$$\boldsymbol{j}(x) = -\rho(x)\boldsymbol{A}(x) - \frac{i\hbar}{2m}\sum_s \left\{\psi_s^\dagger(x)\nabla\psi_s(x) - [\nabla\psi_s^\dagger(x)]\psi_s(x)\right\} \tag{2.470}$$

in the existence of the vector potential $\boldsymbol{A}(x)$.

(5) For the discussion of transport properties, it is important to guarantee the macroscopic conservation laws of the total energy, the total number of electrons, the total momentum, and so on. These macroscopic conservation laws have been rigorously discussed by Baym and Kadanoff [602, 603] for the ground state. Their argument may be generalized to the case of an arbitrary excited eigenstate, and the macroscopic conservation laws hold for arbitrary excited states, when the necessary symmetries suggested by Baym and Kadanoff are satisfied for the excited-state Green's functions. In general, however, we expect that the total number of electrons is conserved in our excited-state theory because the local charge conservation is guaranteed by the Ward–Takahashi identity (2.467). Moreover, we expect that the total energy and the total momentum of the electronically excited many-atom systems should be conserved, if the system is isolated. This is simply due to the time-reversal symmetry and the translational symmetry in space of the total Hamiltonian of the many-atom system.

(6) However, the fluctuation-discipation theorem is not guaranteed for arbitrary excited states because it is based on the canonical ensemble, i.e. the thermal equilibrium, at a given temperature. Therefore, the standard linear response theory cannot be applied to arbitrary excited eigenstates without some necessary modifications.

(7) Last, we discuss how DFT is modified for the excited eigenstate. Similar to the ordinary DFT, we introduce a universal function $F$ defined as

$$F = \langle\, \Phi_\gamma^M \mid H_{\text{eg}} \mid \Phi_\gamma^M \,\rangle, \tag{2.471}$$

where $H_{\text{eg}}$ is the Hamiltonian of the interacting electron gas. By definition, $F$ is a unique functional of the eigenstate $\mid \Phi_\gamma^M \rangle$ of the full Hamiltonian $H$. This statement is a direct consequence of the Luttinger–Ward theory, because $F$ is expressed as a unique functional of the Green's function, which is by definition (see (2.439)) a unique functional of the eigenstate $\mid \Phi_\gamma^M \rangle$. To see that $\mid \Phi_\gamma^M \rangle$ and $F$ are unique functionals of the electron density $\rho(\boldsymbol{r})$ and the occupation numbers of all KS levels, we use the *reductio ad absurdum* along the same line as the Hohenberg–Kohn's theory [60]. Assume that two different local potentials $v(\boldsymbol{r})$ and $v'(\boldsymbol{r})$ give the same electron density $\rho(\boldsymbol{r})$, and write the Hamiltonians with the local potentials $v(\boldsymbol{r})$ and $v'(\boldsymbol{r})$ as $H$ and $H'$, respectively. Their eigenvalues and eigenstates are $E_\gamma^M$, $E'^M_\gamma$, $\mid \Psi_\gamma^M \rangle$, and $\mid \Psi'^M_\gamma \rangle$, which satisfy the eigenvalue equations $H \mid \Psi_\gamma^M \rangle = E_\gamma^M \mid \Psi_\gamma^M \rangle$ and $H' \mid \Psi'^M_\gamma \rangle = E'^M_\gamma \mid \Psi'^M_\gamma \rangle$. The excited eigenstate is orthogonal to all lower excited eigenstates as well as the ground state. Therefore, we still have the variational principle such that $E_\gamma^M$ (or $E'^M_\gamma$) has the minimum value when $\mid \Psi_\gamma^M \rangle$ (or $\mid \Psi'^M_\gamma \rangle$) becomes the true excited eigenstate inside the Fock space orthogonal to all lower excited eigenstates and the ground state. Inside this Fock space, we derive [604]

$$\begin{aligned} E'^{M}_{\gamma} &= \langle \Psi'^{M}_{\gamma} \mid H' \mid \Psi'^{M}_{\gamma} \rangle < \langle \Psi^{M}_{\gamma} \mid H' \mid \Psi^{M}_{\gamma} \rangle \\ &= \langle \Psi^{M}_{\gamma} \mid H \mid \Psi^{M}_{\gamma} \rangle + \int [v'(\boldsymbol{r}) - v(\boldsymbol{r})]\, \rho(\boldsymbol{r})\, d\boldsymbol{r} \\ &= E^{M}_{\gamma} + \int [v'(\boldsymbol{r}) - v(\boldsymbol{r})]\, \rho(\boldsymbol{r})\, d\boldsymbol{r} \end{aligned} \tag{2.472a}$$

$$\begin{aligned} E^{M}_{\gamma} &= \langle \Psi^{M}_{\gamma} \mid H \mid \Psi^{<}_{\gamma} \rangle < \langle \Psi'^{M}_{\gamma} \mid H \mid \Psi'^{M}_{\gamma} \rangle \\ &= \langle \Psi'^{M}_{\gamma} \mid H' \mid \Psi'^{M}_{\gamma} \rangle + \int [v(\boldsymbol{r}) - v'(\boldsymbol{r})]\, \rho(\boldsymbol{r})\, d\boldsymbol{r} \\ &= E'^{M}_{\gamma} + \int [v(\boldsymbol{r}) - v'(\boldsymbol{r})]\, \rho(\boldsymbol{r})\, d\boldsymbol{r}. \end{aligned} \tag{2.472b}$$

Adding these two inequalities, we are led to the final inequality

$$E^{M}_{\gamma} + E'^{M}_{\gamma} < E^{M}_{\gamma} + E'^{M}_{\gamma}. \tag{2.473}$$

Thus we find that the initial assumption was wrong and that $v(\boldsymbol{r})$ must be a unique functional of $\rho(\boldsymbol{r})$ when the corresponding state is constructed as the $\gamma$th excited eignstate. Since $v(\boldsymbol{r})$ determines the form of the Hamiltonian $H$ uniquely, its eigenstate $\mid \Psi^{M}_{\gamma} \rangle$ itself is a unique functional of the density $\rho(\boldsymbol{r})$ and the information that this is the excited level $\gamma$. Since the latter information is equivalent to the knowledge of the occupation numbers of all KS states, we conclude that $\mid \Psi^{M}_{\gamma} \rangle$ and therefore $F$ are unique functionals of the electron density $\rho(\boldsymbol{r})$ and the occupation numbers of all KS states only. The variational principles still holds within the Fock space orthogonal to all lower excited states and the ground state, which can be controlled by the choice of the occupation numbers of KS states. Therefore only additional requirement in the extension of the ordinary DFT to the electronic excited state is that we have to impose the constraint on the occupation numbers of all KS states. In other words, the ordinary DFT still holds for arbitrary excited eigenstates if we additionally give the information on the occupation numbers of all KS states. Of course, this statement only influences the exchange-correlation potential in DFT. We expect that there is a close similarity between the DFT for arbitrary excited eigenstates and the QP theory for arbitrary initial excited eigenstates. This argument encourages a direct use of the DFT to an electronic excited state such as in [605], if a reliable exchange-correlation functional is used.

### 2.7.3 *GW* Approximation

The quasiparticle (QP) equation method in many-body perturbation theory (MBPT) [606, 607] is powerful in determining simultaneously QP energy spectra and QP wavefunctions of target materials from first-principles all at once. In this method, we expand the skeleton diagrams, i.e., the diagrams drawn with the ture Green's function lines, for the self-energy in terms of the electron-electron Coulomb interaction $v$, and solve the QP equation, which is equivalent to the Dyson equation, as an eigenvalue problem self-consistently. Hartree–Fock (HF) approach provides the first order approximation. In Hedin's set of equations [606, 607] known as the GW$\Gamma$ approach, the exchange-correlation part of the self-energy is expressed as $\Sigma_\sigma^{\rm xc} = iG_\sigma W\Gamma_\sigma$, where $G_\sigma$ and $\Gamma_\sigma$ are the one-particle Green's function and the vertex function ($\sigma$ is the spin index), while $W = (1 - vP)^{-1}v$ is the dynamically screened Coulomb interaction ($P = -i\sum_\sigma G_\sigma G_\sigma \Gamma_\sigma$ is the polarization function); see Sect. 2.7.5 for more details. A convenient and probably the most commonly used approximation is to assume $\Gamma_\sigma = 1$, which is called the GW approximation. The name of "GW" comes from the Green's function (G) and the interaction line (W). The first-principles calculation of the GW approximation for real (nonuniform) semiconductors was first carried out by Hybertsen and Louie [608, 609], and soon later by Godby et al. [610–612]. The explicit expression for the Green's function is given by

$$G(\boldsymbol{r}, \boldsymbol{r}'; \omega) = \sum_{\boldsymbol{p}\nu} \frac{\psi_{\boldsymbol{p}\nu}(\boldsymbol{r})\psi^*_{\boldsymbol{p}\nu}(\boldsymbol{r}')}{\omega - \varepsilon_{\boldsymbol{p}\nu} - i\delta_{\boldsymbol{p}\nu}}, \tag{2.474}$$

where $\delta_{\boldsymbol{p}\nu} = 0^+$ for $\varepsilon_{\boldsymbol{p}\nu} < \mu$ and $\delta_{\boldsymbol{p}\nu} = 0^-$ for $\varepsilon_{\boldsymbol{p}\nu} > \mu$. The corresponding spectral function is defined as

$$A(\boldsymbol{r}, \boldsymbol{r}'; \omega) = \sum_{\boldsymbol{p}\nu} \psi_{\boldsymbol{p}\nu}(\boldsymbol{r})\psi^*_{\boldsymbol{p}\nu}(\boldsymbol{r}')\delta(\omega - \varepsilon_{\boldsymbol{p}\nu}). \tag{2.475}$$

The calculation of the GW approximation is usually carried out within the RPA. In this case, the vertex function in reciprocal space is approximated by the screened Coulomb interaction, which is given by the matrix inversion

$$W_{\boldsymbol{G},\boldsymbol{G}'}(\boldsymbol{q}, \omega) = [\epsilon_{\boldsymbol{G},\boldsymbol{G}'}(\boldsymbol{q}, \omega)]^{-1}\widetilde{U}(\boldsymbol{q} + \boldsymbol{G}'). \tag{2.476}$$

This matrix inversion corresponds to summing up the ring diagrams [17, 20] appearing in the RPA; see Fig. 2.24. In (2.476), $\widetilde{U}(\boldsymbol{q} + \boldsymbol{G}')$ is given by $4\pi/\Omega|\boldsymbol{q} + \boldsymbol{G}'|^2$ ($\Omega$ is the volume of the unit cell). The dielectric matrix for nonzero wavevectors $\boldsymbol{q} + \boldsymbol{G}$ and frequencies $\omega$ can be calculated in the same way as mentioned in Sect. 2.3.2:

$$\epsilon_{\boldsymbol{G},\boldsymbol{G}'}(\boldsymbol{q}, \omega) = \delta_{\boldsymbol{G},\boldsymbol{G}'} - \widetilde{U}(\boldsymbol{q} + \boldsymbol{G})P_{\boldsymbol{G},\boldsymbol{G}'}(\boldsymbol{q}, \omega), \tag{2.477}$$

$$P_{\boldsymbol{G},\boldsymbol{G}'}(\boldsymbol{q}, \omega) = \sum_{\boldsymbol{k}}\sum_{\lambda}\sum_{\nu} \frac{\langle \boldsymbol{k}, \lambda|\mathrm{e}^{-i(\boldsymbol{q}+\boldsymbol{G})\cdot\boldsymbol{r}}|\boldsymbol{k} + \boldsymbol{q}, \nu\rangle\langle \boldsymbol{k} + \boldsymbol{q}, \nu|\mathrm{e}^{i(\boldsymbol{q}+\boldsymbol{G}')\cdot\boldsymbol{r}'}|\boldsymbol{k}, \lambda\rangle}{\omega - \varepsilon_{\boldsymbol{k}\lambda} + \varepsilon_{\boldsymbol{k}+\boldsymbol{q}\nu} - i\delta_{\boldsymbol{k}\lambda}} \times [f_0(\varepsilon_{\boldsymbol{k}+\boldsymbol{q}\nu}) - f_0(\varepsilon_{\boldsymbol{k}\lambda})]. \tag{2.478}$$

$$W = U + UPW$$

Fig. 2.24 Ring diagrams contributing to the screened Coulomb interaction $W_{GG'}(q,\omega)$ in the random-phase approximation (RPA). Symbolically this equation is solved as $W = (1 - UP)^{-1}U = \varepsilon^{-1}U$. They correspond to the inverse of the dielectric matrix $\epsilon_{GG'}(q,\omega)$ multiplied by $\widetilde{U}(q+G')$ as given in (2.476)

The polarization part (2.478) is a nonlocal Fourier transformtion (from $(\boldsymbol{r}, \boldsymbol{r}')$ space to $(\boldsymbol{q}+\boldsymbol{G}, \boldsymbol{q}+\boldsymbol{G}')$ space) of

$$\begin{aligned} P_{\boldsymbol{r},\boldsymbol{r}'}(\boldsymbol{q},\omega) &= \sum_{\boldsymbol{k}}\sum_{\lambda}\sum_{\nu} \frac{\psi^*_{\boldsymbol{k}\lambda}(\boldsymbol{r})\psi_{\boldsymbol{k}+\boldsymbol{q}\nu}(\boldsymbol{r})\psi^*_{\boldsymbol{k}+\boldsymbol{q}\nu}(\boldsymbol{r}')\psi_{\boldsymbol{k}\lambda}(\boldsymbol{r}')}{\omega - \varepsilon_{\boldsymbol{k}\lambda} + \varepsilon_{\boldsymbol{k}+\boldsymbol{q},\nu} - i\delta_{\boldsymbol{k}\lambda}} \\ &\times [f_0(\varepsilon_{\boldsymbol{k}+\boldsymbol{q},\nu}) - f_0(\varepsilon_{\boldsymbol{k},\lambda})]. \end{aligned} \tag{2.479}$$

Here we should note that $-i\delta_{\boldsymbol{k}\lambda}$ in the denominator of this time-ordered function is different from the retarded form given before in (2.252). They are related to each other via the fluctuation-dissipation theorem. In MBPT, one has to always use the time-ordered function instead of the retarded function. (Sometimes, the explicit evaluation of $P_{\boldsymbol{r},\boldsymbol{r}'}(\boldsymbol{q},\omega)$ with nonzero $\omega$ are totally skipped by using the plasmon-pole approximation [607], in which the screened Coulomb interaction $W$ is approximated by a delta function peaked at the frequency of a plasmon branch. Using the generalized plasmon-pole (GPP) model, one can determine $P_{\boldsymbol{r},\boldsymbol{r}'}(\boldsymbol{q},\omega)$ from the knowledge of electron charge density $\rho(\boldsymbol{G})$ and $P_{\boldsymbol{G},\boldsymbol{G}'}(\boldsymbol{q},0)$ so as to satisfy the Johnson's sum rule [613], as discussed in detail by Hybertsen and Louie [608, 609]. Other plasmon pole models have been proposed by Engel and Farid [614] and by von der Linden and Horsch [615]; see also [616]. However, recently, plasmon pole models (including GPP model) sometimes give results which are quite different from the numerical $\omega$ integration; see, for example, [617–622] for the discussion of TiO and ZnO.)

The screened Coulomb interaction is expressed in real space as

$$W(\boldsymbol{r},\boldsymbol{r}';\omega) = \sum_{\boldsymbol{q},\boldsymbol{G},\boldsymbol{G}'} \mathrm{e}^{i(\boldsymbol{q}+\boldsymbol{G})\cdot\boldsymbol{r}} W_{\boldsymbol{G},\boldsymbol{G}'}(\boldsymbol{q},\omega)\mathrm{e}^{-i(\boldsymbol{q}+\boldsymbol{G}')\cdot\boldsymbol{r}'}. \tag{2.480}$$

Introducing a suitable spectral decomposition, one may rewrite this function $W$ in a following form:

$$W(\boldsymbol{r},\boldsymbol{r}';\omega) = U(\boldsymbol{r},\boldsymbol{r}') + \int \frac{2\omega' B(\boldsymbol{r},\boldsymbol{r}';\omega')}{\omega^2 - (\omega' - i\eta)^2} d\omega'. \tag{2.481}$$

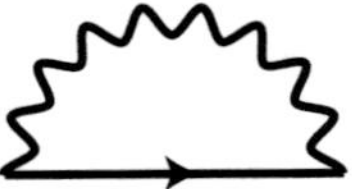

**Fig. 2.25** Diagrammatic representation of the GW approximation. The solid and wavy lines represent, respectively, the Green's function $G(\boldsymbol{r}, \boldsymbol{r}'; t - t')$ and the screened Coulomb interaction $W(\boldsymbol{r}, \boldsymbol{r}'; t - t')$

Here, $\eta$ is a positive and infinitesimal number.

From a very general argument given in previous sections, one finds that the quasiparticle (QP) energies and wavefunctions are obtained by solving the QP equation

$$\left(-\frac{1}{2}\nabla^2 + V_{\rm ext} + V_{\rm H}\right)\psi_{\boldsymbol{k}\lambda}(\boldsymbol{r}) + \int d\boldsymbol{r}\,\Sigma(\boldsymbol{r}, \boldsymbol{r}'; \varepsilon_{\boldsymbol{k}\lambda})\psi_{\boldsymbol{k}\lambda}(\boldsymbol{r}') = \varepsilon_{\boldsymbol{k}\lambda}\psi_{\boldsymbol{k}\lambda}, \quad (2.482)$$

where $V_{\rm ext}$ and $V_{\rm H}$ represent, respectively, the external potential due to the nuclei and the average Coulomb (Hartree) potential due to the electrons, which is expressed by the electron–electron Coulomb interaction $U(\boldsymbol{r} - \boldsymbol{r}') = 1/|\boldsymbol{r} - \boldsymbol{r}'|$ and the electron charge density $\rho(\boldsymbol{r}')$ as $\int U(\boldsymbol{r} - \boldsymbol{r}')\rho(\boldsymbol{r}')d\boldsymbol{r}'$. The newly appeared term in (2.482) is $\Sigma(\boldsymbol{r}, \boldsymbol{r}'; \varepsilon_{\boldsymbol{k}\lambda})$, which represents the electron self-energy. In the GWA, the self-energy is expressed as

$$\Sigma(\boldsymbol{r}, \boldsymbol{r}'; \omega) = i\int_{-\infty}^{\infty} \frac{d\omega'}{2\pi} \mathrm{e}^{-i\eta\omega'} G(\boldsymbol{r}, \boldsymbol{r}'; \omega - \omega')W(\boldsymbol{r}, \boldsymbol{r}'; \omega'), \quad (2.483)$$

where $\eta = 0+$. This self-energy corresponds to the diagram represented by Fig. 2.25, where the solid line is the Green's function (2.474) and the wavy line is the screened Coulomb interaction (2.480) which is diagrammatically given by Fig. 2.24.

Hedin and Lundqvist [607] divided the contribution to the contour integral of (2.483) in the complex $\omega'$ plane into two parts; one is the Coulomb–hole part, which comes from the poles of $W$ only and is given by

$$\Sigma_{\rm COH}(\boldsymbol{r}, \boldsymbol{r}'; \omega) = \sum_{\boldsymbol{k},\nu} \psi_{\boldsymbol{k}\nu}(\boldsymbol{r})\psi^*_{\boldsymbol{k}\nu}(\boldsymbol{r}') \int_0^{\infty} \frac{B(\boldsymbol{r}, \boldsymbol{r}'; \omega')}{\omega - \omega' - \varepsilon_{\boldsymbol{k}\nu}} d\omega', \quad (2.484)$$

and the other is the screened–exchange part, which comes from the poles of $G$ only and is given by

$$\Sigma_{\rm SEX}(\boldsymbol{r}, \boldsymbol{r}'; \omega) = -\sum_{\boldsymbol{k},\nu}^{O} W(\boldsymbol{r}, \boldsymbol{r}'; \varepsilon_{\boldsymbol{k}\nu} - \omega)\psi_{\boldsymbol{k}\nu}(\boldsymbol{r})\psi^*_{\boldsymbol{k}\nu}(\boldsymbol{r}'). \quad (2.485)$$

The summation runs over all states in (2.484), while it runs over only occupied states in (2.485). Hedin and Lundqvist [607] discussed a possible, static approximation, by simply ignoring the $\omega$-dependence of the screened Coulomb interaction $W$. One would then obtain a local interaction

$$\Sigma_{\text{COH}}(\boldsymbol{r}, \boldsymbol{r}') \sim \frac{1}{2}\delta(\boldsymbol{r} - \boldsymbol{r}')\left[W(\boldsymbol{r}, \boldsymbol{r}'; \omega = 0) - U(\boldsymbol{r}, \boldsymbol{r}')\right] \tag{2.486}$$

for the Coulomb-hole part. Note that the sum over all states in this part resumes the delta function via the completeness relation. This term (2.486) represents the static interaction of the quasiparticle with the induced potential due to the rearrangement of the electrons around the quasiparticle. On the other hand, for the screened–exchange part, one would obtain

$$\Sigma_{\text{SEX}}(\boldsymbol{r}, \boldsymbol{r}') \sim -W(\boldsymbol{r}, \boldsymbol{r}'; \omega = 0)\sum_{\boldsymbol{k},\nu}^{O}\psi_{\boldsymbol{k}\nu}(\boldsymbol{r})\psi^*_{\boldsymbol{k}\nu}(\boldsymbol{r}'), \tag{2.487}$$

which represents the statically screened exchange interaction. (Why this term corresponds to the exchange interaction multiplied by $W(\omega = 0)$ can be seen very easily, if we evaluate $\int \Sigma_{\text{SEX}}(\boldsymbol{r}, \boldsymbol{r}')\psi_\lambda(\boldsymbol{r}')d\boldsymbol{r}'$ and compare it with (2.22).) The approximation introduced here is called the (static) COHSEX approximation. The validity of this COHSEX approximation depends, of course, on whether the effect of the dynamical screening is important or not. The COHSEX approximation overestimates, in general, the magnitude of the electron self-energy, as was pointed out by Hybertsen and Louie [608].

Godby et al. [610–612] suggested another useful way to divide the integral (2.483), without introducing any approximation, into following two parts: One is the bare exchange term

$$\begin{aligned}\Sigma_{\text{x}}(\boldsymbol{r}, \boldsymbol{r}') &= iU(\boldsymbol{r} - \boldsymbol{r}')\int_{-\infty}^{\infty}\frac{d\omega}{2\pi}\text{e}^{-i\eta\omega}G(\boldsymbol{r}, \boldsymbol{r}'; -\omega)\\ &= iU(\boldsymbol{r} - \boldsymbol{r}')\int_{-\infty}^{\infty}\frac{d\omega}{2\pi}\text{e}^{i\eta\omega}G(\boldsymbol{r}, \boldsymbol{r}'; \omega)\end{aligned} \tag{2.488}$$

which appears as the lowest-order correction in perturbation theory and is diagrammatically represented by Fig. 2.2b. In (2.488), $U(\boldsymbol{r} - \boldsymbol{r}')$ denotes $1/|\boldsymbol{r} - \boldsymbol{r}'|$. It is a good exercise for the reader to derive (2.22) from $\int \Sigma_{\text{x}}(\boldsymbol{r}, \boldsymbol{r}')\psi_\lambda(\boldsymbol{r}')d\boldsymbol{r}'$. Thus, the matrix elements of this term sandwiched by the LDA Bloch functions can be accurately evaluated through the relation

$$\begin{aligned}\langle\psi^*_{\boldsymbol{k}\lambda}|\Sigma_{\text{x}}|\psi_{\boldsymbol{k}\lambda'}\rangle &= \int\int \psi^*_{\boldsymbol{k}\lambda}(\boldsymbol{r})\Sigma_{\text{x}}(\boldsymbol{r}, \boldsymbol{r}')\psi_{\boldsymbol{k}\lambda'}(\boldsymbol{r}')d\boldsymbol{r}d\boldsymbol{r}'\\ &= -\sum_{p\nu}^{O}\int\int\frac{\psi^*_{\boldsymbol{k}\lambda}(\boldsymbol{r})\psi^*_{p\nu}(\boldsymbol{r}')\psi_{p\nu}(\boldsymbol{r})\psi_{\boldsymbol{k}\lambda'}(\boldsymbol{r}')}{|\boldsymbol{r} - \boldsymbol{r}'|}d\boldsymbol{r}d\boldsymbol{r}'.\end{aligned} \tag{2.489}$$

The other is the correlation term

$$\Sigma_c(\boldsymbol{r}, \boldsymbol{r}'; \omega) = i \int_{-\infty}^{\infty} \frac{d\omega'}{2\pi} \mathrm{e}^{-i\eta\omega'} G(\boldsymbol{r}, \boldsymbol{r}'; \omega - \omega') \Big[ W(\boldsymbol{r}, \boldsymbol{r}'; \omega') - U(\boldsymbol{r} - \boldsymbol{r}') \Big], \tag{2.490}$$

which involves all the rest terms which start from the second-order in $U$. Since $W(\boldsymbol{r}, \boldsymbol{r}'; \omega')$ approaches $U(\boldsymbol{r} - \boldsymbol{r}')$ as $|\omega'|$ increases, the integrand vanishes as fast as $1/|\omega'|^2$ in the limit $|\omega'| \to \infty$, and the $\omega'$ integral converges well. However, one cannot easily perform the $\omega'$ integral along the real axis, since $W$ and $G$ have a strong structure on this axis. In order to avoid this difficulty, Godby et al. restricted the values of $\omega$ to small imaginary numbers and changed the contour of the $\omega'$ integral from real axis to imaginary axis [612]. This is possible because the integrand does not have a pole inside the two quadrants of complex $\omega'$ plane corresponding to (Re $\omega' >$ 0, Im $\omega' <$ 0) and (Re $\omega' <$ 0, Im $\omega' >$ 0). They expanded the matrix elements of $\Sigma_c(\omega)$ in Taylor series of $\omega$, and evaluated these coefficients from integrations at several values of $\omega$ on imaginary axis. Then, by analytic continuation, the resulting Taylor series is used to estimate the matrix elements for real values of $\omega$. In this way, without invoking the plasmon-pole approximation, they succeeded in estimating the band-gap energy for several typical semiconductors and obtained good agreements with experiments. As they pointed out in their paper, however, the above integration method is applicable to the energy region close to the Fermi level only.

More recently, Ishii [339] and Zhang et al. [622] showed that this integration method employed by Godby et al. [612] can be extended rather easily to real numbers of $\omega$ by slightly modifying the contour. For real values of $\omega$, the integrand of (2.489) has no pole inside two quadrants of the complex $\omega'$ plane, corresponding to (Re $\omega' >$ max(0,$\omega$), Im $\omega' <$ 0) and (Re $\omega' <$ min(0,$\omega$), Im $\omega' >$ 0). Therefore, the contour along the real $\omega'$ axis from $-\infty$ to $+\infty$ for the integral (2.490) can be replaced by the contour $C$ shown in Fig. 2.26 [339, 622]. If we choose the LDA Bloch functions $\psi_{\boldsymbol{k},\lambda}$ as basis functions, the matrix elements of the correlation part of the self-energy are expressed as

$$\begin{aligned} \langle \boldsymbol{k}, \lambda | \Sigma_c(\boldsymbol{r}, \boldsymbol{r}'; \omega) | \boldsymbol{k}, \lambda' \rangle &= \sum_{\nu} \sum_{\boldsymbol{q},\boldsymbol{G},\boldsymbol{G}'} \langle \boldsymbol{k}, \lambda | \mathrm{e}^{i(\boldsymbol{q}+\boldsymbol{G})\cdot\boldsymbol{r}} | \boldsymbol{k} - \boldsymbol{q}, \nu \rangle \\ \times\ i \int_C \frac{d\omega'}{2\pi} \mathrm{e}^{-i\eta\omega'} & \frac{[W_{\boldsymbol{G},\boldsymbol{G}'}(\boldsymbol{q}, \omega') - (4\pi/\Omega G^2)\delta_{\boldsymbol{G},\boldsymbol{G}'}]}{\omega - \omega' - \varepsilon_{\boldsymbol{k}-\boldsymbol{q}\nu} - i\delta_{\boldsymbol{k}-\boldsymbol{q}\nu}} \\ \times\ \langle \boldsymbol{k} - \boldsymbol{q}, \nu | \mathrm{e}^{-i(\boldsymbol{q}+\boldsymbol{G}')\cdot\boldsymbol{r}'} & | \boldsymbol{k}, \lambda' \rangle, \end{aligned} \tag{2.491}$$

where the contour $C$ must pass through $\omega$ as is illustrated in Fig. 2.26. Using this method combined with the all-electron mixed-basis approach, Ishii et al. calculated the quasiparticle energies of argon [339] and alkali metal clusters [644], and Zhang et al. calculated $TiO_2$ with and without Nb impurities [651], and obtained fairly a good agreement with experiments.

It has been recognized, from several calculations for semiconductors and ionic crystals, that the solution $\psi_{\boldsymbol{k}\lambda}$ of (2.482) coincides to an accuracy of 0.1% with the

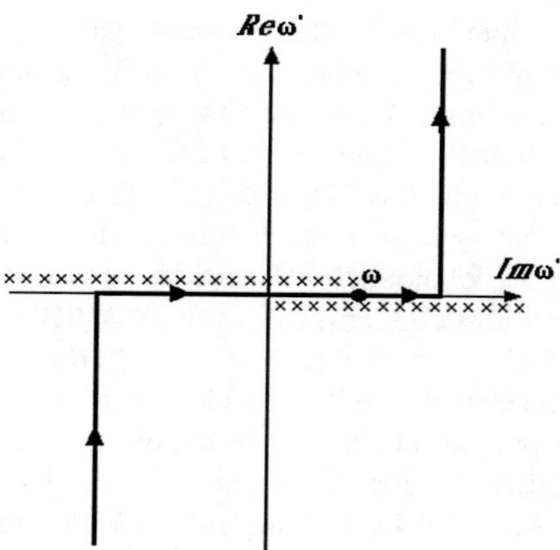


**Fig. 2.26** Contour $C$ of the $\omega'$ integration in (2.491)

wavefunction obtained within the LDA. Therefore, the quasiparticle energy in the GWA can be estimated from

$$\begin{aligned}\varepsilon_{\boldsymbol{k}\lambda}^{\mathrm{GWA}} &= \varepsilon_{\boldsymbol{k}\lambda}^{\mathrm{LDA}} + \int d\boldsymbol{r}d\boldsymbol{r}'\psi_{\boldsymbol{k}\lambda}^{\mathrm{LDA}*}(\boldsymbol{r}) \\ &\times \left[\Sigma(\boldsymbol{r},\boldsymbol{r}';\varepsilon_{\boldsymbol{k}\lambda}^{\mathrm{GWA}}) - \mu_{\mathrm{xc}}^{\mathrm{LDA}}(\boldsymbol{r})\delta(\boldsymbol{r}-\boldsymbol{r}')\right]\psi_{\boldsymbol{k}\lambda}^{\mathrm{LDA}}(\boldsymbol{r}'). \end{aligned} \tag{2.492}$$

Since the right-hand side involves $\varepsilon_{\boldsymbol{k}\lambda}^{\mathrm{GWA}}$ implicitly, this equation must be solved iteratively. This may be performed by linearizing the self-energy as a function of $\varepsilon_{\boldsymbol{k}\lambda}^{\mathrm{GWA}}$ around $\varepsilon_{\boldsymbol{k}\lambda}^{\mathrm{LDA}}$ as follows [608, 609]:

$$\varepsilon_{\boldsymbol{k}\lambda}^{\mathrm{GWA}} = \varepsilon_{\boldsymbol{k}\lambda}^{\mathrm{LDA}} + Z_{\boldsymbol{k}\lambda}\langle \boldsymbol{k},\lambda|\left[\Sigma(\boldsymbol{r},\boldsymbol{r}';\varepsilon_{\boldsymbol{k}\lambda}^{\mathrm{LDA}}) - \mu_{\mathrm{xc}}^{\mathrm{LDA}}(\boldsymbol{r})\delta(\boldsymbol{r}-\boldsymbol{r}')\right]|\boldsymbol{k},\lambda\rangle, \tag{2.493}$$

$$Z_{\boldsymbol{k}\lambda} = \left[1 - \frac{\partial}{\partial\omega}\langle \boldsymbol{k},\lambda\,|\,\Sigma(\omega)\,|\,\boldsymbol{k},\lambda\rangle\right]^{-1}_{\omega=\varepsilon_{\boldsymbol{k}\lambda}^{\mathrm{LDA}}}. \tag{2.494}$$

The above approximation that uses only the diagonal elements $\Sigma_{\boldsymbol{k}\lambda} = \langle \boldsymbol{k},\lambda|\Sigma|\boldsymbol{k},\lambda\rangle$ in evaluating the quasiparticle energies leads to errors less than 0.05 eV for silicon and LiCl as was discussed by Hybertsen and Louie [608]. For example, the direct band gap of silicon at the $\Gamma$ point has been calculated to be 3.35 eV in very good agreement with experimental value 3.4 eV [608]. (Note that the original band gap obtained by the LDA is 2.5 eV.) This approach is now generally called as the one-shot GW approximation ($G_0W_0$).

To conclude, the GW approximation treats the electron correlations in nonuniform systems on the basis of many-body theory. It is much more accurate but much more time-consuming to calculate than the LDA. Photoelectron spectra of semiconductors and insulators are usually well reproduced using this theory. There are a lot of GWA calculations reported so far; for example, within the jellium model, simple metallic clusters were treated by Saito et al. [623]; with the pseudopotential–PW method combined with the generalized plasmon-pole (GPP) model, which was originally devised by Hybertsen and Louie [608, 609], various standard semiconductors [608, 609, 624], AlN and GaN [625], $\beta$-$C_3N_4$ crystal [626], six II-VI compounds [627], H:Si(111)–$(1 \times 1)$ surface [628], $C_{60}$ crystals [466], several ionic, noble-gas and molecular solids [629], and $ZrO_2$ solid [630] were examined; with the pseudopotential–PW method combined with Godby's method (see (2.488)–(2.490)), various standard semiconductors [610–612], cubic GaN [631], and Si and Ge (including a method to calculate charge densities with the GW approximation) [632] were investigated; with the LCAO method using GTOs combined with the pseudopotential approach and the plasmon-pole model, Rohlfing and co-authors have studied various group-IV and III-V semiconductors [633], Si and Ge surface (including discussion of core-level shift) [634], $\beta$-SiC(001) surface [635], CdS, ZnS, GaN (including effect of semicore d electrons) [636], and HgSe [637]; with a product basis approach in the LMTO formalism and a full $\omega$-dependence calculation for $W$, Aryasetiawan and co-authors have investigated Si and Ni [638], Ge, GaAs, ZnSe, NiO [639] Na and Al (including cumulant expansion) [641] MgO (including energy-loss spectrum and MgO(100) surface) [640] and Ni, NiO, Gd, ZnSe, GaAs [643].

Using all-electron mixed basis approach, Ishii et al. calculated the QP energies of alkali metal clusters [645], silicon clusters [646], germanium clusters [647], GaAs clusters and crystal [648], silicon, carbon and SiC crystals [649], and Noguchi et al. calculated the QP energies of alkali metal clusters [707] and aluminum clusters [650]. Figure 2.27 shows the QP energy spectra of aluminum clusters by Noguchi et al. [650]. In the case of the fcc Ar crystal, the band gap is estimated to be 13.62 eV by the pseudopotential approach with the plasmon-pole approximation [629] and 14.06 eV by the all-electron mixed-basis approach without the plasmon-pole approximation [339], which should be compared with the experimental band gap 14.19 eV. (The original band gap obtained by the LDA is 8.02 eV.)

So far, we discussed the ordinary QP energiescalculated for the initial electronic ground state. As has been demonstrated in earlier sections (Sects. 2.7.1 and 2.7.2), the QP theory holds not only for the initial ground state but also for any of the initial excited eigenstates [13]. To see this, let us consider first a hydrogen atom with $N = 1$ and for the case $M = 0$ as the simplest case. Its ($M = 0$)-electron excited state is none other than a proton. Then, the lowest-energy (empty) QP state having $M+1 = 1$ electron is the hydrogen 1s orbital. The self-consistent calculation surely gives this result exactly, because there is no occupied electron. Note that there is no Hartree term, no exchange term, and no higher-order terms. This is even so in the LDA or GGA of DFT. The same thing holds not only for the 1s state but also for any excited eigenstates of the hydrogen atom. Similar situation occurs also for an alkali metal atom like lithium, sodium, and potassium, if we can ignore the core contributions. In this sense, the excited-state formalism has a chance to serve an extremely good result compared to the ordinary ground-state formalism in some special cases.

Next we consider a general isolated atom A or molecule M. It is obvious that the ionization potential (IP) of A or M is identical to the electron affinity (EA) of its cation $A^+$ or $M^+$, provided that the atomic geometries of M and $M^+$ are the same

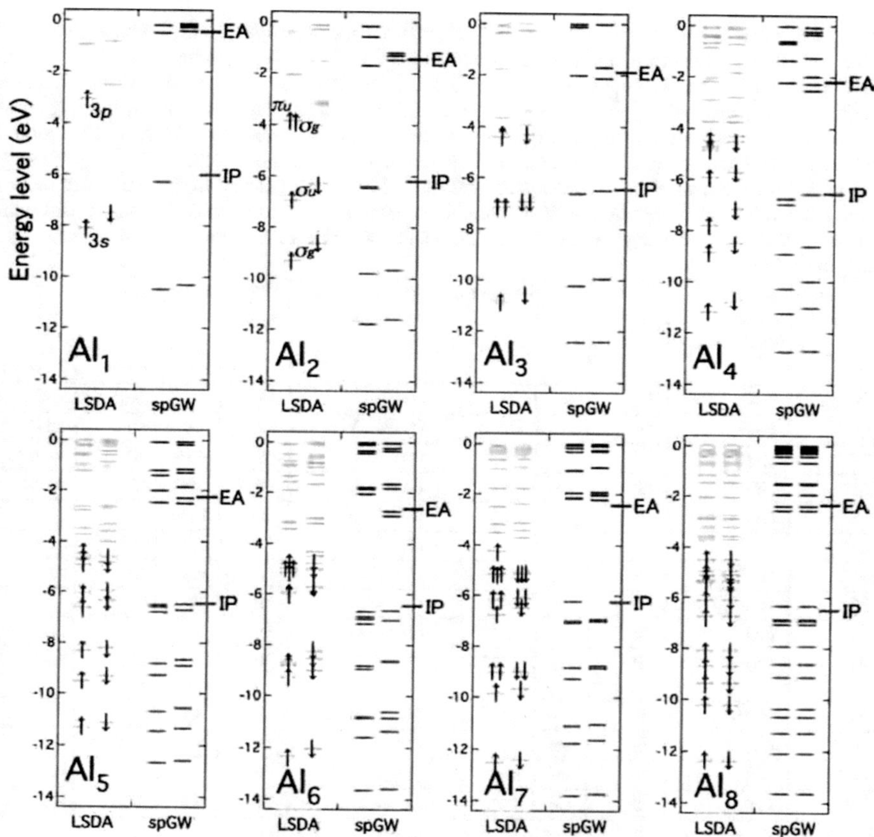


**Fig. 2.27** Spin-dependent QP energy spectra of aluminum clusters obtained by the one-shot GW approximation [650]. Corresponding LSDA Kohn-Sham energies are showen for comparison. Vertical arrows denote the spin directions of the occupied electrons. The experimental ionization potential and electron affinity with negative signs are indicated as IP and EA, respectively, on the right line of each frame

(This is not a bad approximation for typical molecules). Similarly, there are several other identities between the electron removal and absorption energies of cationic, anionic and photoabsorbed states of A or M. These identities allow us to check the validity of the theory.

For this purpose, we use the standard one-shot GW approximation ($G_0W_0$) by beginning with the LDA for both the ground-state and excited-state configurations of isolated atoms and diatomic molecules [13]. We use the all-electron mixed basis approach, FCC unit cell (edge length is 18 Å for Li, $Li_2$, and Al, and 12 Å for Be, $N_2$, and $O_2$). Cutoff energies for PWs, the Fock exchange and the correlation part of the self-energy (as well as the polarization part) are set as 11.05 Ry, 44.22 Ry, 11.05 Ry for Li, $Li_2$, and Al, 19.65 Ry, 78.61 Ry, 19.65 Ry for Be and $N_2$, and 27.71 Ry,

**Table 2.5** $G_0W_0$ QP energies of neutral (A), cationic ($A^+$), anionic ($A^-$), and photoabsorbed ($A^*$) states of lithium, aluminum, and beryllium atoms in units of eV [13]. The right atom/ion has one more electron than the left atom/ion; and the left and the right have $N(-1)$ and $N(+1)$ electron configurations, respectively ($^{2S+1}L$ states are also listed); In the same row, the QP energies of the forward, →, electron absorption process and the reverse, ←, electron release process should coincide with each other. The values inside the parentheses indicate the usual $G_0W_0$ QP energy calculated for the neutral atom at the ground state. EC in the superscript in the Reference values means the value calculated by the EOM coupled cluster singles and doubles (EOM-CCSD) with aug-cc-pV5Z basis set using Gaussian 09 [652]. All the other reference values$^{a-g}$ are experimental data [653–658]

| Atom/Ion | $N(-1)$ electron configuration | $^{2S+1}L$ state | | Atom/Ion | $N(+1)$ electron configuration | $^{2S+1}L$ state | QP energies (eV) | | |
|---|---|---|---|---|---|---|---|---|---|
| | | | | | | | → | ← | References |
| $Li^+$ | $(1s)^2$ | $^1S$ | ↔ | Li | $(1s)^2(2s)$ | $^2S$ | 5.1 | (5.6) | 5.39[a] |
| $Li^+$ | $(1s)^2$ | $^1S$ | ↔ | $Li^*$ | $(1s)^2(2p)$ | $^2P$ | 3.0 | 3.2 | 3.54[b] |
| $Li^{+*}$ | $(1s)(2s)$ | $^1S$ | ↔ | Li | $(1s)^2(2s)$ | $^2S$ | 63.9 | (63.4) | 64.46[b] |
| $Li^{+*}$ | $(1s)(2p)$ | $^1P$ | ↔ | $Li^*$ | $(1s)^2(2p)$ | $^2P$ | 65.0 | 66.0 | 65.76[b] |
| Li | $(1s)^2(2s)$ | $^2S$ | ↔ | $Li^-$ | $(1s)^2(2s)^2$ | $^1S$ | (0.6) | 0.3 | 0.61[c] |
| $Al^+$ | $(3s)^2$ | $^1S$ | ↔ | Al | $(3s)^2(3p)$ | $^2S$ | 5.7 | (5.6) | 5.98[d] |
| $Al^+$ | $(3s)^2$ | $^1S$ | ↔ | $Al^*$ | $(3s)^2(4s)$ | $^2P$ | 2.6 | 3.1 | 2.84[e] |
| $Al^{+*}$ | $(3s)(3p)$ | $^1S$ | ↔ | Al | $(3s)^2(3p)$ | $^2P$ | 12.6 | (10.2) | 13.40[e] |
| Al | $(3s)^2(3p)$ | $^2P$ | ↔ | $Al^-$ | $(3s)^2(3p)^2$ | $^3S$ | (0.3) | - | 0.43[f] |
| Al | $(3s)^2(3p)$ | $^2P$ | ↔ | $Al^{-*}$ | $(3s)^2(3p)(4s)$ | $^3P$ | (-0.1) | 0.5 | 0.2[EC] |
| $Al^*$ | $(3s)^2(4s)$ | $^2S$ | ↔ | $Al^{-*}$ | $(3s)^2(3p)(4s)$ | $^3P$ | 3.5 | 2.7 | 3.3[EC] |
| $Be^+$ | $(2s)$ | $^2S$ | ↔ | Be | $(2s)^2$ | $^1S$ | 9.3 | (9.3) | 9.32[f] |
| $Be^+$ | $(2s)$ | $^2S$ | ↔ | $Be^*$ | $(2s)(2p)$ | $^1P$ | 4.2 | 5.4 | 4.40[g] |
| $Be^+$ | $(2s)$ | $^2S$ | ↔ | $Be^*$ | $(2s)(2p)$ | $^3P$ | 6.4 | 6.2 | 6.59[g] |

[a][653], [b][653, 654], [c][655], [d][656], [e][656, 657], [f][658], [g][654, 658]

122.83 Ry, 30.71 Ry for $O_2$, respectively. In the summation in the correlation part, 400 levels are included for Li, Al, and Be; 3000, 6000, and 7000 levels are included, respectively, for $Li_2$, $N_2$, and $O_2$. We use the generalized plasmon pole model [608, 609] to avoid the $\omega$ integration. The energy dependence of the self-energy is dealt by introducing the $Z$ factor as in (2.493) and (2.494). The computational time scales as $O(N_{\text{emp}}N_{\text{occ}}N_G^2)$ where $N_{\text{emp}}$, $N_{\text{occ}}$, and $N_G$ are the numbers of empty states, occupied states, and $\boldsymbol{G}$ vectors needed for the non-local Fourier components of the polarization function and the correlation part of the self-energy.

There is no essential difference between the QP theory for the initial neutral ground state and that for initial (neutral or charged) excited eigenstates; the GW calculations for the latter initial excited eigenstates can be simply performed by just changing the order of levels after the diagonalization. However, in order to compare the results, we will call the former neutral ground-state case the ordinary QP (OQP) theory and the latter excited-state case the extended QP (EQP) theory.

**Table 2.6** $G_0W_0$ QP energies of neutral (M), cationic ($M^+$), and photoabsorbed ($M^*$) states of nitrogen, oxygen, and lithium molecules in units of eV [13]. The right molecule has one more electron than the left molecule; and the left and the right have $N-1$ and $N$ electron configurations, respectively (electronic states are also listed); In the same row, the QP energies of the forward, →, electron absorption process and the reverse, ←, electron release process should coincide with each other and also with the reference values [659–662]. The values inside the parentheses indicate the usual $G_0W_0$ QP energy calculated for the neutral atom at the ground state

| Molecule | State | | Molecule | State | QP energies (eV) | | |
|---|---|---|---|---|---|---|---|
| | | | | | → | ← | References |
| $N_2^+$ | $^2\Sigma_g^+$ | ↔ | $N_2$ | $^1\Sigma_g^+$ | 15.69 | (15.40) | 15.58[a] |
| $N_2^+$ | $^2\Sigma_g^+$ | ↔ | $N_2^*$ | $^1\Pi_g$ | 7.28 | 7.21 | 6.99[b] |
| $O_2^+$ | $^2\Pi_g$ | ↔ | $O_2$ | $^3\Sigma_g^-$ | 12.16 | (12.35) | 12.30[c] |
| $O_2^+$ | $^2\Pi_g$ | ↔ | $O_2^*$ | $^3\Pi_g$ | 3.80 | 3.73 | 4.17[d] |
| $Li_2^+$ | $^2\Sigma_g^+$ | ↔ | $Li_2$ | $^1\Sigma_g^-$ | 5.30 | (5.32) | 5.11[e] |
| $Li_2^+$ | $^2\Sigma_g^+$ | ↔ | $Li_2^*$ | $^1\Sigma_u^-$ | 3.74 | 3.95 | 3.35[f] |

[a][659], [b][659, 660], [c] [661], [d][660, 661], [e][662], [f][660, 662]

The calculated results of the QP energies of several QP states are listed in Table 2.5 for Li, Al, and Be, and in Table 2.6 for $N_2$, $O_2$, and $Li_2$, together with the reference values. EC in Table 2.5 means the value calculated by the EOM coupled cluster singles and doubles (EOM-CCSD) with aug-cc-pV5Z basis set using Gaussian 09 [652]. References [653–662] are all experimental data. We find good agreements between the EQP energies of the forward (→) electron absorption process and the backward (←) electron release process. Maximum deviation from the reference value is 1.0 eV in the case $Be^+ \leftarrow Be^*$ for atoms, and 0.6 eV in the case $Li_2^+ \leftarrow Li_2^*$ for molecules, which are smaller than the deviations in the OQP theory (values inside parentheses in Tables 2.5 and 2.6); 3.2 eV in the case $Al^{+}* \leftarrow Al$. The average deviations are 0.4 eV and 0.3 eV, respectively, for atoms and molecules; they are the same orders for both the OQP and EQP theories (values outside parentheses in Tables 2.5 and 2.6). This means that there is no specific deviation in the EQP theory, demonstrating the validity of the EQP theory. The resulting QP energy spectra of $Li^{+}*$ calculated by $G_0W_0$ are shown in Fig. 2.28. It is seen from this figure that the agreement between the theory and experiment is very good already in the present approximation.

All the discussions given above were based on the one-shot GW approximation ($G_0W_0$). In this approximation, the QP energies are evaluated within first-order perturbation theory as (2.492)–(2.494), although the ideal GW approximation has to solve the self-consistent equation (2.482). This treatment, however, allows us to estimate the band gap of various semiconductors and insulators which is generally in good agreement with the experimental values. It is well known that the self-consistent treatment of the GW approximation (hereafter we write it $GW$ in contrast to $G_0W_0$) usually overestimates the energy gap [586, 665], while $G_0W_0$ usually gives a better energy gap. This is due to the fact that $GW$ does not satisfy the Pauli principle or the Ward–Takahashi identity [599–601], which is equivalent to the gauge invariance (continuity equation for the electron density), i.e., the local conservation of

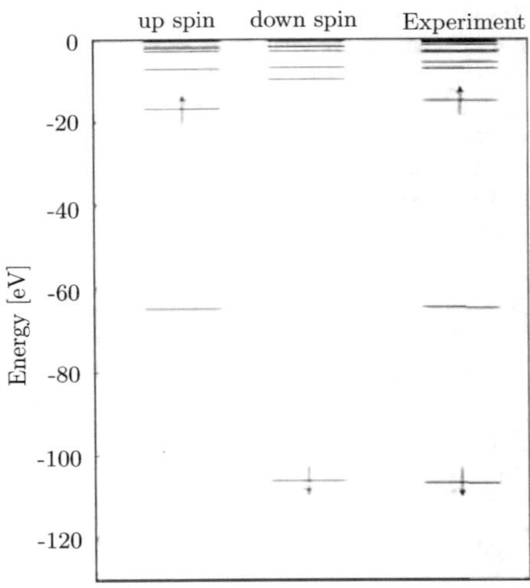


**Fig. 2.28** QP energy spectra of $Li^{+*}$ calculated by $G_0W_0$ in units of eV. Experimental data are taken from [653, 654, 663, 664]. Arrows represent electrons (spin) occupying the level

electron number (or charge) is violated in $GW$. This violation can be partially cured by linearizing the energy dependence of the self-energy as is demonstrated in [587, 592]. On the other hand, it is possible to calculate the total energy within $GW$ using the Luttinger–Ward functional (2.466) as described in [87, 587, 666–668]. although the force acting each nucleus cannot be simply evaluated in $GW$. The virial theorem in $GW$ is discussed by van Leeuwen et al. [667], and it is at least roughly satisfied in $GW$ as investigated in [87]. Furthermore, there is a great advantage in $GW$ because the ring diagrams in the RPA, which become most dominant in the long wavelength limit, are exactly included on the formulation. Therefore, this approximation is expected to become exact in the description of long range interaction. For example, experimentally, He dimer is known to have a quite long average interatomic distance ($\sim$52 Å) [669]. Self-consistent GW approach ($GW$) was used to investigate this van der Waals potential between helium atoms and partially succeeded in reproducing the very long distance behavior by additionally solving the Schrödinger equation for the two nuclei in the extrapolated $GW$ adiabatic potential (Born–Oppenheimer surface) [670].

The great success in the calculation of quasiparticle energies with the GWA is mainly due to the fact that the quasiparticle energies are sensitive to the long-range screened interaction, which can be treated properly by the RPA, but rather insensitive to the other short-range interactions. However, there are many situations where the short-range interactions become more important; for example, in photoelectron spectra observed by photoemission experiments, there are often satellite peaks at a few eV below the main peak [671], which is due to the creation of localized holes after a photoelectron is emitted. The creation of more than two localized holes is an essentially many-body phenomenon associated with the short-range interactions, which cannot be treated within the RPA. Therefore, a natural extension of the GWA is to go beyond the RPA and to include the effect of short-range correlation; see below.

### 2.7.4 Exchange and Coulomb Holes

Since the Hartree–Fock approximation estimates the exchange term correctly, the investigation of this term may give information on the correlation term also. Two electrons with the same spin cannot come near each other, because of Pauli's exclusion principle. Therefore, there is a clear dip in the pair distribution function for electrons with parallel spins $g^{\uparrow\uparrow}(\boldsymbol{r}, \boldsymbol{r}')$ in the region of short distances. This dip is called the "exchange hole". The exchange term describes effectively the reduction, of the electron–electron interaction energy due to the existence of the exchange hole. It is found that the volume of the exchange hole, i.e. the spatial integral of $g^{\uparrow\uparrow}(\boldsymbol{r}, \boldsymbol{r}') - \rho(\boldsymbol{r}')$ with respect to $\boldsymbol{r}'$, exactly equals unity corresponding to the amount of one electron. Similarly, one may study the effect of electron correlation on the behavior of two electrons with antiparallel spins. Generally, the screening effect is important at large distances, while it is known to the electron gas that there is also a dip called the "Coulomb hole" in the pair distribution function of electrons with antiparallel spins $g^{\uparrow\downarrow}(\boldsymbol{r}, \boldsymbol{r}')$. The origin of the Coulomb hole is due to the strong repulsion between electrons. It is found, however, that the volume of the Coulomb hole exactly equals zero, i.e. the spatial integral of $g^{\uparrow\downarrow}(\boldsymbol{r}, \boldsymbol{r}') - \rho(\boldsymbol{r}')$ with respect to $\boldsymbol{r}'$ is identically zero at arbitrary $\boldsymbol{r}$. Therefore, there should be an enhancement of the electron population at medium distances of $|\boldsymbol{r} - \boldsymbol{r}'|$ to compensate for the Coulomb hole. As a result, for isolated atoms and clusters, the exact exchange potential falls off as $-1/r$ for large distances $r$, while the exact correlation potential falls off faster than $O(1/r^2)$. Unfortunately, the standard LDA or LSDA cannot yield this expected dependence, although the exact correlation potential of DFT falls off as $O(1/r^4)$ for isolated atoms [106]. This is related to a fact that one cannot describe correctly the van der Waals interaction within the LDA and the LSDA. (A consideration in the case of the GGA was given by [672].) Even in the case of the GW approximation, the particle number conservation is violated if one does not solve the quasiparticle equation self-consistently as was demonstrated by Schindlmayr [673].

To take into account the effects of both the exchange hole and the Coulomb hole rigorously in a systematic manner, the diagrammatic expansion up to infinite order has so far been analyzed by many authors. In particular, for low electron densities, it is known that the "ladder diagrams" become more important, and the system approaches the behavior discussed by Yasuhara [674] and Kimball [675] in the low-density limit; see also [676–678]. Inclusion of ladder diagrams in real atomic systems will be described in detail in Sect. 2.7.6.

In order to guarantee the Pauli principle in each order of the perturbation theory, it is necessary to consider the corresponding direct and exchange terms always as a pair. Such a perturbation expansion has been performed in the field of quantum chemistry and is known as the Møller–Plesset (MP) perturbation theory [48]. However, the method is to obtain the perturbation corrections to the total energy to a finite order by beginning with the Hartree–Fock approximation, and cannot give the QP energies. In order to calculate the QP energy spectra all at once in a single calculation, it is necessary to adapt the method to the Green's function formalism and solve the Dyson equation self-consistently. This can be possible by constructing the self-energy with the skeleton diagrams up to the second order in the electron-electron Coulomb interaction, i.e., the first four terms in Fig. 2.22. This method can be called the self-consistent second order approach. Another merit of this approach compared to the GW method is that one can calculate the force acting on each atom in addition to the total energy because of its structural simplicity.

### 2.7.5 Vertex Correction

One way to improve the self-consistent GW approximation ($GW$) is to go beyond the RPA and include vertex corrections [679, 680]. There are some recent reports on non-self-consistent GW calculations including the second-order screened exchange (SOSEX) by Ren et al. [681]. and the GW$\Gamma^1$ method (i.e., GW$^{\mathrm{TC-TC}}$ + single-shot vertex correction for the self-energy with the static approximation) by Grüneis et al. [682]. All these authors uses the KS, HF, or HSE wavefunctions throughout the calculations.

Here we will explain the recent self-consistent GW$\Gamma$ calculation [592], which involves the self-consistent treatment of the vertex $\Gamma$ and satisfies the Ward–Takahashi identity [599–601] to the first order in $v$ or $W$. It remarkably improves the QP energies and the optical gap; the optical gap is explained in later section. In this method, the fully self-consistent GW$\Gamma$ one-particle Green's function, i.e., self-consistent QP energies and wavefunctions in the GW$\Gamma$ scheme, are obtained.

Except for the $G_0W_0$ and $GW$ calculations, we use our recent technique to linearize the energy dependence of the self-energy $\Sigma_\sigma(\varepsilon_n)$ [587] to avoid the non-hermitian problem caused by the energy dependence and to perform fully self-consistent calculations. The important point of this technique is that $\Lambda_\sigma = \lim_{(\omega,\boldsymbol{q})\to 0} \Gamma_\sigma(\boldsymbol{r}_1, \boldsymbol{r}_2, \boldsymbol{q}; \mu+\omega, \mu) = 1 - \partial\Sigma_\sigma(\omega)/\partial\omega|_{\omega=\mu}$ is the vertex function and the Ward–Takahashi identity in the limit $(\omega, \boldsymbol{q}) \to 0$. (Here, $\mu = (\varepsilon_{\mathrm{HOMO}} + \varepsilon_{\mathrm{LUMO}})/2$.) The QP equation is given by $H_\sigma \,|\, n\sigma\,\rangle = \varepsilon_{n\sigma}\Lambda_\sigma \,|\, n\sigma\,\rangle$ with $H_\sigma = \hat{T} + \hat{v}_{\mathrm{nuc}} + \Sigma_\sigma(\mu) + \mu(\Lambda_\sigma - 1)$. Then, with the lower triangular matrix $L_\sigma$ in the Choleski decomposition $\Lambda_\sigma = L_\sigma L_\sigma^\dagger$, the renormalized QP states are given by $|\, \widetilde{n\sigma}\,\rangle = L_\sigma^\dagger \,|\, n\sigma\,\rangle$, which satisfy $\widetilde{H}_\sigma \,|\, \widetilde{n\sigma}\,\rangle = \varepsilon_{n\sigma} \,|\, \widetilde{n\sigma}\,\rangle$ with $\widetilde{H}_\sigma = L_\sigma^{-1} H_\sigma L_\sigma^{-1\dagger}$ as well as orthogonality and completeness conditions. As well, the renormalized Green's function is given by $\widetilde{G}_\sigma(\omega) = L_\sigma^\dagger G_\sigma(\omega) L_\sigma$; see [587] for more details.

(Theorem 1) In this linearized formulation, we can additionally introduce the vertex part $\Gamma_\sigma^{(1)}(\boldsymbol{r}_1, \boldsymbol{r}_2, \boldsymbol{q}; \varepsilon+\omega, \varepsilon)$ at the first order either in $v$ (Fig. 2.29a, which is called the LGW$\Gamma_v$ method), or in $W$ (Fig. 2.29b, which is called call the LGW$\Gamma_W$ method). These vertex parts depend fully on the energy and momentum transfers $\omega$ and $\boldsymbol{q}$ at the center (cross in these figures).

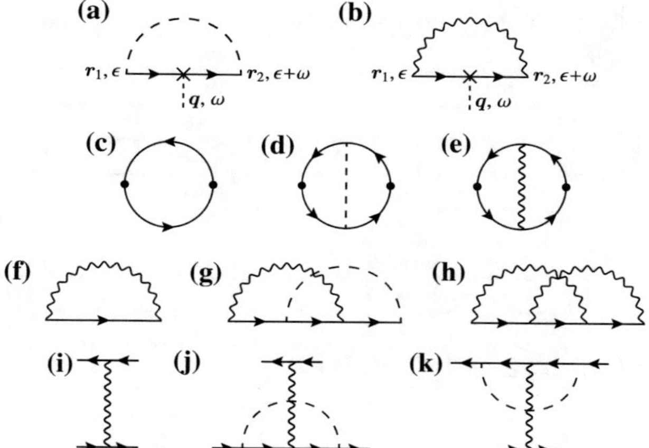


**Fig. 2.29** Skeleton diagrams of **a**, **b** the first-order vertex part $\Gamma_\sigma^{(1)}$, **c–e** the polarization part $P$, **f–h** the exchange-correlation part of the self-energy $\Sigma_\sigma^{\rm xc}$, and **i–k** the interaction kernel $\widetilde{I}^{\sigma_1\sigma_1'}$ for the BSE. **c**, **f**, **i** are usual diagrams without vertex correction, **a**, **d**, **g**, **j**, **k** involves the first-order vertex in $v$ (dotted line), and **b**, **e**, **h** involves the first-order vertex in $W$ (wavy line)

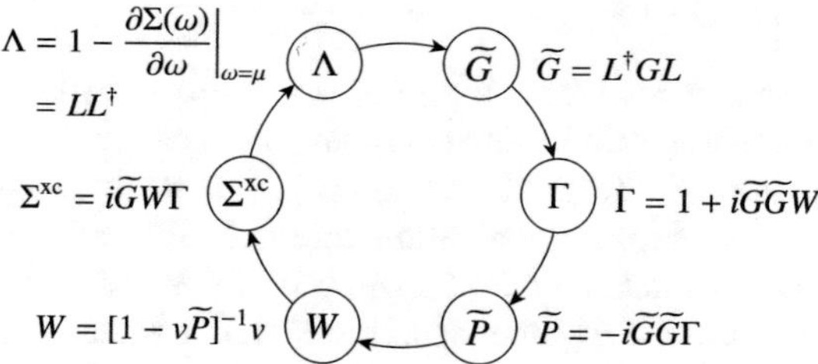


**Fig. 2.30** Flow chart of the self-consistent LGW$\Gamma$ method. $W$ in $\Gamma$ is replaced by $v$ for the self-consistent LGW$\Gamma_v$ method

Then, the polarization function and the self-energy include the skeleton diagrams shown in Fig. 2.29c–h: c and f represent the diagrams without vertex; d, g and e, h are the corresponding vertex corrections at the first order in $v$ (the LGW$\Gamma_v$ method) and $W$ (the LGW$\Gamma$ method), respectively. Figure 2.30 illustrates the flow chart of the self-consistent LGW$\Gamma_W$ method.

(Theorem 2) The present LGW$\Gamma_v$ and LGW$\Gamma_W$ methods satisfy the general Ward–Takahashi identity for arbitrary $\omega$ and $\boldsymbol{q}$, which is equivalent to the gauge invariance (continuity equation for the electron density) [599–601], up to the first order in $v$ and $W$, respectively, because we calculate the vertex correction to this order.

The contribution to the polarization function and self-energy coming from the first-order vertex correction Fig. 2.29e, h are given by

$$\widetilde{P}^{(1)}_{\boldsymbol{G}\boldsymbol{G}'}(\omega) = \frac{1}{\Omega}\int\frac{d\omega'}{2\pi}e^{i\omega'\eta}\int\frac{d\omega''}{2\pi}e^{i\omega''\eta}\sum_{\sigma}\sum_{ijkl}\sum_{\boldsymbol{G}''\boldsymbol{G}'''}\frac{W_{\boldsymbol{G}''\boldsymbol{G}'''}(\omega'')}{(\omega'-\varepsilon_{i\sigma}-i\eta_{i\sigma})}$$
$$\times\frac{\langle\widetilde{i\sigma}\,|\,e^{-i\boldsymbol{G}\cdot\boldsymbol{r}}\,|\,\widetilde{j\sigma}\rangle\langle\widetilde{j\sigma}\,|\,e^{-i\boldsymbol{G}''\cdot\boldsymbol{r}''}\,|\,\widetilde{k\sigma}\rangle\langle\widetilde{k\sigma}\,|\,e^{i\boldsymbol{G}'\cdot\boldsymbol{r}'}\,|\,\widetilde{l\sigma}\rangle\langle\widetilde{l\sigma}\,|\,e^{i\boldsymbol{G}'''\cdot\boldsymbol{r}'''}\,|\,\widetilde{i\sigma}\rangle}{(\omega+\omega'-\varepsilon_{j\sigma}-i\eta_{j\sigma})(\omega'+\omega''+\omega-\varepsilon_{k\sigma}-i\eta_{k\sigma})(\omega'+\omega''-\varepsilon_{l\sigma}-i\eta_{l\sigma})}, \tag{2.495}$$

**Table 2.7** Ionization potential (IP) and electron affinity (EA) of Na and $Na_3$ (in eV)

| | Na | | $Na_3$ | |
|---|---|---|---|---|
| | IP | EA | IP | EA |
| $G_0W_0$ | 5.15 | 0.41 | 4.10 | 1.14 |
| $GW$ | 5.40 | 0.33 | 4.64 | 0.51 |
| LGW | 5.23 | 0.42 | 4.48 | 0.66 |
| LGW$\Gamma_v$ | 5.01 | 0.60 | 4.08 | 1.04 |
| LGW$\Gamma_W$ | 5.12 | 0.58 | 4.04 | 1.15 |
| MRDCI | 4.97[a] | 0.44[b] | 3.76[c] | 1.07/1.17[b] |
| Expt. | 5.14[d] | 0.55[e] | 3.97[f] | 1.02/1.16[g] |

[a][683], [b][684], [c][685], [d][686], [e][687], [f][688], [g][689]

and

$$(\alpha\,|\,\Sigma_\sigma^{(2)}(\omega)\,|\,\beta) = \int \frac{d\omega'}{2\pi} e^{i\omega'\eta} \int \frac{d\omega''}{2\pi} e^{i\omega''\eta} \sum_\sigma \sum_{ijk} \sum_{\boldsymbol{G}\boldsymbol{G}'} \sum_{\boldsymbol{G}''\boldsymbol{G}'''} W_{\boldsymbol{G}\boldsymbol{G}'}(\omega') W_{\boldsymbol{G}''\boldsymbol{G}'''}(\omega'')$$
$$\times \frac{(\alpha\,|\,e^{i\boldsymbol{G}\cdot\boldsymbol{r}}\,|\,\widetilde{i\sigma}\,)\langle\,\widetilde{i\sigma}\,|\,e^{i\boldsymbol{G}'\cdot\boldsymbol{r}'}\,|\,\widetilde{j\sigma}\,\rangle\langle\,\widetilde{j\sigma}\,|\,e^{-i\boldsymbol{G}''\cdot\boldsymbol{r}''}\,|\,\widetilde{k\sigma}\,\rangle\langle\,\widetilde{k\sigma}\,|\,e^{-i\boldsymbol{G}'''\cdot\boldsymbol{r}'''}\,|\,\beta)}{(\omega+\omega'-\varepsilon_{i\sigma}-i\eta_{i\sigma})(\omega+\omega'+\omega''-\varepsilon_{j\sigma}-i\eta_{j\sigma})(\omega+\omega''-\varepsilon_{k\sigma}-i\eta_{k\sigma})}, \qquad (2.496)$$

respectively, where $W_{\boldsymbol{G}\boldsymbol{G}'}(\omega)$ is the dynamically screened Coulomb interaction, $\eta$ is a positive infinitesimal ($\eta_{i\sigma}$ is $\eta$ for occupied $i\sigma$ and $-\eta$ for empty $i\sigma$), $\Omega$ is the volume of the unit cell, and $(\alpha\,|, \,|\,\beta)$ denote basis functions.

The resulting ionization potential (IP) and the electron affinity (EA) of Na and $Na_3$ calculated by the $G_0W_0$, $GW$, LGW, LGW$\Gamma_v$, and LGW$\Gamma_W$ methods are listed in Table 2.7 together with the MRDCI values [683–685] and experimental values [686–689]. The calculated IP and EA are just the minus of the HOMO and LUMO energies, respectively. Comparing the calculated and experimental values, we find that $G_0W_0$ overestimates both IP and EA, while GW overestimates IP and underestimates EA. LGW improves $GW$ [587], but is not perfect. In contrast, LGW$\Gamma_v$ and LGW$\Gamma_W$ almost perfectly improves both IP and EA. Among all, the best agreement with the experimental values is achieved by LGW$\Gamma_W$ (or LGW$\Gamma_v$ for EA of $Na_3$). From Table 2.7, we find that LGW$\Gamma_W$ gives better IP and EA than MRDCI. Similar agreement with the experimental values was also obtained for $B_2$ and $C_2H_2$ [592].

Thus, the self-consistent treatment of $\Gamma$ is required to obtain consistently good results for both PES and PAS. Its computational cost scales as $O(N^2M^3)$, where $N$ and $M$ are the numbers of basis functions and empty states, respectively, if the plasmon-pole model is used in the calculation of the vertex part only. This should be compared with the $O(L^6)$ scaling required in the MRDCI calculations ($L$ is the number of electrons). In this respect, the LGW$\Gamma_W$ approach may have an advantage in the calculations of larger systems and crystal systems.

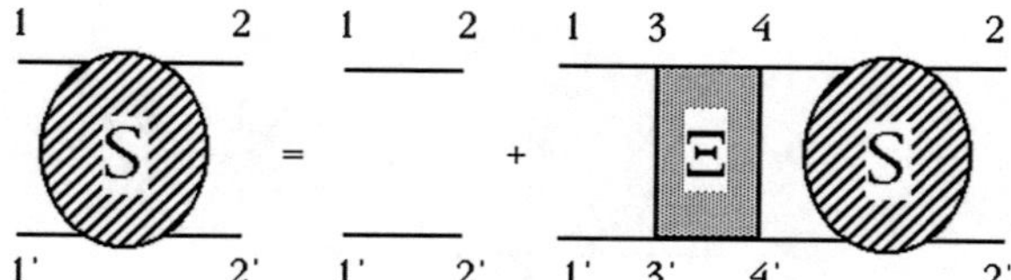


**Fig. 2.31** Diagrammatic representation of the Bethe–Salpeter equation given by (2.497)

## 2.7.6 *Inclusion of Ladder Diagrams*

Many spectroscopic properties involving two-hole or two-particle excitations cannot be described correctly within the GWA, which are based on the RPA, as has been pointed out by many authors [641, 642, 690–692], and also as was noted at the end of Sect. 2.7.3. This can be possible also to include vertex corrections but into the two-particle Green's function (instead of the one-particle Green's function, or the self-energy). As we have seen in Sect. 2.3.2, photoabsorption spectra are obtained from the imaginary part of the electric susceptibility $\chi(\boldsymbol{q}, \omega)$ in the limit $\boldsymbol{q} \to 0$. Since a two-particle Green's function can generally be related to the electric susceptibility, a way to attack the above problem is given by summing the ladder diagrams, which become important in particular at short distances, in the two-particle Green's function. The equation which includes the sum of ladder diagrams is called the Bethe-Salpeter equation [318, 597, 693–696] for a two-particle Green's function $S(1, 1'; 2, 2') = S(\boldsymbol{r}_1, t_1, \boldsymbol{r}_{1'}, t_{1'}; \boldsymbol{r}_2, t_2, \boldsymbol{r}_{2'}, t_{2'})$:

$$S(1, 1'; 2, 2') = S_0(1, 1'; 2, 2') + S_0(1, 1'; 3, 3')\Xi(3, 3'; 4, 4')S(4, 4'; 2, 2') \tag{2.497}$$

Here $S_0(1, 1'; 2, 2') = G(1', 2')G(2, 1)$ is the unperturbed function, in which disconnected term $-G(1, 1')G(2, 2')$ is excluded (we introduced the notation $G(1, 2) = G(\boldsymbol{r}_1, \boldsymbol{r}_2; t_1 - t_2)$ to express the one-particle Green's function). The Bethe–Salpeter equation (2.497) is diagrammatically represented by Fig. 2.31, in which the solid lines represent the one-particle Green's function, and the circle indicated by S and the square indicated by $\Xi$ represent, respectively, the two-particle Green's function $S$ and the interaction $\Xi$.

In the formulation introduced first by Sham and Rice [694] and later used by Hanke and Sham [695] and others [690, 696–698], the interaction $\Xi$ stands for the irreducible electron–hole interaction and contains two different contributions

$$\Xi(1, 1'; 2, 2') = -i\delta(1, 1')\delta(2, 2')U(1, 2) + i\delta(1, 2)\delta(1', 2')W'(1, 1'). \tag{2.498}$$

(Here, more complicated vertex corrections, etc., are neglected. Sham and Rice [694] proved that this simplification is justified in the effective-mass approximation.) The first term, $U(1, 2)$, is given by the bare Coulomb interaction $1/|\boldsymbol{r}_1 - \boldsymbol{r}_2|$, and the

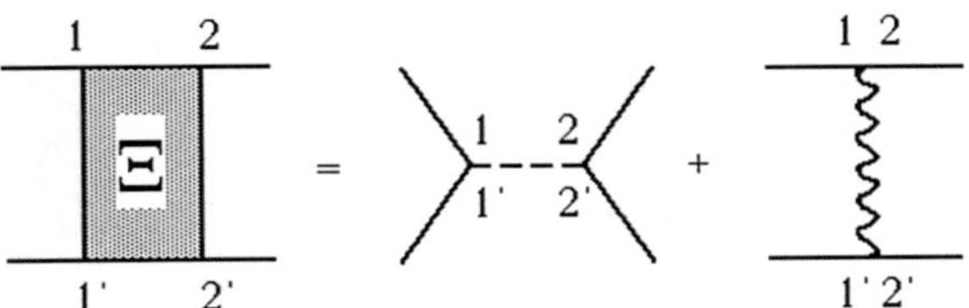


**Fig. 2.32** Diagrammatic representation of the irreducible electron–hole interaction $\Xi$ (hatched square) given by (2.498). The solid, dotted and wavy lines represent, respectively, the one-particle Green's function, the bare Coulomb interaction and the screened Coulomb interaction

second term, $W'(1,2) = W'(\boldsymbol{r}_1, \boldsymbol{r}_2; t_1 - t_2)$, represents the screened Coulomb attraction between an electron and a hole, which is given by the functional derivative of the self-energy $\Sigma$ in the GW approximation with respect to the one-particle Green's function $G$. Usually, the latter term is further approximated by $W$, which is given by the Fourier transformation of (2.480) with (2.476) from $\omega$ to $t_1 - t_2$; this approximation is equivalent to neglecting a term $G\delta W/\delta G$, as was discussed by Strinati [696]. Equation (2.498) is represented diagrammatically by Fig. 2.32, in which the dotted and wavy lines represent, respectively, the bare and screened Coulomb interaction. The inclusion of the first term of (2.498) is equivalent to summing up all the ring diagrams as in standard RPA calculations. This is similar to evaluating the matrix inversion to obtain $W$ in (2.476) and (2.477).

To solve (2.498), one has to invert a four-point function $I - S_0\Xi$, where $I$ is the identity operator. In spectral methods using a plane wave expansion, the indices of basis functions might be used as a straightforward substitution for the four space arguments of the four-point functions. Although such a procedure is sometimes prohibitive because the size of the matrix may become quite large, it became possible to do it recently because of the increase of today's computer power.

In order to compute X-ray photoemission spectra of Ni from quasi-particle energies, Springer et al. [699] used Aryasetiawan's LMTO-basis formalism for the GW calculation (see also [638–641, 643]) to put $\Xi = W$ in the Bethe–Salpeter equation (2.497) (this corresponds to summing up ladder diagrams only) and calculated the direct and exchange contributions to the self-energy $\Sigma$ in their $T$-matrix formulation within the static approximation, where the $\omega$ dependence in $W(\boldsymbol{r}, \boldsymbol{r}'; \omega)$ is completely ignored. In their calculation of the matrix inversion of the four-point kernel, they expanded the Bloch wavefunctions in localized functions of the linear muffin-tin orbitals (LMTO). More recently, Noguchi et al. applied the $T$-matrix method together with the all-electron mixed basis approach to the calculation of double excitation spectra [650, 650, 700, 701] and Auger spectra [702] of hydrocarbon molecules. Because this method treats the two-electron wavefunctions, it is possible to identify the Coulomb hole by drawing the two-particle wavefunctions [701]. Moreover, this method can treat the Auger spectra of the two electrons emitted from the valence band in the Auger process [702], because another (Auger) electron is emitted from the valence band together with the electron emitted from the core state, when one valence electron falls down to the core hole created by a X-ray photon

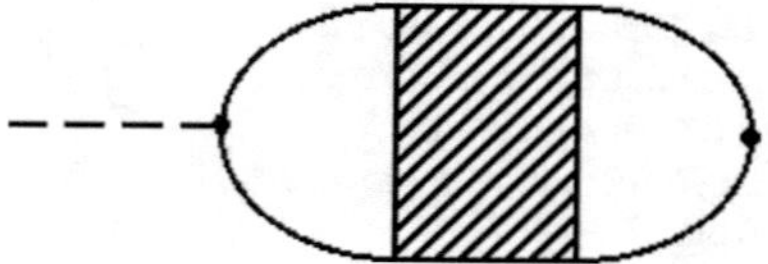

**Fig. 2.33** Diagrammatic representation of the electric susceptibility $\chi(\boldsymbol{q}, \omega)$ given by the product of $-i\widetilde{U}(q)$ and the Fourier transform of $S(\boldsymbol{r}, t, \boldsymbol{r}, t; \boldsymbol{r}', t', \boldsymbol{r}', t')$

absorption; therefore Auger spectra represent the spectra of two interacting valence electrons, not the spectra of core electrons!

Strinati [696] has derived a simplified eigenvalue equation which can be used to bypass the matrix inversion of the Bethe–Salpeter equation with $\Xi$ given by (2.498). (His scheme allows one to evaluate not only an absorption spectrum but also an Auger decay rate, in principle.) Recently, Rohlfing and Louie [697] used the Strinati scheme within the plasmon-pole approximation and localized basis set of Gaussian-type orbitals (see also [633, 634, 636] for their GWA calculation) to obtain the electric susceptibility and the optical absorption spectrum of hydrogenated Si clusters from the two-particle Green's function. They discussed also optical excitations in conjugated polymers [703]. The electric susceptibility $\chi(\boldsymbol{q}, \omega)$ is given by a product of $-i\widetilde{U}(q)$ $(\widetilde{U}(q) = 4\pi/\Omega q^2)$ and the Fourier transform of $S(\boldsymbol{r}, t, \boldsymbol{r}, t; \boldsymbol{r}', t', \boldsymbol{r}', t')$ from $(\boldsymbol{r} - \boldsymbol{r}'; t - t')$ space into $(\boldsymbol{q}, \omega)$ space. This is diagrammatically represented by Fig. 2.33. The photo absorption spectra can be obtained from the imaginary part of the electric susceptibility $\chi(\boldsymbol{q}, \omega)$ in the limit $\boldsymbol{q} \to 0$ as has been mentioned already.

Instead of using a localized basis set, the LDA Bloch functions $\psi_{\boldsymbol{k}\lambda}$ themselves can be adopted as basis functions as was suggested by Albrecht et al. [698]. This enables one to treat the problem by the usual pseudopotential–PW expansion method. Since only a limited number of electron-hole pairs is expected to contribute to each excitation, not very many Bloch functions are needed in the basis. Using this idea and the static approximation (ignorance of the $\omega$-dependence in $W(\boldsymbol{r}, \boldsymbol{r}'; \omega)$), Albrecht et al. calculated the matrix inverse of $I - S_0\Xi$ by means of spectral decomposition, in a quasi-particle level [698]. In this way, they succeeded in obtaining the photoabsorption spectra of Si crystal which is in very good agreement with experiments.

In these ways, the Bethe–Salpeter equation has been solved for the two-particle Green's function in order to treat the excitonic effect in photoabsorption [697, 698] and photoemission spectra [699].

However, quite recently, it is also pointed out that the photoabsorption spectra (PAS) obtained by solving the Bethe–Salpeter Equation (BSE) with the $G_0W_0$ QP energies and the Kohn–Sham (KS) wavefunctions are often significantly red shifted for small molecules by about 1 eV [704, 705]. Use of the HSE functional or the self-consistent GW calculation improves the results but not perfect [705, 706]. In the case of spin-polarized sodium atom (Na) and trimer ($Na_3$), the PAS calculated with $G_0W_0$+ BSE is also extremely bad, although the $G_0W_0$ QP energies are reasonably good [707]. The calculated and experimental [708] optical gaps are 1.32 eV and 2.10 eV, respectively, for Na, and the calculated and experimental [709] PAS for $Na_3$ are shown in Fig. 2.34. These calculated results are far off from the experimental data;

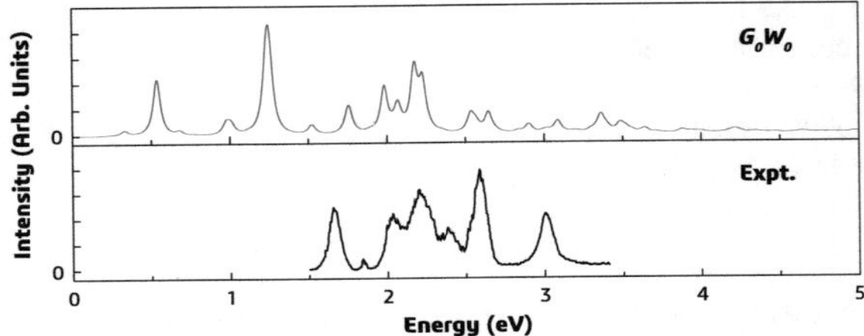


**Fig. 2.34** Photoabsorption spectra of $Na_3$ calculated by the $G_0W_0$ method. Experimental data are taken from [709]

**Table 2.8** The quasiparticle (QP) gap $E_g^{QP}$ (=IP−EA in Table 2.7) and the optical gap $E_g^{Opt}$ (corresponding to S → P transition) of Na and $Na_3$ (in eV). $E_g^{Opt}$ of $Na_3$ corresponds to the $P_1$ peak in Fig. 2.35. MRDCI values for $E_g^{QP}$ are taken from Table 2.7

| | Na | | $Na_3$ | |
|---|---|---|---|---|
| | $E_g^{QP}$ | $E_g^{Opt}$ | $E_g^{QP}$ | $E_g^{Opt}$ |
| $G_0W_0$ | 4.74 | 1.32 | 2.96 | 0.53 |
| $GW$ | 5.07 | 2.23 | 4.13 | 1.91 |
| LGW | 4.81 | 2.18 | 3.82 | 1.92 |
| LGW$\Gamma_v$ | 4.41 | 2.00 | 3.04 | 1.57 |
| LGW$\Gamma$ | 4.54 | 2.16 | 2.89 | 1.60 |
| MRDCI | 4.53 | 1.98[a] | 2.69 | 1.61[a] |
| Expt. | 4.59 | 2.10[b] | 2.95 | 1.65[c] |

[a] [684], [b] [708], [c] [709]

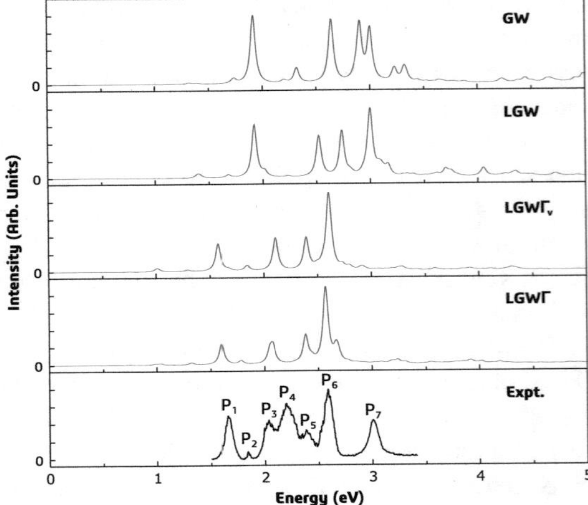


**Fig. 2.35** Photoabsorption spectra of $Na_3$ calculated by $GW$, LGW, LGW$\Gamma_v$, and LGW$\Gamma$. Experimental data are taken from [709]

For the even number clusters of sodium, the $G_0W_0$+ BSE result is not so bad [690, 710].

Recently, the BSE has been solved for the one-shot second order approach to alkaline-earth-metal atoms and spin-unpolarized small molecules [711]. The BSE was also solved for the spin-polarized LGW$\Gamma_v$ and LGW$\Gamma_W$ approaches for Na atom and $Na_3$, $B_2$, and $C_2H_2$ molecules [592]. Table 2.8 lists the QP energy gap $E_g^{QP}$ =IP−EA and the optical gap $E_g^{Opt}$ of PAS corresponding to S → P transition, of Na and $Na_3$ calculated by $G_0W_0$, $GW$, LGW, LGW$\Gamma_v$, and LGW$\Gamma_W$ as well as the corresponding MRDCI data [684] and experimental data [708, 709] (MRDCI data for $E_g^{QP}$ are taken from Table 2.7). Although $G_0W_0$ gives reasonable $E_g^{QP}$, it significantly underestimates the experimental $E_g^{Opt}$. GW overestimates the experimental $E_g^{QP}$ and $E_g^{Opt}$; The deviation from the experimental values is larger for $Na_3$ than Na. If we compare LGW with $GW$, $E_g^{QP}$ is somewhat improved by LGW, but $E_g^{Opt}$ is not. In contrast, LGW$\Gamma_v$ and LGW$\Gamma_W$ give excellent $E_g^{QP}$ and $E_g^{Opt}$. For LGW$\Gamma_W$, the difference between the theoretical and experimental values is less than 0.07 eV both for Na and $Na_3$. The resulting accuracy of LGW$\Gamma_W$ is comparable to that of MRDCI for $E_g^{Opt}$, but the former is better than the latter for $E_g^{QP}$. Figure 2.35 shows the PAS of $Na_3$ calculated by GW, LGW, LGW$\Gamma_v$, and LGW$\Gamma$ as well as the experimental spectra [709]. The overall spectral shapes are similar in all these methods, although the peak positions are almost constantly shifted by the amount indicated by the difference between the calculated and experimental $E_g^{Opt}$'s in Table 2.8 and the peak heights somewhat change after the inclusion of the vertex correction. Obviously, $GW$ and LGW overestimate the peak positions, while LGW$\Gamma_v$ and LGW$\Gamma$ give good peak positions except for $P_4$ and $P_7$. (LGW$\Gamma_v$ has a small peak at 2.9 eV, which may correspond to $P_7$.) The remaining discrepancy between the theory and experiment in the case of $Na_3$ can be attributed to the ignorance of isomers and atomic vibration effects.

### 2.7.7 *Further Remarks: Cusp Condition and Parquet Diagrams*

Another way of obtaining information about the electron correlation is to refer to some existing exact solutions for simple many-electron systems. The easiest case is that of two-electron systems in free space. The use of the center-of-mass system translates this problem into a simple one-body problem, which can be solved exactly in power series along the same lines as the solution of the hydrogen atom but with a positive Coulomb potential. (We notice that even the more complicated two-electron system in a quantum dot which is expressed by a smooth well potential, $V(\boldsymbol{r}_1, \boldsymbol{r}_2) = \gamma^2(r_1^2 + r_2^2)$, can be solved exactly in a power series expansion [712].) The form of the wavefunction when two electrons come close to each other should be unique and universal, i.e. it is essentially the same in any system with an arbitrary number of electrons and does not depend on the details of the external field. This imposes a constraint on the many-electron wavefunction at a short distance. This constraint is called the cusp condition [713, 714]. It has been demonstrated that the expectation value of the Hamiltonian calculated for a many-electron wavefunction

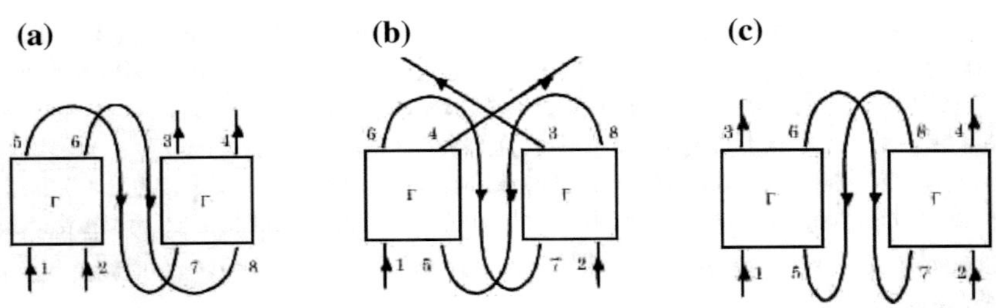


**Fig. 2.36** The basic multiplications: **a** $\Gamma s \Gamma$, **b** $\Gamma u \Gamma$, and **c** $\Gamma c \Gamma$

which does not satisfy the cusp condition cannot give a correct answer, because of the divergence of the electron–electron interaction $e^2/r$ at short distances. Therefore, the cusp condition is very important.

On the other hand, one can consider the many-electron wavefunction on the basis of a more orthodox variational principle. The variational function is given by, e.g., the Jastrow function, which will be described in Sect. 6.1. Another possibility is the use of the Gutzwiller function, which is given by applying the projection operator to a Slater determinant and thus removing the up-spin and down-spin electrons occupying the same Wannier state. The Gutzwiller ansatz is known to be very accurate for 1D systems [715, 716], but turned out to be not so appropriate for 2D systems [717, 718].

Probably, the most recent progress in many-body perturbation theory (MBPT) is the attempt to sum up parquet diagrams [719, 720], although they are based on models and not ab initio calculations. The parquet diagrams were earlier used in elementary particle and nuclear physics, and later they were introduced in solid state physics by Dominicis and Martin [721]; see also the papers by Winer [722], Bickers and White [723], and Lande and Smith [724, 725].

In parquet theory, if we ignore the left and right vertex corrections, there are three ways of connecting two vertices by two solid lines (Green's function) as shown in Fig. 2.36. All these diagrams can be separated into two disconnected parts if we cut these two solid lines. Such diagrams are called non-simple or reducible. The diagram (a) is non-simple in 12-channel (or $s$-channel) because it is separated into (12) and (34) pairs. Similarly, the diagrams (b) and (c) are non-simple, respectively, in 14-channel (or $u$-channel) and in 13-channel (or $c$-channel). If the diagram cannot be separated into (12) and (34) pairs by cutting any two of the internal solid lines, the diagram is called simple or irreducible in 12-channel (or $s$-channel). Similarly, we can define diagrams, which are simple (irreducible) in 14-channel (or $u$-channel) or in 13-channel (or $c$-channel). The diagram (a), which is non-simple in 12-channel ($s$-channel), is simple in 14- and 13-channels, because it cannot be separated into (14) and (23) pairs or (13) and (24) paris: Either (12) or (34) pair is still (indirectly) connected in one $\Gamma$ even if two internal solid lines are removed from the other $\Gamma$. Similarly, the diagram (b) is non-simple in 14-channel ($u$-channel) but simple in 12- and 13-channels, and the diagram (c) is non-simple in 13-channel ($c$-channel)

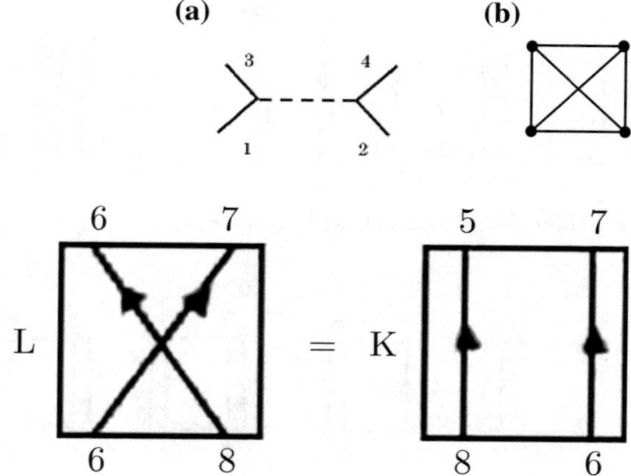


**Fig. 2.37** **a** the lowest order $I_{1,2;\,3,4}$ is the bare Coulomb interaction given by (2.37) and **b** the next order of $I_{1,2;\,3,4}$

**Fig. 2.38** $L_{6,8;\,5,7}$ is identical to $K_{8,6;\,5,7}$. Note that the first two (in the figure, the lower two) indices are interchanged

but simple in 12- and 14-channels. Hereafter, the sum of all diagrams, which are non-simple in 12-channel, is denoted by $S$. Similarly, all non-simple diagrams in 14-channel and in 13-channel are denoted by $U$ and $C$, respectively. Obviously, they are all different diagrams constituting the vertex functions. Only missing is a kind of diagrams which are simple in all 12-, 14- and 13-channels. Such diagrams are called irreducible interaction and denoted by $I$. Thus the sum of all vertex diagrams $\Gamma$ is given by a sum of $I$, $S$, $U$, and $C$:

$$\Gamma_{1,2;\,3,4} = I_{1,2;\,3,4} + S_{1,2;\,3,4} + U_{1,2;\,3,4} + C_{1,2;\,3,4}. \tag{2.499}$$

As a very good approximation, we can replace $I_{1,2;3,4}$ by the bare Coulomb interaction (Fig. 2.37a)

$$I_{1,2;\,3,4} \approx \frac{1}{|\boldsymbol{r}_1 - \boldsymbol{r}_2|}\delta(\boldsymbol{r}_1 - \boldsymbol{r}_3)\delta(\boldsymbol{r}_2 - \boldsymbol{r}_4)\delta(t_1 - t_2)\delta(t_1 - t_3)\delta(t_2 - t_4), \tag{2.500}$$

because the next order correction for $I_{1,2;\,3,4}$ is given by Fig. 2.37b, which is non-simple for all channels. If we use this simplest $I$ in $\Gamma$ in Fig. 2.36, we see that (a) expresses a particle-particle ladder, (b) expresses a particle-hole ladder, and (c) expresses a RPA ring. The sum of all diagrams, which are simple in 12-channel ($s$-channel), is given by $I + U + C = \Gamma - S$. Similarly, the sum of all diagrams, which are simple in 14-channel ($u$-channel) or in 13-channel ($c$-channel), is given by $I+S+C = \Gamma-U$ or $I+S+U = \Gamma-C$. Then, the sum of all non-simple diagrams in 12-channel ($s$-channel), $S$, is given by Fig. 2.36a with the left $\Gamma$ replaced by $\Gamma-S$. Similarly, the sum of all non-simple diagrams in 14-channel ($u$-channel), $U$, or in 13-channel ($c$-channel), $C$, is given by Fig. 2.36b or c with the left $\Gamma$ replaced by $\Gamma - U$ or $\Gamma - C$. Thus we have the basic parquet equations

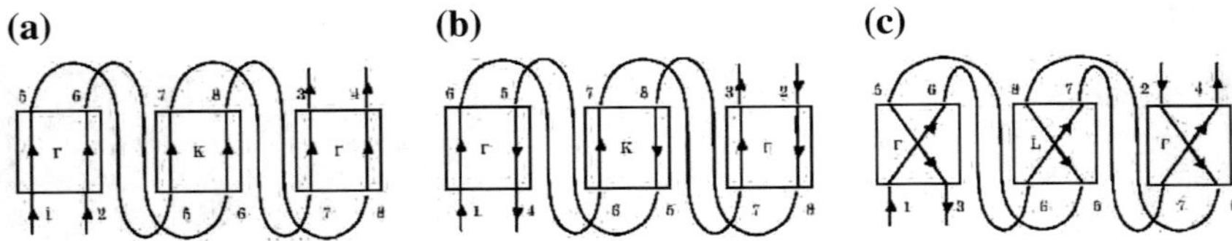


**Fig. 2.39** Multiplications used in (2.503)a–c

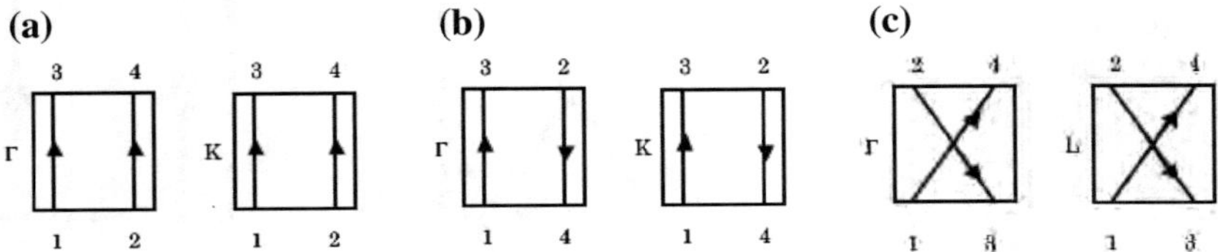


**Fig. 2.40** The diagrammatic representation of $\Gamma_{1,2;\,3,4}$ and $K_{1,2;\,3,4}$ in each term of (2.503); **a–c** correspond to the first, second, and third terms, respectively

$$S_{1,2;\,3,4} = \sum_{5,6}\sum_{7,8}(\Gamma_{1,2;\,5,6} - S_{1,2;\,5,6})K_{5,6;\,7,8}\Gamma_{7,8;\,3,4}, \tag{2.501a}$$

$$U_{1,2;\,3,4} = \sum_{5,6}\sum_{7,8}(\Gamma_{1,5;\,6,4} - S_{1,5;\,6,4})K_{6,8;\,7,5}\Gamma_{7,2;\,3,8}, \tag{2.501b}$$

$$C_{1,2;\,3,4} = \sum_{5,6}\sum_{7,8}(\Gamma_{1,5;\,3,6} - S_{1,5;\,3,6})L_{6,8;\,5,7}\Gamma_{7,2;\,8,4}, \tag{2.501c}$$

$$\Gamma_{1,2;\,3,4} = I_{1,2;\,3,4} + S_{1,2;\,3,4} + U_{1,2;\,3,4} + C_{1,2;\,3,4}, \tag{2.501d}$$

where we put $K_{5,6;\,7,8} = G(5,7)G(6,8)$ and $L_{6,8;\,5,7} = K_{8,6;\,5,7} = G(8,5)G(6,7)$. That is, we define $L$ by interchanging the first two indices in $K$; see Fig. 2.38. Multiplications in the r.h.s. of (2.503)a–c are diagrammatically given in Fig. 2.39, and $\Gamma_{1,2;\,3,4}$ and $K_{1,2;\,3,4}$ in each term of (2.503)a–c are given in Fig. 2.40. The important fact is that (2.499)–(2.503) give a closed set of equations provided that $I_{1,2;\,3,4}$ is given by (2.500). (Note that we totally omitted the vertex corrections, but they can be incorporated by a simple extension of the formulae given above; see [722–725].) Therefore, if one could solve it, the resulting four point vertex function would include almost all important diagrams and become almost exact. Maiti and Chubukov [719], and Lange et al. [720] treated these parquet diagrams with the renormalization group method to discuss superconductivity or X-ray spectra. Although there approach is not ab initio, it is highly desirable to extend their approach to ab initio calculations and solve the parquet equation at least approximately to obtain quite accurate correlation energies.

# References


1. A. Messiah, *Mécanique Quantique* (Dunod, Paris, 1959); English translation, *Quantum Mechanics* (North-Holland, Amsterdam, 1961)
2. L.I. Schiff, *Quantum Mechanics* (McGraw-Hill, New York, 1955, 1968)
3. P.A.M. Dirac, *The Principles of Quantum Mechanics* (Oxford University Press, London, 1958)
4. L.D. Landau, E.M. Lifshitz, *Quantum Mechanics; Non-relativistic Theory* (Pergamon, Oxford, 1965)
5. V.B. Berestetskii, E.M. Lifshitz, L.P. Pitaevskii, *Relativistic Quantum Theory*, Part I and II (Pergamon, Oxford, 1971)
6. W. Pauli, Z. Phys. **31**, 765 (1925)
7. J.C. Slater, Phys. Rev. **34**, 1293 (1929)
8. E. Fermi, Z. Phys. **36**, 902 (1926)
9. P.A.M. Dirac, Proc. R. Soc. A **112**, 661 (1926)
10. S.N. Bose, Z. Phys. **26**, 178 (1924)
11. A. Einstein, Sitzber. Preuss. Akad. Wiss. (1924), p. 261; Sitzber. Preuss. Akad. Wiss. (1925), p. 3
12. F. Hund, Z. Phys. **51**, 759 (1928); Z. Phys. **73**, 1 (1931)
13. K. Ohno, S. Ono, T. Isobe, J. Chem. Phys. **146**, 084108 (2017)
14. D.R. Hartree, Proc. Camb. Philos. Soc. **24**, 89 (1928); Mem. Manchester Philos. Soc. **77**, 91 (1932–1933)
15. V. Fock, Z. Phys. **61**, 126 (1930)
16. R.P. Feynman, Science **153**, 699 (1966)
17. R.P. Feynman, *Statistical Mechanics Frontiers in Physics*. Lecture Note Series, vol. 36 (Benjamin, Reading, 1972)
18. J. Goldstone, Proc. R. Soc. A **239**, 267 (1957)
19. M. Gell-Mann, K.A. Brueckner, Phys. Rev. **106**, 364 (1957)
20. D. Pines, *The Many Body Problem, Frontiers in Physics*. Lecture Note Series, vol. 6 (Benjamin/Cummings, Reading, 1961)
21. P. Nozières, D. Pines, Nuovo Cimento [X] **9**, 470 (1958)
22. J.C. Slater, Phys. Rev. **36**, 57 (1930); Phys. Rev. **42**, 33 (1932)
23. S.F. Boys, Proc. R. Soc. A **200**, 542 (1950)
24. W.J. Hehre, R.F. Stewart, J.A. Pople, J. Chem. Phys. **51**, 2657 (1970); R. Ditchfield, W.J. Hehre, J.A. Pople, J. Chem. Phys. **54**, 724 (1971)
25. J.A. Pople, M. Head-Gordon, D.J. Fox, K. Raghavachari, L.A. Curtiss, J. Chem. Phys. **90**, 5622 (1989)
26. B. Delley, J. Chem. Phys. **92**, 508 (1990)
27. A. Szabo, N.S. Ostlund, *Modern Quantum Chemistry: Introduction to Advanced Electronic Structure Theory* (McGraw-Hill, New York, 1989)
28. A.A. Hasanein, M.W. Evans, *Quantum Chemistry Volume 2: Computational Methods in Quantum Chemistry*. Contemporary Chemical Physics, vol. 5 (World Scientific, Singapore, 1996)
29. C.C.J. Roothaan, Rev. Mod. Phys. **23**, 69 (1951); Rev. Mod. Phys. **32**, 179 (1960)
30. C.C.J. Roothaan, J. Chem. Phys. **19**, 1445 (1951)
31. K. Rüdenberg, J. Chem. Phys. **19**, 1459 (1951)
32. P. Fulde, *Electron Correlations in Molecules and Solids*. Solid-State Sciences, vol. 100 (Springer, Berlin, 1991, 1993, 1995)
33. P.W. Anderson, *Basic Notions of Condensed Matter Physics, Frontiers in Physics*. Lecture Note Series, vol. 55 (Benjamin/Cummings, Menlo Park, 1984)
34. N.F. Mott, *Metal–Insulator Transitions* (Taylor & Francis, London 1974, 1990)
35. A. Shukla, M. Dolg, P. Fulde, H. Stoll, Phys. Rev. B **57**, 1471 (1998)
36. S. Kalvoda, M. Dolg, H.-J. Flad, P. Fulde, H. Stoll, Phys. Rev. B **57**, 2127 (1998)
37. M. Albrecht, A. Shukla, M. Dolg, P. Fulde, H. Stoll, Chem. Phys. Lett. **285**, 174 (1998)
38. K. Rościszewski, K. Doll, B. Paulus, P. Fulde, H. Stoll, Phys. Rev. B **57**, 2127 (1998)

39. W. Heitler, F. London, Z. Phys. **44**, 455 (1927)
40. L. Pauling, E.B. Wilson, *Introduction to Quantum Mechanics* (McGraw-Hill, New York, 1935)
41. N.F. Mott, Can. J. Phys. **34**, 1356 (1956)
42. P.W. Anderson, Phys. Rev. **109**, 1492 (1958)
43. P.-O. Löwdin, Phys. Rev. **97**, 1474 (1955); Adv. Chem. Phys. **2**, 207 (1959)
44. P.L. Brillouin, J. Phys. **3**, 373 (1932). in French
45. R.J. Buenker, S.D. Peyerimhoff, Theor. Chim. Acta (Berl.) **39**, 217 (1975)
46. B.O. Roos, P.R. Taylor, P.E.M. Siegbahn, Chem. Phys. **48**, 157 (1980)
47. J. Olsen, Int. J. Quantum Chem. **111**, 3267 (2011)
48. C. Møller, M.S. Plesset, Phys. Rev. **46**, 618 (1934)
49. J.A. Pople, J.B. Binkley, R. Seeger, Int. J. Quantum Chem. Symp. **10S**, 1 (1976)
50. R. Krinshnan, J.A. Pople, Int. J. Quantum Chem. Symp. **14**, 91 (1978)
51. K. Wolinski, P. Pulay, J. Chem. Phys. **90**, 3647 (1989)
52. O. Sinanoğlu, Adv. Chem. Phys. **6**, 315 (1964); R K. Nesbet, Adv. Chem. Phys. **9**, 321 (1965)
53. J. Čižek, Adv. Chem. Phys. **14**, 35 (1969); J. Paldus, J. Čižek, Adv. Chem. Phys. **9**, 105 (1975)
54. M.A. Robb, in *Computational Techniques in Quantum Chemistry*, ed. by G.H.F. Dierchksen, R.T. Sutcliffe, A. Veilard (Reidel, Boston, 1975), p. 435
55. R.J. Bartlett, Ann. Rev. Phys. Chem. **32**, 359 (1981)
56. R.J. Bartlett, C.E. Dykstra, J. Paldus, in *Advanced Theories and Computational Approaches to the Electronic Structure of Molecules* ed. by C.E. Dykstra (Reidel, Dordrecht, 1984), p. 127
57. C. Kittel, *Quantum Theory of Solids* (Wiley, New York, 1972)
58. W.A. Harrison, *Solid State Theory* (McGraw-Hill, New York, 1970; Dover, New York, 1979)
59. W.A. Harrison, *Electronic Structure and the Properties of Solids* (Freeman, San Francisco, 1980)
60. P. Hohenberg, W. Kohn, Phys. Rev. **136**, 864 (1964)
61. D.M. Ceperley, B.J. Alder, Phys. Rev. Lett. **45**, 566 (1980)
62. N.D. Mermin, Phys. Rev. **137**, 1441 (1965)
63. E. Runge, E.K.U. Gross, Phys. Rev. Lett. **52**, 997 (1984)
64. P. Hessler, J. Park, K. Burke, Phys. Rev. Lett. **82**, 378 (1999)
65. J. Harris, Phys. Rev. B **31**, 1770 (1985)
66. H.M. Polatoglou, M. Methfessel, Phys. Rev. B **37**, 10403 (1988)
67. W.M.C. Foulkes, R. Haydock, Phys. Rev. B **39**, 12520 (1989)
68. A.J. Read, R.J. Needs, J. Phys.: Condens. Matter **1**, 7565 (1989)
69. M.W. Finnis, J. Phys.: Condens. Matter **2**, 331 (1990)
70. M. Pearson, E. Smargiassi, P.A. Madden, J. Phys.: Condens. Matter **5**, 3221 (1993)
71. W. Kohn, L.J. Sham, Phys. Rev. **140**, 1133 (1965)
72. U. von Barth, L. Hedin, J. Phys. C: Solid State Phys. **5**, 1629 (1972)
73. O. Gunnarsson, B.I. Lundqvist, Phys. Rev. B **13**, 4274 (1976)
74. T.A. Koopmans, Physica **1**, 104 (1933)
75. J.C. Slater, T.M. Wilson, J.H. Wood, Phys. Rev. **179**, 28 (1969)
76. R. Gaspar, Acta Phys. Acad. Sci. Hung. **3**, 263 (1954)
77. E.A. Kmetko, Phys. Rev. A **1**, 37 (1970)
78. V.J. Moruzzi, J.F. Janak, A.R. Williams, *Calculated Electron Properties of Metals* (Pergamon, New York, 1978)
79. J.P. Perdew, A. Zunger, Phys. Rev. B **23**, 5048 (1981)
80. W. Kohn, Y. Meir, D.E. Makarov, Phys. Rev. Lett. **80**, 4153 (1998)
81. M. Dion, H. Rydberg, E. Schröder, D.C. Langreth, B.I. Lundqvist, Phys. Rev. Lett. **92**, 246401 (2004)
82. O.A. von Lilienfeld, I. Tavernelli, U. Rothlisberger, D. Sebastiani, Phys. Rev. Lett. **93**, 153004 (2004)
83. A.D. Becke, E.R. Johnson, J. Chem. Phys. **127**, 154108 (2007)
84. A. Tkatchenko, M. Scheffler, Phys. Rev. Lett. **102**, 073005 (2009)

85. J. Klimeš, D.R. Bowler, A. Michaelides, J. Phys.: Condens. Matter **22**, 022201 (2010)
86. F.W. Averill, G.S. Painter, Phys. Rev. B **24**, 6795 (1981)
87. R. Kuwahara, K. Ohno, J. Chem. Phys. **141**, 084108 (2014)
88. D.C. Langreth, J.P. Perdew, Phys. Rev. B **21**, 5469 (1980)
89. D.C. Langreth, M.J. Mehl, Phys. Rev. B **28**, 1809 (1983)
90. J.P. Perdew, Y. Wang, Phys. Rev. B **33**, 8800 (1986)
91. J.P. Perdew, Phys. Rev. B **33**, 8822 (1986)
92. A.D. Becke, Phys. Rev. A **38**, 3098 (1988)
93. J.P. Perdew, Y. Wang, Phys. Rev. B **45**, 13244 (1992)
94. J.P. Perdew, J.A. Chevary, S.H. Vosko, K.A. Jackson, M.R. Pederson, D.J. Singh, C. Fiolhais, Phys. Rev B **46**, 6671 (1992)
95. J.P. Perdew, K. Burke, M. Ernzorhof, Phys. Rev. Lett. **77**, 3865 (1996); (Errata) **78**, 1396 (1997)
96. A.D. Becke, J. Chem, Phys. **84**, 4524 (1986)
97. J.P. Perdew, S. Kurth, A. Zupan, P. Blaha, Phys. Rev. Lett. **82**, 2544 (1999)
98. J.P. Perdew, A. Ruzsinszky, G.I. Csonka, O.A. Vydrov, G.E. Scuseria, L.A. Constantin, X. Zhou, K. Burke, Phys. Rev. Lett. **100**, 136406 (2008)
99. P. Bagno, O. Jepsen, O. Gunnarsson, Phys. Rev. B **40**, 1997 (1989)
100. T. Asada, K. Terakura, Phys. Rev. B **46**, 13599 (1992)
101. L. Vocadlo, G.A. Dewijs, G. Kress, M. Gillan, G.D. Price, Faraday Discuss. **106**, 205 (1997)
102. D.R. Hamann, Phys. Rev. B **55**, R10157 (1997)
103. R. Kuwahara, private communication; see also: R. Kuwahara, Y. Kudo, T. Morisato, K. Ohno, J. Phys. Chem. A **115**, 5147 (2011)
104. J. Janak, Phys. Rev. B **18**, 7165 (1978)
105. F. Gygi, A. Baldereschi, Phys. Rev. Lett. **62**, 2160 (1989)
106. C.O. Almbladh, U. von Barth, Phys. Rev. B **31**, 3231 (1985)
107. D. Vogel, P. Krüger, J. Pollmann, Phys. Rev. B **52**, R14316 (1995); Phys. Rev. B **54**, 5495 (1996); Phys. Rev. B **55**, 12836 (1997)
108. V.I. Anisimov, J. Zaanen, O.K. Andersen, Phys. Rev. B **44**, 943 (1991)
109. A.I. Liechtenstein, V.I. Anisimov, J. Zaanen, Phys. Rev. B **52**, R5467 (1995)
110. M. Cococcioni, in *Correlated Electrons: From Models to Materials*. Modelling and Simulation, vol. 2, ed. by E. Pavaini, E. Koch, F. Anders, M. Jarrell (Forschungszentrum Jülich, 2012), pp. 4.1–4.40. ISBN 978-3-89336-796-2
111. E. Clementi, S.J. Chakravorty, J. Chem. Phys. **93**, 2591 (1990)
112. A.D. Becke, J. Chem. Phys. **98**(2), 1372–1377 (1993)
113. D.C. Langreth, J.P. Perdew, Phys. Rev. B **15**, 2884 (1977)
114. M. Levy, Phys. Rev. A **43**, 4637 (1991)
115. A.D. Becke, J. Chem. Phys. **88**, 1053 (1988)
116. A.D. Becke, J. Chem. Phys. **98**(7), 5648–5652 (1993)
117. K. Kim, K.D. Jordan, J. Phys. Chem. **98**, 10089 (1994)
118. C. Lee, W. Yang, R.G. Parr, Phys. Rev. B **37**, 785 (1988)
119. S.H. Vosko, L. Wilk, M. Nusair, Can. J. Phys. **58**, 1200 (1980)
120. J.P. Perdew, M. Ernzerhof, K. Burke, J. Chem. Phys. **105**, 9982 (1996)
121. J. Heyd, G.E. Scuseria, M. Ernzerhof, J. Chem. Phys. **118**, 8207 (2003)
122. Y. Zhao, D.G.A. Truhlar, J. Chem. Phys. **125**, 194101 (2006)
123. R. Peverati, D.G. Truhlar, J. Phys. Chem. Lett. **3**, 117 (2012)
124. R. Peverati, D.G. Truhlar, Phys. Chem. Chem. Phys. **14**, 13171 (2012)
125. H.S. Yu, X. He, D.G. Truhlar, J. Chem. Theory Comput. **12**, 1280 (2016)
126. Y. Zhao, N.E. Schultz, D.G. Truhlar, J. Chem. Phys. **123**, 161103 (2005)
127. Y. Zhao, N.E. Schultz, D.G. Truhlar, J. Chem. Theory Comput. **2**, 364 (2006)
128. Y. Zhao, D. Truhlar, Theor. Chem. Acc. **120**, 215 (2008)
129. Y. Zhao, D.G. Truhlar, J. Phys. Chem. A **110**, 13126 (2006)
130. Y. Zhao, D.G. Truhlar, J. Chem. Theory Comput. **4**, 1849 (2008)
131. H.S. Yu, X. He, S.L. Li, D.G. Truhlar, Chem. Sci. **7**, 5032 (2016)

132. R. Peverati, D.G. Truhlar, J. Phys. Chem. Lett. **2**, 2810 (2011)
133. R. Peverati, D.G. Truhlar, Phys. Chem. Chem. Phys. **14**, 16187 (2012)
134. N. Mardirossian, M. Head-Gordon, J. Chem. Theory Comput. **12**, 4303 (2016)
135. C. Kittel, *Introduction to Solid State Physics*, 7th edn. (Wiley, New York, 1996)
136. I. Souza, N. Marzari, D. Vanderbilt, Phys. Rev B **65**, 035109 (2001)
137. D.J. Chadi, M.L. Cohen, Phys. Rev. B **8**, 5747 (1973)
138. H.J. Monkhorst, J.D. Pack, Phys. Rev. B **13**, 5188 (1976)
139. M. Orchin, H.H. Jaffé, *Symmetry, Orbitals and Spectra* (Wiley, New York, 1974)
140. M. Tinkham, *Group Theory and Quantum Mechanics* (McGraw-Hill, New York, 1964)
141. J.F. Cornwell, *Group Theory and Electronic Energy Bands in Solids* (North-Holland, Amsterdam, 1969)
142. J.C. Slater, G.F. Koster, Phys. Rev. **94**, 1498 (1954)
143. D.E. Ellis, G.S. Painter, Phys. Rev. B **2**, 2887 (1970)
144. A. Rosén, D.E. Ellis, H. Adachi, F.W. Averill, J. Chem. Phys. **65**, 3629 (1976)
145. H. Adachi, M. Tsukada, C. Satoko, J. Phys. Soc. Jpn. **45**, 875 (1978)
146. C. Satoko, M. Tsukada, H. Adachi, J. Phys. Soc. Jpn. **45**, 1333 (1978)
147. H. Adachi, S. Shiokawa, M. Tsukada, C. Satoko, S. Sugano, J. Phys. Soc. Jpn. **45**, 1528 (1979)
148. H.-P. Cheng, R.S. Berry, Phys. Rev. B **43**, 10647 (1991)
149. G. Kresse, J. Joubert, Phys. Rev. B **59**, 1758 (1999)
150. C. Herring, Phys. Rev. **57**, 1169 (1940)
151. J. Callaway, Phys. Rev. **97**, 933 (1955)
152. E. Brown, J.A. Krumhansl, Phys. Rev. **109**, 30 (1958)
153. T.O. Woodruff, in *Solid State Physics*, vol. 4, ed. by F. Seitz, D. Turnbull (Academic Press, New York, 1957), p. 367
154. T.O. Woodruff, Phys. Rev. **103**, 1159 (1956)
155. R.A. Deegan, W.D. Twose, Phys. Rev. **164**, 993 (1967)
156. R.N. Euwema, D.J. Stukel, Phys. Rev. B **1**, 4692 (1970)
157. A.B. Kunz, Phys. Lett. A **27**, 401 (1968); Phys. Rev. **180**, 934 (1969)
158. F.A. Butler, F.K. Bloom, E. Brown, Phys. Rev. **180**, 744 (1969)
159. P. Bendt, A. Zunger, Phys. Rev. B **26**, 3114 (1982)
160. R.N. Euwema, J. Quantum Chem. **5**, 61 (1971); Phys. Rev. B **4**, 4332 (1971)
161. S.G. Louie, K.-M. Ho, M.L. Cohen, Phys. Rev. B **19**, 1774 (1979)
162. K.M. Ho, C. Elsässer, C.T. Chan, M. Fähnle, J. Phys.: Condens. Matter **4**, 5189 (1992); J. Phys.: Condens. Matter **4**, 5207 (1992)
163. A.B. Gordienko, Y.N. Zhuravlev, A.S. Poplavnoi, Phys. Status Solidi B **198**, 707 (1996)
164. S. Koslmeier, C. Elsässer, B. Meyer, M.W. Finnis, Phys. Status Solidi A **166**, 417 (1998)
165. C. Friedli, N.W. Ashcroft, Phys. Rev. B **16**, 662 (1977)
166. H. Kamiyama, K. Ohno, Y. Maruyama, Y. Kawazoe, in *Physics and Chemistry of Finite Systems: From Clusters to Crystals*, vol. II, ed. by P. Jena, S.N. Khama, B.K. Rao (Kluwer Academic, Amsterdam, 1992), p. 1335
167. Y. Maruyama, H. Kamiyama, K. Ohno, Y. Kawazoe, in *Advanced Materials '93 I/B*, ed. by M. Homma et al., Trans. Mat. Res. Soc. Jpn. **14B**, 1215 (1994)
168. K. Ohno, Y. Maruyama, Y. Kawazoe, Phys. Rev. B **53**, 4078 (1996)
169. K. Ohno, Y. Maruyama, K. Esfarjani, Y. Kawazoe, N. Sato, R. Hatakeyama, T. Hirata, M. Niwano, Phys. Rev. Lett. **76**, 3590 (1996)
170. T. Ohtsuki, K. Masumoto, K. Ohno, Y. Maruyama, Y. Kawazoe, K. Sueki, K. Kikuchi, Phys. Rev. Lett. **77**, 3522 (1996)
171. Y. Maruyama, Ph.D thesis, Tohoku University, 1996
172. K. Ohno, F. Mauri, S.G. Louie, Phys. Rev. B **56**, 1009 (1997)
173. K. Ohno, Y. Maruyama, H. Kamiyama, E. Bei, K. Shiga, Z.-Q. Li, K. Esfarjani, Y. Kawazoe, in *Mesoscopic Dynamics of Fracture: Computational Materials Design*, vol. 1, ed. by H. Kitagawa, T. Aihara, Jr., Y. Kawazoe, *Advances in Materials Research* (Springer, Berlin, 1998), p. 210
174. F. Herman, S. Skillman, *Atomic Structure Calculations* (Prentice-Hall, New Jersey, 1963)

175. S. Ono, Y. Noguchi, R. Sahara, Y. Kawazoe, K. Ohno, Comput. Phys. Commun. **189**, 20 (2015)
176. T. Ohtsuki, K. Ohno, K. Shiga, Y. Kawazoe, Y. Maruyama, K. Masumoto, Phys. Rev. Lett. **81**, 967 (1998)
177. T. Ohtsuki, K. Ohno, K. Shiga, Y. Kawazoe, Y. Maruyama, K. Masumoto, Phys. Rev. B **60**, 1531 (1999)
178. T. Ohtsuki, K. Ohno, K. Shiga, Y. Kawazoe, Y. Maruyama, K. Masumoto, J. Chem. Phys. **112**, 2834 (2000)
179. K. Shiga, K. Ohno, T. Ohtsuki, Y. Kawazoe, Mat. Trans. JIM **42**(11), 2189 (2001)
180. T. Ohtsuki, K. Ohno, K. Shiga, Y. Kawazoe, Y. Maruyama, K. Shikano, K. Masumoto, Phys. Rev. B **64**, 125402 (2001)
181. T. Ohtsuki, K. Ohno, K. Shiga, Y. Kawazoe, H. Yuki, Phys. Rev. B **65**, 73402 (2002)
182. T. Ohtsuki, K. Ohno, Phys. Rev. B **72**, 153411 (2005)
183. C.M.J. Wijers, H.G. Bekker, N.E. Christensen, Phys. Rev. B **52**, 7767 (1995)
184. J.C. Dobrowolski, G. Karpinska, A.P. Mazurek, Polish J. Chem. **69**, 1059 (1995)
185. S. Rabii, N.A.W. Holzwarth, G. Loupias, V. Codazzi, D. Guerard, J. Phys. Chem. Solids. **57**, 921 (1996)
186. W.H. Press, S.A. Teukolsky, W.T. Vettering, B.R. Fiannery, *Numerical Recipes in FORTRAN: The Art of Scientific Computing*, 2nd edn. (Cambridge University Press, New York, 1992)
187. V.J. Moruzzi, C.B. Sommers, *Calculated Electronic Properties of Ordered Alloys: A Handbook* (World Scientific, Singapore, 1995)
188. D.A. Papaconstantopoulos, *Handbook of The Band Structure of Elemental Solids* (Plenum, New York, 1986)
189. W.A. Harrison, *Pseudopotentials in the Theory of Metals* (Benjamin/Cummings, Menlo Park, 1966)
190. G.B. Bachelet, D.R. Hamann, M. Schlüter, Phys. Rev. Lett. **26**, 4199 (1982)
191. D.R. Hamann, Phys. Rev. B **40**, 2980 (1989)
192. L. Kleinman, D.M. Bylander, Phys. Rev. Lett. **48**, 1425 (1982)
193. A.M. Rappe, K.M. Rabe, E. Kaxiras, J.D. Joannopoulos, Phys. Rev. B **41**, 1227 (1990)
194. N. Troullier, J.L. Martins, Phys. Rev. B **43**, 1993 (1991)
195. X. Gonze, R. Strumpf, M. Scheffler, Phys. Rev. B **41**, 12264 (1990); Phys. Rev. B **44**, 8503 (1991)
196. D. Vanderbilt, Phys. Rev. B **41**, 7892 (1990)
197. P.E. Blöchl, Phys. Rev. B **50**, 17953 (1994); see also, P.E. Blöchl, Phys. Rev. B **41**, 5414 (1990)
198. J.C. Slater, Phys. Rev. **51**, 846 (1937)
199. J.C. Slater, Phys. Rev. **92**, 603 (1953)
200. J.C. Slater, *Quantum Theory of Molecules and Solids*, vol. 2 (McGraw-Hill, New York, 1965)
201. T.L. Loucks, *Augmented Plane Wave Method* (Benjamin/Cummings, New York, 1967)
202. O.K. Andersen, Phys. Rev. B **12**, 3060 (1975)
203. J. Korringa, Physica **13**, 392 (1947)
204. W. Kohn, J. Rostoker, Phys. Rev. **94**, 1111 (1954)
205. J. Kübler, V. Eyert, in *Materials Science and Technology*, vol. 3A (Electronic and Magnetic Properties of Metals and Ceramics, Part I), ed. by R.W. Cahn, P. Haasen, E.J. Kramer (VCH, Weinheim, 1992)
206. J.M. Ziman, Proc. Phys. Soc. **86**, 337 (1965)
207. J.C. Slater, Phys. Rev. **145**, 599 (1966)
208. J. Callaway, *Quantum Theory of the Solid State* (Academic Press, New York, 1974) (Parts A and B)
209. E. Wimmer, H. Krakauer, M. Weinert, A.J. Freeman, Phys. Rev. B **24**, 864 (1981)
210. M. Weinert, E. Wimmer, A.J. Freeman, Phys. Rev. B **26**, 4571 (1982)
211. B.I. Min, S. Massidda, A.J. Freeman, Phys. Rev. B **38**, 1970 (1988)
212. N. Hamada, M. Hwang, A.J. Freeman, Phys. Rev. B **41**, 3620 (1990)
213. R. Wu, L. Zhong, L.-J. Chen, A.J. Freeman, Phys. Rev. B **54**, 7084 (1996)

214. A.B. Shick, D.L. Novikov, A.J. Freeman, Phys. Rev. B **56**, R14259 (1997)
215. P. Blaha, K. Schwarz, P. Sorantin, S.B. Trickey, Comput. Phys. Commun. **59**, 399 (1990)
216. B. Kohler, S. Wilke, M. Scheffler, R. Kouba, C. Ambrosch-Draxl, Comput. Phys. Commun. **94**, 31 (1996)
217. P.H. Dederichs, T. Hoshino, B. Drittler, K. Abraham, R. Zeller, Physica **172**, 203 (1991)
218. O.K. Andersen, R.G. Wooly, Mol. Phys. **26**, 905 (1973)
219. H. Skriver, *The LMTO Method* (Springer, Berlin, 1984)
220. J.M. Wills, B.R. Cooper, Phys. Rev. B **36**, 3809 (1987)
221. D. Price, B.R. Cooper, Phys. Rev. B **39**, 4945 (1989)
222. S.Y. Savrasov, Phys. Rev. B **54**, 16470 (1996)
223. S.A. Ostanin, E.I. Salamatov, V.I. Kormilets, Phys. Solid State **39**, 148 (1997)
224. S.A. Ostanin, E.I. Salamatov, J. Phys.: Condens. Matter **9**, 7063 (1997)
225. E. Kulatov, H. Ohta, T. Arioka, S. Halikov, L. Vinokurova, J. Phys.: Condens. Matter **9**, 9043 (1997)
226. A.R. Williams, J. Kübler, C.D. Gelatt, Phys. Rev. B **19**, 6094 (1979)
227. G. Binnig, H. Rohrer, Helv. Phys. Acta 55, 726 (1982); Rev. Mod. Phys. **59**, 615 (1987)
228. J. Bardeen, Phys. Rev. Lett. **6**, 57 (1961)
229. J. Tersoff, D.R. Hamann, Phys. Rev. Lett. **50**, 1998 (1983); Phys. Rev. B **31**, 805 (1985)
230. M. Tsukada, K. Kobayashi, N. Issiki, H. Kageshima, Surf. Sci. Rep. **13**, 265 (1991)
231. T. Hashizume, K. Motai, X.-D. Wang, N. Shinohara, Y. Saito, Y. Maruyama, K. Ohno, Y. Kawazoe, Y. Nishina, H.W. Pickering, Y. Kuk, T. Sakurai, Phys. Rev. Lett. **71**, 2959 (1993)
232. Y. Maruyama, K. Ohno, Y. Kawazoe, Phys. Rev. B **52**, 2070 (1995)
233. S. Wakoh, J. Yamashita, J. Phys. Soc. Jpn. **21**, 1712 (1966)
234. J.W.D. Connolly, Phys. Rev. **159**, 415 (1967)
235. C.L. Fu, A.J. Freeman, T. Oguchi, Phys. Rev. Lett. **54**, 2700 (1985)
236. S. Blügel, Solid State Commun. **84**, 621 (1992)
237. T. Asada, S. Blügel, Phys. Rev. Lett. **79**, 507 (1997)
238. O. Gunnarsson, J. Phys. F **6**, 587 (1976)
239. J. Janak, Phys. Rev. B **16**, 255 (1977)
240. K. Terakura, N. Hamada, T. Oguchi, T. Asada, J. Phys. F **12**, 1661 (1982)
241. F. Mauri, S.G. Louie, Phys. Rev. Lett. **76**, 4246 (1996)
242. E.L. Shirley, X.J. Zhu, S.G. Louie, Phys. Rev. B **56**, 6648 (1997)
243. J. Zaanen, C. Westra, G.A. Sawatzky, Phys. Rev. B **33**, 8060 (1986)
244. J. Zaanen, G.A. Sawatzky, Phys. Rev. B **33**, 8074 (1986)
245. J. Schwitalla, H. Ebert, Phys. Rev. Lett. **80**, 4586 (1998)
246. L.X. Benedict, E.L. Shirley, R.B. Bohn, Phys. Rev. Lett. **80**, 4514 (1998)
247. W. Kutzelnigg, U. Fleischer, M. Scindler, in *NMR Basic Principles and Progress*, vol. 23, ed. by P. Diehl et al. (Springer, Berlin, 1990), p. 165
248. A M. Lee, N.C. Handy, S.M. Colwell, J. Chem. Phys. **103**, 10095 (1995)
249. F. Mauri, B.G. Pfrommer, S.G. Louie, Phys. Rev. Lett. **77**, 5300 (1996); Phys. Rev. Lett. **79**, 2340 (1997)
250. Y.-G. Yoon, B.G. Pfrommer, F. Mauri, S.G. Louie, Phys. Rev. Lett. **80**, 3388 (1998)
251. A.J. Freeman, R.E. Watson, Phys. Rev. Lett. **5**, 498 (1960)
252. A.J. Freeman, R.B. Frankel (eds.), *Hyperfine Interactions* (Academic Press, New York, 1967)
253. H. Katayama-Yoshida, A. Zunger, Phys. Rev. Lett. **53**, 1256 (1984); Phys. Rev. B **31**, 7877 (1985)
254. M. Matsumoto, K. Tomimoto, S. Wakoh, J. Phys. Soc. Jpn. **62**, 2734 (1993)
255. J. Moscovici, G. Loupias, S. Rabii, S. Erwin, A. Rassat, C. Fabre, Europhys. Lett. **31**, 87 (1995)
256. D. Ayma, M. Rérat, A. Lichanot, J. Phys.: Condens. Matter **10**, 557 (1998)
257. B. Králik, P. Delaney, S.G. Louie, Phys. Rev. Lett. **80**, 4253 (1998). and references therein
258. E. Engel, R.M. Dreizler, in *Density Functional Theory II*, , vol. 181, ed. by R.F. Nalewajski. Topics in Current Chemistry (Springer, Berlin, 1996), p. 1
259. W. Yang, Phys. Rev. Lett. 66, 1438 (1991); Phys. Rev. A **44**, 7823 (1991)

260. W. Yang, T.-S. Lee, J. Chem. Phys. **103**, 5674 (1995)
261. C. Lee, W. Yang, J. Chem. Phys. **96**, 2408 (1992)
262. J.-P. Lu, W. Yang, Phys. Rev. B **49**, 11421 (1994)
263. D. York, J.P. Lu, W. Yang, Phys. Rev B **49**, 8526 (1994)
264. T. Zhu, W. Pan, W. Yang, Phys. Rev. B **53**, 12713 (1996)
265. W. Kohn, Phys. Rev. **115**, 809 (1959)
266. S. Baroni, P. Giannozzi, Europhys. Lett. **17**, 547 (1992)
267. R. Haydock, V. Heine, M.J. Kelly, J. Phys. C **5**, 2845 (1972)
268. A.P. Horsfield, A.M. Bratkovsky, M. Fearn, D.G. Pettifor, M. Aoki, Phys. Rev. B **53**, 12694 (1996)
269. D.M.C. Nicholson, G.M. Stocks, Y. Wang W.A. Shelton, Z. Szotek, W.M. Temmerman, Phys. Rev. B **50**, 14686 (1994)
270. Y. Wang, G.M. Stocks, W.A. Shelton, D.M.C. Nicholson, Z. Szotek, W.M. Temmerman, Phys. Rev. Lett. **75**, 2867 (1995)
271. I.A. Abrikosov, A.M.N. Niklasson, S.I. Simak, B. Johansson, A.V. Ruban, H.L. Skriver, Phys. Rev. Lett. **76**, 4203 (1996)
272. I.A. Abrikosov, S.I. Simak, B. Johansson, A.V. Ruban, H.L. Skriver, Phys. Rev. B **56**, 9319 (1997)
273. X.-P. Li, R.W. Nunes, D. Vanderbilt, Phys. Rev. B **47**, 10891 (1993)
274. W. Kohn, Phys. Rev. Lett. **76**, 3168 (1996)
275. M.S. Daw, Phys. Rev. B **47**, 10895 (1993)
276. S.-Y. Qiu, C.Z. Wang, K.M. Ho, C.T. Chan, J. Phys.: Condens. Matter **6**, 9153 (1994)
277. R.W. Nunes, D. Vanderbilt, Phys. Rev. Lett. **73**, 712 (1994)
278. C.S. Jayanthi, S.Y. Wu, J. Cocks, N.S. Luo, Z.L. Xie, M. Menon, G. Yang, Phys. Rev. B **57**, 3799 (1998)
279. W. Hierse, E.B. Stechel, Phys. Rev. B **50**, 17811 (1994)
280. W. Hierse, E.B. Stechel, Phys. Rev. B **54**, 16515 (1996)
281. E. Hernandez, M.J. Gillan, Phys. Rev. B **51**, 10157 (1995)
282. E. Hernandez, M.J. Gillan, C.M. Goringe, Phys. Rev. B **53**, 7147 (1996)
283. A.E. Carlsson, Phys. Rev. B **51**, 13935 (1995)
284. D.J. Singh, H. Krakauer, C. Haas, A.Y. Liu, Phys. Rev. B **46**, 13065 (1992)
285. G. Galli, M. Parrinello, Phys. Rev. Lett. **69**, 3547 (1992)
286. F. Gygi, Europhys. Lett. **19**, 617 (1992); Phys. Rev. B **48**, 11692 (1993); *ibid***51**, 11190 (1995)
287. D.R. Hamann, Phys. Rev. B 51, 7337 (1995); Phys. Rev. B **51**, 9508 (1995)
288. J.R. Chelikowsky, N. Troullier, Y. Saad, Phys. Rev. Lett. **72**, 1240 (1994)
289. E.L. Briggs, D.J. Sullivan, J. Bernholc, Phys. Rev. B **52**, R5471 (1995)
290. F. Mauri, G. Galli, R. Car, Phys. Rev. B **47**, 9973 (1993)
291. F. Mauri, G. Galli, Phys. Rev. B **50**, 4316 (1994)
292. P. Ordejón, D.A. Drabold, M.P. Grumbach, R.M. Martin, Phys. Rev. B **48**, 14646 (1993)
293. P. Ordejón, D.A. Drabold, R.M. Martin, M.P. Grumbach, Phys. Rev. B **51**, 1456 (1995)
294. J. Kim, F. Mauri, G. Galli, Phys. Rev. B **52**, 1640 (1995)
295. P. Ordejón, E. Artacho, J.M. Soler, Phys. Rev. B **53**, R10441 (1996)
296. S. Itoh, P. Ordejón, D.A. Drabold, R.M. Martin, Phys. Rev. B **53**, 2132 (1996)
297. W. Kohn, Chem. Phys. Lett. **208**, 167 (1993)
298. W. Kohn, Int. J. Quantum Chem. **56**, 229 (1995)
299. E.B. Stechel, A.R. Williams, P.J. Feibelman, Phys. Rev. B **49**, 10088 (1994)
300. D.R. Jennison, P.A. Schultz, M.P. Sears, J. Chem. Phys. **106**, 1856 (1997)
301. J.F. Annett, Comput. Mater. Sci. **4**, 23 (1995)
302. Z. Gonze, Phys. Rev. **54**, 4383 (1996)
303. O.F. Sankey, D.J. Niklewski, Phys. Rev. B **40**, 3979 (1989)
304. G.B. Adams, O.F. Sankey, Phys. Rev. Lett. **67**, 867 (1991)
305. G.B. Adams, J.B. Page, O.F. Sankey, Phys. Rev. B **50**, 17471 (1994)
306. M.-H. Tsai, J.D. Dow, O.F. Sankey, Phys. Rev. B **46**, 10464 (1992)
307. S.H. Yang, D.A. Drabold, J.B. Adams, A. Sachdev, Phys. Rev. B **47**, 1567 (1993)

308. S.H. Yang, D.A. Drabold, J.B. Adams, Phys. Rev. B **48**, 5261 (1993)
309. S.H. Yang, D.A. Drabold, J.B. Adams, P. Ordejón, K. Glassford, J. Phys.: Condens. Matter **9**, L39 (1997)
310. A.A. Demkov, J. Ortega, O.F. Sankey, M.P. Grumbach, Phys. Rev. B **52**, 1618 (1995)
311. A.A. Demkov, O.F. Sankey, J. Gryko, P.F. McMillan, Phys. Rev. B **55**, 6904 (1997)
312. D.A. Drabold, P.A. Fedders, S. Klemm, O.F. Sankey, Phys. Rev. Lett. **67**, 2179 (1991)
313. J.R. Smith, T. Hong, D. Srolovitz, Phys. Rev. Lett. **72**, 4021 (1994)
314. J. Fritsch, J.B. Page, K.E. Schmidt, G.B. Adams, Phys. Rev. B **57**, 9745 (1998)
315. S.L. Adler, Phys. Rev. **126**, 413 (1962)
316. R. Kubo, J. Phys. Soc. Jpn. **12**, 570 (1957)
317. R. Landauer, Philos. Mag. **21**, 863 (1970)
318. A.L. Fetter, J.D. Walecka, *Quantum Theory of Many-Particle Systems* (McGraw-Hill, New York, 1971)
319. S. Doniach, E.H. Sondheimer, *Green's Functions for Solid State Physics, Frontiers in Physics*. Lecture Note Series, vol. 44 (Benjamin/Cummings, Reading, 1974)
320. M. Itoh, Phys. Rev. B **45**, 4241 (1992)
321. H. Tanaka, M. Itoh, Phys. Rev. Lett. **81**, 3727 (1998)
322. H. Ebert, Phys. B **1989**, 175 (1989)
323. H. Ebert, H. Akai, J. Appl. Phys. **67**, 4798 (1990)
324. S.B. Halilov, Y.A. Uspenskii, J. Phys.: Condens. Matter **2**, 6137 (1990)
325. E.T. Kulatov, Y.A. Uspenskii, S.V. Halilov, Phys. Lett. A **195**, 267 (1994); J. Magn. Magn. Mat. **145**, 39 (1995)
326. Y.A. Uspenskii, E.T. Kulatov, S.V. Halilov, Phys. Rev. B **54**, 474 (1996)
327. H. Miyazawa, T. Oguchi, Materia **37**, 560 (1998). in Japanese
328. P.M. Oppeneer, P.M. Maurer, J. Sticht, J. Kübler, Phys. Rev. B **45**, 10924 (1992)
329. P.M. Oppeneer, J. Sticht, T. Maurer, J. Kübler, Z. Phys. B **88**, 309 (1992)
330. A. Cebollada, D. Weller, J. Sticht, G.R. Harp, R.F.C. Farrow, R.F. Marks, R. Savoy, J.C. Scott, Phys. Rev. B **50**, 3419 (1994)
331. V. Nathan, A.H. Guenther, S.S. Mitra, J. Opt. Soc. Am. B **2**, 294 (1985)
332. M. Murayama, T. Nakayama, Phys. Rev. B **49**, 5737 (1994)
333. S.V. Nair, T. Takagahara, Phys. Rev. B **53**, R10516 (1996)
334. J. Li, C.-G. Duan, Z.-Q. Gu, D.-S. Wang, Phys. Rev. B **57**, 2222 (1998)
335. A.I. Shkrebtii, J.L.P. Hughes, J.E. Sipe, O. Pulci, Thin Solid Films **313–314**, 574 (1998)
336. W. Shockley, Phys. Rev. **78**, 173 (1950); **90**, 491 (1953)
337. G. Dresselhaus, A.F. Kip, C. Kittel, Phys. Rev. **98**, 368 (1955)
338. E.O. Kane, J. Phys. Chem. Solids **1**, 249 (1957)
339. S. Ishii, Master thesis, Tohoku University (1999)
340. P.S. Epstein, Phys. Rev. **28**, 695 (1926); G. Wentzel, Z. Phys. **38**, 518 (1926); I. Waller. Z. Phys. **38**, 635 (1926)
341. R.H. Orcutt, R.H. Cole, J. Chem. Phys. **46**, 697 (1967)
342. D. Gugan, G.W. Michel, Mol. Phys. **39**, 783 (1980)
343. K.D. Bonin, V.V. Kresin, *Electric-Dipole Polarizabilities of Atoms, Molecules and Clusters* (World Scientific, Singapore, 1997)
344. P. Giannozzi, S. de Gironcoli, P. Pavone, S. Baroni, Phys. Rev. B **43**, 7231 (1991)
345. J.C. Slater, J. Chem. Phys. **57**, 2389 (1972)
346. R. Yu, H. Krakauer, Phys. Rev. B **49**, 4467 (1994)
347. S. Baroni, P. Giannozzi, A. Testa, Phys. Rev. Lett. **58**, 1861 (1987)
348. S. Baroni, S. de Gironcoli, A.D. Corso, P. Giannozzi, Rev. Mod. Phys. **73**, 515 (2001)
349. S. de Gironcoli, Phys. Rev. B **51**, 6773 (1995)
350. F. Giustino, Rev. Mod. Phys. **89**, 015003 (2017)
351. H. Fröhlich, Adv. Phys. **3**, 325 (1954)
352. C. Verdi, F. Giustino, Phys. Rev. Lett. **115**, 176401 (2015)
353. J. Sjakste, N. Vast, M. Calandra, F. Mauri, Phys. Rev. B **92**, 054307 (2015)

354. M. Bernardi, D. Vigil-Fowler, J. Lischner, J.B. Neaton, S.G. Louie, Phys. Rev. Lett. **112**, 257402 (2014)
355. M. Bernardi, D. Vigil-Fowler, C.S. Ong, B. Neaton, S.G. Louie, Proc. Nat. Acad. Sci. USA (PNAS) **112**, 5291 (2015)
356. J.A. Moriarty, Phys. Rev. B **38**, 3199 (1988)
357. D.K. Blat, N.E. Zein, V.I. Zinenko, J. Phys.: Condens. Matter **3**, 5515 (1991)
358. R. Bauer, O. Schütt, P. Pavone, W. Windl, D. Strauch, Phys. Rev. B **51**, 210 (1995)
359. K.-P. Bohnen, A. Eichler, J. Hafner, Surf. Sci. **368**, 222 (1996)
360. P. Söderlind, J.A. Moriarty, J.M. Wills, Phys. Rev. B **53**, 14063 (1996)
361. K. Parlinski, Z.-Q. Li, Y. Kawazoe, Phys. Rev. Lett. **78**, 4063 (1997)
362. M. Sluiter, M. Weinert, Y. Kawazoe, Europhys. Lett. **43**, 183 (1998)
363. R.D. King-Smith, D. Vanderbilt, Phys. Rev. B **49**, 5828 (1994)
364. S. de Gironcoli, S. Baroni, R. Resta, Phys. Rev. Lett. **62**, 2853 (1989)
365. G. Sághi-Szabó, R.E. Cohen, H. Krakauer, Phys. Rev. Lett. **80**, 4321 (1998)
366. F. London, H. London, Proc. R. Soc. Lond. A **149**, 71 (1935); Physica **2**, 341 (1935)
367. F. London, *Superfluids I: Macroscopic Theory of Superconductivity* (Dover, New York, 1950, 1961)
368. D.A. Greenwood, Proc. Phys. Soc. **71**, 585 (1958)
369. R. Kaschner, M. Schöne, G. Seifert, G. Pastore, J. Phys.: Condens. Matter **8**, L653 (1996)
370. F. Kirchhoff, J.M. Holender, M.J. Gillan, Phys. Rev. B **54**, 190 (1996)
371. J.M. Holender, M.J. Gillan, J. Non-Cryst. Solids **205/207**, 866 (1996)
372. P.L. Silvestrelli, A. Alavi, M. Parrinello, Phys. Rev. B **55**, 15515 (1997)
373. R. Landauer, IBM. J. Res. Dev. **1**, 223 (1957)
374. A.D. Stone, Phys. Rev. Lett. **54**, 2692 (1985)
375. L. Chico, L.X. Benedict, S.G. Louie, M.L. Cohen, Phys. Rev. B **54**, 2600 (1996)
376. M.S. Green, J. Chem. Phys. **22**, 398 (1954)
377. R. Tamura, M. Tsukada, Solid State Commun. **101**, 601 (1997)
378. J.M. Luttinger, Phys. Rev. **6A**, A1505 (1964)
379. G. Seifert, R. Kaschner, M. Schöne, G. Pastore, J. Phys.: Condens. Matter **10**, 1175 (1998)
380. N. Ashcroft, N. Mermin, *Solid State Physics* (Saunders College, Philadelphia, 1976)
381. R. Juneja, T. Pandey, A.K. Singh, Chem. Mater. **29**, 3723 (2017)
382. A. Marcolongo, P. Umari, S. Baroni, Nat. Phys. **12**, 80 (2015)
383. R. Peierls, Ann. Phys. **3**, 1055 (1929)
384. C. Herring, Phys. Rev. **95**, 954 (1954)
385. P.G. Klemens, Proc. Phys. Soc. Sect. A **68**, 1113 (1955)
386. R.A. Cowley, Rep. Prog. Phys. **31**, 123 (1968)
387. K. Fuchs, Math. Proc. Camb. Philos. Soc. **34**, 100 (1938)
388. E.H. Sondheimer, Adv. Phys. **50**, 499 (1952)
389. M. Maldovan, J. Appl. Phys. **111**, 024311 (2012)
390. Z. Aksamija, I. Knezevic, Phys. Rev. B **90**, 0355419 (2014)
391. S. Tamura, Phys. Rev. B **27**, 858 (1983)
392. L. v. Wei, A. Henry, New J. Phys. **18**, 013028 (2016)
393. P.B. Allen, J.L. Feldman, Phys. Rev. B **48**, 12581 (1993)
394. D.P. Sellan, E.S. Landry, J.E. Turney, A.J.H. Mcgaughey, C.H. Amon, Phys. Rev. B **81**, 214305 (2010)
395. K. Esfarjani, H. Stokes, Phys. Rev. B **77**, 1 (2008)
396. M. Petersilka, U.J. Gossmann, E.K.U. Gross, Phys. Rev. Lett. **76**, 1212 (1996)
397. I. Vasiliev, S. Öğüt, J.R. Chelikowsky, Phys. Rev. Lett. **82**, 1919 (1999)
398. G. Kotliar, S.Y. Savrasov, K. Haule, V.S. Oudovenko, O. Parcollet, C.A. Marianetti, Rev. Mod. Phys. **78**, 865–951 (2006)
399. A. Georges, G. Kotliar, W. Krauth, M.J. Rozenberg, Rev. Mod. Phys. **68**, 13–125 (1996)
400. R.A. Winer, Phys. Rev. Lett. **24**, 1071 (1970)
401. N.E. Bickers, S.R. White, Phys. Rev. B **43**, 8044 (1991)
402. A. Lande, R.A. Smith, Phys. Lett. **131B**, 253 (1983)

403. A. Lande, R.A. Smith, Phys. Rev. A **45**, 921 (1983)
404. G. Kotliar, S.Y. Savrasov, G. Pálsson, G. Biroli, Phys. Rev. Lett. **87**, 186401 (2001)
405. S. White, Phys. Rev. Lett. **69**, 2863 (1992)
406. W. Metzner, D. Vollhardt, Phys. Rev. Lett. **62**, 324 (1989)
407. E. Müller-Hartmann, Z. Phys. B **74**, 507 (1989)
408. S. Biermann, F. Aryasetiawan, A. Georges, Phys. Rev. Lett. **90**, 086402 (2003)
409. S. White, Phys. Rev. Lett. **69**, 2863 (1992)
410. S.R. White, Phys. Rev. B **48**, 10345 (1993)
411. S.R. White, Phys. Rep. **301**, 187 (1998)
412. U. Schollwöck, Rev. Mod. Phys. **77**, 259 (2005)
413. S.R. White, A.E. Feiguin, Phys. Rev. Lett. **93**, 076401 (2004)
414. M. Levin, C.P. Nave, Phys. Rev. Lett. **99**, 120601 (2007)
415. Z.Y. Xie, J. Chen, M.P. Qin, J.W. Zhu, L.P. Yang, T. Xiang, Phys. Rev. B **86**, 045139 (2012)
416. G.M. Crosswhite, D. Bacon, Phys. Rev. A **78**, 012356 (2008)
417. U. Schollwöck, Ann. Phys. **326**, 96 (2011)
418. G. Evenbly, G. Vidal, J. Stat. Phys. **145**, 891 (2011)
419. A.M. Goldsborough, R.A. Römer, Phys. Rev. B **89**, 214203 (2014)
420. M. Born, Z. Phys. **40**, 167 (1927); M. Born, J.R. Oppenheimer, Ann. Phys. **84**, 457 (1927)
421. A. Gross, M. Scheffler, Phys. Rev. B **57**, 2493 (1998)
422. R. Car, M. Parrinello, Phys. Rev. Lett. **55**, 2471 (1985)
423. M.C. Payne, M.P. Teter, D.C. Allan, T.A. Arias, J.D. Joannopoulos, Rev. Mod. Phys. **64**, 1045 (1992)
424. T. Oguchi, T. Sasaki, Prog. Theor. Phys. Suppl. **103**, 97 (1991)
425. J.P. Ryckaert, G. Ciccotti, H.J.C. Berendsen, J. Comput. Phys. **23**, 327 (1977)
426. M. Scheffler, J.P. Vigneron, G.B. Bachelet, Phys. Rev. B **31**, 6541 (1985)
427. M.C. Payne, M. Needels, J.D. Joannopoulos, Phys. Rev. B **37**, 8137 (1988); M.P. Teter, M.C. Payne, D.C. Allen, Phys. Rev. B **40**, 12255 (1989); see also [417]
428. E.R. Davidson, J. Comput. Phys. **17**, 87 (1975)
429. B. Liu, in *Numerical Algorithms in Chemistry: Algebraic Methods*, ed. by C. Moler, I. Shavitt (Lawrence Berkeley, Berkeley, 1978), pp. 49–52
430. R.M. Parrish, E.G. Hohenstein, T.J. Martínez, J. Chem. Theory Comput. **12**, 3003 (2016)
431. G.P. Francis, M.C. Payne, J. Phys.: Condens. Matter **2**, 4395 (1990)
432. P. Pulay, in *Modern Theoretical Chemistry*, vol. 4, ed. by H.F. Schaefer (Plenum, New York, 1977); P. Pulay, J. Chem. Phys**79**, 2491 (1983); G.P. Srivasfava et al., Adv. Phys. **36**, 475 (1987)
433. K.M. Ho, C.L. Fu, B.N. Harmon, Phys. Rev. B **28**, 6687 (1983)
434. T. Kishi, S. Itoh, J. Phys.: Condens. Matter **7**, 6197 (1995)
435. T. Kishi, S. Itoh, Surf. Sci. **358**, 186 (1996)
436. J.M. Soler, A.R. Williams, Phys. Rev. B **40**, 1560 (1989)
437. J.M. Soler, A.R. Williams, Phys. Rev. B **42**, 9728 (1990)
438. J.M. Soler, A.R. Williams, Phys. Rev. B **47**, 6784 (1993)
439. R. Yu, D. Singh, H. Krakauer, Phys. Rev. B **43**, 6411 (1991); D. Singh. Phys. Rev. B **40**, 5428 (1989)
440. R. Yu, D. Singh, H. Krakauer, Phys. Rev. B **45**, 8671 (1992)
441. S. Goedecker, K. Mashke, Phys. Rev. B **45**, 1597 (1992)
442. T. Oguchi, in *Interatomic Potential and Structural Stability*, ed. by K. Terakura, H. Akai (Springer, Berlin, 1993), p. 33
443. T. Oguchi, T. Sasaki, in *Computer Aided Innovation of New Materials II*, ed. by M. Doyama, J. Kihara, M. Tanaka, R. Yamamoto (North-Holland, Amsterdam, 1993), p. 107
444. H.G. Krimmel, J. Ehmann, C. Elsässer, M. Fähnle, J.M. Soler, Phys. Rev. B **50**, 8846 (1994)
445. J.F. Janak, Phys. Rev. B **9**, 3985 (1974)
446. T. Oguchi, N. Hamada, J. Phys. Soc. Jpn. **66**, 2751 (1997)
447. I. Ohsaki, T. Oguchi, J. Phys. Soc. Jpn. **67**, 978 (1998)

448. B. Kohler, S. Wilke, M. Scheffler, R. Kouba, C. Ambroschdraxl, Comput. Phys. Commun. **94**, 31 (1996)
449. I.G. Batilev, J.A. Leiro, T. Oguchi, Philos. Mag. Lett. **73**, 385 (1996)
450. R.Q. Wu, L.P. Zhong, L.J. Chen, A.J. Freeman, Phys. Rev. B **54**, 7084 (1996)
451. J. Sarnthein, K. Schwarz, P.E. Blöchl, Phys. Rev. B **53**, 9084 (1996)
452. K. Schwarz, E. Nusterer, P. Margl, Int. J. Quantum Chem. **61**, 369 (1997)
453. H. Ness, A.J. Fisher, Phys. Rev. B **55**, 10081 (1997)
454. N.A.W. Holzwarth, G.E. Matthews, A.R. Tackett, R.B. Dunning, Phys. Rev. B **57**, 11827 (1998)
455. N.A.W. Holzwarth, G.E. Matthews, R.B. Dunning, A.R. Tackett, Y. Zeng, Phys. Rev. B **55**, 2005 (1997)
456. R.O. Jones, G. Seifert, Phys. Rev. Lett. **79**, 443 (1997)
457. K. Raghavachari, V. Logovinsky, Phys. Rev. Lett. **55**, 2853 (1985)
458. S. Saito, S. Ohnishi, C. Satoko, S. Sugano, Z. Phys. D **14**, 237 (1989)
459. J.-Y. Yi, D.J. Oh, J. Bernholc, R. Car, Chem. Phys. Lett. **174**, 461 (1990)
460. J.-Y. Yi, D.J. Oh, J. Bernholc, Phys. Rev. Lett. **67**, 1594 (1991)
461. A. Jackson, Phys. Rev. B **47**, 9715 (1993)
462. C. Massobrio, A. Pasquarello, R. Car, Chem. Phys. Lett. **238**, 215 (1995)
463. J.H. Weaver, D.M. Poirier, in *Solid State Physics*, , vol. 48, ed. by H. Ehrenreich, F. Spaepen (Academic Press, Boston, 1994), p. 1
464. M.S. Dresselhaus, G. Dresselhaus, P.C. Eklund, *Science of Fullerenes and Carbon Nanotubes* (Academic Press, San Diego, 1996)
465. H.W. Kroto, J.R. Heath, S.C. O'Brien, R.F. Curl, R.E. Smaley, Nature **318**, 162 (1985)
466. W. Krätschmer, L.D. Lamb, K. Fostiropoulos, D.R. Huffman, Nature **347**, 354 (1990)
467. S. Iijima, I. Ichihashi, Y. Ando, Nature **356**, 776 (1992)
468. S. Saito, A. Oshiyama, Phys. Rev. Lett. **66**, 2637 (1991)
469. N. Troullier, J.L. Martins, Phys. Rev. B **46**, 1754 (1992)
470. E.L. Shirley, S.G. Louie, Phys. Rev. Lett. **71**, 133 (1993)
471. B.-L. Gu, Y. Maruyama, J.-Z. Yu, K. Ohno, Y. Kawazoe, Phys. Rev. B **49**, 16202 (1994)
472. K. Ohno, Y. Maruyama, M. Takahashi, J.-Z. Yu, B.-L. Gu, Y. Kawazoe, Surf. Rev. Lett. **3**, 735 (1996)
473. K. Ohno, J.-Z. Yu, Y. Maruyama, Y. Kawazoe, T. Takahashi, Chem. Phys. Lett. **255**, 367 (1996)
474. Q.M. Zhang, J.Y. Yi, J. Bernhole, Phys. Rev. Lett. **66**, 2633 (1991)
475. B.P. Feuston, W. Andreoni, M. Parrinello, E. Clementi, Phys. Rev. B **44**, 4056 (1991)
476. D.H. Robertson, D.W. Brenner, C.T. White, J. Phys. Chem. **96**, 6133 (1992)
477. S. Serra, S. Sanguinetti, L. Colombo, J. Chem. Phys. **102**, 2151 (1995)
478. J.R. Chelikowsky, Phys. Rev. Lett. **67**, 2970 (1991); Phys. Rev. B **45**, 12062 (1992)
479. C. Xu, G.E. Scuseria, Phys. Rev. Lett. **72**, 669 (1994)
480. G. Galli, F. Mauri, Phys. Rev. Lett. **73**, 3471 (1994)
481. A. Alavi, J. Kohanoff, M. Parrinello, D. Frenkel, Phys. Rev. Lett. **73**, 2599 (1994)
482. M.P. Grumbach, D. Hohl, R.M. Martin, R. Car, J. Phys.: Condens. Matter **6**, 1999 (1994)
483. H. Shinohara, H. Sato, M. Ohkohchi, Y. Ando, T. Kodama, T. Shida, T. Kato, Y. Saito, Nature **357**, 52 (1992)
484. M. Takata, B. Umeda, E. Nishibori, M. Sakata, Y. Saito, M. Ohno, H. Shinohara, Nature **377**, 46 (1995)
485. Z. Wan, J.F. Christian, S.L. Anderson, Phys. Rev. Lett. **69**, 1352 (1992)
486. E.E.B. Campbell, R. Tellgmann, N. Krawez, I.V. Hertel, J. Phys. Chem. Solids **57**, 1763 (1997)
487. S. Aoyagi, E. Nishibori, H. Sawa, K. Sugimoto, M. Takata, Y. Miyata, R. Kitaura, H. Shinohara, H. Okada, T. Sakai, Y. Ono, K. Kawachi, K. Yokoo, S. Ono, K. Omote, Y. Kasama, S. Ishikawa, T. Komuro, H. Tobita, Nat. Chem. **2**, 678 (2010)
488. K. Ohno, A. Manjanath, Y. Kawazoe, R. Hatakeyama, F. Misaizu, E. Kwon, H. Fukumura, H. Ogasawara, Y. Yamada, C. Zhang, N. Sumi, T. Kamigaki, K. Kawachi, K. Yokoo, S. Ono, and Y. Kasama, to appear

489. A. Seeger, D. Schumacher, W. Schilling, J. Diehl (eds.), *Vacancies and Interstitials in Metals* (North-Holland, 1970)
490. A.M. Stoneham, *Theory of Defects in Solids* (Clarendon, Oxford, 1975)
491. R.W. Cahn, P. Haasen (eds.), *Physical Metallurgy* (Elsevier, Amsterdam, 1983)
492. J. Kondo, Prog. Theor. Phys. **32**, 37 (1964)
493. K.G. Wilson, Rev. Mod. Phys. **47**, 773 (1975)
494. N. Andrei, K. Furuya, J. Lowenstein, Rev. Mod. Phys. **55**, 331 (1983)
495. P. Schlottmann, Phys. Rev. Lett. **80**, 4975 (1998)
496. D.J. Chadi, K.J. Chang, Phys. Rev. Lett. **61**, 873 (1988); Phys. Rev. B **39**, 10063 (1989)
497. E. Yamaguchi, K. Shiraishi, T. Ohno, J. Phys. Soc. Jpn. **60**, 3093 (1991)
498. M. Saito, A. Oshiyama, O. Sugino, Phys. Rev. B **47**, 13205 (1993)
499. A. Zunger, in *Solid State Physics*, vol. 39, ed. by H. Ehrenreich, D. Turnbull (Academic Press, Orlando, 1986), p. 275
500. K. Sato, H. Katayama-Yoshida, Jpn. J. Appl. Phys. **39**, L555 (2000)
501. L.A. Girifalco, *Atomic Migration in Crystals* (Blaisdell, New York, 1964)
502. G.E. Murch, A.S. Nowick (eds.), *Diffusion in Crystalline Solids* (Academic Press, London, 1984)
503. R. Thomson, in *Solid State Physics*, vol. 39, ed. by H. Ehrenreich, D. Turnbull (Academic Press, Orlando, 1986), p. 1
504. H. Kitagawa, T. Aihara Jr., Y. Kawazoe (eds.), *Mesoscopic Dynamics of Fracture: Computational Materials Design*. Advances in Materials Research, vol. 1 (Springer, Berlin, 1998)
505. F. Buda, G.L. Chiarotti, R. Car, M. Parrinello, Phys. Rev. Lett. **63**, 294 (1989)
506. M.J. Gillan, J. Phys.: Condens. Matter **1**, 689 (1989)
507. U. Breier, V. Schott, M. Fähnle, Phys. Rev. B **55**, 5772 (1997)
508. B. Meyer, M. Fahnle, Phys. Rev. B **56**, 13595 (1997)
509. J. Mayer, B. Meyer, J.S. Oehrens, G. Beser, N. Bornsen, M. Fehnle, Intermetallics **5**, 597 (1997)
510. P.H. Dederichs, T. Hoshino, B. Drittler, K. Abraham, R. Zeller, Phys. B **172**, 203 (1991)
511. M. Asato, T. Hoshino, T. Asada, R. Zeller, P.H. Dederichs, J. Magn. Magn. Mater. **177–181**, 1403 (1998)
512. T. Hoshino, N. Papanikolaou, R. Zeller, P.H. Dederichs, M. Asato, T. Asada, N. Stefanou, Comput. Mater. Sci. **14**, 99 (1999)
513. M. Needels, M.C. Payne, J.D. Joannopoulos, Phys. Rev. Lett. **58**, 1765 (1987)
514. S. Ihara, S.L. Ho, T. Uda, M. Hirao, Phys. Rev. Lett. **65**, 1909 (1990)
515. I. Štich, J. Kohanoff, K. Terakura, Phys. Rev. B **54**, 2642 (1996)
516. K. Kobayashi, Y. Morikawa, K. Terakura, S. Blügel, Phys. Rev. B **45**, 3469 (1992)
517. J. Yamauchi, K. Kobayashi, M. Tsukada, Surf. Sci. **306**, 42 (1994)
518. A.J. Dyson, P.V. Smith, Surf. Sci. **375**, 45 (1997)
519. E. Kim, C.W. Oh, Y.H. Lee, Phys. Rev. Lett. **79**, 4621 (1997)
520. A. Oshiyama, Phys. Rev. Lett. **74**, 130 (1995)
521. M.R. Jarvis, R. Pérez, F.M. van Bouwelen, M.C. Payne, Phys. Rev. Lett. **80**, 3428 (1998)
522. A. Gross, M. Bockstedte, M. Scheffler, Phys. Rev. Lett. **79**, 701 (1997)
523. A. Alavi, P. Hu, T. Deutsch, P.L. Silvestrelli, J. Hutter, Phys. Rev. Lett. **80**, 3650 (1998)
524. D.M. Newns, Phys. Rev. **178**, 1123 (1969)
525. H. Suhl, Phys. Rev. B **11**, 2011 (1975)
526. J.G. Kirkwood, J. Chem. Phys. **3**, 300 (1935)
527. D. Frenkel, A.J.C. Ladd, J. Chem. Phys. **81**, 3188 (1984)
528. D. Frenkel, B. Smit, *Understanding Molecular Simulation: From Algorithms to Applications* (Academic Press, San Diego, 1996)
529. M. Watanabe, W.P. Reinhardt, Phys. Rev. Lett. **65**, 3301 (1990)
530. O. Sugino, R. Car, Phys. Rev. Lett. **74**, 1823 (1995)
531. G.A. de Wijs, G. Kresse, M.J. Gillan, Phys. Rev. B **57**, 8223 (1998)
532. G. Cicotti, D. Frenkel, I.R. McDonald, *Simulation of Liquids and Solids* (North-Holland, Amsterdam, 1987)

533. K. Hoshino, F. Shimojo, J. Phys.: Condens. Matter **8**, 9315 (1996)
534. R.M. Wentzcovitch, Phys. Rev. B **50**, 10358 (1994)
535. A.Y. Liu, R.M. Wentzcovitch, Phys. Rev. B **50**, 10362 (1994)
536. R.M. Wentzcovitch, N.L. Ross, G.D. Price, Phys. Earth Planet. Inter. **90**, 101 (1995)
537. R.M. Wentzcovitch, D.A. Hughjones, R.J. Angel, G.D. Price, Phys. Chem. Miner. **22**, 453 (1995)
538. P. Focher, G.L. Chiarotti, M. Bernasconi, E. Tosatti, M. Parrinello, Europhys. Lett. **26**, 345 (1994)
539. M. Bernasconi, G.L. Chiarotti, P. Focher, S. Scandolo, E. Tosatti, M. Parrinello, J. Phys. Chem. Solids **56**, 501 (1995)
540. M. Bernasconi, M. Parrinello, G.L. Chiarotti, P. Focher, E. Tosatti, Phys. Rev. Lett. **76**, 2081 (1996)
541. M. Benoit, M. Bernasconi, M. Parrinello, Phys. Rev. Lett. **76**, 2934 (1996)
542. M. Bernasconi, G.L. Chiarotti, P. Focher, M. Parrinello, E. Tosatti, Phys. Rev. Lett. **78**, 2008 (1997)
543. J. Kohanoff, S. Scandolo, G.L. Chiarotti, E. Tosatti, Phys. Rev. Lett. **78**, 2783 (1997)
544. M. Benoit, D. Marx, M. Parrinello, Comput. Mat. Sci. **10**, 88 (1998)
545. S. Biermann, D. Hohl, D. Marx, J. Low. Temp. Phys. **110**, 97 (1998)
546. J. Bardeen, L.N. Cooper, J.R. Schrieffer, Phys. Rev. **108**, 1175 (1957)
547. R.E. Prange, S.M. Girvin (eds.), *The Quantum Hall Effect* (Springer, Berlin, 1987, 1990) (See, in particular, a chapter written by R. B. Laughlin)
548. T. Ando, Prog. Theor. Phys. Suppl. **84**, 69 (1985)
549. S. Forte, Rev. Mod. Phys. **64**, 193 (1992)
550. P. Weinberger, P.M. Levy, J. Banhart, L. Szunyogh, B. Újfalussy, J. Phys.: Condens. Matter **8**, 7677 (1996)
551. J. Mathon, Phys. Rev. B **56**, 11810 (1997)
552. P. Zahn, J. Binder, I. Mertig, R. Zeller, P.H. Dederichs, Phys. Rev. Lett **80**, 4309 (1998)
553. A. Sakuma, J. Magn. Magn. Mater. **102**, 127 (1991)
554. S. Matter, Z. Phys. B **87**, 91 (1992)
555. B.I. Min, Phys. Rev. B **46**, 8232 (1992)
556. S. Ishida, K. Kitawase, S. Fujii, S. Asano, J. Phys.: Condens. Matter **4**, 765 (1992)
557. R. Coehoon, G.H.O. Daalderrop, H.J.F. Jansen, Phys. Rev. B **48**, 3830 (1993)
558. T. Oguchi, in *Modelling and Simulation for Materials Design* (Proc. of CMSMD'96), ed. by S. Nishijima, H. Onodera (National Research Institute for Metals, Tsukuba, 1996), p. 55
559. H. Lin, R.S. Markiewicz, L.A. Wray, L. Fu, M.Z. Hasan, A. Bansil, Phys. Rev. Lett. **105**, 036404 (2010)
560. G.Y. Guo, H. Ebert, Hyperfine Interact. **97/98**, 11 (1996)
561. P. Hüsser, H.U. Suter, E.P. Stoll, P.F. Meier, Phys. Rev. B **61**, 1567 (2000)
562. M.S. Bahramy, M.H.F. Sluiter, Y. Kawazoe, Phys. Rev. B **73**, 045111 (2006)
563. W. Shockley, H.J. Queisser, J. Appl. Phys. **32**, 510 (1961)
564. L. Yu, A. Zunger, Phys. Rev. Lett. **108**, 068701 (2012)
565. F. Zheng, H. Takenaka, F. Wang, N.Z. Koocher, A.M. Rappe, J. Phys. Chem. Lett. **6**, 31 (2015)
566. N. Samadian, R. Sainz, B. Partoens, D. Lamoen, J. Appl. Phys. **120**, 085707 (2016)
567. N.C. Giebink, G.P. Wiederrecht, M.R. Wasielewski, S.R. Forrest, Phys. Rev. B **83**, 195326 (2011)
568. K. Seki, A. Furube, Y. Yoshida, Appl. Phys. Lett. **103**, 253904 (2013); see also: S.R. Forrest, N.C. Giebink, Appl. Phys. Lett. **104**, 146101 (2014)
569. S. Ono, R. Kuwahara, K. Ohno, J. Appl. Phys. **116**, 054305 (2014)
570. T. Yamamoto, K. Watanabe, Phys. Rev. Lett. **96**, 255503 (2006)
571. K. Esfarjani, M. Zebarjadi, Y. Kawazoe, Phys. Rev. B **73**, 085406 (2006)
572. T. Pandey, A.K. Singh, Phys. Chem. Chem. Phys. **17**, 16917 (2015)
573. J.R. Chelikowsky, S.G. Louie (eds.), *Quantum Theory of Real Materials* (Kluwer Academic, Amsterdam, 1996)

574. J.-P. Connerade, *Correlations in Clusters and Related Systems: New Perspectives on the Many-Body Problem* (World Scientific, Singapore, 1996)
575. K. Terakura, H. Akai (eds.), *Interatomic Potential and Structureal Stability* (Springer, Berlin, 1993)
576. G.L. Malli (ed.), *Relativistic and Electron Correlation Effects in Molecules and Solids* (Plenum, New York, 1994)
577. R.M. Martin, *Electronic Structure: Basic Theory and Practical Methods* (Cambridge University Press, Cambridge, 2008)
578. P. Soven, Phys. Rev. **156**, 809 (1967)
579. D.W. Taylor, Phys. Rev **156**, 1017 (1967)
580. B. Velicky, S. Kirkpatrick, H. Ehrenreich, Phys. Rev. **175**, 747 (1968)
581. F. Yonezawa, T. Matsubara, Prog. Theor. Phys. **35**, 357 (1966)
582. J.M. Ziman, *Models of Disorder* (Cambridge University Press, Cambridge, 1979)
583. F. Ducastelle, *Order and Phase Stability in Alloys* (North-Holland, Amsterdam, 1991)
584. A. Gonis, *Green Functions for Ordered and Disordered Systems* (North-Holland, Amsterdam, 1992)
585. Y. Marechal, J. Chem. Phys. **83**, 247 (1985)
586. A.A. Abrikosov, L.P. Gorkov, I.E. Dzyaloschinski, *Methods of Quantum Field Theory in Statistical Physics* (Dover, New York, 1963)
587. E.M. Lifshitz, L.P. Pitaevskii, *"Statistical Physics Part 2" L. D. Landau and E. M. Lifshitz, Course of Theoretical Physics*, vol. 9 Translated from the Russian by J. B. Sykes and M. J. Kearsley, (Pergamon Press, Oxford, 1980) Chapter 1, Section 8
588. U. Saalman, R. Schmidt, Z. Phys. D **38**, 153 (1996)
589. R. McWeeny, Rev. Mod. Phys. **32**, 335 (1960)
590. M. Head-Gordon, J.C. Tully, J. Chem. Phys. **103**, 10137 (1995) (and references therein)
591. E.K.U. Gross, J.F. Dobson, M. Petersilka, in *Density Functional Theory II*. Topics in Current Chemistry; Relativistic and Time-Dependent Extensions, vol. 181, ed. by R.F. Nalewajski (Springer, Berlin, 1996), p. 81
592. P.-O. Löwdin, Adv. Phys. **5**, 1 (1956)
593. O. Goscinski, P. Linder, J. Math. Phys. **11**, 1313 (1970)
594. M. Shishkin, M. Marsman, G. Kresse, Phys. Rev. Lett. **99**, 246403 (2007)
595. R. Kuwahara, K. Ohno, Phys. Rev. A **90**, 032506 (2014)
596. R. Kuwahara, Y. Noguchi, K. Ohno, Phys. Rev. B **94**, 121116(R) (2016)
597. V.M. Galitskii, A.B. Migdal, Sov. Phys. JETP **7**, 96 (1958); Russian original: ZhETF **34**, 139 (1958)
598. P. Sánchez-Friera, R.W. Godby, Phys. Rev. Lett. **85**, 5611 (2000)
599. K. Morokuma, S. Iwata, Chem. Phys. Lett. **16**, 192 (1972)
600. J.M. Luttinger, Phys. Rev. **121**, 942 (1961)
601. P. Nozières, *Theory of Interacting Fermi Systems* (Benjamin, New York, 1964; Addison-Wesley, New York, 1997)
602. J.M. Luttinger, J.C. Ward, Phys. Rev. **118**, 1417 (1960)
603. J.C. Ward, Phys. Rev. **78**, 182 (1950)
604. Y. Takahashi, Il Nuovo Cimento **6**, 371 (1957)
605. J.R. Schrieffer, *Theory of Superconductivity*. Advanced Book Classic (Westview Press, 1999)
606. G. Baym, L.P. Kadanoff, Phys. Rev. **124**, 287 (1961)
607. G. Baym, Phys. Rev. **127**, 1391 (1962)
608. K. Ohno, Mater. Trans. **48**, 649 (2007)
609. F. Mauri, R. Car, Phys. Rev. Lett. **75**, 3166 (1995)
610. L. Hedin, Phys. Rev. **139**, A796 (1965)
611. L. Hedin, S. Lundqvist, in *Solid State Physics*, vol. 23, ed. by F. Seitz, D. Turnbull, H. Ehrenreich (Academic Press, New York, 1969), p. 1
612. M.S. Hybertsen, S.G. Louie, Phys. Rev. Lett. **55**, 1418 (1985); Phys. Rev. B **36**, 5390 (1986)
613. M.S. Hybertsen, S.G. Louie, Phys. Rev. B **38**, 4033 (1988)

614. R.W. Godby, M. Schlüter, L.J. Sham, Phys. Rev. Lett. **56**, 2415 (1986); Phys. Rev. B **35**, 4170 (1987)
615. R.W. Godby, M. Schlüter, L.J. Sham, Phys. Rev. B **36**, 6497 (1987)
616. R.W. Godby, M. Schlüter, L.J. Sham, Phys. Rev. B **37**, 10159 (1988)
617. D.L. Johnson, Phys. Rev. B **9**, 4475 (1974)
618. G.E. Engel, B. Farid, Phys. Rev. B **46**, 15812 (1992)
619. W. von der Linden, P. Horsch, Phys. Rev. B **37**, 8351 (1988)
620. W.G. Aulbur, L. Jönsson, J.W. Wilkins, *Solid State Phyics*, vol. 54 (Academic Press, San Diego, 2000)
621. W. Kang, M.S. Hybertsen, Phys. Rev. B **82**, 085203 (2010)
622. C. Friedrich, M.C. Müller, S. Blügel, Phys. Rev. B **83**, 081101 (2011)
623. C. Friedrich, M.C. Müller, S. Blügel, Phys. Rev. B **84**, 039906 (2011)
624. M. Stankovski, G. Antonius, D. Waroquiers, A. Miglio, H. Dixit, K. Sankaran, M. Giantomassi, X. Gonze, M. Côté, G.-M. Rignanese, Phys. Rev. B **84**, 241201 (2011)
625. J.A. Berger, L. Reining, F. Sottile, Phys. Rev. B **85**, 085126 (2012)
626. M. Zhang, S. Ono, N. Nagatsuka, K. Ohno, Phys. Rev. B **93**, 155116 (2016)
627. S. Saito, S.B. Zhang, S.G. Louie, M.L. Cohen, Phys. Rev. B **40**, 3643 (1989)
628. Z. Zhu, S.G. Louie, Phys. Rev. B **43**, 14142 (1991)
629. A. Rubio, J.L. Corkill, M.L. Cohen, Phys. Rev. B **49**, 1952 (1994)
630. J.L. Corkill, M.L. Cohen, Phys. Rev. B **48**, 17622 (1993)
631. O. Zakharov, A. Rubio, Z. Blase, M.L. Cohen, S.G. Louie, Phys. Rev. B **50**, 10780 (1994)
632. X. Blase, X.J. Zhu, S.G. Louie, Phys. Rev. B **49**, 4973 (1994)
633. E.L. Shirley, Phys. Rev. B **58**, 9579 (1998)
634. B. Kralik, E.K. Chang, S.G. Louie, Phys. Rev. B **57**, 7027 (1998)
635. M. Palummo, L. Reining, R.W. Godby, C.M. Bertoni, N. Börnsen, Europhys. Lett. **26**, 607 (1994)
636. M.M. Rieger, R.W. Godby, Phys. Rev. B **58**, 1343 (1998)
637. M. Rohlfing, P. Krüger, J. Pollmann, Phys. Rev. B **48**, 17791 (1993)
638. M. Rohlfing, P. Krüger, J. Pollmann, Phys. Rev. B **52**, 1905 (1995); Phys. Rev. B **54**, 13759 (1996); Phys. Rev. B **56**, 2191 (1997)
639. M. Sabisch, P. Kruger, A. Mazur, M. Rohlfing, J. Pollmann, Phys. Rev. B **53**, 13121 (1996)
640. M. Rohlfing, P. Krüger, J. Pollmann, Phys. Rev. Lett. **75**, 3489 (1995); Phys. Rev. B **57**, 6485 (1998)
641. M. Rohlfing, S.G. Louie, Phys. Rev. Lett. **57**, R9392 (1998)
642. F. Aryasetiawan, Phys. Rev. B **49**, 16214 (1994)
643. F. Aryasetiawan, O. Gunnarsson, Phys. Rev. Lett. **74**, 3221 (1995); Phys. Rev. B **54**, 17564 (1996)
644. U. Schonberger, F. Aryasetiawan, Phys. Rev. B **52**, 8788 (1995)
645. F. Aryasetiawan, F. Hedin, K. Karlsson, Phys. Rev. Lett. **77**, 2268 (1996)
646. F. Aryasetiawan, K. Karlsson, Phys. Rev. B **54**, 5553 (1996)
647. F. Aryasetiawan, Phys. B **237**, 321 (1997)
648. S. Ishii, K. Ohno, Y. Kawazoe, S.G. Louie, Phys. Rev. B **63**, 155104 (2001)
649. S. Ishii, K. Ohno, Y. Kawazoe, S.G. Louie, Phys. Rev. B **65**, 245109 (2002)
650. S. Ishii, K. Ohno, V. Kumar, Y. Kawazoe, Phys. Rev. B **68**, 195412 (2003)
651. E. Kikuchi, S. Ishii, K. Ohno, Phys. Rev. B **74**, 195410 (2006)
652. E. Kikuchi, S. Iwata, S. Ishii, K. Ohno, Phys. Rev. B **76**, 075325 (2007)
653. S. Ishii, S. Iwata, K. Ohno, Mater. Trans. **51**, 2150 (2010)
654. Y. Noguchi, K. Ohno, I. Solovyev, T. Sasaki, Phys. Rev. B **81**, 165411 (2010)
655. M. Zhang, S. Ono, K. Ohno, Phys. Rev. B **92**, 035205 (2015)
656. M.J. Frisch et al., Gaussian 09, Revision A.02, Gaussian Inc., 2009
657. B.A. Bushaw, W. Nörtershäuser, G.W.F. Drake, H.-J. Kluge, Phys. Rev. A **75**, 052503 (2007)
658. A. Kramida, W.C. Martin, J. Phys. Chem. Ref. Data **26**, 1185 (1997)
659. G. Haeffler, D. Hanstorp, I. Kiyan, A.E. Klinkmüller, U. Ljungblad, D.J. Pegg, Phys. Rev. A **53**, 4127 (1996)

660. E.S. Chang, J. Phys. Chem. Ref. Data **19**, 119 (1990)
661. W.C. Martin, R. Zalubas, J. Phys. Chem. Ref. Data **8**, 817 (1979)
662. R. Belgang, D. Sxhmidt, P.J. West, J. Phys. (Paris) Colloques **44**, C7-229 (1983)
663. T. Trickl, E.F. Cromwell, Y.T. Lee, A.H. Kung, J. Chem. Phys. **91**, 6006 (1989)
664. K.P. Huber, G. Herzberg, *Molecular Spectra and Molecular Structure*, vol. IV. Constants of Diatomic Molecules (Van Nostrand Reinhold, New York, 1979)
665. K. Kimura, S. Katsumata, Y. Achiba, T. Yamazaki, S. Iwata, *Ionization Energies, Ab Initio Assignments, and Valence Electronic Structure for 200 Molecules in the Handbook of HeI Photoelectron Spectra of Fundamental Organic Compounds* (Japan Scientific Society Press, Tokyo, 1981)
666. M.W. McGeogh, R.E. Schlier, Chem. Phys. Let. **99**, 347 (1983)
667. G.W. Erickson, J. Phys. Chem. Ref. Data **6**, 831 (1977)
668. G.W.F. Drake, Can. J. Phys. **66**, 586 (1988)
669. T. Kotani, M. van Schilfgaarde, S.V. Faleev, Phys. Rev. B **76**, 165106 (2007)
670. N.E. Dahlen, U. von Barth, Phys. Rev. B **69**, 195102 (2004)
671. N.E. Dahlen, R. van Leeuwen, J. Chem. Phys. **122**, 164102 (2005)
672. A. Stan, N.E. Dahlen, R. van Leeuwen, Europhys. Lett. **76**, 298 (2006)
673. R.E. Grisenti, W. Schöllkopf, J.P. Toennies, G.C. Hegerfeldt, T. Köhler, M. Stoll, Phys. Rev. Lett. **85**, 2284 (2000)
674. T. Shoji, R. Kuwahara, S. Ono, K. Ohno, Phys. Chem. Chem. Phys. **18**, 24477 (2016)
675. S. Hüfner, *Photoelectron Sectroscopy, Solid State Sciences*, vol. 82 (Springer, Berlin, 1996)
676. H. Chermette, A. Lembarki, H. Razafinjanahary, F. Rogemond, in *Advances in Quantum Chemistry, Volume 33, Density Functional Theory*, ed. by P. Löwdin et al. (Academic, San Diego, 1999), p. 105
677. A. Schindlmayr, Phys. Rev. B **56**, 3528 (1997)
678. H. Yasuhara, Solid State Commun. **11**, 1481 (1972); J. Phys. Soc. Jpn. **36**, 361 (1974); Physica **78**, 420 (1974)
679. J.C. Kimball, Phys. Rev. B **14**, 2371 (1976)
680. H. Suehiro, Y. Ousaka, H. Yasuhara, J. Phys. C: Solid State Phys. **19**, 4247 (1986); J. Phys. C: Solid State Phys. **19**, 4263 (1986)
681. H. Yasuhara, H. Suehiro, Y. Ousaka, J. Phys. C: Solid State Phys. **21**, 4045 (1988)
682. H. Yasuhara, Y. Takada, Phys. Rev. B **43**, 7200 (1991)
683. A. Fleszar, W. Hanke, Phys. Rev. B **56**, 10228 (1997)
684. A. Schindlmayr, R.W. Godby, Phys. Rev. Lett. **80**, 1702 (1998)
685. X. Ren, P. Rinke, G.E. Scuseria, M. Scheffler, Phys. Rev. B **88**, 035120 (2013)
686. A. Grüneis, G. Kresse, Y. Hinuma, F. Oba, Phys. Rev. Lett. **112**, 096401 (2014)
687. R.J. Buenker, S. Krebs, in *Recent Advances in Multireference Methods*. Recent Advances in Computational Chemistry, vol. 4, ed. by K. Hirao (World Scientific, Singapore, 1999), pp. 1–29
688. V. Bonačić-Koutecký, P. Fantucci, J. Koutecký, J. Chem. Phys. **93**, 3802 (1990)
689. V. Bonačić-Koutecký, I. Boustani, M. Guest, J. Koutecký, J. Chem. Phys. **89**, 4861 (1988)
690. A. Herrmann, E. Schumacher, L. Wöste, J. Chem. Phys. **68**, 2327 (1978)
691. H. Hotop, W.C. Lineberger, J. Phys. Chem. Ref. Data **14**, 731 (1985)
692. A. Herrmann, S. Leutwyler, E. Schumacher, L. Woste, Helv. Chim. Acta **61**, 453 (1978)
693. K.M. McHugh, J.G. Eaton, G.H. Lee, H.W. Sarkas, L.H. Kidder, J.T. Snodgrass, M.R. Manaa, K.H. Bowen, J. Chem. Phys. **91**, 3792 (1989)
694. G. Onida, L. Reining, R.W. Godby, R. DelSole, W. Andreoni, Phys. Rev. Lett. **75**, 818 (1995)
695. S. Albrecht, G. Onida, L. Reining, Phys. Rev. B **55**, 10278 (1997)
696. W.D. Schone, A.G. Eguiluz, Phys. Rev. Lett. **81**, 1662 (1998)
697. H.L. Frisch, F.C. Collins, J. Chem. Phys. **21**, 2158 (1953)
698. L.J. Sham, T.M. Rice, Phys. Rev. **144**, 708 (1966)
699. W. Hanke, L.J. Sham, Phys. Rev. Lett. **43**, 387 (1979); Phys. Rev. B **21**, 4656 (1980)
700. G. Strinati, Phys. Rev. Lett. **49**, 1519 (1982); Phys. Rev. B **29**, 5718 (1984)
701. M. Rohlfing, S.G. Louie, Phys. Rev. Lett. **80**, 3320 (1998)

702. S. Ablrecht, L. Reining, R.D. Sole, G. Onida, Phys. Rev. Lett. **80**, 4510 (1998)
703. M. Springer, F. Aryasetiawan, K. Karlsson, Phys. Rev. Lett. **80**, 2389 (1998)
704. Y. Noguchi, Y. Kudo, S. Ishii, K. Ohno, J. Chem. Phys. **123**, 144112 (2005)
705. Y. Noguchi, S. Ishii, K. Ohno, J. Chem. Phys. **125**, 114108 (2006)
706. Y. Noguchi, S. Ishii, K. Ohno, I. Solovyev, T. Sasaki, Phys. Rev. B **77**, 035132 (2008)
707. M. Rohlfing, S.G. Louie, Phys. Rev. Lett. **82**, 1959 (1999)
708. D. Hirose, Y. Noguchi, O. Sugino, Phys. Rev. B **91**, 205111 (2015)
709. D. Jacquemin, I. Duchemin, X. Blase, J. Chem. Theory Comput. **11**, 3290 (2015)
710. F. Bruneval, S.M. Hamed, J.B. Neaton, J. Chem. Phys. **142**, 244101 (2015)
711. Y. Noguchi, S. Ishii, K. Ohno, T. Sasaki, J. Chem. Phys. **129**, 10104 (2008)
712. J.E. Sansonetti, J. Phys. Chem. Ref. Data **37**, 1659 (2008)
713. C.R.C. Wang, S. Pollack, T.A. Dahlseid, G.M. Koretsky, M.M. Kappes, J. Chem. Phys. **96**, 7931 (1992)
714. Y. Noguchi, K. Ohno, Phys. Rev. A **81**, 045201 (2010)
715. D. Zhang, S.N. Steinmann, W. Yang, J. Chem. Phys. **139**, 154109 (2013)
716. J.-L. Zhu, Z.-Q. Li, J.-Z. Yu, K. Ohno, Y. Kawazoe, Phys. Rev. B **55**, 15819 (1997)
717. T. Kato, Commun. Pure Appl. Math. **10**, 151 (1957)
718. D.P. Chong, D.M. Schrader, Mol. Phys. **16**, 137 (1969)
719. T.A. Kaplan, P. Horsch, P. Fulde, Phys. Rev. Lett. **49**, 889 (1982)
720. P. Horsch, T.A. Kaplan, J. Phys. C **16**, L1203 (1983)
721. P. Horsch, T.A. Kaplan, Bull. Am. Phys. Soc. **30**, 513 (1985)
722. C. Gros, R. Joint, T.M. Rice, Phys. Rev. B **36**, 381 (1987)
723. S. Maiti, A.V. Chubukov, AIP Conf. Proc. **1550**, 3 (2013)
724. P. Lange, C. Drukier, A. Sharma, P. Kopietz, J. Phys. A: Math. Theor. **48**, 395001 (2015)
725. C. De Dominicis, P.C. Martin, J. Math. Phys. **5**, 31 (1964)
726. Y. Noda, K. Ohno, S. Nakamura, Phys. Chem. Chem. Phys. **18**, 13294 (2016)
727. T. Okugawa, K. Ohno, Y. Noda, S. Nakamura, J. Phys.: Cond. Mat. **30**, 075502 (2018)

# Chapter 3
# Tight-Binding Methods

## 3.1 Introduction

Despite recent major developments in algorithms and computer hardware, the simulation of large systems of particles by ab initio methods is still limited to about a hundred particles. For treating larger systems by molecular dynamics, one can use either tight-binding (TB) or classical molecular- dynamics methods. The TB method has the advantage of being quantum mechanical; therefore one has, in addition to its higher accuracy, information about the electronic structure of the system. In the field of quantum chemistry, other semi-empirical methods, such as MNDO (modified neglect of differential overlap), also exist. These are, in their nature, very similar to Hartree–Fock methods, but the computations of the Hamiltonian and overlap matrix elements are based on semi-empirical formulae.

The TB method is, in this respect, similar to the MNDO method, since the above-mentioned matrix elements are also calculated according to formulae obtained from fits to ab initio data. In this method, one adopts a localized-orbital basis set (the so-called Linear Combination of Atomic Orbitals or LCAO). The analytic form of these basis functions is not known; one can, however, calculate the matrix elements of the Hamiltonian and overlap matrices in this basis and the electronic eigenvalues and total energy can be deduced from these matrix elements. Next, the forces on the atoms are calculated by taking derivatives of the total energy with respect to their coordinates with the help of the well-known Hellmann–Feynman theorem. From the forces, it is then possible to perform molecular-dynamics simulations by using Verlet method or predictor–corrector method. In such simulations, as we will see later, the time-consuming part of the calculation resides in finding the eigenvalues and eigenvectors of the Hamiltonian and in computing the forces on the particles.

In the following sections, we will describe the formalism of the tight-binding theory, methods to calculate the spectrum and the forces, and the way the parametrization is done. Then a few of its applications to simulation of semiconductor systems based on carbon and silicon, and of transition-metal systems such as copper will be given.



K. Ohno et al., *Computational Materials Science*,
https://doi.org/10.1007/978-3-662-56542-1_3

The chapter will be concluded with comments about the applicability and limitations of this theory.

## 3.2 Tight-Binding Formalism

The TB theory can be deduced from a Kohn–Sham type of Hamiltonian; therefore it is a one-particle mean-field theory. It was widely used after the pioneering work of Slater and Koster [1], who devised it for band structure calculations of simple crystals. Later, with the advent of the density functional (DF) theory, and recent advances in computer hardware, its place was taken by ab initio calculations, until very recently, when there was a need to treat large systems, and researchers in this field came back to the TB theory with new and more accurate parametrizations based on ab initio calculations, to be able to make more quantitative predictions.

The TB Hamiltonian is composed of on-site terms and hopping terms as follows

$$H = \sum_{i\alpha} \varepsilon_{i\alpha} c^{\dagger}_{i\alpha} c_{i\alpha} + \sum_{ij,\alpha\beta} t_{i\alpha,j\beta}\; c^{\dagger}_{i\alpha} c_{j\beta}. \tag{3.1}$$

Here the indices $i$ and $j$ refer to the atoms and the Greek letters $\alpha$ and $\beta$ to the orbitals on these atoms. The on-site term $\varepsilon_{i\alpha}$ includes the atomic contribution, the effect of the neighboring atoms (crystal or molecular field) and, to some extent, the exchange–correlation potential. Strictly speaking, it is not the atomic eigenvalue, and includes on average, the effect of the nearest neighbors. This is in fact the source of one of the major problems with this theory, which is its *transferability*. In other words, if this on-site term includes the effect of the four nearest neighbors in a diamond structure, for example, it would not be accurate to use it in an FCC system, where the atoms have 12 nearest neighbors, unless the correction due to these nearest neighbors is included explicitly.

The second term represents hopping between the nearest neighbors and is parametrized as a function of the nearest-neighbor distance and according to the type of orbitals $\alpha$ and $\beta$ involved. It includes a contribution from the kinetic energy, the potential due to the ions and other electrons, and again some part of the exchange–correlation potential. The matrix elements of $H$ can be read explicitly from this equation: The diagonal terms are $\varepsilon_{i\alpha}$, and the off-diagonal terms are $t_{i\alpha,j\beta}$. These terms decay to zero as the distance between the atoms $i$ and $j$ becomes large, and in practice they are set to zero after a cutoff distance. We see then that the matrix of the Hamiltonian in the $|i\alpha\rangle$ basis may contain many zeros if the size of the system one is treating is larger than the cutoff radius.

These parameters $\varepsilon_{i\alpha}$ and $t_{i\alpha,j\beta}$ are usually determined by a fit of the TB eigenvalues to the ab initio ones [2, 3]. One could also try to calculate them directly by performing the same calculation for a localized basis set exactly [4–6]. In the older versions of the TB method, the universal or canonical parameters of Harrison [7] were used; but since they were "universal", these parameters were either not very transferable or not accurate enough (or both), and thus they were either slightly modified [8, 9] or replaced altogether. In all these cases, there is a price to pay, and that is of **transferability**. Anderson has shown [10] that there exists a pseudoatomic Hamiltonian that has as its eigenstates the basis states used here, but this Hamiltonian is not an atomic one and depends yet again on the neighboring atoms, hence the prefix "pseudo". From Anderson's analysis, one can understand why such a simple approach as the TB method (or Hückel's method) can work so well without the need to include the overlap matrix. This approach is in fact very similar to the well-known and widely used pseudopotential method, where one constructs a pseudopotential which gives eigenstates that are smooth (nodeless) and have the same eigenvalues as the true Hamiltonian. Below is the expression for the pseudo-Hamiltonian of atom $i$

$$H_i|i\alpha\rangle = \left( T + V_i + \sum_{j \neq i, \beta} V_j - |j\beta\rangle\langle j\beta|V_j \right) |i\alpha\rangle = \varepsilon_{i\alpha}|i\alpha\rangle. \tag{3.2}$$

The difference is that here, instead of projecting out the core states, one projects out the states of the neighboring atoms which overlap with the atomic basis functions. This is where the dependence on the environment comes from in this atomic pseudo-Hamiltonian (see (3.2)). From the above equation, one can see that the more the states are localized on the atom and the less they overlap with the neighboring orbitals, the more this formalism is transferable, since it is less dependent on the environment. There would be a large degree of cancellation between the third and the fourth terms on the right-hand side of (3.2). Therefore, atomic eigenstates, since they have a long exponential tail, are not appropriate in such treatments. In some parametrizations [5, 11], a harmonic potential is added to the atomic Hamiltonian in order to obtain a more localized basis set. For more details on this issue, we refer the reader to papers by Anderson [10] and by Heine and Bullet [12].

### 3.2.1 *Tight-Binding Parametrization*

In the usual TB formalisms, the total energy is written as a sum of two (or more) terms. The first term is the band energy as shown in (3.22), and the second, an empirical repulsive potential, is usually of the two-body type, meaning that it is written as a sum of interactions between pairs of atoms. The justification for this can be found in [13], although the repulsive part is not always parametrized in a pair form. As an example, one can cite the orthogonal parametrization used for carbon and silicon [2, 14]. This parametrization was first formulated by Goodwin et al. for silicon [15], and then improved by the Iowa state group [2, 14]. The hopping matrix elements of the Hamiltonian and the repulsive pair potential assume the same form

$$H^{\alpha\beta}(R_{ij}) = V^{\alpha\beta}(r_0) f(R_{ij}),$$
$$f(r) = \begin{cases} \left(\frac{r_0}{r}\right)^n \exp\left\{n\left[-\left(\frac{r}{r_c}\right)^{n_c} + \left(\frac{r_0}{r_c}\right)^{n_c}\right]\right\} & r < r_1 \\ c_0 + c_1(r - r_1) + c_2(r - r_1)^2 + c_3(r - r_1)^3 & r \geq r_1. \end{cases} \quad (3.3)$$

Above some distance $r_1$, the exponential form is replaced by a polynomial which tends smoothly to 0 at the cutoff radius. The repulsive energy is then written as

$$E_{\text{rep}} = \sum_i g\left[\sum_j \phi(R_{ij})\right], \quad (3.4)$$

where $g$ is an embedding function which can be fitted by a polynomial, and $\phi$ is of the same exponential form as that in (3.3) but with different $r_c$, $n_c$ and $n$ parameters.

To obtain these parameters, one first minimizes the difference between the TB band structure and first-principles band structure with respect to the adopted parameters. This could be done for several structures in order to get the best transferability. Next, the repulsive parameters are fixed by fitting the TB total-energy curves of several crystalline phases to available experimental or accurate theoretical curves. Determining the parameters really depend on what kind of structures or systems one wants to simulate by the TB method. Sometimes some of the parameters are obtained by fitting to the dimer bond length or binding energy, or other available experimental quantities. When studying bulk properties, defect formation energies or phonon frequencies, one fits the TB parameters to the bulk phases. In the case of small clusters, however, one might want to fit to smaller systems, or to bulk phases with similar (small) coordination numbers to ensure the best transferability. In all cases, having a nonorthogonal parametrization (i.e. including the overlap matrix) even though computationally heavier, is preferable. The reason is that an orthogonal basis set that is equivalent to a nonorthogonal one (i.e. one that reproduces the same eigenvalues for a given structure) has orbital wavefunctions (called Wannier functions in this case) that have a larger extent. Their overlap, however, is zero because of their oscillatory tail. This long oscillatory tail will have a nonzero overlap in other structures with different coordination numbers and nearest-neighbor distances. Therefore, the orthogonality, and consequently the transferability, will be lost. Generally, the best transferability is achieved with the most localized basis sets [10].

More recently, Mehl and Papaconstantopoulos [3] have proposed tight-binding parametrizations for most of the elements. In this case, the total energy is just the sum of the occupied eigenvalues. The repulsive potential is included in the diagonal elements of the Hamiltonian matrix, and this results in a shift of the eigenvalues, simulating the effect of a repulsive interaction. This approach might be valid for bulk crystals, for which they indeed obtain very good agreement with the ab initio data to which they have fitted the parameters. For small clusters, where the coordination number is different from atom to atom, however, this approach might not work as well. We have tested their parameters for copper clusters, and have found that in some cases the total energy *increases* in some small region when an atom approaches a cluster, whereas physically, this interaction should always be attractive. The main shortcoming of their formulation is that it does not incorporate charge transfer effects which are important for clusters and surfaces. This parametrization has, however, the merit of being accurate for reproducing the band structure, cohesive energy, vacancy formation energy and phonon frequencies of bulk systems. In general, TB basis sets that are environment-independent are not very accurate for performing molecular-dynamics simulations. A more recently developed parametrization is due to Tang et al. [16]. This parametrization claims to take three-center integrals into account, and also uses an optimized basis set which depends on the environment of each atom. It has been applied to carbon, silicon and germanium, where it has many more parameters than their previous parametrization. The results for the total energy and eigenvalues of bulk systems in several phases are in much better agreement with LDA calculations than previous parametrizations. For surfaces or clusters, another problem which occurs is the charge transfer phenomenon, i.e. the total charge on every atom, unlike the case in a crystal, is not the same. As will be discussed in Sect. 3.4, one needs to add a Hubbard term to the Hamiltonian in order to avoid excessive charge transfer and include screening in a self-consistent manner.

### 3.2.2 Calculation of the Matrix Elements

In the usual TB parametrizations, the distance dependence of the overlap and hopping matrix elements (Slater–Koster parameters) between orbitals at different sites is given as previously shown in (3.3). The index of each of these functions contains the two orbital types involved and their hybridization. For example, $H^{sp\sigma}$ is the hopping matrix element between an $s$ orbital on the first atom and a $p$ orbital on the second atom forming a $\sigma$ bond. The Hamiltonian matrix, however, is composed of matrix elements such as $H_{ij}^{sp_x}$ representing the hopping element between the $s$ orbital of atom $i$ and the $p_x$ orbital of atom $j$. This must then be expressed in terms of the standard Slater–Koster (SK) parameters, which are hybridizations either along a bond ($\sigma$ type) or perpendicular to the bond ($\pi$ type). For example, if one considers atom $j$ at the origin, and an atom $i$ at $\mathbf{r}$ making an angle of $\theta$ with the $x$ direction, the matrix element $H^{sp\sigma}$ can be expressed as

$$H^{sp\sigma} = H_{ij}^{sp_x} cos\theta + H_{ij}^{sp_y} sin\theta. \tag{3.5}$$

This is due to the fact that the $p\sigma$ orbital is a linear combination of $p_x$ and $p_y$ orbitals. This relation must be inverted to obtain the actual Hamiltonian matrix elements $H_{ij}^{sp_x}, \ldots$ as a function of the SK parameters. These relationships are given in the original paper of Slater and Koster [1] and the book by Harrison [7] for $s$, $p$ and $d$ orbitals, where the Cartesian matrix elements are given in terms of the direction cosines and the SK parameters. Below, we give the relationships for an $sp$ system such as carbon or silicon. If we take $l = x/r$, $m = y/r$ and $n = z/r$, then we have

$$\langle is|H|js\rangle = H_{ss\sigma}(1-\delta_{ij}) + E_s\delta_{ij}, \tag{3.6}$$
$$\langle is|H|jp_x\rangle = l\, H_{sp\sigma}, \tag{3.7}$$
$$\langle is|H|jp_y\rangle = m\, H_{sp\sigma}, \tag{3.8}$$
$$\langle is|H|jp_z\rangle = n\, H_{sp\sigma}, \tag{3.9}$$
$$\langle ip_x|H|js\rangle = -l\, H_{sp\sigma}, \tag{3.10}$$
$$\langle ip_y|H|js\rangle = -m\, H_{sp\sigma}, \tag{3.11}$$
$$\langle ip_z|H|js\rangle = -n\, H_{sp\sigma}, \tag{3.12}$$
$$\langle ip_x|H|jp_x\rangle = [l^2\,(H_{pp\sigma} - H_{pp\pi}) + H_{pp\pi}](1-\delta_{ij}) + E_p\delta_{ij}, \tag{3.13}$$
$$\langle ip_y|H|jp_y\rangle = [m^2\,(H_{pp\sigma} - H_{pp\pi}) + H_{pp\pi}](1-\delta_{ij}) + E_p\delta_{ij}, \tag{3.14}$$
$$\langle ip_z|H|jp_z\rangle = [n^2\,(H_{pp\sigma} - H_{pp\pi}) + H_{pp\pi}](1-\delta_{ij}) + E_p\delta_{ij}, \tag{3.15}$$
$$\langle ip_x|H|jp_y\rangle = lm\,(H_{pp\sigma} - H_{pp\pi}), \tag{3.16}$$
$$\langle ip_y|H|jp_z\rangle = mn\,(H_{pp\sigma} - H_{pp\pi}), \tag{3.17}$$
$$\langle ip_z|H|jp_x\rangle = nl\,(H_{pp\sigma} - H_{pp\pi}). \tag{3.18}$$

### 3.2.3 *Total Energy*

The total-energy and force calculations follow ab initio formulae. In density functional formalism, as seen in the previous chapter, the total energy can be written as

$$E_{\rm DF} = 2\sum_\lambda n_\lambda \varepsilon_\lambda + V_{\rm ii} - \frac{1}{2}\int \frac{\rho(r)\,\rho(r')}{|r-r'|}\mathrm{d}^3r\mathrm{d}^3r' + E_{\rm xc}[\rho] - \int v_{\rm xc}(r)\rho(r)\mathrm{d}^3r, \tag{3.19}$$

where $V_{\rm ii}$ is the ion–ion repulsive interaction and $n_\lambda$ is the occupation number of the eigenstate $|\lambda\rangle$ (it is equal to 0 for unoccupied, and to 1 for occupied levels; $n_\lambda = \theta(E_F - \varepsilon_\lambda)$ where $\theta$ is the Heaviside or step function); the factor of 2 accounts for the spin degeneracy. In the TB method, one can write the total energy as the sum of the band energy (sum of the electronic eigenvalues up to the Fermi level), and a short-range repulsive interaction replacing the repulsive interaction between the core electrons and the term correcting the double counting of the electronic Coulomb repulsion. As said previously, the exchange–correlation is included in some average sense in the band term, and the correction to it can also be included in the empirical repulsive term. A comment about this term is in order here. For atoms separated from each other, it can be shown by using Gauss's law, in the limit of nonoverlapping atomic charge densities, that the electron double-counting correction term cancels the ion–ion repulsive Coulomb interaction exactly; and in the limit of very short distances, the former is finite and the latter becomes infinitely large, hence the name "repulsive" used for this empirical term. A detailed discussion regarding TB parametrizations of this form can be found in the paper by Foulkes and Haydock [13]. Therefore, the TB total energy can be, in general, written as follows

$$E_{\mathrm{TB}} = 2\sum_{\lambda} n_{\lambda}\, \varepsilon_{\lambda} + E_{\mathrm{repulsive}}(R_1, \ldots, R_N). \tag{3.20}$$

The band term is usually attractive and is responsible for the cohesion of molecules and solids, the only exception being for closed-shell atoms such as Helium, where there is no energy gain even after hybridization of $1s$ orbitals. In general, as atoms get closer to each other, the hybridization becomes stronger, and since the lowest orbitals are occupied, the energy of the bonding orbitals decreases, hence the attractive nature of the band term. This band term can also be written as a function of the density and Hamiltonian matrices: expanding the eigenstates in the localized basis $|\lambda\rangle = \sum_i c_i^{\lambda}\, |i\rangle$, this term can be rewritten as follows

$$E_{\mathrm{band}} = 2\sum_{\lambda} n_{\lambda}\, \varepsilon_{\lambda} = 2\sum_{\lambda} n_{\lambda} \langle\lambda|H|\lambda\rangle \tag{3.21}$$

$$= 2\sum_{\lambda}\sum_{ij} n_{\lambda}\, c_i^{\lambda *}\, c_j^{\lambda}\, H_{ij} = 2\sum_{ij} \rho_{ij}\, H_{ij} = 2\mathrm{Tr}(\rho\, H), \tag{3.22}$$

where the elements of the density matrix are defined as

$$\rho_{ij} = \sum_{\lambda} n_{\lambda}\, c_i^{\lambda *}\, c_j^{\lambda}. \tag{3.23}$$

All ground-state quantities can in fact be expressed with the density matrix, since it contains as much information as the eigenvectors.

### 3.2.4 Forces

For the computation of the forces on each atom, one can use within the density functional formalism, the Hellmann–Feynman theorem. This theorem states that in the expression of the total energy, one only needs to take derivatives with respect to the atomic coordinates when the latter appear explicitly, even though the ground-state density depends on these coordinates as well. This result comes from the following general identity, in which $A$ is the many-body Hamiltonian and $\psi$ any of its eigenfunctions

$$\frac{\mathrm{d}}{\mathrm{dx}}\langle\psi|A|\psi\rangle = \langle\psi|\frac{\mathrm{d}A}{\mathrm{dx}}|\psi\rangle. \tag{3.24}$$

In a nonorthogonal TB formalism, and in the matrix notation, the Schrödinger equation is written as follows

$$\sum_j H_{ij}\, c_j^\lambda = \varepsilon_\lambda \sum_j S_{ij}\, c_j^\lambda, \tag{3.25}$$

where $S$ represents the overlap matrix.

To compute the force on an atom located at $R$, one must take the derivative of (3.20) with respect to $R$: $F_i = -\partial E/\partial R_i$. The repulsive term is a classical potential, and taking its derivative is straightforward; thus we discuss only the band term; $\sum_\lambda n_\lambda\, \varepsilon_\lambda$. From (3.25) one can solve for the eigenvalue after multiplying by the *normalized* eigenvector $c$ and summing

$$\varepsilon_\lambda = \sum_{ij} c_i^{\lambda *}\, H_{ij}\, c_j^\lambda. \tag{3.26}$$

After taking the derivative with respect to $R$, owing to the orthonormality of the eigenstates: $\sum_{ij} c_i^{\lambda *}\, S_{ij}\, c_j^{\lambda'} = \delta_{\lambda,\lambda'}$, one obtains

$$\frac{\mathrm{d}\varepsilon_\lambda}{\mathrm{d}R} = \sum_{ij} c_i^{\lambda *} \left( \frac{\mathrm{d}H_{ij}}{\mathrm{d}R} - \varepsilon_\lambda \frac{\mathrm{d}S_{ij}}{\mathrm{d}R} \right) c_j^\lambda. \tag{3.27}$$

From this relation, one can obtain the contribution of the band term to the force

$$F_{\mathrm{band}} = -2 \sum_{ij} \left( \frac{\mathrm{d}H_{ij}}{\mathrm{d}R}\, \rho_{ij} - \frac{\mathrm{d}S_{ij}}{\mathrm{d}R}\, w_{ij} \right), \tag{3.28}$$

where $w$ is the energy-weighted density matrix: $w_{ij} = \sum_\lambda n_\lambda\, \varepsilon_\lambda\, c_i^{\lambda *}\, c_j^\lambda$. In this way, if the matrix elements of the Hamiltonian and overlap matrices are known explicitly as a function of the atomic coordinates $R$, one can calculate exactly the forces on the atoms (exactly in the sense that the force would be the exact derivative of the total energy) provided that the vectors $c$ are the exact eigenvectors of $H$.

For finite temperature simulations, or metallic systems which require many k-points in order to converge, one needs to replace $n_\lambda$ by the Fermi-Dirac distribution function $n_\lambda = f_{FD}(\varepsilon_\lambda) = 1/[1 + \exp\beta(\varepsilon_\lambda - \mu)]$. In this case, a small correction needs to be added to the force due to the dependence of the chemical potential $\mu$ and $f_{FD}(\varepsilon_\lambda)$ on the positions of the particles. Equation (3.27) remains correct of course, but instead of $n_\lambda \mathrm{d}\varepsilon_\lambda/\mathrm{d}R$, one needs to compute $\mathrm{d}(n_\lambda \varepsilon_\lambda)/\mathrm{d}R$. This yields

$$\frac{\mathrm{d}(n_\lambda \varepsilon_\lambda)}{\mathrm{d}R} = n_\lambda \frac{\mathrm{d}\varepsilon_\lambda}{\mathrm{d}R} + n_\lambda(1 - n_\lambda)\beta\varepsilon_\lambda \left[ \frac{\sum_\mu n_\mu(1 - n_\mu)\frac{\mathrm{d}\varepsilon_\mu}{\mathrm{d}R}}{\sum_\mu n_\mu(1 - n_\mu)} - \frac{\mathrm{d}\varepsilon_\lambda}{\mathrm{d}R} \right] \tag{3.29}$$

The contribution of the additional term is important only around the Fermi level where $n_\lambda(1 - n_\lambda)$ is nonzero, i.e., in metallic and open shell systems.

## 3.3 Methods to Solve the Schrödinger Equation for Large Systems

The straightforward way to solve the Schrödinger equation for electrons is usually the direct, "brute-force" diagonalization. This is a so-called $O(N^3)$ task, which means that the CPU time needed to perform the diagonalization is proportional to the cube of the number of atoms, or equivalently the number of (valence) electrons involved, denoted here by $N$. The alternative way to find the eigenvalues and eigenfunctions of the electronic Hamiltonian is to minimize the total energy with respect to the electronic degrees of freedom, by steepest-descent (SD) or conjugate-gradient (CG) methods [17]. Although faster than direct diagonalization, these are also $O(N^3)$ methods, since they require orthogonalization of the eigenfunctions. In the case of a band or tridiagonal matrix, the task can be reduced to $O(N^2)$.

### 3.3.1 *The Density Matrix $O(N)$ Method*

Recently, order-$N$ methods have been put forth [18–21]. In these methods, one assumes that electronic properties of the system, such as the charge density or the density matrix, are not affected by changes in the potential far enough away from the point considered. One also assumes that such changes in the potential are done at *constant chemical potential*, otherwise one could lower the potential far away so much that all the electrons would flow there to fill up the potential well. In this way, one only needs to include the few nearest shells in order to compute the charge density at a given point. The range must be determined by checking the convergence of the result as the number of included shells is increased. The physical reason for this assumption is screening: electrons tend to screen any perturbation in the system by satisfying Poisson's equation: $\Delta V = -4\pi\rho$.

In semiconductors, one does not need to include more than a few shells, but in metals, one has to cover a few screening lengths. This could, however, lead to very large cutoff distances because of the well-known Friedel oscillations of the charge (and potential) around an impurity. (In a three-dimensional metal, this oscillation is of order $\cos(2k_F r)/r^3$, and thus long-ranged).

One of the implementations of this order-$N$ method is performed by adopting the density matrix (DM) formalism, where the total electronic energy is written as $2\text{Tr}(\rho H)$ (see (3.23) and (3.22)), and computing the DM by minimizing it. This formulation is based on a variational principle, and hence, for any approximation, will lead to an upper bound of the ground-state energy. The usual approximation used in this case is to truncate the range of the DM, and assume that only DM elements between two atoms within a cutoff radius are nonzero. If we denote by $M$ the number of relevant neighboring sites times the number of orbitals on each atom, then the memory needed to store the nonzero elements of $\rho$ is of size $N \times M$, where $N$ is the number of electrons.

The elements of the DM can be found by minimizing the total energy (or rather, $\Omega = E - \mu N$ if one is working at constant chemical potential) with the constraint that $\rho$ is idempotent: $\rho_{ij} = \sum_k \rho_{ik}\rho_{kj}$. (In here and what follows, we have assumed an orthogonal basis: $S = 1$). Additionally, one must take care to satisfy the Hermitian nature of $\rho$ and the normalization of charge $(2\int_r \rho(r)\mathrm{d}^3r = 2\,\mathrm{Tr}\,\rho = N)$. Kohn has suggested [21] that one could minimize

$$Q[\rho] = E[\rho^2] - \mu N[\rho^2] + \alpha\Big[\int \rho^2(1-\rho^2)\Big]^{1/2}\mathrm{d}^3r$$

with respect to the matrix $\rho$ by the CG method, for example. Here $\rho^2$ represents the square of the trial density matrix $\rho$, which would be identical to $\rho$ for the final idempotent result; the last term is a penalty functional ($\alpha > 0$) of $\rho$ which reduces to its minimum, i.e. zero, for idempotent matrices since their eigenvalues are either 0 or 1. Furthermore, Kohn has shown that one reaches the correct ground-state density and energy if the positive coefficient $\alpha$ is large enough, typically of the order of the valence bandwidth times $N\sqrt{\mu}$. This cost function involves the fourth power of the DM and might end up a bit more time-consuming that the following methods proposed by Li et al. [18] and Daw [19], where one iterates to more idempotent density matrices by taking $\tilde{\rho} = 3\rho^2 - 2\rho^3$ as the physical density matrix. As in Kohn's method, the matrix elements of $\rho$ are taken as the variational degrees of freedom. The minimization condition is imposed on $\Omega = E - \mu N = 2\mathrm{Tr}[(3\rho^2 - 2\rho^3)(H - \mu)]$ if one works at constant chemical potential. Using this form of the density matrix, if $\rho$ has eigenvalues in $[-0.5, 1.5]$, then the physical DM's eigenvalues are strictly in $[0, 1]$. Furthermore, in the limit of a large cutoff radius, the potential $\Omega = E - \mu N$ tends to its exact value from above since this is a variational formulation. The minimization with respect to $\rho$ can be done by steepest-descent (SD) or conjugate-gradient (CG) methods, since the gradient of the potential with respect to the matrix elements of $\rho$ is readily calculated

$$\frac{\mathrm{d}\Omega}{\mathrm{d}\rho_{ij}} = 6(\rho_{ik}H'_{kj} + H'_{ik}\rho_{kj}) - 4[(\rho^2)_{ik}H'_{kj} + \rho_{ik}H'_{kl}\rho_{lj} + H'_{ik}(\rho^2)_{kj}], \qquad (3.30)$$

where $H' = H - \mu$, and summation over repeated indices is implied. Generalizations for nonorthogonal basis functions have also been obtained [22, 23]. The CG method can easily be implemented as follows. In each iteration $n$ one performs a linear minimization of $\Omega[\rho]$ in the conjugate direction $F^{[n]}$

$$\rho^{[n+1]} = \rho^{[n]} + \alpha^{[n]}F^{[n]}. \qquad (3.31)$$

This search can be performed analytically since the function to be minimized is a third-degree polynomial in $\alpha^{[n]}$. After $\rho^{[n+1]}$ is calculated, one proceeds to compute the steepest-descent direction $G_{ij}^{[n+1]} = -\partial\Omega/\partial\rho_{ij}^{[n+1]}$ at this point. Then, the conjugate direction is defined as $F^{[n+1]} = G^{[n+1]} + \gamma^{[n]}F^{[n]}$, where $\gamma^{[n]} = (F^{[n+1]} - F^{[n]}).F^{[n+1]}/F^{[n]}.F^{[n]}$, the dot product for two matrices being defined as

$A.B = \sum_{ij} A_{ij} B_{ij}$. Since the size of all these arrays is proportional to $N$, all matrix multiplications take a CPU time proportional to $N$. Minimizations are done analytically, and a few iterations independent of $N$ are needed to converge to the minimum; therefore the total CPU time is of the order of $N$.

If one is working at constant particle number, one should aim instead to minimize the potential $\Omega = E - \mu(2\text{Tr}\rho - N)$, and consider $\mu$ as a Lagrange multiplier that is to be determined so as to satisfy the electron number conservation at all times. This method was proposed by Millam and Scuseria [24]. According to (3.31), if $2\text{Tr}\rho^{[n]} = N$, it is sufficient to choose $\mu$ in $\Omega$ such that $\text{Tr}F^{[n]} = 0$; this way $2\text{Tr}\rho^{[n+1]} = N$ will also be satisfied. For $G$ to be traceless, it is sufficient to make the derivative of $\Omega$ traceless; this gives

$$\mu = \text{Tr}[2(H\rho^2 + \rho H\rho + \rho^2 H) - 3(\rho H + H\rho)]. \tag{3.32}$$

Then the electron number is conserved at all times and $\Omega$ is just be the total energy, although its derivative is different. This algorithm can be summarized in the flowchart of Fig. 3.1

A more recent development in this field is based on the formal knowledge of the DM [25] as a function of the Hamiltonian matrix: $\rho(H) = \theta(E_\text{F} - H)$. The Fermi energy $E_\text{F}$ is found from $2\text{Tr}\rho(H) = N_\text{e}$. Therefore, the minimization task is avoided altogether! To actually compute matrix elements of $\rho$, one can use its "analytic approximation": $\rho(H) = 1/(1 + \text{e}^{\beta(H-\mu)})$ and expand it on a polynomial basis, since powers of $H$ are easy to calculate, and, furthermore, one can take advantage of the sparsity of $H$ to compute $H^n$ with a CPU time of $O(N^2)$ at the most, since the matrix of $H$ in an LCAO basis is always short-ranged. If, furthermore, the system is a semiconductor or insulator (nonzero gap), the DM itself becomes short-ranged, and the task of computing $\rho$ becomes $O(N)$.

A Taylor expansion would have convergence problems at points far away from the center of the expansion; a Chebyshev expansion, however, has the best convergence properties. One must first shift and rescale $H$ so that its spectrum lies in $[-1, 1]$, and then use

$$\rho_s(H) = \frac{1}{1 + \text{e}^{\beta_s(H_s - \mu_s)}} = \sum_{n=0}^{p-1} a_n(\beta_s, \mu_s) T_n(H_s), \tag{3.33}$$

where the subscript "$s$" refers to scaled quantities, and $T_n$ is the $n$th Chebyshev polynomial: $T_n(x) = \cos(n\cos^{-1} x)$. The coefficients $a_n$ can be calculated once and for all, by using Fast Fourier Transformation (FFT). The only approximation then lies in truncating the summation at $p$. If the guessed difference between the upper and lower eigenvalues is too small, then the series expansion is not convergent (i.e. the eigenvalues of the density matrix would be larger than 1 or negative); and if it is too large, then a large number of terms ($p$) are necessary to converge the sum to the real DM. Therefore, it is important first to get a good approximation to the highest and lowest eigenvalues of $H$, by using the Lanczos method, for example. This method is not variational, but seems more efficient than the previously proposed ones. For

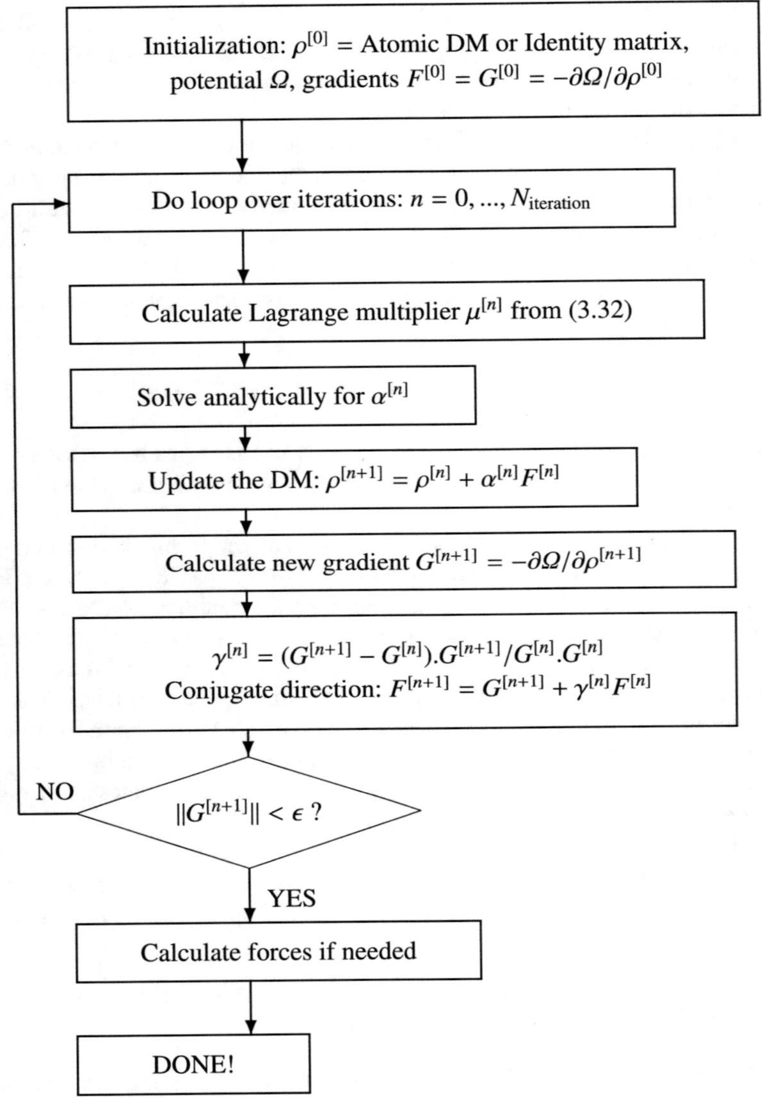


**Fig. 3.1** Flowchart of the conjugate-gradient (CG) minimization algorithm for the $O(N)$ density matrix formalism

more details, the reader can refer to the paper by Baer and Head-Gordon [26] and the references therein.

Note that such methods become preferable when the number of atoms involved is typically more than 100, for carbon crystals, where the density matrix is rather short-ranged. Even in this case, the solution is not exact, and taking larger cutoff radii would increase the memory size, and the CPU time. So, in practice, having an exact density matrix implies a break-even point of even larger than 100 for insulators such as diamond. In conclusion, the order-$N$ DM methods become advantageous for systems of quite large size, depending on the range of $\rho$. They work best for systems with a large gap; typically, the range of the DM scales like $1/\sqrt{E_{\rm gap}}$. In any case, the previously mentioned Chebyshev method leads at the most to an $O(N^2)$ scaling, although it is approximate, since the polynomial expansion must be truncated at some point. At lower temperatures, where the step in the Fermi–Dirac function becomes sharper, one would need to include more terms in the Chebyshev expansion. In the case of a metal with zero gap, the range of the DM scales like $\sqrt{\beta}$, so that it is necessary to compute electronic properties of metallic systems at large temperatures, typically of the order of 0.2 eV or even more. Basically, the range is limited by the deBroglie thermal wavelength of the electrons $\lambda_{\rm thermal} = \sqrt{\hbar^2/2\pi mkT}$ [26], and as $T$ approaches zero, the off-diagonal terms of the DM tend to behave like $\cos(2k_{\rm F}|r-r'|)/|r-r'|^3$. A very recent energy-renormalization-group proposal has been able to make the TB calculations for a metal scale like $N(\log N)^2$ [27], though the coefficient of this term might be very large.

### 3.3.2 The Recursion Method

To compute the density of states (DOS) in bulk or very large systems, Haydock [28] has proposed the recursion method, which is, in its idea, similar to order-$N$ methods, i.e. it relies on the local environment of a given atom, and computes the contribution of neighboring shells, and increases the latter until convergence is reached. This method is based on an LCAO basis as well and, more precisely, calculates the local DOS (LDOS), which contains information about the distribution of the energy levels of the system *on each atom*:

$$LDOS(i, E) = \langle i|\delta(E-H)|i\rangle = \sum_{\lambda} \delta(E-\varepsilon_\lambda)\, |\langle i|\lambda\rangle|^2. \tag{3.34}$$

It can be easily shown that the LDOS is directly related to the imaginary part of the Green's function through

$$LDOS(i, E) = -\frac{1}{\pi}\mathrm{Im}\langle i|\frac{1}{E-H+\mathrm{i}0}|i\rangle, \tag{3.35}$$

where the energy $E$ is given a small positive imaginary part i 0. The recursion method gives a continued-fraction expansion of this diagonal matrix element of the Green's function. The idea is to go from the LCAO basis of the system at hand to a "chain" basis which is orthonormal and satisfies the following

$$H|u_n\rangle = a_n|u_n\rangle + b_{n+1}|u_{n+1}\rangle + b_n|u_{n-1}\rangle, \tag{3.36}$$

where $|u_0\rangle$ is the orbital $|i\rangle$ at which we want to compute the LDOS, and $a_j$ and $b_j$ real numbers. One can start by normalizing $|i\rangle$ to get $|u_0\rangle$ and then use (3.36) with $b_0 = 0$ to obtain $a_0$, $b_1$ and $|u_1\rangle$, and continue by *recursion* to transform the original TB basis to the new orthonormal basis $|u_j\rangle$. As can be easily seen from (3.36), the matrix of the Hamiltonian in this basis is symmetric and tridiagonal (hence the name "chain" used for this representation), with $a_j$ on its diagonal, and $b_j$ above and below it. As an example, for a linear chain with hopping between nearest neighbors, after $n$ iterations the generated $|u_n\rangle$ has only components on sites that are within $n$ unit cells of the original site $|u_0\rangle = |i\rangle$:

$$|u_n\rangle = \sum_{j=-n,n} c_j |i + j\rangle.$$

Therefore, to obtain a tridiagonal matrix of size 100 (i.e. generate up to $|u_{100}\rangle$), one has to go up to 100 shells beyond the central site. In three dimensions, this could involve about 10000 atoms or more.

By using the recursion relation between the matrix elements of the Green's function in the chain basis, one can show that the diagonal matrix element of the retarded Green's function can be written as a continued fraction as a function of $a_n$ and $b_n$ as follows:

$$\langle u_0|\frac{1}{z - H}|u_0\rangle = \frac{1}{z - a_0 - \frac{b_1^2}{z-a_1-\frac{b_2^2}{z-a_2\cdots}}}, \tag{3.37}$$

where $z = E + \mathrm{i}\,0$. There are techniques for terminating the continued fraction, which could have a complex structure for systems with multiple bands and for complicated systems such as binary alloys, for example. We refer the reader to [28] for these subtleties, but basically, once a few shells about an atom are chosen, one can transform the orbitals to the "chain" basis and obtain the coefficients $a_j$ and $b_j$, which are then used in the continued fraction (3.37) in order to compute the LDOS. To get an accurate LDOS, one should of course increase the number of shells until the fraction converges and can be properly terminated. In a simple case where $a_n \to a$ and $b_n \to b$ for large $n$, the terminator can be analytically solved for, since it satisfies

$$G_{nn}(z) \approx 1/[z - a - b^2 G_{nn}(z)]$$

for $n \to \infty$. This is another order-$N$ method in the sense that computing the physical properties of the whole system scales like $N$ (one computes the LDOS at each of the $N$ sites; but each of the computations, does not depend on $N$). Furthermore, each of the LDOS computations is quite fast, though the number of atoms involved in a few shells could quickly reach thousands. From the knowledge of the LDOS, computing the DOS, the forces and the total energy of the system is a straightforward task.

### 3.3.3 Recursive Green's Function Method

This method is very similar to Haydock's recursion method in spirit, but does not involve the calculation of the continued fraction. If the Hamiltonian of a system is written as the sum of the Hamiltonians of its two subsystems and their interaction $H = H_a + H_b + V$, then one can show that the projection of the Green's function of the whole system in the space spanned by the subsystem $a$ for example is:

$$G_a(z) = \frac{1}{z - H_b} \tag{3.38}$$

This relation relates the two GFs of the 2 subsystems, therefore one should solve iteratively for both $G_a$ and $G_b$, by starting with an initial guess function which could be the GF of the noninteracting subsystems. This very simple relation can be generalized for the many subsystems (atoms or a few atom) case as follows:

$$G_i(z) = \sum_{j,k} V_{ij} G'_{jk}(z) V_{ki} \tag{3.39}$$

where $G'_{jk}(z)$ is the Green's function of the system in which state $i$ was removed; $V_{ij} = \langle i|V|j\rangle$, and $i$ and $j$ label the subsystems which could be atoms. In this case, $V$ is a sparse matrix and the $V_{ij}$ need only to be computed for the neighboring atoms. The memory needed for this calculation is therefore proportional to $N$, the number of atoms. The CPU is manifestly of order N as well. The GF must, however, be solved iteratively. In the case of the semiconductors, few iterations are needed before convergence, but for metals presumably more iterations are necessary. This method is in principle exact, there is no need to truncate (DM) nor to approximate the terminator of the continued fraction (previous section). Convergence might, however, be a problem at times. In such cases, one can take a larger imaginary part for the energy variable $z$. As in the previous section, one can obtain all the electronic properties by this method.

Compared to DM methods, one can say that methods based on recursion typically take longer times and more shells to converge, regardless of the materials involved. The convergence properties of DM methods in terms of forces and total energy are found to be much better than those of recursion-based algorithms, especially for insulators and semiconductors, i.e. systems with a gap.

For metallic systems, however, the DM is long-ranged ($\propto 1/R^3$ in 3D) and, basically, will not converge properly unless one is able somehow to renormalize it. In this case, recursion-based methods, since they are able to sum over all shells in some approximate way by properly summing the infinite continued fraction, are more accurate and therefore preferable to DM methods. In addition to the forces and the total energy, recursion-based methods have the advantage of giving also the electronic properties such as the DOS. A detailed comparative study of these methods can be found in reference [29].

## 3.4 Self-consistent Tight-Binding Formalism

Unlike the Hamiltonian in ab initio methods, the tight-binding Hamiltonian is blind with respect to the charge of the system. In other words, it predicts the same eigenvalues and eigenvectors for a system of any charge. The only difference for a charged system is in the occupation numbers, which would yield a different total energy and forces. But adding an electron to the system also changes the Hamiltonian by adding a repulsive Coulomb, and an exchange–correlation potential to the Hamiltonian, as a result of which the electronic eigenvalues will be shifted upwards. Furthermore, for a charge-neutral system, the TB method is unable to properly take into account the screening of an external field. Concerning the transferability, not only do the shape of the wavefunctions and their overlap affect transferability, but also a phenomenon called "charge transfer" could cause an unphysical charge distribution in the system. This is basically due to the environment of a given atom. For example, atoms at the surface usually tend to get more charge than their neighbors, and sometimes this charge transfer can be quite large. This, in turn, affects the potential via Poisson's equation, and thus the Hamiltonian matrix elements. Consequently, this problem must be dealt with self-consistently, just like the problem of charged systems where the effect of the additional Coulomb interaction must be reflected in the Hamiltonian.

To overcome these mentioned problems an ad-hoc Hubbard like term $\frac{1}{2}\sum_i U(n_i - n_i^0)^2$ is usually added to the total energy [2, 15, 30]. This way, any change in the electronic population $n_i^0$ would cost some energy, and thus, large charge transfer effects are avoided. The problem becomes now self-consistent since the added term depends on the charge density itself, and the total energy should be minimized with respect to the occupations $n_i$. The minimization of $E$ is equivalent to adding $U(n_i - n_i^0)$ to the on-site matrix element of the Hamiltonian. The Hubbard coefficient $U$ is usually determined empirically by fitting to the ab initio data of inhomogeneous structures with different environments. The addition of such a term is in fact crucial in ensuring good transferability of a given parametrization, even if it is a nonorthogonal one, since these are two different and unrelated issues. Below, we will show how this method can be implemented and how the "Hubbard" coefficient may be determined, and, as will be shown, the results are very different from the empirically adopted values.

Let us consider the Hamiltonian of a finite system in the neutral configuration and denote it by $H_0$. The addition of an electron will add an electrostatic Coulomb term and an exchange–correlation term to it. In the mean-field, or first-order perturbation theory, one considers the average of the perturbation in an eigenstate to obtain the shift in the corresponding eigenvalue. This first-order shift can be interpreted as the Coulomb interaction between the added electron and the electronic charge distribution of the eigenstate considered. The change in the exchange–correlation potential is neglected.

One ends up, for a negatively charged system, for instance, with the following expression:

$$\delta\varepsilon_\lambda = \int_{r,r'} \frac{\rho_\lambda(r)\delta\rho(r')}{|r-r'|}\, \mathrm{d}^3r\mathrm{d}^3r' = \int_{r,r'} \frac{\rho_\lambda(r)\rho_{\mathrm{LUMO}}(r')}{|r-r'|}\, \mathrm{d}^3r\mathrm{d}^3r'. \tag{3.40}$$

The added charge ($\delta\rho = \rho - \rho^0$) will occupy the lowest unoccupied molecular orbital (LUMO).

This charge will affect the ground-state charge distribution itself, and thus instead of using first-order perturbation theory, one can calculate the matrix elements of the new perturbed Hamiltonian and diagonalize it numerically and self-consistently to obtain the exact shift of the eigenvalues. This approach, allowing for the relaxation of the ground state charge density, will partially include correlation effects as well.

By using the density matrix, the change in the Hamiltonian operator $H$ and its matrix $\mathcal{H}$, respectively, can be written as follows:

$$\delta H(r) = \int_{r'} \frac{\delta\rho(r')}{|r-r'|}\, \mathrm{d}^3r', \tag{3.41}$$

$$\delta\mathcal{H}_{ij}^{\alpha\beta} = \sum_{kl,\gamma\delta} U_{ij,kl}^{\alpha\beta,\gamma\delta} (\rho_{kl}^{\gamma\delta} - \rho^{0}{}_{kl}^{\gamma\delta}), \tag{3.42}$$

where $U$ is defined as

$$U_{ij,kl} = \int_{r,r'} \phi_i^*(r)\phi_j(r) \frac{1}{|r-r'|} \phi_k^*(r')\phi_l(r')\, \mathrm{d}^3r\mathrm{d}^3r'. \tag{3.43}$$

Here, the superscript $^0$ refers to the neutral unperturbed ground state. It can be seen from the above equation that the matrix element is a product of two terms: one is the two-particle Coulomb integral, and the other factor is the change in the density matrix $\rho_{ij}^{\alpha\beta}$. If the latter is diagonal, one recovers the often-used Hubbard term $U(n_i - n_i^0)$, which is the correction to the Hartree potential, i.e. it represents the electrostatic "Coulomb" potential of the induced (or added) charge. The additional Coulomb interaction was the reason why, physically, the above term needed to be added to $H_0$. One can also see it mathematically, as the matrix elements of $H_0$ were defined from a fit to crystalline band structures where there is no charge transfer. Therefore, for a surface or a cluster, or a charged system, there must be a correction to these matrix elements. To first order, this correction is linear in the charge difference $\delta\rho$, and $U$ can be seen as the proportionality constant in this expansion, which, by using Taylor's formula, should be the first derivative of the Hamiltonian with respect to the charge density: $H(\rho) = H(\rho_0) + \mathcal{U}\delta\rho + O(\delta\rho^2)$. Within the density functional formalism, $\mathcal{U}$ is given by

$$\mathcal{U}_{ij,kl}^{\mathrm{DF}} = \int_{r,r'} \phi_i^*(r)\phi_j(r) \left( \frac{\partial v_{xc}(r)}{\partial\rho(r')} + \frac{1}{|r-r'|} \right) \phi_k^*(r')\phi_l(r')\, \mathrm{d}^3r\mathrm{d}^3r'. \tag{3.44}$$

Here the Greek superscripts have been dropped, but $i$, $j$, $k$, $l$ can be thought of as basis indices. Notice that we have still reserved "$U$" for the pure electrostatic Coulomb term

(3.43), the term represented by the calligraphic symbol $\mathcal{U}$ includes the exchange–correlation effects as well.

One can discuss the effect of exchange in an LCAO–HF approach. Because of the small overlap, the exchange contribution is usually very small except when the four orbitals are localized on the same atom. The exchange, therefore, appears only in the on-site terms of $U$ through $U^{\alpha\beta,\alpha\beta}\delta\rho^{\beta\beta}$ and should be included if $U^{\alpha\beta,\alpha\beta}$ is large. In general, its effect becomes important when the off-diagonal terms of the density matrix are large. Such is the case for metallic systems, where the density matrix is long-ranged. In this case $-U^{\alpha\gamma,\beta\delta}_{ik,jl}/2$ must be added to the usual $U^{\alpha\beta,\gamma\delta}_{ij,kl}$ in the Hamiltonian and the total energy to represent the exchange effects.

This treatment is also relevant to the concept of chemical hardness which can be defined as [31]

$$\eta_{\lambda\mu} = \frac{\partial^2 E}{\partial n_\lambda \partial n_\mu},$$

where $n_\lambda$ and $n_\mu$ are the occupation numbers of the eigenstates $\lambda$ and $\mu$. The first derivative is of course the single-particle eigenvalue $\varepsilon_\lambda$, and the second tells us how this eigenvalue changes as the occupation number of another level $\mu$ is changed. This is exactly the shift of the eigenvalues as the system is charged. In other words, we have calculated the chemical hardness by taking the state $\mu$ as being the lowest unoccupied state, and $\lambda$ any other state. This chemical hardness matrix allows one to predict the effect of any charge transfer, (or more generally, any change in the occupation numbers) on the eigenvalues of the system. It is given by

$$\eta_{\lambda\mu} = \sum_{i\alpha,j\beta,k\gamma,l\delta} C^{\lambda\,*}_{i\alpha}\, C^{\lambda}_{j\beta}\, C^{\mu\,*}_{k\gamma}\, C^{\mu}_{l\delta}\, \mathcal{U}^{\alpha\beta,\gamma\delta}_{ij,kl} \tag{3.45}$$

and the shift in the energy levels is given by

$$\varepsilon_\lambda = \varepsilon^0_\lambda + \sum_\mu \eta_{\lambda\mu}\,(n_\mu - n^0_\mu). \tag{3.46}$$

As one can see, $\eta$ involves the eigenstates of the system, and therefore depends on the specific structure of the system considered, although its atomic value has also been used in some literature [4] in order to compute charge transfer in neutral systems.

The screening of an external field can also be treated properly by this method. If one approximates the matrix elements of the external field by

$$\langle i\alpha|V|j\beta\rangle = \int \phi^*_{i\alpha}(r)\, V(r)\, \phi_{j\beta}(r)\, \mathrm{d}^3 r \approx \frac{1}{2} S^{\alpha\beta}_{ij}[V(R_i) + V(R_j)], \tag{3.47}$$

where $S$ is the overlap matrix, the matrix elements of the perturbed Hamiltonian can easily be calculated:

$$\langle i\alpha|\delta H|j\beta\rangle = \langle i\alpha|V|j\beta\rangle + \sum_{kl,\gamma\delta} \mathcal{U}_{ij,kl}^{\alpha\beta,\gamma\delta}(\rho_{kl}^{\gamma\delta} - \rho^{0\gamma\delta}_{kl}). \tag{3.48}$$

The second term, the induced potential, being the change of $H$ due to the change in the charge density, involves the full Coulomb term $\mathcal{U} = \partial H/\partial\rho$, and not only the electrostatic induced potential defined with $U$.

In the usual TB formalisms, the total energy of the unperturbed system is the sum of a band term (sum of the occupied eigenvalues) and an empirical repulsive potential. This potential substitutes for the residual terms in the total energy:

$$E_{\text{rep}}(R_1, \ldots, R_N) = V_{\text{ion-ion}}$$
$$-\frac{1}{2}\int \frac{\rho(r)\rho(r')}{|r-r'|}\,\mathrm{d}^3r\mathrm{d}^3r' + E_{xc}[\rho] - \int v_{xc}[\rho](r)\rho(r)\,\mathrm{d}^3r. \tag{3.49}$$

When an electron is added to the system, this repulsive energy is changed because of the change in the charge density $\rho$. Using the above definition of $E_{\text{rep}}$, one can write the total energy within this formalism as

$$E[\rho, R] = 2\text{Tr}\,\rho\mathcal{H} + E_{\text{rep}}[R] - \frac{1}{2}\int \frac{\rho(r)\rho(r') - \rho^o(r)\rho^o(r')}{|r-r'|}\,\mathrm{d}^3r\mathrm{d}^3r' + \delta(\text{XC}). \tag{3.50}$$

The first term, i.e. the sum of the eigenvalues of the perturbed system, includes the effect of an external potential as well. The last two terms are necessary since they represent the change in the Coulomb energy due to charging effects. Written in the TB basis, and using the symmetry of the kernel $\mathcal{U}$ under exchange of $ij$ and $kl$, this change can be expressed as

$$\Delta E_{\text{Coulomb}} = \frac{1}{2}\sum_{i\alpha, j\beta, k\gamma, l\delta} \delta\rho_{ij}^{\alpha\beta}\,\Sigma\rho_{kl}^{\gamma\delta}\,\mathcal{U}_{ij,kl}^{\alpha\beta,\gamma\delta}, \tag{3.51}$$

where $\delta\rho = \rho - \rho^0$ and $\Sigma\rho = \rho + \rho^0$. It must be noted that this correction also includes corrections to the exchange–correlation energy, since we have used $\mathcal{U}$ and not the electrostatic $U$. The final expression for the total energy then reads

$$E[\rho, R] = 2\text{Tr}\,\rho\,\mathcal{H}^0 + \frac{1}{2}\delta\rho\;\mathcal{U}\,\delta\rho + E_{\text{rep}}[R]. \tag{3.52}$$

### 3.4.1 Parametrization of the Coulomb Integral $\mathcal{U}$

In order to perform such calculations, one has to determine the kernel $\mathcal{U}$. We can approximate the general terms by

$$\mathcal{U}_{ij,kl}^{\alpha\beta,\gamma\delta} = \frac{1}{4} S_{ij}^{\alpha\beta} S_{kl}^{\gamma\delta} (U_{i,k}^{\alpha,\gamma} + U_{j,k}^{\beta,\gamma} + U_{i,l}^{\alpha,\delta} + U_{j,l}^{\beta,\delta}). \tag{3.53}$$

For $U_{i,k}^{\alpha,\gamma}$ one can either use the Ohno–Klopman form [32, 33]:

$$U_{i,k}^{\alpha,\gamma} = \left[ |R_i - R_k|^2 + \frac{1}{4}\left( \frac{1}{u_{\alpha\alpha}} + \frac{1}{u_{\gamma\gamma}} \right)^2 \right]^{-1/2}, \tag{3.54}$$

or simply the following formula:

$$U_{i,k}^{\alpha,\gamma} = [|R_i - R_k|^2 + u_{\alpha\gamma}^{-2}]^{-1/2}, \tag{3.55}$$

where $u$ is the on-site (atomic) Coulomb integral. This way, $U$ has the proper limits for large and small distances $R_{ik}$; namely, it goes to $1/R$ for large $R$ and tends to $u_{\alpha\gamma}$ for small distances, typically of the order of the atomic size. The above approximation has been used for an arbitrary normal basis set; it reduces to a simpler form if the basis is orthonormal. Such a basis has been used for carbon [2] and silicon [14], where each atom has one 2s and three 2p orbitals. One is left with $u_{ss,ss}$, $u_{ss,pp}$, $u_{pp,pp}$, $u_{pp,p'p'}$, $u_{sp,sp}$ and $u_{pp',pp'}$, which can be obtained from an atomic calculation. Here $p'$ and $p$ denote two different $p$-orbitals.

In the case of an atom, the chemical-hardness matrix $\eta$ and the Coulomb integral $\mathcal{U}$ are identical (see (3.45)) since the basis indices are identical to the eigenstate indices. It follows that exchange terms such as $u_{sp,sp}$ and $u_{pp',pp'}$ are identically zero for the atom; the exchange effects are really included in $u$ itself, as can be seen from (3.44). This, fortunately, reduces the number of parameters, and anyway, since the TB parameters are usually derived from a fit to DF calculations where the exchange–correlation potential is local, there is no need to include exchange integrals unless one is doing an HF calculation. Furthermore, using the invariance of $u$ with respect to any rotation, one can easily show that $u_{pp,p'p'} = u_{pp,pp} - 2u_{pp',pp'}$. The only parameters to determine are thus $u_{ss,ss} u_{ss,pp}$ and $u_{pp,pp}$.

The values of these parameters (not including the exchange–correlation effects i.e. including purely the Coulomb interaction in the integral) for carbon can be found in the book by Fulde [33], which is a good reference on correlation calculations for atoms and molecules. For carbon and silicon, using the change in the atomic eigenvalues as a function of occupation numbers, we found the values for the Hubbard parameters given in Table 3.1.

The self-consistency iterations can be performed by using the linear mixing scheme, in which the output charge from the diagonalization is mixed linearly with the input charge of the same iteration to give the input charge of the next iteration. More efficient and sophisticated methods include the use of Broyden's algorithm [34], which seeks the zero of the function $F(\rho) = \rho_{\text{out}} - \rho_{\text{in}}$. This method will be explained in Appendix D.

As an example, we have calculated and compared to ab initio calculations the shift in the eigenvalues of a charged $C_{20}$ bowl shape corannulene molecule. We obtained a rigid shift of 4.17 eV, whereas a calculation with the Gaussian94 program with a

**Table 3.1** Values in eV of the on-site Hubbard terms for carbon and silicon

| Parameter | Carbon from (3.45) | Carbon screened (from [33]) | Carbon (from [33]) | Silicon from (3.45) |
|---|---|---|---|---|
| $u_{ss,ss}$ | 10.83 | 11.43 | 16.2 | 8.16 |
| $u_{ss,pp}$ | 10.37 | – | 14.0 | 7.37 |
| $u_{sp,sp}$ | 0.0 | – | 2.30 | 0.0 |
| $u_{pp,pp}$ | 9.93 | 11.11 | 14.6 | 6.76 |
| $u_{pp,p'p'}$ | 9.93 | – | 13.5 | 6.76 |
| $u_{pp',pp'}$ | 0.0 | – | 0.54 | 0.0 |

3-21G basis (10 orbitals per carbon atom) and the $X_\alpha$ exchange–correlation potential yielded, for the same coordinates, an average shift of 4.24 eV for all the levels [35]. It turns out that the results are affected by about 10 percent when the parameters from [33], as given in the third column of Table 3.1, are used. Furthermore, setting the exchange integrals to zero affects the results by less than 0.1 eV, meaning that the relevant parameters for such calculation are indeed the direct terms, except maybe for metallic systems, where the off-diagonal part of the DM is long-ranged.

For magnetic systems this could be generalized within either an unrestricted Hartree–Fock (UHF) [36, 37] or a DF formalism, where one would have to consider two Hamiltonians, one for each spin. Omitting basis indices and leaving only spin indices, the $\mathcal{U}$s would then be defined by

$$H_\sigma = H_\sigma^0 + \sum_{\sigma\sigma'} \mathcal{U}_{\sigma\sigma'}\, \delta\rho_{\sigma'}, \tag{3.56}$$

where $H_\sigma^0$ is a reference Hamiltonian matrix which does not necessarily need to depend on $\sigma$ (if, for example, it is fitted from a paramagnetic calculation). The same type of interpolation as in (3.55) can then be used and one would again end up with an atomic calculation from which the spin-dependent on-site terms may easily be extracted. One can then also treat systems with magnetic anisotropy by adding an $L \cdot S$ coupling term to $H$ [38]. This would fill up the (spin-)off-diagonal part of the Hamiltonian matrix, but involves no additional difficulty if one assumes a constant coupling constant for this term.

## 3.5 Applications to Fullerenes, Silicon and Transition-Metal Clusters

With recent technological advances in the fields of software and hardware, it is now possible to treat larger molecular systems with high accuracy. Clusters have, in the past decade, attracted much of physicists, materials scientists and chemists' attention. The newly emerged field of cluster science is interesting for at least two reasons. One is that clusters are a novelty, a transition between atoms and solids, with properties certainly different from the latter two. So, regardless of them having any technological applications, they await scientists to discover their properties. The second reason is that, because of their size, between angströms and a micron, they have the potential of making useful nanoscale devices with complex or nontrivial properties. Given their large size, one of the appropriate methods to treat them is the tight-binding method, which can very easily and rapidly calculate their electronic structure. For sizes of a few hundred atoms or more, ab initio methods are unable at this point to tackle this problem, but an $O(N)$ implementation of TB could treat up to 50000 atoms or even more on today's workstations. In this section, we will illustrate how this method has been applied to clusters of various nature, from semiconducting to metallic.

### *3.5.1 Fullerene Collisions*

The TB method, because of its simplicity, can be used to perform structure relaxations and molecular dynamics, which are very time-consuming by ab initio calculations. We have applied it to investigate the properties of fullerene materials. Among these calculations, we can mention simulations of collisions between two $C_{60}$. We found that no bonding takes place, regardless of the incident energy. In fact the binding energy of $C_{60}$ dimer was found to be extremely small. This, however, was a shortcoming of the TB parameters used in the calculation. Snapshots of collision processes can be seen in Fig. 3.2 for three different incident kinetic energies of 2, 45 and 100 eV. The maximum energy at which there is no bond-breaking but just large deformation of the cages is about 40 eV. Even in the second snapshot, where the incident energy of each cage is 45 eV, the cages do not break, and will be reconstructed. At energies of the order of 100 eV, one reaches the limit of stability of the balls: the front atoms move backwards into the volume of the ball, and hence, because of the great stress accumulated in the bent region, the molecules break.

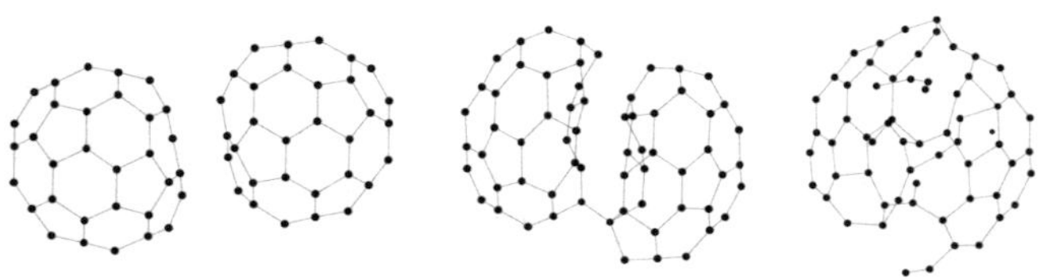

**Fig. 3.2** Collision of two $C_{60}$ molecules at incident energies of 2, 45 and 100 eV, respectively, in the center-of-mass frame

### 3.5.2 $C_{240}$ Doughnuts and Their Vibrational Properties

Some recently observed doughnut shapes in electron microscope images have attracted much interest in these interesting topological shapes. We have investigated their electronic and vibrational properties by the TB method [39, 40]. We have tried two parametrizations, one orthogonal [2] and one nonorthogonal [41]. The latter has been very successful in predicting the spectrum of $C_{60}$ and other fullerene cages since it was obtained after a fit to the band structure of graphite. The configuration of several isomers was relaxed by using the steepest-descent method. The electronic eigenvalues were then calculated by direct diagonalization. The use of the Hellmann–Feynman theorem allowed us to compute the force constants matrix by finite differences of the forces computed analytically. From the force constants matrix one is then able to compute the square of the phonon frequencies (see Appendix B). It is also possible to use this Hessian matrix $H$ to optimize a structure by the Newton–Raphson method: the displacement after the $m+1$th iteration can be obtained from the displacement and force and the Hessian matrix of the $m$th iteration through $X^{[m+1]} - X^{[m]} = H^{-1[m]}F^{[m]}$. From the electronic eigenvalues, one can predict the kinetic stability of a cluster. Large-gap systems, similarly to rare-gas atoms, which have a large energy separation between their occupied and empty levels, have more chance of being chemically inert. Whereas if the gap is small, or a level is half-occupied, it is prone to hybridization with another atom in its vicinity. Thermodynamic stability can be inferred from the total free energy which includes the vibrational, rotational and electronic energy and entropy. The free energy can be easily computed within the harmonic approximation in the low temperature regime. For $C_{240}$ doughnut-shape isomers, we computed the free energy up to 1500 K, where the harmonic approximation might still be valid. To discuss the thermodynamic stability, one must consider the free energy at high temperatures since after all, the clusters are created at those temperatures. A third criterion for stability is the mechanical one: clusters with soft phonon modes have a lower chance of being stable. These three criteria are, of course, not unrelated. It is possible that a cluster with a small gap would be soft, and therefore some deformation could open up the gap and make it more stable.

### 3.5.3 IR Spectra of $C_{60}$ and $C_{60}$ Dimers

Motivated by recent experiments performed on $C_{60}$ dimers [42], we have performed total energy minimizations for several isomers of the $C_{60}$ dimer in the neutral configuration in order to find the ground state structure. From these calculations, one can also deduce the IR absorption spectrum [43]. By comparing our results with the available IR experimental data, we have drawn conclusions regarding the structure of the $C_{60}$ dimer used in the experiments.

IR modes are active, if a vibration in the direction of that mode induces a dipole moment in the system. Likewise, a mode is Raman-active if a vibration in the direction of that mode induces a nonzero polarizability. IR and Raman activities of a vibrational mode of a molecule can be deduced from its symmetry and the character table of the point group of that molecule.

To calculate the IR intensities of each of the vibrational modes, we can use either the finite difference method [44] or the linear-response theory [45] to compute the effective charge tensor. In the finite-difference method, one calculates the change in the dipole moment after a small distortion corresponding to the vibrational mode is applied to the system. The IR intensity is given by the square of this change:

$$I_\lambda \propto |D(\lambda) - D_0|^2 \tag{3.57}$$

where $D_0$ is the permanent dipole moment and $D(\lambda)$ is the dipole moment of the distorted cluster corresponding to the vibrational mode $\lambda$. In the linear-response formalism, one computes the same expression by calculating the difference analytically; in other words, by computing the first-order term in the Taylor expansion of the dipole moment:

$$I_\lambda \propto \sum_{\alpha=1,3} \mid \sum_{i=1,N} \nabla_{i\beta} D^\alpha e_i^\beta(\lambda) \mid^2, \tag{3.58}$$

where $e_i^\beta(\lambda)$ is the $\beta$ component of the polarization vector of mode $\lambda$ on atom $i$. The effective charge tensor $Z = \nabla D$ is calculated from linear-response (or perturbation) theory:

$$\nabla_{i\beta} D^\alpha = Z_i^{\alpha\beta} = \int \nabla_{i\beta}\rho(r) r^\alpha \, \mathrm{d}^3 r - Z_i \, \delta_{\alpha\beta}. \tag{3.59}$$

Here $Z_i$ is the nuclear charge, and $\rho(r)$ the total electronic charge density:

$$\rho(r) = 2 \sum_{\mu}^{\mathrm{occ}} |\psi_\mu(r)|^2. \tag{3.60}$$

Assuming an "infinitesimal" displacement corresponding to the phonon mode, one can calculate the change in the electronic eigenstates by using first-order perturbation theory. In the tight-binding molecular-dynamics (TBMD) formalism, the matrix elements of the Hamiltonian are known analytically; thus one can easily calculate the matrix elements of the perturbation $V$ by using the Hellmann–Feynman theorem. The final result for the effective charge matrix can be written as follows:

$$Z_i^{\alpha\beta} = \sum_{\mu}^{\mathrm{occ}} \sum_{\nu}^{\mathrm{unocc}} \frac{4}{E_\mu - E_\nu} \langle \nu | r^\alpha | \mu \rangle \langle \mu | \nabla_i^\beta H | \nu \rangle - Z_i \delta_{\alpha\beta}, \tag{3.61}$$

where the first sum is over "*occ* upied", and the second sum over "*unocc* upied" orbitals. This general expression can be used within any one-particle formalism

(HF, LDA, TB, etc). The wavefunctions $|\psi_\nu\rangle$ are the eigenstates of the unperturbed one-particle Hamiltonian with eigenvalues $E_\nu$. For the matrix elements of $\boldsymbol{r}$, the TB basis functions being orthonormal, one can use the following approximation:

$$\langle\nu|\,\boldsymbol{r}\,|\mu\rangle = \sum_{i\gamma}\sum_{j\beta} c_{i\gamma}^{\nu\,*} c_{j\beta}^{\mu}\, \delta_{ij}\Big(\delta_{\gamma\beta}\boldsymbol{R}_i + \boldsymbol{S}_{\gamma\beta}\Big), \tag{3.62}$$

where $S^x_{\gamma\beta} = \int \psi_\gamma(r)\, x\, \psi_\beta(r)\mathrm{d}^3r = S$ for $\gamma = 2s$ and $\beta = 2p_x$, and zero otherwise. We enforce the sum rule on the effective charge tensor $Z$ to fix the value of $S$ by minimizing the left-hand side of (3.63), since in a perturbation treatment, the sum rule is not satisfied anymore. The sum rule

$$\sum_i Z_i^{\alpha\beta} = 0 \tag{3.63}$$

comes from the fact that the dipole moment should not change for a random but uniform displacement of the atoms (see (3.59).) The parameter $S = 3.87$ was found by imposing the above sum rule on $C_{60}$ and then using the optimized value for the calculation of all other systems, since it should be the same as long as the basis functions used are the same. Among the two methods mentioned, the finite-difference one is more appropriate for TB calculations, because it relies on ground state properties, and has a better chance of being accurate. The linear-response formulation, involving the unoccupied or excited states, would be less accurate since the latter are not well approximated even by ab initio LDA-based methods. Furthermore, the sum should extend over all the excited states, which in the TB case, are quite limited, and hence, the calculation would be inaccurate.

For a comparison between LDA, TB, and MNDO methods, one can refer to [46], or to Table 3.2, in which one can notice the change in the IR intensities even from one LDA to another LDA calculation. The TB results are still better than the MNDO [47], which predicts that the fourth peak is more than four times as intense as the first one, and the third peak almost twice the first one. Besides, to get better agreement with experiment, the MNDO frequencies have been scaled by a factor of 0.87 [47]. A comparison between the TB and experimental results for the spectrum of $C_{60}$ is shown in Fig. 3.3. The frequencies agree reasonably well, but as far as the intensities of the $C_{120}$ isomers (discussed below) are concerned, one should compare the change of the peak intensities relative to that of $C_{60}$, between theory and experiment.

Among the nine isomers of different geometry that we considered, it was found that the IR spectrum of the "dumbbell" (two $C_{60}$ units connected by a $C_4$ square ring) isomer with $D_{2h}$ symmetry agrees best with experiment on undoped photopolymerized $C_{60}$ films, even though there are many isomers of $C_{120}$ that have a larger bandgap and a much lower binding energy. The two curves are displayed in Fig. 3.3. For the sake of comparison with experiment, the peaks have been broadened with a Gaussian of width 5 cm$^{-1}$; the plotted intensities are renormalized to the intensity of the first peak. For the calculation of the IR spectrum, because of the limitations in treating

**Table 3.2** Comparison between experimental and calculated IR intensities ($I$) for $C_{60}$, the latter obtained by TB, MNDO and two different LDA calculations. The columns headed $\omega$ show the frequencies in $cm^{-1}$ and the columns headed $I$ their intensities in arbitrary units

| TB (present) | | MNDO [47] | | LDA [48] | | LDA [46] | | Experiment [42] | |
|---|---|---|---|---|---|---|---|---|---|
| $\omega$ | $I$ | $\omega$ | $I$ | $\omega$ | $I$ | $\omega$ | $I$ | $\omega$ | $I$ |
| 486 | 1.00 | 504 | 1.00 | 527 | 1.00 | 516 | 1.00 | 527 | 1.00 |
| 600 | 0.08 | 626 | 0.35 | 586 | 0.63 | 557 | 0.65 | 576 | 0.30 |
| 1199 | 0.18 | 1179 | 1.87 | 1218 | 0.36 | 1202 | 0.59 | 1183 | 0.22 |
| 1571 | 1.74 | 1419 | 4.35 | 1462 | 0.57 | 1589 | 0.41 | 1429 | 0.24 |

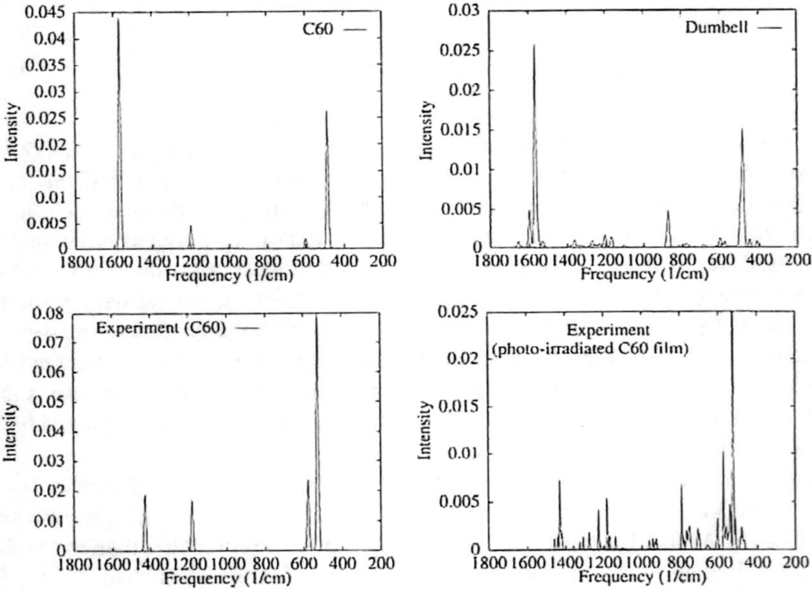


**Fig. 3.3** Calculated (top) and experimental (bottom) IR intensities of $C_{60}$ and the dumbbell structure of $C_{120}$

120 atoms with DF–LDA methods, we have used the tight-binding approach, which for larger fullerenes is still tractable on large computers. This method is successful in predicting the ground state geometry of carbon clusters as well as their vibrational frequencies, in some instances as accurately as ab initio methods. But it is not as accurate in the computation of the dipole moment, since the charge density is not really known. One can, however, make judgements on the agreement of the experimental IR data with the calculated results by comparing the shifts in the frequencies and the *relative* change of the IR intensities with respect to $C_{60}$.

### 3.5.4 Simulated Annealing of Small Silicon Clusters

There has been much interest in studying the properties of small Si clusters theoretically [8, 49–52] and experimentally [53, 54]. Using a recently developed adaptive simulated annealing (ASA) algorithm [55], we have performed systematic total-energy minimizations for small silicon clusters of sizes from 3 to 30 [56]. The total energy and forces are calculated with the generalized tight-binding parametrization by Kwon et al. [14] (see (3.3) in Sect. 3.2.1) In general, the LDA cohesive energy curves of cubic and diamond structures are reproduced accurately with this orthogonal scheme.

In the ASA algorithm, one starts the molecular-dynamics simulation at a high temperature well above melting, and lowers it after about 1000 MD steps, according to the following equation (see also Appendix D):

$$\frac{\mathrm{d}T}{\mathrm{d}t} = -\nu T/\epsilon\sqrt{C}, \tag{3.64}$$

where $\nu$ is a velocity scale taken as a constant ($\nu = 0.1$–$1$), which controls the cooling rate, $C$ is the heat capacity of the system, indicating the fluctuations in the average configurational (or potential) energy, and $\epsilon$ represents the relaxation time, and is related to the second-highest eigenvalue of the transition matrix by the following relation: $\epsilon = -log(q_2)$ (for more details see footnote in Sect. 5.2.3). The thermodynamic properties of the system ($C$ and $\epsilon$) are computed from the information gathered during the 1000 steps. The MD time step was 1.018 fs and the velocity-Verlet algorithm was used to update the atomic coordinates. The annealing was done in a hard-wall spherical box with a radius increasing with decreasing temperature. The runs were made performed on five processors in parallel (an ensemble of five different samples, all evolving at the same temperature each simulated on a processor of a Hitachi SR-2201 supercomputer). After reaching a temperature of about 300 K, we stopped the Simulated Annealing (SA) algorithm, and minimized the structure by a velocity scaling technique which was very similar to the steepest-descent algorithm. For computations of the vibrational spectra or IR intensities, one needs a fully relaxed set of coordinates; in this case, the relaxation was performed by computing the force-constants matrix and using the Newton–Raphson method. The final ground-state structures were reached after about 100000–200000 molecular dynamics time steps.

The binding energies per atom, as well as the bandgaps of the annealed structures, are presented in Fig. 3.4. It is found that most of the larger structures are based on the icosahedral geometry of $Si_{12}$, where all atoms are fivefold coordinated. In fact, the average coordination number per atom for the studied clusters turned out to be 5 in more than 70% of the atoms.

For the stable, high-cohesive-energy clusters, one can also calculate the vibrational spectra, and predict the modes that can be detected in IR and Raman spectroscopy, similarly to the above study of $C_{60}$ dimers. Because of the importance of $Si_{12}$ in the optimized structures that were found, we have also calculated and displayed their vibrational frequencies in Fig. 3.5.

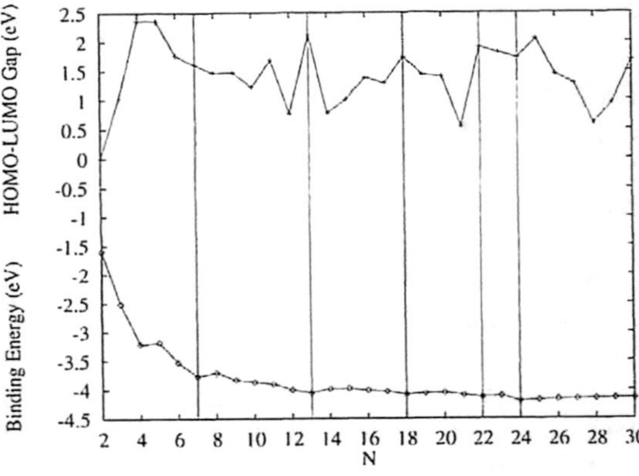


**Fig. 3.4** SA results for the gap and binding energy of $Si_N$ clusters

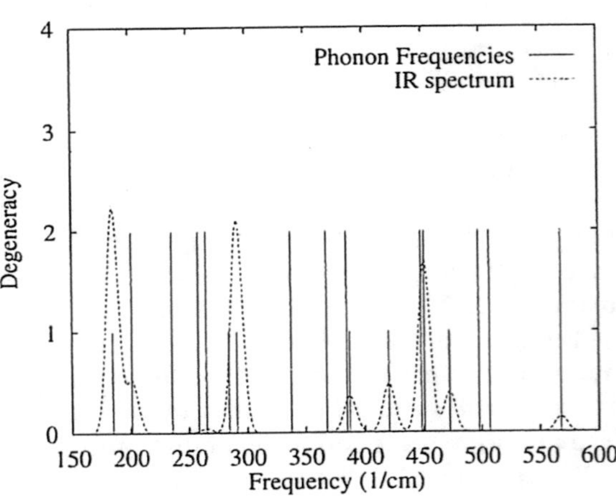


**Fig. 3.5** Vibrational frequencies ($cm^{-1}$) of $Si_{12}$. The dashed curve shows the intensity of the IR peaks

For small Si clusters, the TB calculations were in good agreement with more sophisticated ab initio calculations. The same parametrization was also applied to study the Si(100) surface reconstruction. It was found that parallel buckled dimers were lower in energy than antiparallel buckled dimers, in disagreement with experimental findings. This energy difference being very small, one can conclude that the tight-binding method was not appropriate for such sensitive calculations, and in general, one should not expect always to obtain good agreement with experiment or ab initio calculations. After all, in this case, we have only used four orbitals per atom, and this incomplete basis, even with ab initio methods, can not give accurate results for total energies or vibrational frequencies. Besides the small number of basis functions, the Hamiltonian matrix elements are universal in the sense that for any environment the hopping and on-site energies are always the same. These simplifications, not unexpectedly, lower the accuracy, and while performing such calculations, one must always bear in mind how simplified they are, and what one can expect from a given TB parametrization.

### 3.5.5 Titanium and Copper Clusters

The same SA algorithm was also applied to find ground-state configuration of small copper and titanium clusters. In this case, we used the recently developed parameters by Mehl and Papaconstantopoulos [3]. This nonorthogonal parametrization, as mentioned earlier, has incorporated the repulsive interaction into the diagonal part of the Hamiltonian matrix, which is now environment-dependent. Usually, the on-site terms are just a constant near the atomic eigenvalues. But in this case, they are given by

$$h_{i\alpha} = a_{i\alpha} + b_{i\alpha}\rho_i^{2/3} + c_{i\alpha}\rho_i^{4/3} + d_{i\alpha}\rho_i^2, \tag{3.65}$$

where the constants $a, b, c$ and $d$ depend only on the atom and the orbital type, whereas the "density" $\rho_i = \sum_{j \neq i} exp(-\gamma^2 R_{ij})$ counts the number of nearest neighbors to the atom $i$. As usual, the overlap and hopping terms are decreasing functions of distance, parametrized in this case with a Gaussian function. This parametrization has been able to reproduce ground-state bulk properties of copper in excellent agreement with FLAPW calculations. We adopted this parameter set to search for ground-state configurations of small copper clusters.

For $Cu_3$ up to $Cu_7$ we found planar structures. This is presumably because more compact three-dimensional forms, having a smaller volume, would have electrons with higher kinetic energy. There is a competition between kinetic energy (which favors large volume, or in other words, flat clusters) and bonding energy (which favors compact clusters). For $n = 8$ atoms, we find a transition to a 3D structure, namely a structure with tetrahedral symmetry. Experimentally, 8, 14 and 20 were found to be magic numbers for copper clusters. For all these systems, we find a ground state having $T_d$ symmetry. Furthermore, we have compared the stability of icosahedral versus FCC clusters. Starting with FCC $Cu_{55}$, we found that it deforms continuously to an icosahedral shape (Fig. 3.6).

As for titanium, it was found that from $n = 4$ the structures become compact and three-dimensional. No outstanding magic numbers were found, except for $n = 10$ and 13, where the binding energy per atom was slightly higher than the neighboring values of $n$, and the HOMO–LUMO gaps also were larger than for $n = 9, 11, 12, 14$. In general the optimized structures were compact and of icosahedral shape, capped with additional atoms. The gaps were very small since the outermost shell is $3d$, so this small gap is opened in the middle of the $d$ band, which is narrow itself. The occurrence of the gaps is due to the Jahn–Teller distortions of the clusters. Unlike the copper clusters which displayed strong magic numbers with gaps of the order of 1–2 eV, due to the shell effect, one can say that titanium clusters are metallic regardless of their size, and do not display any specific stability.

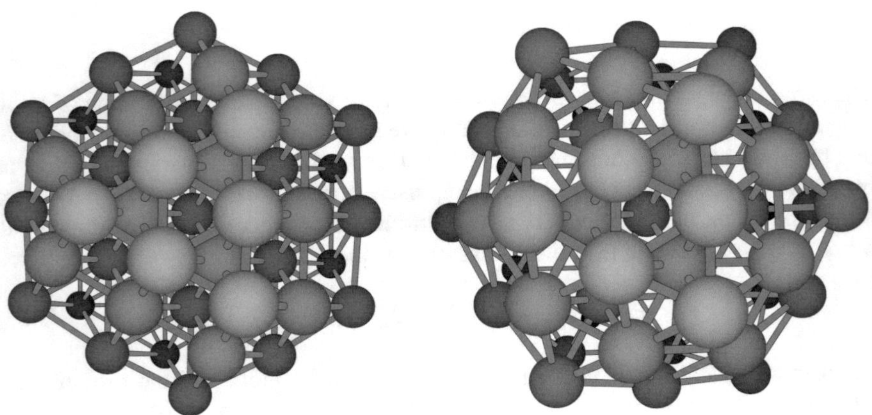

**Fig. 3.6** Initial (cuboctahedron from FCC lattice on the left) and final (icosahedron on the right) configurations of a $Cu_{55}$ cluster

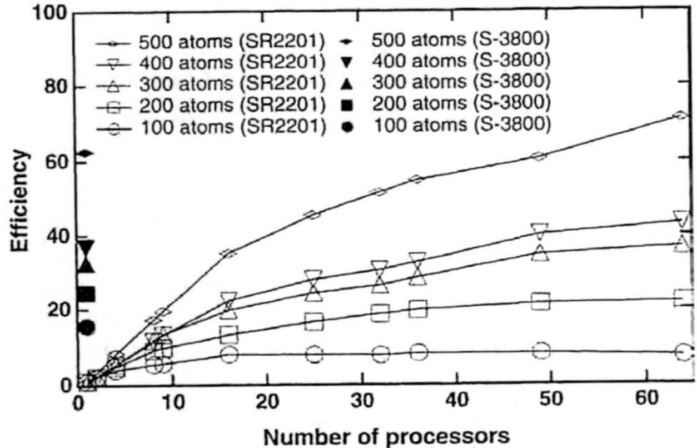


**Fig. 3.7** Performance of the parallel Hitachi SR2201 running a TB code as a function of particle number and processors, compared to that of Hitachi S-3800 vector machine (8 GFLOPS/CPU)

## 3.6 Conclusions

As explained in Chap. 2, by performing MD simulations one can obtain thermodynamic quantities as well as the response functions of a system. These kinds of simulations, however, take very long times, and ab initio methods are unable at this point to calculate such data for moderately large systems. In this case, TB becomes a powerful tool in computing physical and chemical properties of a system by MD methods.

As mentioned earlier, one cannot expect from TB calculations a very accurate description of a system, namely the stability of two nearly degenerate phases may not be given correctly by TB, as was the case for parallel buckled dimers on the Si(001) surface. With today's supercomputers with a maximum memory of several gigabytes, it is possible to perform TB calculations on systems of up to 1000,000 particles. Using an $O(N)$ algorithm will push this point further up to about tens of millions of particles. Such programs have also been implemented on parallel machines. The performance on a Hitachi SR2201 is displayed in Fig. 3.7 for different numbers of processors.

## References


1. J.C. Slater, G.F. Koster, Phys. Rev. **94**, 1498 (1954)
2. C. Xu, C.Z. Wang, C.T. Chan, K.M. Ho, J. Phys. Condens. Matter **4**, 6047 (1992)
3. M. Mehl, D. Papaconstantopoulos, Phys. Rev. B **54**, 4519 (1996)
4. F. Liu, Phys. Rev. B **52**, 10677 (1995)
5. D. Porezag, Th Frauenheim, Th Köhler, G. Seifert, R. Kaschner, Phys. Rev. B **51**, 12947 (1995)
6. A. Taneda, K. Esfarjani, Z.Q. Li, Y. Kawazoe, Comput. Mat. Sci. **9**, 343 (1998)
7. W.A. Harrison, *Electronic Structure and the Properties of Solids* (Dover, New York, 1980)
8. M. Menon, K.R. Subbaswamy, Phys. Rev. Lett. **67**, 3487 (1991); Int. J. Mod. Phys. **6**, 3839 (1992)
9. A.P. Sutton, M.W. Finnis, D.G. Pettifor, Y. Ohta, J. Phys. C. **21**, 35 (1988)
10. P.W. Anderson, Phys. Rev. Lett. **21**, 13 (1968); Phys. Rev. **181**, 25 (1969)
11. A.P. Horsefield, Phys. Rev. B **56**, 6594 (1997). References therein
12. V. Heine, Solid State Phys. **35**, 47 (1980); D.W. Bullet, Solid State Phys. **35**, 173 (1980)
13. M. Foulkes, R. Haydock, Phys. Rev. B **39**, 12520 (1989)
14. I. Kwon, R. Biswas, C.Z. Wang, K.M. Ho, C.M. Soukoulis, Phys. Rev. B **49**, 7242 (1994)
15. L. Goodwin, A.J. Skinner, D.G. Pettifor, Europhys. Lett. **9**, 701 (1989)
16. M.S. Tang, C.Z. Wang, C.T. Chan, K.M. Ho, Phys. Rev. B **53**, 979 (1996)
17. M.C. Payne, M.P. Teter, D.C. Allan, T.A. Arias, J.D. Joannopoulos, Rev. Mod. Phys. **64**, 1045 (1992)
18. X.P. Li, R.W. Nunes, D. Vanderbilt, Phys. Rev. B **47**, 10891 (1993)
19. M.S. Daw, Phys. Rev. B **47**, 10895 (1993)
20. P. Ordejon, D.A. Drabold, R.M. Martin, M.P. Grumbach, Phys. Rev. B **51**, 1456 (1995)
21. W. Kohn, Phys. Rev. Lett. **76**, 3168 (1996)
22. R.W. Nunes, D. Vanderbilt, Phys. Rev. B **50**, 17611 (1994)
23. W. Hierse, E.B. Stechel, Phys. Rev. B **50**, 17811 (1994)
24. J.M. Millam, G.E. Scuseria, J. Chem. Phys. **106**, 5569 (1997)
25. S. Goedecker, L. Colombo, Phys. Rev. Lett. **73**, 122 (1994); S. Goedecker, M. Teter, Phys. Rev. B **51**, 9455 (1995)
26. R. Baer, M. Head-Gordon, J. Chem. Phys. **107**, 10003 (1997)
27. R. Baer, M. Head-Gordon, Phys. Rev. B **58**, 15296 (1998)
28. R. Haydock, V. Heine, M.J. Kelly, J. Phys. Chem. **5**, 2845 (1972); R. Haydock, Solid State Phys. **35**, 215 (1980)
29. D.R. Bowler, M. Aoki, C.M. Goringe, A.P. Horsefield, D.G. Pettifor, Model. Sim. Matls. Sci. Eng. **5**, 199 (1997)
30. D. Tomańek, M. Schlüter, Phys. Rev. B **54**, 4519 (1986); Phys. Rev. Lett. **65**, 1306 (1991)
31. M. Teter, Phys. Rev. B **48**, 5031 (1993). References therein
32. K. Ohno, Theor. Chim. Acta **2**, 219 (1964); G. Klopman, J. Am. Chem. Soc. **86**, 4450 (1964)
33. P. Fulde, *Electron Correlations in Molecules and Solids*, vol. 100. Solid-State Series (Springer, Berlin, 1993), p. 151

34. C.G. Broyden, Math. Comput. **19**, 577 (1965); D. Vanderbilt, S.G. Louie, Phys. Rev. B **30**, 6118 (1984)
35. K. Esfarjani, Y. Kawazoe, J. Phys. Condens. Matter **10**, 8257 (1998)
36. J. Guevara, F. Parisi, A.M. Llois, M. Weissmann, Phys. Rev. B **55**, 13283 (1997)
37. P. Villaseñor-Gonzales, J. Dorantes-Dávila, H. Dreyssé, G.M. Pastor, Phys. Rev. B **55**, 15084 (1997)
38. J. Dorantes-Dávila, H. Dreyssé, G.M. Pastor, Phys. Rev. B **55**, 15033 (1997)
39. Y. Hashi, K. Esfarjani, S. Itoh, S. Ihara, Y. Kawazoe, Trans. Mat. Res. Soc. Jpn. **20**, 486 (1996)
40. K. Esfarjani, Y. Hashi, S. Itoh, S. Ihara, Y. Kawazoe, Z. Phys. D **41**, 73 (1997)
41. S. Saito, S. Okada, S. Sawada, N. Hamada, Phys. Rev. Lett. **75**, 685 (1995)
42. J. Onoe, K. Takeuchi, Phys. Rev. B **54**, 6167 (1996)
43. K. Esfarjani, Y. Hashi, J. Onoe, K. Takeuchi, Y. Kawazoe, Phys. Rev. B **57**, 223 (1998)
44. A.P. Smith, G.F. Bertsch, Phys. Rev. B **53**, 7002 (1996)
45. P. Giannozzi, W. Andreoni, Phys. Rev. Lett. **76**, 4915 (1996)
46. G.F. Bertsch, A.P. Smith, K. Yabana, Phys. Rev. B **52**, 7876 (1995)
47. D. Bakowies, W. Thiel, Chem. Phys. **151**, 309 (1991), R.E. Stratton, M. Newton, J. Phys. Chem. **92**, 2141 (1988)
48. P. Giannozzi, S. Baroni, J. Chem. Phys. **100**, 8537 (1994)
49. D. Tomańek, M. Schlüter, Phys. Rev. Lett. **56**, 1055 (1986); Phys. Rev. B **36**, 1208 (1987)
50. R. Biswas, D.R. Hamann, Phys. Rev. Lett. **55**, 2001 (1985); Phys. Rev. B **36**, 6434 (1987)
51. K. Raghavachari, V. Logovinsky, Phys. Rev. Lett. **55**, 2853 (1985); K. Raghavachari, J. Chem. Phys. **83**, 3520 (1985); J. Chem. Phys. **84**, 5672 (1986); K. Raghavachari, C.M. Rohlfing, Chem. Phys. Lett. **143**, 428 (1988); J. Chem. Phys. **89**, 2219 (1988)
52. M.R. Pederson, A.A. Quong, Phys. Rev. Lett. **74**, 2319 (1995)
53. J.L. Elkind, J.M Alford, F.D. Weiss, R.T. Laaksonen, R.E. Smalley, J. Chem. Phys **87**, 2397 (1987); S. Maruyama, L.R. Anderson, R.E. Smalley, J. Chem. Phys **93**, 5349 (1990); J.M. Alford, R.T. Laaksonen, R.E. Smalley, J. Chem. Phys. **94**, 2618 (1991)
54. M.F. Jarrold, J.E. Brower, K.M. Creegan, J. Chem. Phys **90**, 3615 (1989); M.F. Jarrold, U. Ray, K.M. Creegan, J. Chem. Phys. **93**, 224 (1990); U. Ray, M.F. Jarrold, J. Chem. Phys **94**, 2631 (1991)
55. B. Andersen, J.M. Gordon, Phys. Rev. E **50**, 4346 (1994)
56. K. Esfarjani, Y. Hashi, Y. Kawazoe, *Nuclus 97 Conference Proceedings*, Tsukuba, Japan (1998), p. 403

# Chapter 4
# Empirical Methods and Coarse-Graining

## 4.1 Introduction

In the previous two chapters, we learned that atomistic- and electronic-scale simulations can be performed by means of ab initio methods or semi-empirical methods such as a tight-binding method. However, we learned also that these are, at present, still restricted in their capability with respect to both the number of atoms and the simulation timescale. In order to study longer-timescale phenomena of systems composed of larger numbers of particles, it becomes necessary to introduce much easier but still atomic scale methods. If such methods are more or less based on the ab initio or semi-empirical total energies and they are not simplified too much, one may rely on these methods as a substitute. The important issue here is how to reduce the amount of necessary computation in such methods, and how to introduce parametrizations or fittings into the interatomic potentials without losing too much accuracy and reliability. In principle, it is possible to construct realistic classical potentials based on ab initio calculations. A possible methodology here is to determine classical potentials by, for example, fitting them to contour maps of the total energy, which may be obtained with an ab initio method by changing the position of one atom while fixing the coordinates of all other atoms. In this book, we do not go into details of the applications of classical molecular dynamics, since readers can refer to many good reviews. Instead, a few examples are given of how classical potentials can be constructed from ab initio theories.

The next topic in this chapter is concerned with phenomena on a much larger space scale and longer timescale, such as the diffusion of particles in a bulk solid. Even with classical molecular dynamics methods with very simple potentials, the maximum number of particles which can be handled at present is about one million and the simulation timescale is of the order of 10 ns, because of the limitation of computational power. These numbers might seem quite large. However, real materials are composed of $\sim 10^{23}$ atoms and molecules, and sometimes it becomes necessary to perform far-larger-scale simulations. Examples include phase transitions, solidification from the liquid state, crystal growth processes, melting, segregation and


K. Ohno et al., *Computational Materials Science*,
https://doi.org/10.1007/978-3-662-56542-1_4

diffusion. One possible strategy for such purposes is to accelerate force calculations required at each MD step, as will be described in detail in Appendix A. For far larger-scale phenomena, one may introduce alternative approaches using very simplified coarse-grained models such as lattice gas models. Lattice gas models, if they are simple enough, can be treated with the methods used commonly in statistical mechanics, such as mean-field theory, the cluster variation method (CVM) and Monte Carlo (MC) methods.

A question arising now is how to provide a link between the ab initio, semi-empirical or classical potentials and the potential parameters suitable for lattice gas models. In order to determine realistic potentials for lattice gas models from ab initio, semi-empirical or empirical potentials, there are several possible methodologies. One is the well-established method called the Connolly–Williams (CW) approximation, which is simply the "inverse" method to the usual cluster expansion technique. The other is a useful method, called the "potential renormalization" (PR), which may have a capability to take the effect of off-lattice motions such as vibrations or atomic displacements into account.

## 4.2 Reduction to Classical Potentials

Interatomic potentials among atoms or ions can, in principle, be determined at the absolute zero of temperature by estimating the ground-state total energy of the electronic system with given atomic positions. Of course, theoretically, this is the most reliable way. However, for large systems, this becomes very time-consuming and heavy on computer resources. In such cases, it is convenient to fit the ab initio total energies to simple analytic potential functions, which we call the molecular-dynamics potentials. Once such an MD potential is determined with high accuracy and reliability, a classical MD simulation or, alternatively, Monte Carlo (MC) methods for classical particles (see Sect. 5.4.1) can be used for atomistic simulations instead of directly using initio methods. A classical MD simulation is similar to but much easier than an ab initio MD one. The classical MD is described in Appendix A, and may be performed along the lines described there for the structural optimization, several techniques described in Appendix D can be used as well.

As the simplest example, the long-range intermolecular potential between ions is dominated by the Coulomb interaction. In the case of intermolecular potentials, there are also multipole potentials depending on the electron charge density distribution; the intermolecular potentials can be represented in the form of a multipole expansion, except for a term representing the short-range core repulsion. The interaction between atoms which are electrically neutral is not so simple. The problem of how to fit the ab initio energy to simple MD potential functions is generally very complicated and not unique at all. Here we will only describe a few special cases. The detailed description of the many other empirical potentials [1, 2] is beyond the scope of this book.

### 4.2.1 Polar Systems

In the case of ionic crystals, the total energy can be estimated accurately by summing the Coulomb interaction energies between the ions. When the ions, with have a charge of $\pm Z$ (in units of $e$), are located at the lattice points $\boldsymbol{R}_i$ ($i = 1, 2, \ldots, N$), the total electrostatic energy of this system is given by

$$E = \frac{1}{2} \sum_{ij}^{N} \frac{\pm Z^2}{|\boldsymbol{R}_i - \boldsymbol{R}_j|}. \tag{4.1}$$

This can be conveniently rewritten as

$$E = -\frac{NZ^2}{2} \frac{\alpha}{a}, \tag{4.2}$$

where $a$ is a lattice constant. The coefficient $\alpha$ is called a "Madelung constant", which is a dimensionless number and depends on the lattice and the way of ionic alignments but not on the spatial scale $a$. The Madelung constant can be determined by identifying (4.2) with the Madelung energy, which is evaluated by means of the Ewald sum as described in Appendix C; see (C.10). For example, $\alpha = 1.7476$ for a NaCl-type crystal and $\alpha = 1.6381$ for a zinc-sulfide-type (ionic) structure. For more general crystal structures or noncrystalline ionic solids, one has to calculate the electrostatic energy of the system by means of the Ewald sum for every given set of atomic coordinates.

In another kind of polar systems, namely one composed of polar molecules, electrostatic interactions also play an essential role. In such a case, the interactions are not the direct Coulomb interactions but the multipole interactions. If the charge distribution $\{q_n(\boldsymbol{r}_n)\}$ is given around, say, the $j$th molecule, the electrostatic potential at a position $\boldsymbol{R}_i$ is calculated as

$$V(\boldsymbol{R}_i) = \sum_{n} \frac{q_n(\boldsymbol{r}_n)}{|\boldsymbol{R}_i - \boldsymbol{r}_n|}. \tag{4.3}$$

Let us write the center of the $j$th molecule as $\boldsymbol{R}_j$. If one is interested only in the cases where the distance $\boldsymbol{R}_{ij} = (X_{ij}^{(1)}, X_{ij}^{(2)}, X_{ij}^{(3)}) = \boldsymbol{R}_i - \boldsymbol{R}_j$ is large enough, (4.3) can be expanded in the form of a multipole expansion:

$$V(\boldsymbol{R}_{ij}) = \frac{1}{R_{ij}} \sum_{n} q_n + \frac{1}{R_{ij}^2} \sum_{\alpha=1}^{3} \frac{X_{ij}^{(\alpha)}}{R_{ij}} \sum_{n} q_n x_n^{(\alpha)} +$$
$$\frac{1}{R_{ij}^3} \left[ \frac{1}{2} \sum_{\alpha=1}^{3} \left( \frac{3X_{ij}^{(\alpha)^2}}{R_{ij}^2} - 1 \right) \sum_{n} q_n x_n^{(\alpha)^2} + \sum_{\alpha>\beta}^{3} \frac{3X_{ij}^{(\alpha)} X_{ij}^{(\beta)}}{R_{ij}^2} \sum_{n} q_n x_n^{(\alpha)} x_n^{(\beta)} \right], \tag{4.4}$$

provided that charge-distribution points $\boldsymbol{r}_n = (x_n^{(1)}, x_n^{(2)}, x_n^{(3)})$ are measured from the center of the $j$th molecule, $\boldsymbol{R}_j$.

This may be written in a more compact form:

$$V(\boldsymbol{R}_{ij}) = \frac{Q_j}{R_{ij}} + \sum_\alpha P_j^{(\alpha)} \frac{\partial}{\partial X_{ij}^{(\alpha)}} \frac{1}{R_{ij}} + \frac{1}{2!} \sum_{\alpha\beta} \Theta_j^{(\alpha\beta)} \frac{\partial^2}{\partial X_{ij}^{(\alpha)} \partial X_{ij}^{(\beta)}} \frac{1}{R_{ij}} + \cdots, \quad (4.5)$$

where we have defined

$$Q_j = \sum_n q_n, \quad P_j^{(\alpha)} = \sum_n q_n x_n^{(\alpha)}, \quad \Theta_j^{(\alpha\beta)} = \sum_n q_n x_n^{(\alpha)} x_n^{(\beta)} \quad (4.6)$$

Here, $Q_j$ is the total charge of the $j$th molecule, $\boldsymbol{P}_j = (P_j^{(1)}, P_j^{(2)}, P_j^{(3)})$ is its dipole moment, and $\overleftrightarrow{\Theta}_j = \{\Theta_j^{(\alpha\beta)}\}$ is a tensor of the second rank, called the quadrupole moment.

If each molecule has no net charge, i.e. $Q_j = 0$, then the most dominant energy may be due to the dipole–dipole interaction between molecules, which can be expressed as

$$H_{ij} = \frac{(\boldsymbol{P}_i \cdot \boldsymbol{P}_j) R_{ij}^2 - 3(\boldsymbol{P}_i \cdot \boldsymbol{R}_{ij})(\boldsymbol{P}_j \cdot \boldsymbol{R}_{ij})}{R_{ij}^5}. \quad (4.7)$$

If the charge distribution of a molecule is not altered, i.e. not affected by the environmental configuration of other molecules, the potential form discussed here can be used without change. However, in the usual cases, the charge distribution depends on the environment. If such an effect is not negligible, one should consider it explicitly. One such effect is the induction of multipoles [3]. The potential due to a single molecule at $\boldsymbol{R}_n$ is given by (4.5), which may be written conveniently for the present purpose as

$$V(\boldsymbol{R}_{ij}) = \frac{1}{R_{ij}^3} \boldsymbol{R}_{ij} \cdot \boldsymbol{P}_j - \frac{1}{R_{ij}^5} \boldsymbol{M}_{ij} \cdot \boldsymbol{\Theta}_j, \quad (4.8)$$

where we have assumed $\Theta_j^{(\alpha\beta)} = 0$ unless $\alpha = \beta$. The symbol $\boldsymbol{\Theta}_j$ here stands for a vector whose components are given by $\Theta_j^{(\alpha\alpha)}$ $(\alpha = 1, 2, 3)$, while $\boldsymbol{M}_{ij}$ is a vector with components $R_{ij}^2 - 3X_{ij}^{(\alpha)^2}$. Then the electric field at $\boldsymbol{R}_i$ caused by this $j$th molecule is given by

$$\begin{aligned} \boldsymbol{E}(\boldsymbol{R}_{ij}) &= -\nabla_i V(\boldsymbol{R}_{ij}) \\ &= -\frac{\boldsymbol{P}_j}{R_{ij}^3} + \frac{3}{R_{ij}^5} (\boldsymbol{R}_{ij} \cdot \boldsymbol{P}_j) \boldsymbol{R}_{ij} + \frac{\boldsymbol{N}_{ij}}{R_{ij}^5} - \frac{5}{2R_{ij}^7} (\boldsymbol{M}_{ij} \cdot \boldsymbol{\Theta}_j) \boldsymbol{R}_{ij}, \end{aligned} \quad (4.9)$$

where $\boldsymbol{N}_{ij}$ is a new vector which has components $N_{ij}^{(1)} = (\Theta_j^{(2)} + \Theta_j^{(3)} - 2\Theta_j^{(1)})X_{ij}^{(1)}$, and $N_{ij}^{(2)}$, $N_{ij}^{(3)}$ are given by cyclic permutation of 1, 2, 3. If there is an $i$th molecule at $\boldsymbol{R}_i$, whose polarizability (electric susceptibility) is given by $\chi_i$, a polarization is induced in this molecule by the electric field (4.9). The induced energy there is given by

$$E_{ij} = -\frac{1}{2}\chi_i \boldsymbol{E}^2(\boldsymbol{R}_{ij}). \tag{4.10}$$

Note that the polarizability of the $i$th molecule, $\chi_i$, can be estimated from first principles by means of the method presented in Sect. 2.3.2. The mean energy in which we are interested is obtained by taking the mean of $\boldsymbol{E}^2$ over all orientations of the $j$th molecule. Thus one finds that the induced interaction between molecules $i$ and $j$ is

$$E_{ij} = -\chi_i \left( \frac{P_j^2}{R_{ij}^6} + \frac{3}{2}\frac{\tau_j^2}{R_{ij}^8} \right) - \chi_j \left( \frac{P_i^2}{R_{ij}^6} + \frac{3}{2}\frac{\tau_i^2}{R_{ij}^8} \right), \tag{4.11}$$

where $P_j$ and $\tau_j$ are defined as $\langle (\boldsymbol{R}_{ij} \cdot \boldsymbol{P}_j)^2 \rangle_{\text{av}} = R_{ij}^2 P_j^2/3$ and $\langle \boldsymbol{N}_{ij}^2 \rangle_{\text{av}} = 2R_{ij}^2 \tau_j^2$, respectively, and we have taken into account the fact that an electric moment is also induced in the $n$th molecule by the field of the $m$th molecule.

The second-order forces between polar molecules caused by the interaction (4.10) may appear physically by the following mechanisms:

(1) The rotating charges in the molecules have a tendency to rotate in phase.
(2) The instantaneous distortion in the molecules has a tendency to occur in phase.

It is clear that these two causes are not limited in their operation to polar molecules but affect nonpolar ones as well. Such an effect is discussed in the next section.

### 4.2.2 Van der Waals Potential

For inert-gas elements such as helium, neon, argon and krypton, it is well known that the Lennard–Jones (LJ) potential,

$$V_{ij}(R_{ij}) = U_0 \left[ \left( \frac{R_0}{R_{ij}} \right)^{12} - 2 \left( \frac{R_0}{R_{ij}} \right)^{6} \right], \tag{4.12}$$

where $R_{ij}$ is the interatomic distance, and $U_0$ and $R_0$ are the binding energy and equilibrium distance of the two atoms, respectively, can be used as a good approximation. The second term of the LJ potential describes the van-der-Waals-type attractive interaction between closed-shell atoms (or molecules), neither of which has any intrinsic electric (or magnetic) dipole moment.

The quantum mechanical origin of the van der Waals (attractive) part is a virtual dipole (or, more generally, multipole) induction due to the virtual process of excitation of electrons to empty levels. This is calculated simply by using second-order perturbation theory in quantum mechanics. Note that the energy correction from the ground state estimated in second-order perturbation theory is always negative because of its energy denominator. Therefore, the resulting interatomic interaction is attractive. On the other hand, the short-range repulsive part in the LJ potential can be considered to be a result of interaction or higher-order terms in the perturbation expansion of the energy shift. For very short distances, however, the origin of the interparticle repulsion can be considered more directly as the result of the overlap of the electron cores, which will be described in the next section.

Now we estimate the van der Waals interaction between two atoms in terms of second-order perturbation theory. Generally, two atoms also interact via the dipole–dipole interaction caused by the induced dipoles. That is, we have an interaction of the form given by $H_{ij}$ in (4.7) already in our starting Hamiltonian. If we assume that the two atoms $i$ and $j$ are located on the $z$ axis, we have

$$H_{ij} = \frac{x_i x_j + y_i y_j - 2z_i z_j}{R_{ij}^3}, \tag{4.13}$$

where $\boldsymbol{r}_i = (x_i, y_i, z_i)$ and $\boldsymbol{r}_j = (x_j, y_j, z_j)$ indicate electron positions inside the $i$ and $j$ atoms, respectively. Of course, this interaction becomes small in the long-distance limit. We start from this limit, where the description of two isolated atoms becomes accurate, and treat the dipole–dipole interaction (4.13) as a perturbation. According to second-order perturbation theory, the shift of the energy caused by this perturbation is given by

$$\Delta E_{ij} = \sum_{\lambda_i \lambda_j}^{\mathcal{O}} \sum_{\lambda'_i \lambda'_j}^{\mathcal{E}} \frac{|\langle \lambda_i, \lambda_j | H_{ij} | \lambda'_i, \lambda'_j \rangle|^2}{\varepsilon_{\lambda_i} + \varepsilon_{\lambda_j} - \varepsilon_{\lambda'_i} - \varepsilon_{\lambda'_j}}, \tag{4.14}$$

where an average is taken over the unperturbed states which characterize the system of two isolated atoms $i$ and $j$; $\mathcal{O}$ and $\mathcal{E}$ mean that the summations should be over all the occupied and empty states, respectively. The electronic states of the unperturbed system can be characterized fully by the principal, angular-momentum and magnetic quantum numbers; here $\lambda_k = n_k, l_k, m_k$ and $\lambda'_k = n'_k, l'_k, m'_k$ for $k = i, j$. Since the denominator does not depend on the magnetic quantum numbers at all, the summation over $m_k, m'_k$ can be performed without regard for the denominator. This summation cancels exactly the cross terms arising from the square in the numerator, so that (4.13) can be rewritten as [3, 4]

$$\Delta E_{ij} = \frac{2}{3R_{ij}^6} \sum_{n_i l_i n_j l_j}^{\mathcal{O}} \sum_{n'_i l'_i n'_j l'_j}^{\mathcal{E}} \frac{|\langle n_i l_i | \boldsymbol{r}_i | n'_i l'_i \rangle|^2 |\langle n_j l_j | \boldsymbol{r}_i | n'_j l'_j \rangle|^2}{\varepsilon_{n_i l_i} + \varepsilon_{n_j l_j} - \varepsilon_{n'_i l'_i} - \varepsilon_{n'_j l'_j}}. \tag{4.15}$$

In optical transitions of atoms or molecules, there is a well-known constant called the oscillator strength $f_{\lambda,\lambda'}$, defined by (2.264) in Sect. 2.3.2. Using (2.264) in (4.15),

**Table 4.1** Lennard–Jones parameters for rare-gas elements

| *Element* | He | Ne | Ar | Kr | Xe |
|---|---|---|---|---|---|
| $U_0/k_B$ [K] | 10.3 | 35.8 | 119 | 167 | 255 |
| $R_0$ [Å] | 2.91 | 3.09 | 3.83 | 4.13 | 4.57 |
| $c/2R_0^6 k_B$ [K] from (4.18) | 7.3 | 19.4 | 63.6 | 78.1 | 92.7 |

we have

$$\Delta E_{ij} = -\frac{c}{R_{ij}^6}, \tag{4.16}$$

where

$$c = -\frac{2}{3} \sum_{n_i l_i n_j l_j}^{\mathcal{O}} \sum_{n'_i l'_i n'_j l'_j}^{\mathcal{E}} \frac{f_{n_i l_i, n'_i l'_i} f_{n_j l_j, n'_j l'_j}}{(\varepsilon_{n_i l_i} - \varepsilon_{n'_i l'_i})(\varepsilon_{n_j l_j} - \varepsilon_{n'_j l'_j})} \times \frac{1}{(\varepsilon_{n_i l_i} + \varepsilon_{n_j l_j} - \varepsilon_{n'_i l'_i} - \varepsilon_{n'_j l'_j})}, \tag{4.17}$$

which was first derived by F. London [4]. If we replace the denominator of (4.14) approximately by the constant $-I_i I_j (I_i + I_j)$ ($I_i, I_j$ are roughly of the order of the first ionization energies of each atom), summations in the numerator can be performed by using the $f$-sum rule (2.267), and we obtain a crude estimation of the parameter $c$ in (4.16) for the van der Waals potential energy [4, 5]:

$$c \sim \frac{3}{2} \frac{I_i I_j}{I_i + I_j} \chi_i(\mathbf{0}, 0) \chi_j(\mathbf{0}, 0), \tag{4.18}$$

where $\chi_i(\mathbf{0}, 0)$, $\chi_j(\mathbf{0}, 0)$ are the polarizabilities of the $i$, $j$ atoms, given within the same approximation by (2.269).

The values for the parameters $U_0$ and $R_0$ in the LJ potential (4.12) determined experimentally from the second virial coefficients, as well as the parameter $c/2R_0^6$ ($= U_0$) for the van der Waals potential (4.16) estimated by (4.18) are listed in Table 4.1. The theoretical values underestimate $U_0$ considerably but this is due to that fact that they are the values only for the $R_{ij}^{-6}$ interaction term, although there are many other higher-order contributions, which can also be estimated in terms of similar equations [3, 4].

### 4.2.3 *Potential for Covalent Bonds*

For estimating the ground-state total energy of an electronic system with given atomic positions, one may use, for example, an ab initio pseudopotential LCAO method or a semi-empirical tight-binding method. On the basis of such an analysis, Abell [6] discussed a universal feature of the potentials between neutral atoms in particular, at short ranges. Although he confined himself to the pair interactions only, he showed that for most atoms the repulsive and attractive pair interactions can be parametrized as simple exponentials and all but nearest-neighbor interactions can be ignored without losing much accuracy.

The expression for the wavefunction $\psi_\lambda(\boldsymbol{r})$ of the λth level in terms of the localized Wannier basis $\phi_m(\boldsymbol{r})$ is

$$\psi_\lambda(\boldsymbol{r}) = \sum_m C_{m,\lambda}\phi_m(\boldsymbol{r}), \tag{4.19}$$

where the $C_{m,\lambda}$ are determined by the secular equation. The two-center approximation gives an expression for the binding energy in terms of bond densities and pair interactions, the latter of which, for a given atomic species, depend only on the pair separation and not on the overall chemical environment. The simplest pairwise potential described by exponentials is the Morse potential

$$V(r) = A\exp(-2\lambda r) - B\exp(-\lambda r). \tag{4.20}$$

The universal form for the pairwise potentials proposed by Abell (the Abell potential) is just its simple generalization

$$V(r) = A\exp(-\theta r) - B\exp(-\lambda r). \tag{4.21}$$

The covalent bonding in semiconductors and other materials has a strong angle dependence and cannot be expressed only by pair potentials. Tersoff [7] used Abell's idea for the interatomic interactions associated with the covalent bonding, and extended the potential to three-body interactions:

$$V = \sum_{i<j} f_c(r_{ij})[A_{ij}\exp(-\lambda_{ij}r_{ij}) - b_{ij}\exp(-\mu_{ij}r_{ij})], \tag{4.22}$$

where $r_{ij}$ is the distance between atoms $i$ and $j$, and $f_c(r_{ij})$ is a cutoff function given by

$$f_c(r_{ij}) = \begin{cases} 1 & r_{ij} < R_{ij} \\ \frac{1}{2} + \frac{1}{2}\cos[\pi(r_{ij} - R_{ij})/(S_{ij} - R_{ij})] & R_{ij} \le r_{ij} \le S_{ij} \\ 0 & r_{ij} > S_{ij} \end{cases} \tag{4.23}$$

The parameter $b_{ij}$ represents the bond strength and the many-body effect and is given by

$$b_{ij} = B_{ij}\chi_{ij}(1 + \beta_i^{n_i}\xi_{ij}^{n_i})^{-1/2n_i}, \tag{4.24}$$

**Table 4.2** Parameters for Tersoff's potential for carbon, silicon and germanium

| | C | Si | Ge |
|---|---|---|---|
| $A$ [eV] | $1.3936 \times 10^3$ | $1.8308 \times 10^3$ | $1.769 \times 10^3$ |
| $B$ [eV] | $3.4670 \times 10^2$ | $4.7118 \times 10^2$ | $4.1923 \times 10^2$ |
| $\lambda$ [Å$^{-1}$] | 3.4879 | 2.4799 | 2.4451 |
| $\mu$ [Å$^{-1}$] | 2.2119 | 1.7322 | 1.7047 |
| $\beta$ | $1.5724 \times 10^{-7}$ | $1.1000 \times 10^{-6}$ | $9.0166 \times 10^{-7}$ |
| $n$ | $7.2751 \times 10^{-1}$ | $7.8734 \times 10^{-1}$ | $7.5627 \times 10^{-1}$ |
| $c$ | $3.8049 \times 10^4$ | $1.0039 \times 10^5$ | $1.0643 \times 10^5$ |
| $d$ | 4.384 | 16.217 | 15.652 |
| $h$ | −0.57058 | −0.59825 | −0.43884 |
| $R$ [Å] | 1.8 | 2.7 | 2.8 |
| $S$ [Å] | 2.1 | 3.0 | 3.1 |
| $\chi$ | $\chi_{\text{C-Si}} = 0.9776$ $\chi_{\text{Si-Ge}} = 1.00061$ | | |

where

$$\xi_{ij} = \sum_{k \neq i,j} f_c(r_{ik})\omega_{ik} g(\theta_{ijk}), \tag{4.25}$$

$$g(\theta_{ijk}) = 1 + (c_i/\mathrm{d}_i)^2 - c_i^2/[\mathrm{d}_i^2 + (h_i - \cos\theta_{ijk})^2], \tag{4.26}$$

$$\omega_{ik} = \exp[\mu_{ik}^3 (r_{ij} - r_{ik})^3]. \tag{4.27}$$

Here $\theta_{ijk}$ is the bond angle between the $ij$ bond and the $ik$ bond; $\beta_i, n_i, c_i, d_i, h_i$ are parameters, given in Table 4.2 for carbon, silicon and germanium. The parameters $A_{ij}$, $B_{ij}$, $\lambda_{ij}$, $\mu_{ij}$, $R_{ij}$, $S_{ij}$ should be determined for different atomic species. Often one assumes that they are simply given by

$$\begin{aligned} A_{ij} &= (A_i A_j)^{1/2}, \\ B_{ij} &= (B_i B_j)^{1/2}, \\ R_{ij} &= (R_i R_j)^{1/2}, \\ S_{ij} &= (S_i S_j)^{1/2}, \\ \lambda_{ij} &= (\lambda_i + \lambda_j)/2, \\ \mu_{ij} &= (\mu_i + \mu_j)/2; \end{aligned} \tag{4.28}$$

$A_i$, $B_i$, $R_i$, $S_i$, $\lambda_i$ and $\mu_i$ are also given in Table 4.2.

Chelikowsky and Phillips [8] also determined a potential function for atomic systems which have covalent bonding so as to fit ab initio calculations. Similarly to Tersoff's potential, the Chelikowsky–Phillips potential is composed of a pair part and a three-body part. However, in contrast to Tersoff's potential, their potential is not based on the Morse potential nor its simple generalization, but is based on a

Gaussian-type potential. As a characteristic, their potential has an additional term representing a quantum effect which allows the transition of a bond from a dangling bond to a back bond.

How to go beyond pair potentials is described in detail, with more emphasis on fitting to experiments, in an article by Carlsson [9].

### *4.2.4 Embedded-Atom Potential*

Especially for transition metals and noble metals, in which d electrons contribute to the bonding, there are several shortcomings in the use of pair potentials. It is known that the use of pair potentials always leads to the Cauchy relation $C_{12} = C_{44}$. However, the elastic constants experimentally observed do not satisfy this relation. Moreover, there is an experimental evidence that the vacancy formation energy ($E_{\mathrm{VF}}$) is not always equal to the cohesive energy ($E_{\mathrm{C}}$), but the use of pair potentials always leads to $E_{\mathrm{VF}} = E_{\mathrm{C}}$. Besides these faults, there are several additional problems. The pair potentials cannot explain the properties of the metal surfaces. Near the surface, the interlayer distance between the first and second layers is experimentally smaller than the bulk interlayer distance, but becomes bigger when one uses pair potentials. Pair potentials cannot reproduce the experimentally observed surface reconstruction. Since $E_{\mathrm{C}}/k_{\mathrm{B}}T_{\mathrm{m}}$ is small, the use of pair potentials gives an unrealistically high vaporization rate near the melting temperature.

In order to solve these problems, Daw and Baskes [10] proposed an embedded-atom method whose potential is composed of pairwise terms and many-body terms, which are expressed in terms of the electron density $n_i$ at each atomic center:

$$V = \sum_{i>j} \Phi(r_{ij}) + \sum_{i}^{N} F(n_i), \tag{4.29}$$

where $\Phi(r_{ij})$ is the pair potential and $F(n_i)$ represents the energy required for embedding the atom at a point of electron density $n_i$. Their idea of introducing this form of potential was originally based on the pseudoatom method due to Stott and Zaremba [11]. Later, this model was reformulated from a more fundamental point of view by Jacobsen et al. [12], who proposed the effective medium theory. This model is essentially the same as the glue model proposed by Ercolessi et al. [13].

In the embedded-atom potential, the electron density $n_i$ is given by the superposition of electron densities of the surrounding atoms:

$$n_i = \sum_{j=1}^{N} \rho_j^{\mathrm{atom}}(r_{ij}). \tag{4.30}$$

Here, $\rho_j^{\text{atom}}$ denotes the electron density of the $j$th atom, which may be determined, for example, by solving the Hartree–Fock equation. However, since the electron densities of an isolated atom and of an atom in a solid are generally different, $\rho_j^{\text{atom}}$ should not simply be set equal to the density of the corresponding isolated atom. Usually, it is modified empirically to

$$\rho_j^{\text{atom}}(r) = N_s \rho_s^{\text{atom}}(r) + (N - N_s)\rho_d^{\text{atom}}(r), \tag{4.31}$$

where $N$ is the number of valence electrons of the atom, and $\rho_s^{\text{atom}}$ and $\rho_d^{\text{atom}}$ denote, respectively, the electron densities of s and d electrons of the isolated atom.

The embedded-atom method has been successfully applied to many FCC and BCC metals and other transition metals whose d-band is almost fully occupied. Finnis and Sinclair [14] proposed to fit the functions appearing in (4.29) to the second-moment approximation in the tight-binding method. Since the analytic polynomial form of the Finnis–Sinclair potential is simple and convenient, its parameters can also be determined empirically so as to fit experimental data. However, the Finnis–Sinclair model appears to be less convenient for direct conversion from pure metals to alloys. In this model, the expression for the total energy of the system is the same as in the embedded-atom method, but the many-body term is given by the square root of a sum of two-body potentials, i.e. $F_i = -\sqrt{\cdots}$, and both $\phi$ and $\rho_j^{\text{atom}}$ are dependent on the species of both atoms $i$ and $j$ and thus they are not directly transferable from a pure metal to an alloy. This difficulty was partly overcome by Ackland and Vitek [15] who determined a Finnis–Sinclair potential suitable for the Cu–Ag–Au system. Johnson [16] proposed an exponential-type analytic function for the embedded-atom potential by considering only the nearest-neighbor interactions in FCC metals. Johnson [17] also gave a convenient way to treat the interactions between atoms of different species. Recently, Baskes extended the embedded-atom potential to include the angle dependence of electron charge distributions [18].

## 4.3 Cluster Expansion

As shown in the previous chapters, atomistic calculations are still limited to a few thousand particles, and for all practical purposes, to systems of nanometer size at the most. To perform thermodynamic calculations, one needs to treat much larger systems. In practice, with the present computing facilities, this task is out of the reach of today's supercomputers. For such calculations, one then resorts to statistical methods such as the Monte-Carlo method, which will be detailed in the next chapter. These methods allow one to calculate thermodynamic properties of large systems by adopting simple Ising-type models for interactions between coarse-grained degrees of freedom of the system.

### 4.3.1 Lattice Gas Model

In this model, the real system is replaced by a much simpler one where the degrees of freedom are discretized and placed on a regular lattice. Such models are referred to as lattice gas or Ising models. Properties of some of these systems, for simple specific interactions have been calculated exactly [19] (for example, the 2D Ising model with nearest-neighbor interactions), by using field-theoretic methods [20], by using the cluster variation method [21–24] (CVM) or by Monte Carlo methods. The advantage of the latter is that it can be used for any system in any number of dimensions as exactly as the statistical and numerical errors of the computations allow. All the above methods go beyond the mean field, and therefore include correlations between the degrees of freedom.

The input to all these coarse-grained models is the "coupling constants" of the theory. For the Ising model, for example, it would be the exchange coupling between the spins: the parameter $J_{ij}$ in the Ising Hamiltonian

$$H = -\sum_{<ij>} J_{ij}\sigma_i\sigma_j \tag{4.32}$$

where the sum is over the sites $i$ and $j$ (usually nearest neighbors), $\sigma_i$ being the spin degree of freedom, taking the discrete values $+1$ and $-1$.

All systems of interacting particles can be reduced to or approximated by such a Hamiltonian. In the case of a lattice gas modeling a binary alloy, $\sigma_i$ would be the occupation number of the site $i$, being either 0 or 1, (even though its thermal average is a real number in between), and the "coupling" $J_{ij}$ represents the pair-interaction between the two atoms at sites $i$ and $j$.

In what follows, we will discuss how the transition from microscopic quantities to their coarse-grained equivalents in the Ising model is achieved.

### 4.3.2 The Connolly–Williams Approximation

In this approximation, the total energy of a thermodynamic system is written as a cluster expansion:

$$E = E_0 + \sum_i J_i\langle\sigma_i\rangle + \sum_{ij} J_{ij}\langle\sigma_i\sigma_j\rangle + \sum_{ijk} J_{ijk}\langle\sigma_i\sigma_j\sigma_k\rangle + \dots, \tag{4.33}$$

where the sites $i, j, k$ are on a regular infinite lattice. The brackets $\langle\cdots\rangle$ represent the ground-state average (at zero temperature) or thermal average (at finite temperatures). $J_i$ is the conjugate field to $\sigma_i$. For a magnetic system, this would be the magnetic field at the site $i$ due to some external applied field or a random quantity representing the disorder or randomness present in the system. As explained before, $J_{ij}$ and the higher-order terms $J_{ijk}$, etc., represent pair, and three-body and higher interactions, respectively.

One can perform accurate total energy calculations of the real system for different orderings, i.e. for different decorations of the lattice characterized by the set $\sigma_i$ ($i = 1, \ldots, N$). From these total energies, it is possible to go back to the couplings by means of a singular value decomposition (SVD) [25]. One solves a set of $p$ linear equations with $q$ unknowns, $p$ being the number of computed total energies and $q$ the number of coupling constants needed:

$$E^{\lambda} = E_o + \sum_i J_i S_i^{\lambda} + \sum_{ij} J_{ij} S_{ij}^{\lambda} + \sum_{ijk} J_{ijk} S_{ijk}^{\lambda} + \ldots, \quad \lambda = 1, \ldots, p. \quad (4.34)$$

The SVD algorithm solves this ill-defined problem in a least-squares sense, i.e. it minimizes the difference between the computed energies and those obtained from (4.33), with respect to the couplings $J$. Clearly, the number of calculated total energies must be enough to yield a mathematically sound solution. The number of $J$s adopted must also be chosen appropriately, so that they are enough to reproduce physically sensible energies, and, at the same time, not redundant. One can either have $p = q$ as Connolly and Williams originally did [26] or have $p > q$, in which case the system is overdetermined. Other constraints such as symmetry considerations can also be imposed on the couplings by modifying the cost function, using Lagrange multipliers, or by direct elimination and substitution. In all cases, the solution must be checked against a set of total energies calculated for an arbitrary ordered decoration of the lattice not used in the SVD fitting procedure.

For this calculation, one must of course compute the values of the correlation functions $S_{i_1 \ldots i_n}^{\lambda} = \langle \sigma_{i_1} \ldots \sigma_{i_n} \rangle^{\lambda}$ of the ordered structure $\lambda$. This calculation is quite simple since each structure is ordered, meaning the occupations are fixed at a constant number.

The error one makes in such treatment resides in (1) the accuracy with which the total energies are computed and (2) the error made by truncating the expansion of the total energy. The first error is usually small in the case of nonmagnetic alloys ("small" meaning "within the DFT functional accuracy"). For systems containing magnetic elements, calculations more sophisticated than the simple LDA or GGA are often necessary. The cluster truncation error can be checked by comparing the energies of other structures with the value obtained from (4.34) after the SVD procedure has been performed. If one can reproduce the energies of these other ordered structures accurately, this means the truncation at $n$ clusters was enough; if not, more terms are needed in the expansion (4.34) and, accordingly, more total energies need to be computed in order to get a good fit. In some cases, it could happen that the expansion is not convergent. Then one needs to adopt another type of expansion (with respect to different variables), or another approach altogether.

This method has been applied to compute phase diagrams of binary alloys by first-principles methods [24]. The Connolly–Williams expansion included clusters up to octahedral ones. The relative errors in the formation energies did not exceed 20%. In terms of the assessment of phase diagrams, this method is not as accurate as empirical

CALPHAD methods, but has the advantage of being first-principles and therefore could have predictive power *outside* the range where thermodynamic quantities have been fitted in the past, and it is able to predict new phases. The total energy must be evaluated as accurate as possible, in order to describe subtle energy differences of the order of mRyd per atom or less between two different phases. The SVD algorithm has basically no errors inherent in it, other than numerical inaccuracies which are orders of magnitude smaller than the errors in the computation of formation energies for example.

## 4.4 Potential Renormalization

Numerically, to get a phase equilibrium as fast as possible or to study a behavior near thermal equilibrium, a simple coarse-grained lattice gas model is, in the usual cases, more tractable than off-lattice models. Lattice gas models are often also used for describing systems far from equilibrium, for example, nucleation and crystal growth, atomic diffusion in the solid state, defects and crack propagation. In standard lattice gas models, the interparticle interaction is assumed to be a nearest-neighbor (n.n.) pair interaction and the particle occupancy at each lattice point is just 0 or 1. However, such an assumption is not very realistic and in many cases too much of a simplification. For example, a lattice gas model overestimates the order–disorder phase-transition temperature [27] because it neglects the internal entropy on a length scale shorter than the lattice constant. Of course, it is possible to introduce many-body interactions in the lattice gas model, but the same problem still remains when one uses directly the bare interparticle interaction, which was originally applicable only to off-lattice systems. Then the question arises: is it possible to *renormalize* the bare potential so as to fit the lattice model? If such a potential-renormalization scheme could be formulated in a closed way, one might expect that the lattice gas model with the renormalized potential would have the same properties as the original off-lattice model with the bare potential. One possible strategy to find such a renormalization scheme is to assume that the lattice system and the off-lattice system give the same local partition function at a given temperature.

Here, we will discuss one systematic way [28] to perform the potential renormalization when the bare potential is short-ranged. The short-range potential is, of course, an assumption, but many preexisting interatomic potentials which have been used in molecular dynamics are sufficiently short-ranged.

### 4.4.1 *Basic Idea: Two-Step Renormalization Scheme*

The renormalization of Hamiltonian parameters was first introduced in the theory of quantum electrodynamics by Dyson [29] who used the many-time perturbation expansion scheme due to Tomonaga [30], Schwinger [31] and Feynman [32]. The basic idea of the renormalization here is to renormalize bare (infinite) interaction parameters to get physical (finite) interactions. Renormalization theory has been the standard technique in elementary-particle physics, and is well established in the grand unified theory. Later, the renormalization group theory was also introduced into the theory of critical phenomena in condensed-matter physics [33], although there is no difficulty of divergence in condensed-matter physics. The "elementary domain" first discussed by Yukawa [34] to avoid the ultraviolet divergence, and the lattice gauge theory [35] have a similarity to the discretization of the configurational space in the lattice gas model. The idea of potential renormalization which will be described here is, therefore, somewhat similar to that developed in elementary-particle physics, and also similar to an approximate real-space renormalization group theory developed by Migdal [36] and Kadanoff and Houghton [37] for spin systems.

The basic idea of potential renormalization in a classical $N$-particle system is to determine a new potential function so that, when we discretize the configurational space we do not change the partition function. The original partition function is expressed in terms of the bare potential $U(\boldsymbol{x}_1, \ldots, \boldsymbol{x}_N)$ as follows:

$$\begin{aligned} Z &= \frac{1}{h^{3N} N!} \int \mathrm{d}\boldsymbol{x}_1 \cdots \mathrm{d}\boldsymbol{x}_N \exp\Big[-\frac{U(\boldsymbol{x}_1, \ldots, \boldsymbol{x}_N)}{k_\mathrm{B} T}\Big] \\ &\quad \times \int \mathrm{d}\boldsymbol{p}_1 \cdots \mathrm{d}\boldsymbol{p}_N \exp\Big(-\sum_{i=1}^{N} \frac{p_i^2}{2mk_\mathrm{B}T}\Big) \\ &= \Big(\frac{2\pi m k_\mathrm{B} T}{h^2}\Big)^{3N/2} \frac{1}{N!} \int \mathrm{d}\boldsymbol{x}_1 \cdots \mathrm{d}\boldsymbol{x}_N \exp\Big[-\frac{U(\boldsymbol{x}_1, \ldots, \boldsymbol{x}_N)}{k_\mathrm{B} T}\Big]. \end{aligned} \tag{4.35}$$

We assume that this partition function can be reproduced by its discretized version:

$$Z = \Big(\frac{2\pi m k_\mathrm{B} T}{h^2}\Big)^{3N/2} \frac{1}{N!} \sum_{i_1} \cdots \sum_{i_N} \exp\Big[-\frac{W(i_1, \ldots, i_N)}{k_\mathrm{B} T}\Big]. \tag{4.36}$$

In (4.36), $i_k$ ($k = 1, \ldots, N$) runs over all the $M$ lattice points, and $W(i_1, \ldots, i_N)$ is the desired renormalized potential, which should be determined from the knowledge of the bare potential.

Imagine a regular lattice composed of two sublattices, A and B, in configurational space. Then one can discretize the configurational space into the Wigner–Seitz (WS) cells, which are polyhedrons centered at sublattice points. A WS cell centered at an A sublattice point $\boldsymbol{R}_k$ is defined as a zone surrounded by planes; each of the planes is perpendicular to a line from the point $\boldsymbol{R}_k$ toward its neighboring B sublattice point, and intersects the mid point of this line. A WS cell centered at a B sublattice point is defined in a similar way. Let us call a WS cell centered at an A or B sublattice point an A-cell or B-cell, respectively. The two sublattices should be chosen so that two different A-cells do not have any common (surface) plane nor do two different B-cells. If there is inevitably a common plane, the common area should be sufficiently

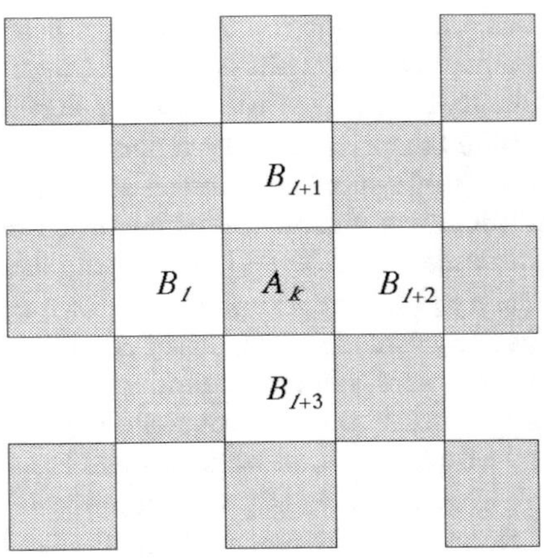


**Fig. 4.1** A sublattice division in the case of a simple cubic lattice

small. Such a sublattice division is possible, for example, in a simple cubic (SC) lattice; see Fig. 4.1.

The bare potential $U(\boldsymbol{x}_1, \ldots, \boldsymbol{x}_N)$ in (4.36) is assumed to be short-ranged and three-body,

$$U(\boldsymbol{x}_1, \ldots, \boldsymbol{x}_N) = \sum_{j}^{N} \sum_{j'(\neq j)} \Big[u_2(\boldsymbol{x}_j, \boldsymbol{x}_{j'}) + \sum_{j''(\neq j, j')} u_3(\boldsymbol{x}_j, \boldsymbol{x}_{j'}, \boldsymbol{x}_{j''})\Big], \tag{4.37}$$

although the generalization to the case of an $n$-body potential is straightforward. Here, $u_2$ and $u_3$ denote two-body and three-body parts, respectively. The cutoff length of the potential is assumed to be shorter than the lattice constant $a$. Note that there is no restriction on the total number of particles $N$ or on the total number of lattice points $M$, which is equal to the total number of WS cells; one may have either $N \leq M$ or $N > M$. That is, more than one particle can be present simultaneously in the same WS cell, or some WS cells may have no particles inside.

In order to map onto the standard lattice gas model, in which one lattice point (or one WS cell) has at most one particle inside, it is necessary to assume that $N \ll M$. In the case $N \simeq M$ or $N > M$, the present way of discretizing the space introduces the generalized lattice gas model, in which many particles can occupy the same lattice point simultaneously, as if it represented an interacting Bose system.

In what follows, the two-step renormalization scheme is described.

### 4.4.2 The First Step

Consider one A-cell, labeled $k$ and centered at $\boldsymbol{R}_k$, and its neighboring B-cells, labeled $l, l+1, \ldots, l+h_{\mathrm{B}}-1$ (here, $h_{\mathrm{B}}$ is the number of neighboring B-cells). Remember that only B-cells are neighbors of an A-cell. Let the number of particles in the $k$th A-cell be $n_k$, and the number in the $(l+i)$th B-cell $(i = 0, 1, \ldots, h_B - 1)$ be $m_{l+i}$; moreover, let the particle positions inside the $(l+i)$th B-cell be $\boldsymbol{x}_{l+i,j}$ $(j = 1, \ldots, m_{l+i})$.

It is possible to evaluate the integral of the configurational exponential with respect to the particle positions inside the $k$th A-cell:

$$\begin{aligned} Z_k &= \int_{k\text{th A-cell}} \mathrm{d}\boldsymbol{x}_{k,1} \cdots \mathrm{d}\boldsymbol{x}_{k,n_k} \\ &\times \exp\Bigg\{ -\frac{1}{k_B T} \sum_{j=1}^{n_k} \Bigg[ \Bigg( \sum_{\substack{j'=\\j+1}}^{n_k} u_2(\boldsymbol{x}_{k,j}, \boldsymbol{x}_{k,j'}) + \sum_{\substack{j''=\\j'+1}}^{n_k} u_3(\boldsymbol{x}_{k,j}, \boldsymbol{x}_{k,j'}, \boldsymbol{x}_{k,j''}) \Bigg) \\ &+ \sum_{i=0}^{h_B-1} \sum_{j'=1}^{m_{l+i}} \Bigg( u_2(\boldsymbol{x}_{k,j}, \boldsymbol{x}_{l+i,j'}) + \sum_{\substack{j''=\\j+1}}^{n_k} u_3(\boldsymbol{x}_{k,j}, \boldsymbol{x}_{l+i,j'}, \boldsymbol{x}_{k,j''}) \\ &+ \sum_{\substack{j''\\=j'+1}}^{m_{l+i}} u_3(\boldsymbol{x}_{k,j}, \boldsymbol{x}_{l+i,j'}, \boldsymbol{x}_{l+i,j''}) + \Phi_{AB} \Bigg) + \Phi_{AA} \Bigg] \Bigg\}, \end{aligned} \tag{4.38}$$

where $\Phi_{AB}$ and $\Phi_{AA}$ are the interactions affecting different A-cells and given, respectively, by a summation with respect to the A-cells neighboring the $(l+i)$th B-cell,

$$\Phi_{AB} = \sum_{i'}^{\text{A}-\text{B}_{l+i}} \sum_{j''=1}^{n_{i'}} u_3(\boldsymbol{x}_{k,j}, \boldsymbol{x}_{l+i,j'}, \boldsymbol{x}_{i',j''}), \tag{4.39}$$

and a summation with respect to the A-cells in the vicinity of the $k$th A-cell,

$$\begin{aligned} \Phi_{AA} &= \sum_{i'}^{\text{A}-\text{A}_k} \sum_{j'=1}^{n_{i'}} \Bigg[ u_2(\boldsymbol{x}_{k,j}, \boldsymbol{x}_{i',j'}) + \sum_{j''=j}^{n_k} u_3(\boldsymbol{x}_{k,j}, \boldsymbol{x}_{k,j''}, \boldsymbol{x}_{i',j'}) \\ &\quad \sum_{j''=j'}^{n_{i'}} u_3(\boldsymbol{x}_{k,j}, \boldsymbol{x}_{i',j'}, \boldsymbol{x}_{i',j''}) \Bigg]. \end{aligned} \tag{4.40}$$

These two terms are called the *edge interaction*; see Fig. 4.2.

If the edge interaction is zero or negligibly small, this integral $Z_k$ depends on the particle positions only in the neighboring B-cells. Then one may evaluate the *trace* of the $k$th A-cell by equating the result to

$$Z_k \equiv \Omega_A \exp\left[ -\frac{\epsilon_k^A + v_k(\{\boldsymbol{x}_{l+i,j}\}; \boldsymbol{R}_k)}{k_B T} \right], \tag{4.41}$$

where $\Omega_A$ denotes the volume of an A-cell; $\epsilon_k^A$ is the renormalized on-site energy at the $k$th A-cell depending only on the occupation numbers $n_k, m_l, m_{l+1}, \ldots, m_{k+h_B}$,

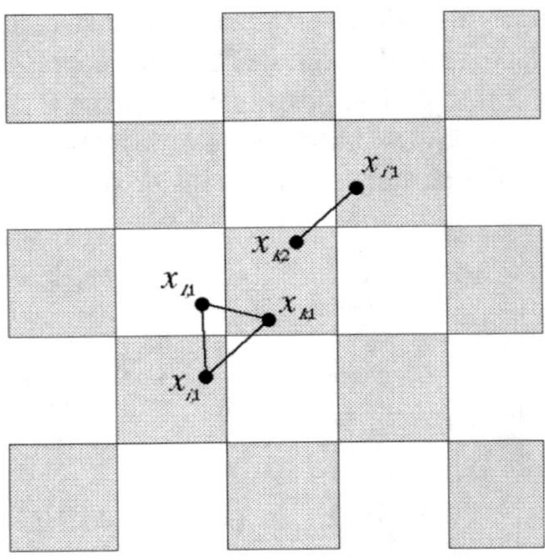

**Fig. 4.2** An illustration of the edge interaction

and $v_k$ is the renormalized potential, depending on all the particle positions in the neighboring B-cells $(l, \ldots, l + h_B - 1)$ and also on $\boldsymbol{R}_k$. The terms $\Phi_3$ and $\Phi_2$ can be treated perturbatively, as is shown in the next subsection.

Thus, one may evaluate the trace of the $k$th A-cell independently for a given set of particle positions $\boldsymbol{x}_{l+i,j}$ for $j = 1, \ldots, m_{l+i}$ in the $h_B$ neighboring B-cells $(l + i = l, \ldots, l + h_B - 1)$, and for a given number of particles in the $k$th A-cell, $n_k$. In this way, one may, in principle, evaluate the trace of the A sublattice, independently, for a given set of particle numbers $\{n_i\}$ in the A-cells and a given set of particle positions $\{\boldsymbol{x}_{i,j}\}$ $(j = 1, \ldots, m_i)$ in the B-cells.

But for practical applications, it is convenient to introduce an approximate parametrization of the renormalized potential by assuming that $v_k$ is a sum of terms, each depending only on the particle positions in *one* neighboring B-cell, say the $(l + i)$th B-cell (and, of course, on $n_k$ and $\boldsymbol{R}_k$), as

$$v_k = \sum_{i=0}^{h_B - 1} v_{k,l+i}(\boldsymbol{x}_{l+i,1}, \ldots, \boldsymbol{x}_{l+i,m_{l+i}}; \boldsymbol{R}_k). \tag{4.42}$$

Then the first step of the potential renormalization is to find the parametrized potential $v_{k,l+i}$ as a function of $n_k$, $\boldsymbol{R}_k$ and the $m_{l+i}$ particle positions $\boldsymbol{x}_{l+i,1}, \ldots, \boldsymbol{x}_{l+i,m_{l+i}}$.

## 4.4.3 The Second Step

The second step is to evaluate the trace of the B sublattice. The potential acting on the particles inside the $l$th B-cell is given by

$$\sum_{j=1}^{m_l}\Biggl\{\sum_{\substack{j'=\\j+1}}^{m_l}\Bigl[u_2(\boldsymbol{x}_{l,j},\boldsymbol{x}_{l,j'})+\sum_{\substack{j''=\\j'+1}}^{m_l}u_3(\boldsymbol{x}_{l,j},\boldsymbol{x}_{l,j'},\boldsymbol{x}_{l,j''})\Bigr]$$
$$+\Phi_{\mathrm{BB}}+\sum_{i=0}^{h_{\mathrm{A}}-1}v_{k+i,l}(\boldsymbol{x}_{l,1},\ldots,\boldsymbol{x}_{l,m_l})\Biggr\}. \tag{4.43}$$

In the third term, the summation with respect to $i$ runs over all $h_{\mathrm{A}}$ A-cells $(k,\ldots,k+h_{\mathrm{A}}-1)$ which are neighbors of the $l$th B-cell ($h_{\mathrm{A}}$ stands for the number of the neighboring A-cells); one of them is, of course, the $k$th A-cell. In the second term, $\Phi_{\mathrm{BB}}$ means the interaction affecting different B-cells and given by a summation with respect to B-cells in the vicinity of the $l$th B-cell (the "edge interaction"):

$$\Phi_{\mathrm{BB}} = \sum_{i'}^{\mathrm{B-B}_l}\sum_{j'=1}^{m_{i'}}\Bigl[u_2(\boldsymbol{x}_{l,j},\boldsymbol{x}_{i',j'})+\sum_{j''=j}^{m_l}u_3(\boldsymbol{x}_{l,j},\boldsymbol{x}_{l,j''},\boldsymbol{x}_{i',j'})$$
$$+\sum_{j''=j'}^{m_{i'}}u_3(\boldsymbol{x}_{k,j},\boldsymbol{x}_{i',j'},\boldsymbol{x}_{i',j''})\Bigr]. \tag{4.44}$$

If there is no edge interaction, the second trace can be evaluated easily by integrating the configurational exponential with respect to all the particle positions belonging to the $l$th B-cell and by equating the result to

$$\int_{l\text{th B-cell}}\mathrm{d}\boldsymbol{x}_{l,1}\cdots\mathrm{d}\boldsymbol{x}_{l,m_l}\exp\Biggl\{-\frac{1}{k_{\mathrm{B}}T}\sum_{j=1}^{m_l}\Bigl[\sum_{j'=j+1}^{m_l}\Bigl(u_2(\boldsymbol{x}_{l,j},\boldsymbol{x}_{l,j'})$$
$$+\sum_{j''=j'+1}^{m_l}u_3(\boldsymbol{x}_{l,j},\boldsymbol{x}_{l,j'},\boldsymbol{x}_{l,j''})\Bigr)+\Phi_{\mathrm{BB}}+\sum_{i=0}^{h_{\mathrm{A}}-1}v_{k+i,l}(\boldsymbol{x}_{l,1},\ldots,\boldsymbol{x}_{l,m_l})\Bigr]\Biggr\}$$
$$\equiv \Omega_{\mathrm{B}}\exp\Bigl[-\frac{\epsilon_l^{\mathrm{B}}}{k_{\mathrm{B}}T}\Bigr]. \tag{4.45}$$

Here, $\Omega_{\mathrm{B}}$ denotes the volume of a B-cell, and $\epsilon_l^{\mathrm{B}}$ the on-site energy of the $l$th B-cell, which depends on the occupation numbers of the $h_{\mathrm{A}}$ A-cells ($n_k$, ..., $n_{k+h_{\mathrm{A}}-1}$) and of their neighboring B-cells (including $m_l$, ..., $m_{l+h_{\mathrm{B}}-1}$ as a part).

Thus one may obtain the renormalized potential $W$ in (4.36) in a slightly different form as

$$Z=\left(\frac{2\pi mk_{\mathrm{B}}T}{h^2}\right)^{3N/2}\frac{1}{N!}\sum_{n_1}\cdots\sum_{n_{M_{\mathrm{A}}}}\sum_{m_1}\cdots\sum_{m_{M_{\mathrm{B}}}}\exp\Bigl[-\frac{\sum_{k=1}^{M_{\mathrm{A}}}\epsilon_k^{\mathrm{A}}+\sum_{l=1}^{M_{\mathrm{B}}}\epsilon_l^{\mathrm{B}}}{k_{\mathrm{B}}T}\Bigr], \tag{4.46}$$

where the summations with respect to the occupation numbers $\{n_k\}$ and $\{m_l\}$ are restricted summations so as to guarantee the sum rule

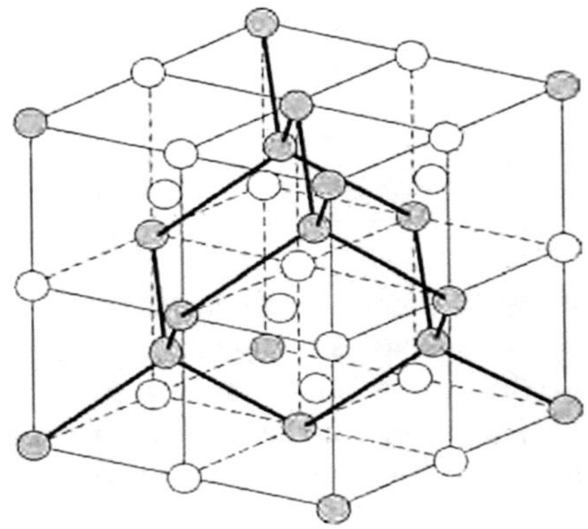

**Fig. 4.3** BCC lattice gas model for Si. The figure shows the regular 50% filling corresponding to the diamond structure

$$\sum_{k=1}^{M_{\mathrm{A}}} n_k + \sum_{l=1}^{M_{\mathrm{B}}} m_l = N. \tag{4.47}$$

In this way, one can derive the renormalized potential, which is not an n.n. pair interaction but inevitably a many-body interaction.

If the bare interaction directly connects an A-cell to another A-cell or a B-cell to another B-cell, i.e. if the edge interactions $\Phi_{\mathrm{AB}}$, $\Phi_{\mathrm{AA}}$ and $\Phi_{\mathrm{BB}}$ in (4.39), (4.40) and (4.44) are not negligible, the two-step renormalization scheme presented here is no longer correct, although it is in principle possible to carry out the first-step renormalization *for the whole A sublattice* independently.

One possible way to take the edge interaction into account may be to carry out a four-step (or multistep) renormalization in place of the two-step renormalization. To do this, one has to decompose the A-sublattice further into two or more sublattices.

### *4.4.4 Application to Si*

The present potential-renormalization scheme should be very suitable to justify the many-body interaction in a lattice gas model for Si, for example. Silicon is a very important material for semiconductor device applications, and investigating its crystal growth process using Monte Carlo simulations is of interest.

For this purpose, the BCC lattice gas model is one possibility for applying the present scheme. A regular diamond structure can be realized by a 50% filling of the BCC lattice positions; see Fig. 4.3. We assume that each lattice point has at most one particle inside.

One may easily obtain the renormalized potential at the first step of renormalization, when all the other atoms are fixed at the ideal positions. Such a treatment would be justified at sufficiently low temperatures. Using Tersoff's short-range three-body potential for Si, for example, one may consider how the potential is renormalized. For

**Table 4.3** Renormalized potential values (eV) at the first step at 1400 K versus coordination number at the n.n. positions. The on-site energy corresponds to the half of this value

| Coordination number | 4 | 3 | 2 |
|---|---|---|---|
| Bare potential at the center | −9.9 | −7.6 | −5.2 |
| Renormalized potential value | −9.1 | −7.0 | −4.7 |

this system, the most important configuration is such that one Si atom is surrounded by four Si atoms at the tetrahedral sites. In the first step renormalization procedure at an assumed temperature of 1400 K (290 K below the melting temperature), the potential value at the tetrahedral center, $-99 \times 10^{-1}$ eV, (the most stable position) becomes $-91 \times 10^{-1}$ eV. If there are only three Si atoms in the neighborhood of the central Si atoms, the potential map is greatly modified. In this case, the renormalized potential becomes $-70 \times 10^{-1}$ eV at the center, which should be compared with the value $-76 \times 10^{-1}$ eV at $T = 0$ K. The results obtained by this first-step renormalization are listed in Table 4.3.

From this renormalization, we see that the potential difference between the 4-coordinated position and the 2-coordinated position is smaller in the renormalized potential ($V_2 - V_4 = 4.4$ eV) than in bare potential ($V_2 - V_4 = 4.7$ eV), i.e. the relative potential depth becomes shallower with respect to a change of the coordination number after renormalization. This shallowing may reduce the melting temperature estimated in the discretized model.

Using an approximate two-step renormalization scheme, the renormalized potentials for all possible first-neighbor configurations concerning occupations were determined [38, 39]. Using the tetrahedral (crystal) configuration only, the Helmholtz free energy of a perfect Si crystal can be determined without performing a simulation. The result is shown in Fig. 4.4. Certainly, the nearest-neighbor distance at the free-energy minimum increases with the temperature, i.e. it exhibits thermal expansion. The thermal expansion coefficient is defined as

$$\alpha = \frac{1}{a}\frac{\mathrm{d}a}{\mathrm{d}T}, \tag{4.48}$$

where $a$ is a lattice constant. From Fig. 4.4, $\alpha$ is estimated to be 7.6 at 500 K, which is about 10% larger than the coefficient obtained by an MD simulation [40]. At 2000 K, $\alpha$ is estimated to be 2.5, which is about half of the MD result. This discrepancy may be attributed to an approximation introduced when the Tersoff potential is renormalized.

Next, in order to carry out a Monte Carlo simulation, the renormalized potentials were also calculated for configurations different from the tetrahedral one. Here, in order to compensate for the unnecessary double counting of the nearest-neighbor three-body interaction due to the approximate replacement by a different kind of three-body interaction as described above, a simple averaging is performed between the resulting renormalized potentials and the renormalized potential for the ideal tetrahedral configuration:

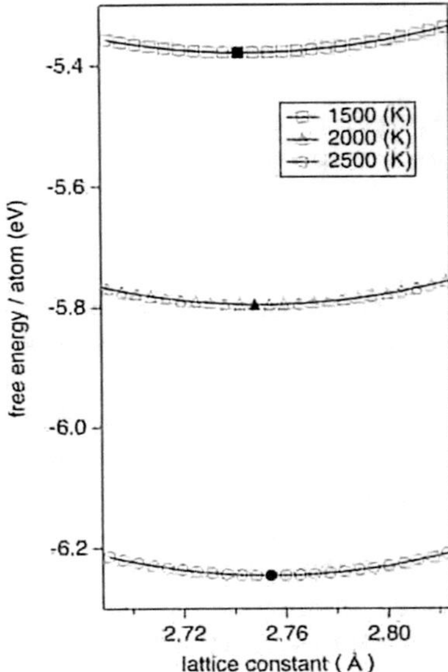


**Fig. 4.4** Helmholtz free energy of a perfect Si crystal [38]

$$\frac{1}{2} \times \frac{(nf^{(n)} + 4f^{(\text{tetra})})}{2}. \tag{4.49}$$

Using renormalized potentials in the BCC lattice gas model, a Monte Carlo simulation was carried out in a long tetrahedral cell at various temperatures under zero pressure. Figure 4.5 plots the density obtained from this simulation together with the density of the pure crystal given in Fig. 4.4. This result shows a broad density change above 2500 K, at which the liquid–solid phase transition (melting transition) takes place. The increase of the density in melts is a well known characteristics of Si. This result is similar to that of the molecular-dynamics study performed by Cook and Clancy [40], who used the same Tersoff potential. In particular, the transition temperatures are close to each other, although they are much higher than the experimental value.

## 4.4.5 Application to FePt Alloys

The potential renormalization theory is applicable not only to a Si system but also to many different systems. $Cu_xAu_{1-x}$ alloys such as a $Cu_3Au$ system were treated with an FCC lattice gas model. For such FCC alloys, in which atoms occupy FCC lattice points, it is convenient to use the tetrahedron approximation [41] in the Connolly–Williams cluster expansion [26]. Then the two-step renormalization scheme can be employed by using a Finnis–Sinclair-type embedded-atom potential as was demonstrated by Sahara et al. [27]; see also the discussion in Sect. 5.4.5.

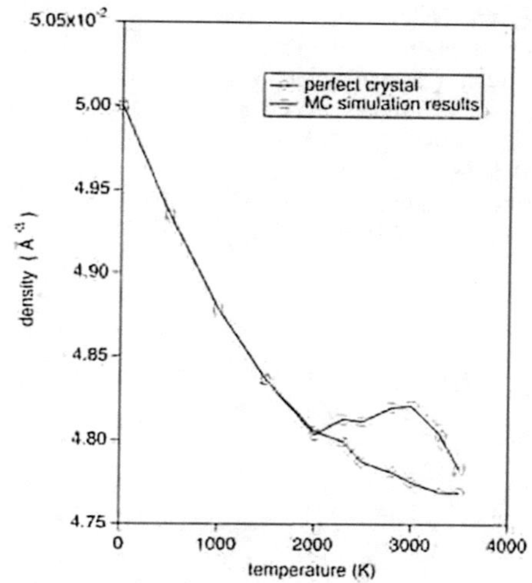


**Fig. 4.5** Density of Si evaluated from a Monte Carlo simulation using the renormalized potential (squares) and density of a perfect Si crystal calculated from Fig. 4.4 (circles) [38, 39]

Next, application of first-principles calculations is explained for a specific example of $Fe_xPt_{1-x}$ alloys. In particular, the FePt system has attracted considerable interest due to its high magneto-crystalline anisotropy and high magnetic coercive force of the $L1_0$ ordered phase. We adopt the tetrahedron approximation [41] in the Connolly–Williams cluster expansion [26]. First, by directly using the tetrahedral 4-atom interaction determined by first principles calculations, one can construct the lattice model with the 4-body interaction. Each simple cubic primitive unit cell includes four atoms (Fe or Pt) at the tetrahedral sites. We call it the basic tetrahedron; it can be either $Fe_2Pt_2$ ($L1_0$), $Fe_4$ (fcc), $Pt_4$ (fcc), $FePt_3$ ($L1_2$), or $Fe_3Pt$ ($L1_2$). However, this direct mapping onto lattice model overestimates the order-disorder transition temperature by a few hundred degrees. In fact, in a previous study [42] using the cluster variation method (CVM) neglecting the off-lattice effects, the transition temperature of the FePt alloy was estimated to be around 1900 K, which is higher than the experimental value (1572 K [43]). To overcome this problem, Mohri et al. calculated the transition temperature with an effect of lattice vibrations using the CVM within the quasi-harmonic approximation via Debye–Grüneisen model and succeeded in obtaining closer value to experimental phase transition temperature [44]. However, the crystal structure of the $L1_0$ ordered phase of the FePt alloy is not rigorously fcc but fct, which cannot be simply treated within the lattice model.

Here, we explain how the potential renormalization theory works in the FePt alloy systems [45]. For the first principles calculations, however, the full two-step renormalization procedure is computationally a bit too heavy. Instead, we use a simpler one-step renormalization scheme. Figure 4.6 schematically illustrates how to perform this one-step renormalization. Using $2 \times 2 \times 2$ simple cubic supercell including 32 atoms, we evaluate the total energy $E(\boldsymbol{R})$ under the condition that all atoms are fixed at the lattice points (black circles in Fig. 4.6) and the total energies $E(\boldsymbol{r}_i)$ under the same condition except a central atom displaced around the lattice

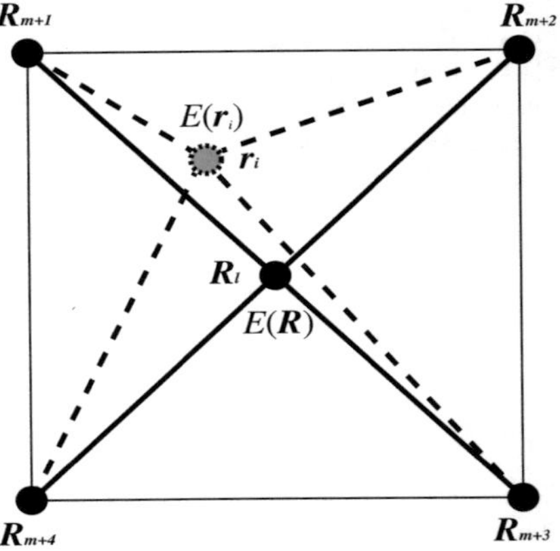


**Fig. 4.6** Schematic illustration representing the potential renormalization. Black circles represent lattice points, while a grey circle and dotted lines show that the central atom is displaced around the lattice point [45]

point (a grey circle in Fig. 4.6). Taking account of the symmetry of the cubic lattice, we displace the central atom inside the one-eighth of the whole volume. The total energy deviation $\Delta E_i^\alpha$ is defined as $\Delta E_i^\alpha = E(\boldsymbol{r}_i) - E(\boldsymbol{R})$, where $i$ and $\alpha$ denote the position and species of the central atom. Calculating altogether $n = 20$ positions of $\boldsymbol{r}_i$ $(i = 1, 2, \ldots, n)$ for each center species $\alpha =$ Fe or Pt, we take the local trace of the Boltzmann distribution to determine the renormalized potential which depends on the temperature. That is, the local partition function is calculated by

$$Z = \sum_i C_\alpha \sum_i^n \exp\left(-\frac{\Delta E_i^\alpha}{k_{\mathrm{B}} T}\right), \tag{4.50}$$

where $C_\alpha$ is the concentration ratio of the center species $\alpha$ (= Fe or Pt) for a given configuration (Fe, Pt, FePt, $FePt_3$, or $Fe_3Pt$). Then, the local free energy correction involving the effects of atomic displacements $\Delta F$ is calculated from this local partition function as

$$\Delta F = -k_{\mathrm{B}} T \log Z. \tag{4.51}$$

We add this $\Delta F$ to the total energy $E(\boldsymbol{R})$ of the perfect crystal in which all atoms were fixed at the lattice points. Finally, all the results are divided by eight, i.e., the number of basic tetrahedrons, to yield the desired renormalized potential of the basic tetrahedron. The calculations were performed by using the Vienna ab-initio simulation package (VASP) [46, 47], using the Vanderbilt ultrasoft pseudopotential and plane waves with the cut-off energy 237.6 eV in the LSDA. Within the first Brillouin zone, $4 \times 4 \times 4$ grid is used for the $k$-point sampling to calculate the self-consistent charge density. The lattice constant is fixed at 3.7 Å, which is the optimized value for the $Fe_2Pt_2$ $L1_0$ ordered phase.

Firstly, Table 4.4 shows the resulting total energies and free energy corrections. The column "before renormalization" lists the total energy $E(\boldsymbol{R})/8$ per basic

**Table 4.4** Total energies and free energy corrections per basic tetrahedron (simple cubic primitive unit cell including four atoms) for five different ordered structures. The columns "before renormalization", "renormalization correction", and "after renormalization" indicate, respectively, $E(\boldsymbol{R})/8$, $\Delta F/8$, and $[E(\boldsymbol{R}) + \Delta F]/8$, in units of eV. From [45]

| Tetrahedron structure | Relative composition | Energy [eV] | | |
|---|---|---|---|---|
| | | Before renormalization, $E(\boldsymbol{R})/8$ | Renormalization correction, $\Delta F/8$ | After renormalization, $[E(\boldsymbol{R}) + \Delta F]/8$ |
| $L1_0$ | $Fe_2Pt_2$ | −33.860 | 0.018 | −33.842 |
| fcc | $Fe_4$ | −35.709 | 0.034 | −35.675 |
| fcc | $Pt_4$ | −27.948 | 0.015 | −27.933 |
| $L1_2$ | $FePt_3$ | −31.484 | 0.015 | −31.469 |
| $L1_2$ | $Fe_3Pt$ | −35.185 | 0.029 | −35.156 |

tetrahedron (i.e., per simple cubic primitive unit cell) in which four atoms exist at the tetrahedral sites. The temperature $T = 1650\,\text{K}$ is assumed for the renormalized potential $\Delta F$. The renormalization correction per basic tetrahedron, $\Delta F/8$, amounts to be $0.015 \sim 0.034\,\text{eV}$, which corresponds to $170 \sim 390\,\text{K}$. In this table, "after renormalization" lists the resulting renormalized potentials per basic tetrahedron, $[E(\boldsymbol{R}) + \Delta F]/8$.

These renormalized total energies are then used to perform the lattice Monte Carlo simulation; for Monte Carlo simulation, see Chap. 5. We randomly select two atoms and exchange them with the probability $\min[1, \exp(\Delta E/k_\text{B}T)]$ according to the Metropolis algorithm (Sect. 5.3.3). Here, $\Delta E$ means the energy gain occurred by the exchange. The energies before and after the exchange are given by the sum of the two on-site energies of the two atoms. Within this exchange algorithm, the on-site energy can be simply calculated as the arithmetic mean of the energies of the eight basic tetrahedrons including the atom at the center, which is equivalent to the treatment of [41] on the basis of the Connolly–Williams approximation [26].

We use 3-dimensional periodic boundary conditions. The size of the system in the present simulation is $20 \times 20 \times 20$ in units of a simple cubic primitive cell of the fcc lattice. The initial configuration is random which is assumed at very high temperature. Then, by quenching the system with keeping the temperature constant, we observe the time evolution of the system. We evaluate the LRO parameter at all times. We judge that the system reaches thermal equilibrium when the LRO parameter converged to some steady value. The LRO parameter is calculated as follows: We divide the fcc lattice into two sub-lattices $a$ and $b$. There are three ways of the division in $x$-, $y$-, and $z$-directions, and we consider all ways. The $A$ and $B$ are, respectively, the numbers of the total atoms of Fe and Pt within a sub-lattice $a$ and $N_a = A + B$. Then, the LRO parameter is defined as the largest value of

$$LRO = \frac{|A - B|}{N_a}, \tag{4.52}$$

among the three ways of the sub-lattice division. Here, the LRO parameter takes the value from 0 to 1, and it takes the value of 0 when the system is completely disordered. In this paper, we compare the patterns obtained by the MC simulations before and after renormalization. We estimate the order-disorder phase transition temperature from the resulting patterns and the LRO parameter of the MC simulations.

Figure 4.7 shows the several typical patterns at 4,000,000 MC steps after quenching the system with keeping the temperature constant. In this figure, Fe and Pt atoms are marked by black diamonds and crosses, respectively. The edge length of the smallest square in this figure corresponds to the half of the lattice constant of the simple cubic primitive unit cell including four atoms. The patterns are the perspective view, seen from one side. In this figure, a point marked by only a cross indicates that a line going through the point and perpendicular to the page consists of Pt only. Therefore, we interpret the areas in which diamonds (Fe) and crosses (Pt) are alternately aligned in one or two directions as the $L1_0$ ordered phase, and, if not, i.e., the areas filled fully by the superposition of diamonds and crosses, the disordered phase. Figure 4.7a, c, and d show patterns obtained after renormalization, and (b) shows a pattern obtained before renormalization. Figure 4.8 shows the LRO parameters at various temperatures before renormalization (dotted curve) and after renormalization (solid curve). In Fig. 4.7a ($T = 1850$ K, after renormalization), the disordered phase extends over a large area in comparison to the ordered phase and the LRO parameter is about 0.1 in Fig. 4.8. That is, the temperature is not enough low to order the system. In Fig. 4.7b ($T = 1970$ K, before renormalization) and c ($T = 1800$ K, after renormalization), it is clear that the area of the ordered phase is much larger than that of (a). In addition, the LRO parameter also represents sharp increase in Fig. 4.8 at these temperatures. From these results, the transition temperatures is estimated to be around $T = 1970$ K before renormalization and around $T = 1800$ K after renormalization. The calculated transition temperature using the renormalization technique becomes 170 K closer to the experimental value. From the figures, it is found that the several ordered phases oriented in different directions appear locally in the whole system, so that the LRO parameter increases only up to about 0.3.

Finally, when the temperature goes below the transition temperature, the area of the ordered phase decreases (in Fig. 4.7d) and the LRO parameter becomes smaller (in Fig. 4.8) than that at the transition temperature. The reason is that we quenched the system from very high temperature, and that the atoms were frozen before getting sufficient thermal energies. This is what observed also in real experimental conditions [48, 49]. If we annealed the system gradually, however, the system could be ordered more perfectly. Such a study will be reported elsewhere in the investigation of FePt clusters.

So far, the treatment of the lattice gas model has been described mainly for solids or liquids. It should be applicable more easily to gas phases. It is also interesting to note that there has been an attempt to apply the lattice gas model to study surface reconstruction, i.e. the reconstruction of atomic alignments at surfaces [50].

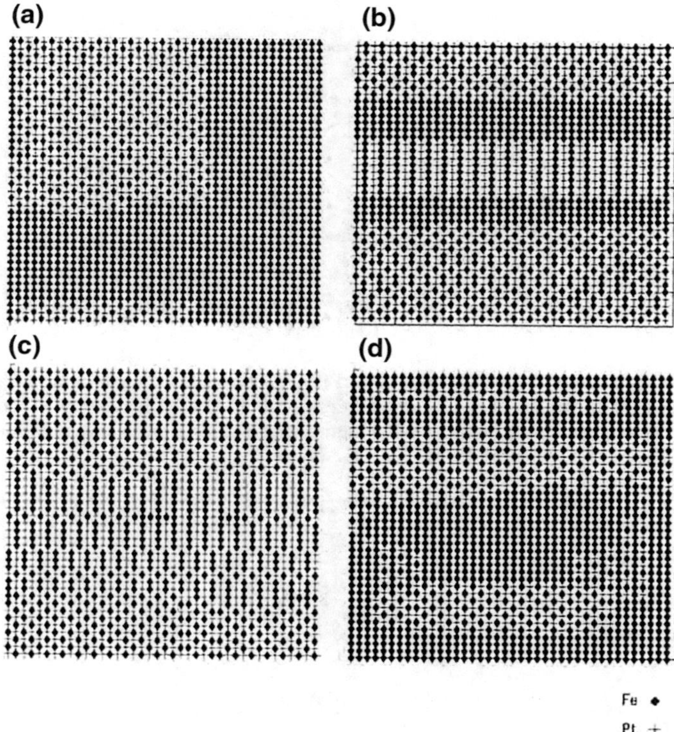


**Fig. 4.7** Perspective view of the resulting patterns of the MC simulation at various temperatures after 4,000,000 MC steps. Fe and Pt atoms are marked by black diamonds and crosses, respectively. The areas where diamonds and crosses are alternately aligned in one or two directions form single domains of the $L1_0$ ordered phase, which perfectly continue in the direction perpendicular to the page, while the areas fully filled by superposition of diamonds and crosses are the disordered phase. **a** is at $T = 1850\,\text{K}$ including the off-lattice effects (after renormalization) and **b** is at $T = 1970\,\text{K}$ without the off-lattice effects (before renormalization). **c** and **d** are at $T = 1800\,\text{K}$ and $T = 1500\,\text{K}$ with the off-lattice effects (after renormalization). From [45]

Another extension is to treat an isolated alloy cluster in the vacuum. To do that, Masatsuji et al. included the vacancy as one species of atoms in the cluster expansion within the tetrahedron approximation using ab initio calculations, and performed MC simulations of FePt alloy clusters [51]. The resulting $L1_0$ order parameter as a function of the cluster size is compared with the experimental result of Miyazaki et al. [52] in Fig. 4.9. There is a good agreement between theory and experiment concerning the critical diameter ($\sim$1.5 nm) to preserve the order.

The lattice MC method, however, cannot treat so many atoms, because it is still an atomic scale simulation. In order to perform mesoscopic simulation, one needs to change the discrete variables such as the occupation number and the atomic species index at each lattice point to the continuous variables such as the density $\phi_X$ of

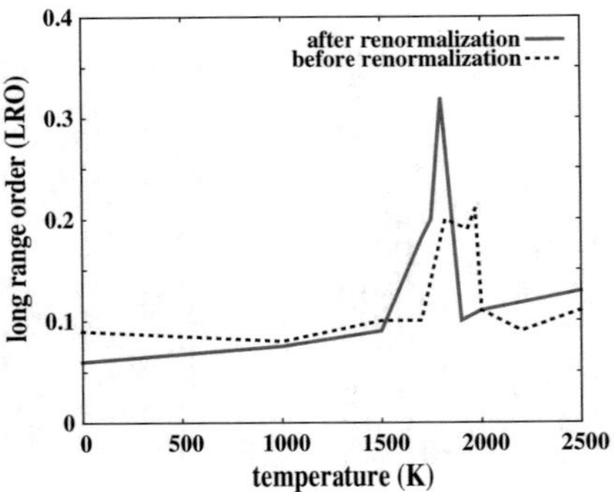


**Fig. 4.8** LRO parameter of the FePt alloy at various temperatures quenched from very high temperature. The solid and dotted curves denote the results obtained after and before renormalization, respectively. The transition temperature using renormalization technique (after renormalization) is about 170 K closer to experimental value 1572 K. From [45]

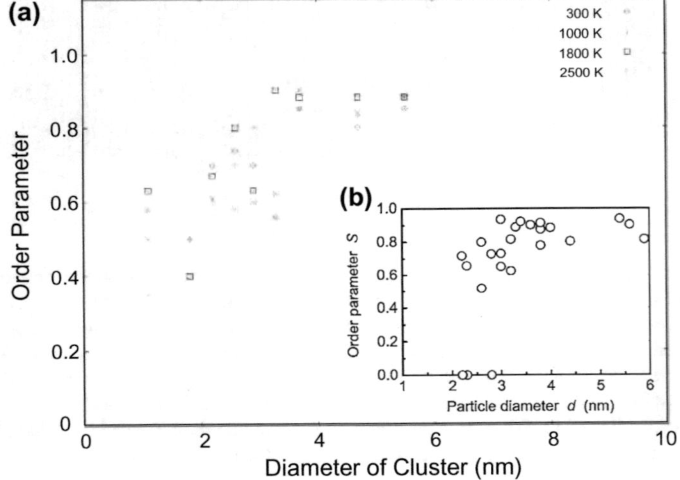


**Fig. 4.9** $L1_0$ order parameter of FePt alloy clusters at various temperatures as a function of the radius of the cluster. **a** The MC result using the ab initio total energies in the tetrahedron approximation including vacancies [51], and **b** the experimental result of Miyazaki et al. [52]

the atomic species $X$ ($X$ refers to, for example, Fe, Pt and vacancy) [53]. In such a treatment, if one divides the space in fine grid and regards each grid point to be a simple cubic unit cell including one tetrahedron, one may use the renormalized potential as the local free energy at each grid point. Then, one may introduce the phase field model from the ab initio basis. For further discussion, see Sect. 5.4.6.

# References


1. S. Erkoc, Phys. Rep. **278**, 79 (1997)
2. R.M. Nieminen, M.J. Puska, M.J. Manninen (eds.), *Many-Atom Interactions in Solids*, vol. 48, Proceedins in Physics (Springer, Berlin, 1990)
3. H. Margenau, Rev. Mod. Phys. **11**, 1 (1939)
4. F. London, Z. Phys. **63**, 245 (1930)
5. L. Pauling, E.B. Wilson, *Introduction to Quantum Mechanics* (McGraw-Hill, New York, 1935)
6. G.C. Abell, Phys. Rev. B **31**, 6184 (1985)
7. J. Tersoff, Phys. Rev. B **39**, 5566 (1989)
8. J.R. Chelikowsky, J.C. Phillips, Phys. Rev. B **41**, 5735 (1990)
9. A.E. Carlsson, in *Solid State Physics*, vol. 43, ed. by H. Ehrenreich, D. Turnbull (Academic Press, San Diego, 1990), p. 1
10. M.S. Daw, M.I. Baskes, Phys. Rev. Lett. 50, 1285 (1983); Phys. Rev. B **29**, 6443 (1984)
11. M.J. Stott, E. Zaremba, Phys. Rev. B **35**, 7423 (1978)
12. K.W. Jacobsen, J.K. Norskov, M.J. Puska, Phys. Rev. B **35**, 7423 (1987)
13. F. Ercolessi, M. Parrinello, E. Tosatti, Philos. Mag. A **58**, 213 (1988)
14. M.W. Finnis, J.E. Sinclair, Philos. Mag. A **50**, 45 (1984)
15. G.J. Ackland, V. Vitek, Phys. Rev. B **41**, 10324 (1990)
16. E.A. Johnson, Phys. Rev. B **37**, 3924 (1988)
17. E.A. Johnson, Phys. Rev. B **39**, 12554 (1989)
18. M.I. Baskes, Phys. Rev. B **46**, 2727 (1992)
19. L. Onsager, Phys. Rev. **65**, 117 (1944)
20. C. Itzykson, J.M. Drouffe, *Statistical Field Theory* (Cambridge University Press, Cambridge, 1989)
21. J. Sanchez, *The Cluster Variation Method* (Oxford University Press, Oxford, 1995)
22. D. de Fontaine, in *Solid State Physics*, vol. 47, ed. by H. Ehrenreich, D. Turnbull (Academic Press, San Diego, 1994), p. 33
23. R. Kikuchi, Phys. Rev. **81**, 988 (1951)
24. M.H.F. Sluiter, Y. Watanabe, D. de Fontaine, Y. Kawazoe, Phys. Rev. B **53**, 6137 (1996). and references therein
25. W.H. Press, S.A. Teukolsky, W.T. Vettering, B.R. Fiannery, *Numerical Recipes in FORTRAN: The Art of Scientific Computing*, 2nd edn. (Cambridge University Press, New York, 1992)
26. J.W. Connolly, A.R. Williams, Phys. Rev. B **27**, 5169 (1983)
27. R. Sahara, H. Ichikawa, H. Mizuseki, K. Ohno, H. Kubo, Y. Kawazoe, J. Chem. Phys. **120**, 9297 (2004)
28. K. Ohno, Sci. Rep. Res. Inst. Tohoku Univ. A **43**, 17 (1997)
29. F.J. Dyson, Phys. Rev. 75, 486 (1949); Phys. Rev. **75**, 1736 (1949)
30. S. Tomonaga, Prog. Theor. Phys. 1, 27 (1946); Phys. Rev. **74**, 224 (1948)
31. J. Schwinger, Phys. Rev. 73, 416 (1948); Phys. Rev. **74**, 1439 (1948)
32. R. Feynman, Rev. Mod. Phys. **20**, 367 (1948); Phys. Rev. **74**, 939, 1430 (1948)
33. J. Zinn-Justin, *Quantum Field Theory and Critical Phenomena* (Oxford Science Publications, Oxford, 1989)
34. H. Yukawa, Prog. Theor. Phys. Suppl. **37/38**, 512 (1966)
35. C. Rebbi, *Lattice Gauge Theories and Monte Carlo Simulations* (World Scientific, Singapore, 1983)
36. A. A. Migdal, Z. Eksper. Theor. Fiz. **69**, 810 (1975); Z. Eksper. Theor. Fiz. **69**, 1457 (1975)
37. L. Kadanoff, A. Houghton, Phys. Rev. B **11**, 377 (1975)
38. R. Sahara, H. Mizuski, K. Ohno, S. Uda, T. Fukuda, Y. Kawazoe, J. Chem. Phys. **110**, 9608 (1999)
39. R. Sahara, H. Mizuseki, K. Ohno, H. Kubo, Y. Kawazoe, J. Cryst. Growth **229**, 610 (2001)
40. S.J. Cook, P. Clancy, Phys. Rev. B **47**, 7686 (1993)
41. K. Terakura, T. Oguchi, T. Mohri, K. Watanabe, Phys. Rev. B **35**, 2169 (1987)

42. T. Mohri, Y. Chen, Mater. Trans. **43**, 2104 (2002)
43. T.B. Massalski (ed.), *Binary Alloy Phase Diagrams* (American Society of Metals, Metals Park, 1986), p. 1093
44. T. Mohri, Y. Chen, J. Alloy. Compod. **383**, 23 (2004)
45. Y. Misumi, S. Masatsuji, R. Sahara, S. Ishii, K. Ohno, J. Chem. Phys. **128**, 234702 (2008)
46. G. Kresse, J. Hafner, Phys. Rev. B **49**, 14251 (1994)
47. G. Kresse, J. Furthmuller, Comput. Mat. Sci. **6**, 15 (1996)
48. S.H. Whang, Q. Feng, Y.-Q. Gao, Acta Materialia **46**, 6485 (1998)
49. C.M. Kuo, P.C. Kuo, H.C. Wu, Y.D. Yao, C.H. Lin, J. Appl. Phys. **85**, 4886 (1999)
50. J. Kanamori, in *Dynamical Processes and Ordering on Solid Surfaces*, vol. 59, Solid-State Science, ed. by A. Yoshimori, M. Tsukada (Springer, Berlin, 1985), p. 154
51. S. Masatsuji, Y. Misumi, S. Ishii, K. Ohno, **49**, 2424 (2008); Y. Misumi, S. Masatsuji, K. Ohno, unpublished
52. T. Miyazaki, O. Kitakami, S. Okamoto, Y. Shimada, Z. Akase, Y. Murakami, D. Shindo, Y.K. Takahashi, K. Hono, Phys. Rev. B **72**, 144419 (2005)
53. S. Bhattacharyya, R. Sahara, K. Ohno (to appear)

# Chapter 5
# Monte Carlo Methods

## 5.1 Introduction

Nature is composed of gross assemblies of huge numbers of atoms and molecules showing a wide variety of phenomena according to the way how they are assembling. The macroscopic behaviors of such systems are rather different from the microscopic laws in the world of atoms and molecules. For example, in the usual cases of macroscopic systems, the motion of the atoms and molecules can be regarded simply as heat. That is, the average kinetic energy of each atom and molecule in a macroscopic system is equal to a quantity measured as the temperature. The other contributor to macroscopic behavior is the cooperative motion of atoms and molecules (or sometimes electrons). Since atoms and molecules interact with each other, their macroscopic assemblies can have cooperative motions. Many examples can be seen in our daily life: spring or rubber elasticity, magnetization in permanent magnets, shape memory alloys, liquid flows, surface tension of liquids, swelling of polymers by water absorption, etc. Sometimes such cooperative motions are frozen as the temperature decreases. In this case, the states which have been realized at higher temperatures become unstable. This phenomenon is called a phase transition [1, 2]. For example, ferromagnets show nonzero magnetization below the Curie temperature, and superconductors show zero resistivity below a certain temperature called the critical temperature. These phase transitions accompany a spontaneous symmetry breakdown (i.e. the breakdown of symmetry in the spin direction in magnets or in the electric gauge in superconductors) and do not produce latent heat or a volume change at the transition temperature. This kind of phase transition is called a second-order phase transition and is distinguished from more usual first-order phase transitions, such as melting, solidification, evaporation and sublimation. At the transition temperature, which is called the critical temperature, of a second-order phase transition, the thermodynamic quantities derived from the second derivative of the free energy, such as the specific heat and the susceptibility, diverge according to a power law. And a spatially fractal structure of ordering in the spin, electronic or atomic alignments appears. The ordering spreads into the whole system as the temperature is decreased.



K. Ohno et al., *Computational Materials Science*,
https://doi.org/10.1007/978-3-662-56542-1_5

Strictly speaking, finite systems cannot undergo a second-order phase transition, and therefore do not show any singularity in the specific heat or other quantities. To simulate such macroscopic phenomena, although one must use finite systems, one has to make the system size as large as possible. The methodology of Monte Carlo simulation is quite powerful in understanding the properties and behaviors of macroscopic systems involving phase transitions [3, 4].

As slightly different subjects, crystal growth from the melt or vapor, and pattern formation as in macromolecular systems are problems which are closely related to a phase transition. Another related type of problem may be that of frustrated systems, where the ground state cannot be uniquely determined, and glass/amorphous systems, where metastable states can be realized. These systems are essentially many-body systems in which searching for the ground state at the absolute zero of temperature is the nontrivial problem. On the other hand, systems such as superconductors or superfluids which exhibit purely quantum effects on macroscopic length scales should be approached with many-body quantum theory. These systems offer a challenging problem which can be treated with Monte Carlo simulations and are currently being studied by many theoreticians. The quantum Monte Carlo method will be described in a later chapter.

## 5.2 Basis of the Monte Carlo Method

In a macroscopic system, it is difficult to treat the motions of all (microscopic) atoms and molecules accurately on the basis of a deterministic model, because it requires much longer space and time intervals than the length and time scales available in the MD technique. Fortunately, in many cases, one is more interested in macroscopic properties (namely, how the whole system behaves) than in microscopic information such as the state of each atom or molecule. In these cases, it is important to study ways to construct a simple and useful model by coarse-graining in both space and time. Coarse-graining procedures have been discussed in the previous chapter, especially from the spatial point of view. If the time evolution of a system is coarse-grained stochastically, one achieves one class of models, called the stochastic models. One of the efficient ways to realize this numerically on a computer is by the Monte Carlo method. Even in a stochastic model, one has to handle directly a complex problem of a many-body system which has quite many degrees of freedom. Except for simplified treatments such as mean-field or classical approximations, most models cannot be solved analytically. Moreover, in many problems, the deviation from the mean-field (or classical) approximation is quite important, and such a deviation is usually called "fluctuation". One might say that a many-body problem is equivalent to a problem of fluctuation. Monte Carlo methods provide a powerful way to solve numerically the problem of fluctuation or relaxation in a stochastic system.

A typical example of the Monte Carlo method is the treatment of colloidal particles (Brownian particles) moving in solution as in Fig. 5.1. Such movement is called "Brownian motion". When one observes it through a microscope, one can identify

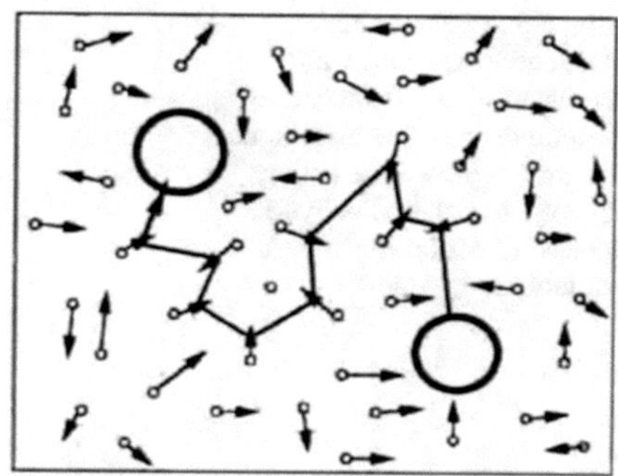

**Fig. 5.1** Brownian motion. The bigger colloidal particle (Brownian particle) moves randomly, colliding with small solvent particles

the position (or velocity) of the Brownian particle only. Applying the coarse-graining procedure, the other degrees of freedom (e.g. the motion of small solvent particles) are removed and, finally, they can be regarded as a random force acting on the Brownian particle. In this section, we introduce the basic idea of the Monte Carlo method on the basis of stochastic processes.

### 5.2.1 Stochastic Processes

Consider a dynamic system, and call a set of variables which is characteristic of the internal state the "dynamic variables". We represent the dynamic variables by $\boldsymbol{q}$. For example, if one considers an assembly of $N$ freely moving molecules in the gas phase (Fig. 5.2), the dynamic variable $\boldsymbol{q}$ is a set of positions $(\mathbf{x}_1, \mathbf{x}_2, \ldots, \mathbf{x}_N)$ and momenta $(\boldsymbol{p}_1, \boldsymbol{p}_2, \ldots, \boldsymbol{p}_N)$. Or, if we consider the thermal motion of the atoms in a solid, which can be regarded as a system of $N$ weakly interacting harmonic oscillators, a set of energies $(\varepsilon_1, \varepsilon_2, \ldots, \varepsilon_N)$ of these oscillators is the dynamic variable $\boldsymbol{q}$ ($\varepsilon_i$ can

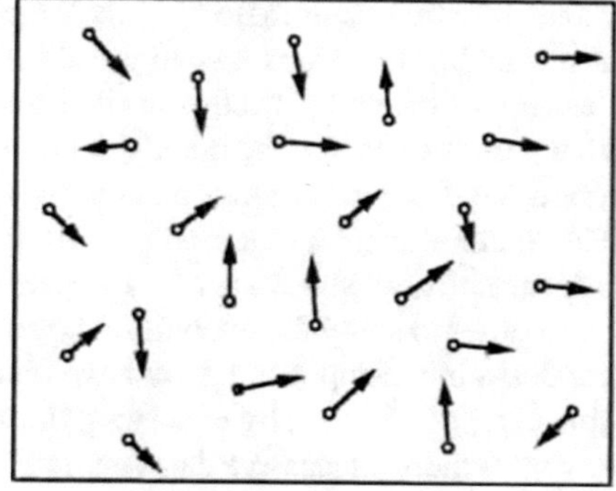

**Fig. 5.2** Motion of gas molecules. The momentum distribution of the gas particles obeys the Boltzmann distribution after a long enough time

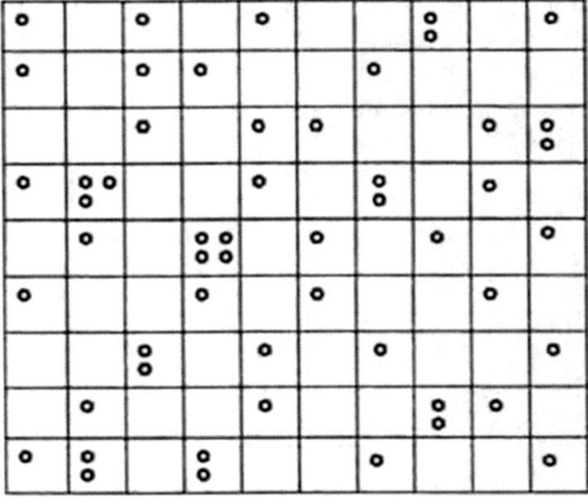

**Fig. 5.3** Einstein model. The energy quanta (white circles) migrate from cell to cell randomly. The histogram of the number of the energy quanta in each cell obeys a Poisson distribution after a suitably long time

be an integer multiple of $h\nu$, ignoring from the zero-point energy $\frac{1}{2}h\nu$; see Fig. 5.3). In the case of a polymer chain in solution, a set of positions of all the segments constructing the polymer is the dynamic variable. In the magnetic spin system, a set of spin variables is the dynamic variable.

Which form of distribution do the dynamic variables take at thermal equilibrium? One of the most important subjects in the Monte Carlo method is to study such a distribution function. For this purpose, it is not important to examine the trajectory of the dynamic variable according to some deterministic equation, but it is important to discuss the value of the dynamic variable at a certain place and a certain time. Then, we observe the system in several regions separately or observe the same region at different times. In the terminology of the Monte Carlo method, these observations correspond to the samplings of the ensemble. The dynamic variables may have different values in different samples even at the same time (or, of course, at different times in the same sample). It is possible to measure the distribution of the dynamic variable via such a sampling at certain time.

Suppose that each cell (or lattice) has some degrees of freedom and that there is a well-defined algorithm for updating the whole system together, such a system is called a "cellular automaton". Such a time evolution does not have a stochastic nature at all, and is therefore essentially different from the Monte Carlo method, which determines the state at the next time step by using random numbers. Therefore, we do not go further into the details of cellular automaton. Recently, there has appeared a new kind of cellular automata which involves randomness in its time evolution. This method can be regarded as a Monte Carlo method with the special constraint that the cells are simultaneously updated at a given time.

In order to study the evolution of dynamic variables, data for their time sequences are desirable. Suppose that we have obtained successive data for the dynamic variables, $\boldsymbol{q}_1, \boldsymbol{q}_2, \ldots, \boldsymbol{q}_L$, by observing the system $L$ times at different time steps, $t_1, t_2, \ldots, t_L$. If these successive data seem to change stochastically with time step, we may call this a "stochastic process". In many cases, in the Monte Carlo methods which are used to the dynamics of a many-body system, the time evolution is considered as a stochastic process. In the Monte Carlo method, one updates the dynamic variables at each time step by using random numbers. In this process, a sample which is obtained in one simulation is composed of a set of $L$ time series of the dynamic variable. The real Monte Carlo simulation repeats the same procedure many times, say $M$ times, and analyzes the resulting $M$ samples in standard ways used in statistical mechanics.

We write the "transition probability" that the system with a dynamic variable $\boldsymbol{q}_i$ at time $t_i$ moves to the state between $\boldsymbol{q}_j$ and $\boldsymbol{q}_j + d\boldsymbol{q}_j$ at a later time $t_j$ as $\Phi(\boldsymbol{q}_i, t_i|\boldsymbol{q}_j, t_j)d\boldsymbol{q}_j$. We call the probability $p(\boldsymbol{q}, t)$ that the system is in the state defined by $\boldsymbol{q}$ at the time $t$ the "distribution function of $\boldsymbol{q}$". Provided that the distribution function at time $t_0$ is given by $p(\boldsymbol{q}, t_0)$, the distribution function $p(\boldsymbol{q}, t)$ at a later time $t$ can be written as

$$p(\boldsymbol{q}, t) = \int p(\boldsymbol{q}_0, t_0)d\boldsymbol{q}_0\Phi(\boldsymbol{q}_0, t_0|\boldsymbol{q}, t). \tag{5.1}$$

Of course, this transition probability should be normalized:

$$\int \Phi(\boldsymbol{q}_i, t_i|\boldsymbol{q}_j, t_j)d\boldsymbol{q}_j = 1, \tag{5.2}$$

for any state $\boldsymbol{q}_i$ and time $t_i < t_j$.

### 5.2.2 *Markov Process*

Most algorithms used in simulating a realistic system by Monte Carlo methods, are based on the following Markov process. In such a stochastic model, if the dynamic variable is given by $\boldsymbol{q}_0$ at time $t_0$, the probability distribution at a later time $t$ does not depend on the trajectory (history) that the dynamic variable has passed through before the time $t_0$. Namely, $\Phi(\boldsymbol{q}_0, t_0|\boldsymbol{q}, t)$ is independent of any information about any time $t'$ before $t_0$. In other words, all the history before the time $t_0$ is contracted into the single piece of information that the system has the dynamic variable $\boldsymbol{q}_0$ at the time $t_0$. As a matter of fact, the transition probability $\Phi(\boldsymbol{q}_0, t_0|\boldsymbol{q}_1, t_1; \boldsymbol{q}, t)d\boldsymbol{q}_1 d\boldsymbol{q}$ that the dynamic variable $\boldsymbol{q}_0$ at time $t_0$ goes through $(\boldsymbol{q}_1, \boldsymbol{q}_1 + d\boldsymbol{q}_1)$ at time $t_1$ and finally arrives at $(\boldsymbol{q}, \boldsymbol{q} + d\boldsymbol{q})$ at time $t$ is given by the product of two probabilities $\Phi(\boldsymbol{q}_0, t_0|\boldsymbol{q}_1, t_1)d\boldsymbol{q}_1$ and $\Phi(\boldsymbol{q}_1, t_1|\boldsymbol{q}, t)d\boldsymbol{q}$:

$$\Phi(\boldsymbol{q}_0, t_0|\boldsymbol{q}_1, t_1; \boldsymbol{q}, t) = \Phi(\boldsymbol{q}_0, t_0|\boldsymbol{q}_1, t_1)\Phi(\boldsymbol{q}_1, t_1|\boldsymbol{q}, t). \tag{5.3}$$

If we integrate (5.3) with respect to the dynamic variable $\boldsymbol{q}_1$ at the intermediate time, we obtain

$$\Phi(\boldsymbol{q}_0, t_0|\boldsymbol{q}, t) = \int \Phi(\boldsymbol{q}_0, t_0|\boldsymbol{q}_1, t_1)d\boldsymbol{q}_1\Phi(\boldsymbol{q}_1, t_1|\boldsymbol{q}, t), \tag{5.4}$$

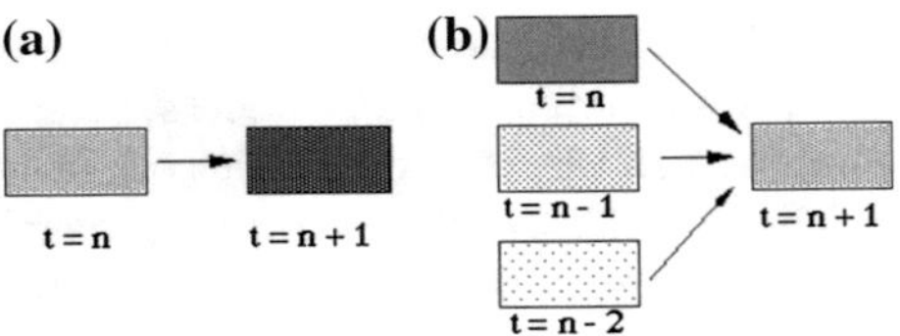


**Fig. 5.4** In the Markov process (**a**), the new state is determined from only information about the present state. In the non-Markov process (**b**), the new state is determined in a way that depends directly on the history of the state in the past

which is called the Chapman–Kolmogorov equation. The stochastic process satisfying (5.3) and (5.4) is called in general a "Markov process" or a "Markov chain". In contrast, a stochastic process depending on the history as shown in Fig. 5.4 is called a "non-Markov process".

For a general dynamic system with many particles and many degrees of freedom, if one can extract just some important degrees of freedom (which we call here dynamic variables) and ignore all the other degrees of freedom, the observed behavior of the system can be described as a simple stochastic process such as a Markov process. That is, by introducing a kind of random updating of the dynamic variable in compensation for neglecting the microscopic details of the real system, one may arrive at a Markov process. To replace the neglected details with a random updating process is identical to the introduction of a heat bath, and this enables us to treat thermal fluctuations. The form and amount of the updating probability are directly related to the temperature of the thermal equilibrium state, as will be described in detail in Sect. 5.3.3.

In order to derive the time-evolution equation of a system obeying a Markov process, we introduce the variables $\Gamma(\boldsymbol{q})$ and $W(\boldsymbol{q}_1; \boldsymbol{q})$ with small values of order $\Delta t$ (such an order is expressed as $O(\Delta t)$, as usual), and expand the transition probability as

$$\Phi(\boldsymbol{q}_1, t|\boldsymbol{q}, t + \Delta t) = [1 - \Gamma(\boldsymbol{q})]\delta(\boldsymbol{q} - \boldsymbol{q}_1) + W(\boldsymbol{q}_1; \boldsymbol{q}) + O(\Delta t^2). \tag{5.5}$$

Using (5.5), we may rewrite the condition (5.4) for the Markov process in the following form:

$$\begin{aligned} &\Phi(\boldsymbol{q}_0, t_0|\boldsymbol{q}, t + \Delta t) - \Phi(\boldsymbol{q}_0, t_0|\boldsymbol{q}, t) \\ = &\int \Phi(\boldsymbol{q}_0, t_0|\boldsymbol{q}_1, t)d\boldsymbol{q}_1 W(\boldsymbol{q}_1; \boldsymbol{q}) - \Phi(\boldsymbol{q}_0, t_0|\boldsymbol{q}, t)\Gamma(\boldsymbol{q}) + O(\Delta t^2). \end{aligned} \tag{5.6}$$

This equation gives a general time-evolution equation for a transition probability obeying a Markov process. As is seen from (5.1) and (5.6), the distribution function $p(\boldsymbol{q}, t)$ of the dynamic variable $\boldsymbol{q}$ also obeys a similar time-evolution equation:

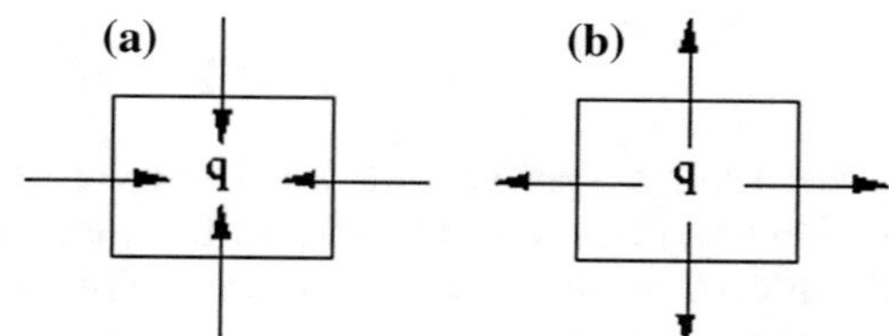

**Fig. 5.5** The change of the distribution function $p(\boldsymbol{q}, t)$ of the state $\boldsymbol{q}$ is equal to the difference of the probabilities of the event (**a**) and the event (**b**). See (5.7) and (5.8)

$$p(\boldsymbol{q}, t + \Delta t) - p(\boldsymbol{q}, t) = \int p(\boldsymbol{q}_1, t) d\boldsymbol{q}_1 W(\boldsymbol{q}_1; \boldsymbol{q}) - p(\boldsymbol{q}, t)\Gamma(\boldsymbol{q}) + O(\Delta t^2). \tag{5.7}$$

When we deal with a physical system which can be described by a Markov process, its basic time-evolution equation (5.7) is called the "master equation". Although this equation could be rewritten in a differential form by taking the continuous limit, here we have left it in the form of a difference equation, which is of course more direct and tractable in Monte Carlo simulations. Here, in order to get $p(\boldsymbol{q}, t)$ normalized, $\Gamma(\boldsymbol{q})$ must be related to $W(\boldsymbol{q}; \boldsymbol{q}_1)$ via the following relation:

$$\Gamma(\boldsymbol{q}) = \int W(\boldsymbol{q}; \boldsymbol{q}_1) d\boldsymbol{q}_1. \tag{5.8}$$

Combining these equations (see also (5.34) later and Fig. 5.5), one finds that the first term of the right-hand side of (5.7) (or (5.34)) describes the process of transitions into a state $\boldsymbol{q}$ (the probability increases), while the second term describes the process of transitions out of this state (the probability decreases). The kernel $W(\boldsymbol{q}_1; \boldsymbol{q})$ is in general an operator, called the stochastic operator.

For example, if we consider a random walk on a $d$-dimensional hypercubic lattice, the kernel $W(\boldsymbol{q}_1; \boldsymbol{q})$, which represents the transition rate, is given by

$$W(\boldsymbol{q}_1; \boldsymbol{q}) = \frac{v}{2d} \sum_{i}^{2d} \delta(\boldsymbol{q} - \boldsymbol{q}_1 + a\boldsymbol{e}_i), \tag{5.9}$$

where $\boldsymbol{e}_i$ denotes one of the $2d$ vectors starting from one lattice site of the $d$-dimensional hypercubic lattice and pointing to an adjacent lattice site; $a$ is the lattice constant and $v$ is the diffusion velocity. Thus the master equation (5.7) becomes a finite-difference version of the diffusion equation:

$$p(\boldsymbol{q}, t + \Delta t) - p(\boldsymbol{q}, t) = v\left[\frac{1}{2d} \sum_{i=1}^{2d} p(\boldsymbol{q} + a\boldsymbol{e}_i, t) - p(\boldsymbol{q}, t)\right] + O(\Delta t^2). \tag{5.10}$$

Now, introducing a "diffusion constant" defined as $D = va^2/2d\Delta t$, and taking the continuous limit $a$, $\Delta t \to 0$ with fixed $D$, we find that (5.10) reduces to the usual diffusion equation,

$$\frac{\partial}{\partial t} p(\boldsymbol{q}, t) = D \nabla^2 p(\boldsymbol{q}, t). \tag{5.11}$$

The diffusion equation (5.11) is a so-called parabolic partial differential equation, with a first derivative with respect to $t$ and a second derivative with respect to the dynamic variable $\boldsymbol{q}$. The more general "Fokker–Planck equation" is also an example of a master equation, and emerges in many physical problems.

In the Einstein model of $N$ harmonic oscillators, we take into account the exchange of the oscillator energy

$$(\varepsilon_1, \ldots, \varepsilon_i, \ldots, \varepsilon_j, \ldots, \varepsilon_N) \to (\varepsilon_1, \ldots, \varepsilon_i' = \varepsilon_i + h\nu, \ldots, \varepsilon_j' = \varepsilon_j - h\nu, \ldots, \varepsilon_N) \tag{5.12}$$

as a result of the weak interaction between oscillators $i$ and $j$. If we assume that this process is a Markov process, the energy distribution function $p(\varepsilon_i, t)$ for the energy $\varepsilon_i$ of the $i$th oscillator ($i$ is arbitrary) has a time dependence characterized by a master equation:

$$\begin{aligned} p(\varepsilon_i, t + \Delta t) - p(\varepsilon_i, t) &= p(\varepsilon_i + h\nu, t) W_- + p(\varepsilon_i - h\nu, t) W_+ \\ &\quad - p(\varepsilon_i, t)(W_- + W_+) + O(\Delta t^2), \end{aligned} \tag{5.13}$$

where $W_-$ represents the probability of removing the energy quantum $h\nu$, while $W_+$ represents the probability of adding $h\nu$. The statistical or physical properties of this system are derived from (5.13). It is possible to add $h\nu$ to the zero energy state, whereas it is not possible to subtract $h\nu$ from the zero-energy state. Therefore, $W_-$ should be greater than $W_+$ for energy conservation. Therefore, we may write $W_- = W_+ \exp \alpha$ with $\alpha > 0$. After a long enough time has passed, the left-hand side of (5.13) approaches 0, and the energy of each oscillator obeys the Poisson distribution $p(\varepsilon_i, t) \sim \exp(-\alpha\varepsilon_i / h\nu)$. It should be checked by the readers that this is the solution of (5.13) with the left-hand side equal to zero. (As will be explained in Sect. 5.3.3, if we write $\alpha = h\nu / k_B T$ according to the principle of detailed balance, we will obtain the Boltzmann distribution $p(\varepsilon_i, t) \sim \exp(-\varepsilon_i / k_B T)$.) A similar thing happens in the kinetic theory of gases: looking at the dynamic variable $\boldsymbol{q}_i = (\mathbf{x}_i, \boldsymbol{p}_i)$ of the $i$th particle ($i$ is arbitrary), we obtain the Boltzmann equation as the master equation:

$$\begin{aligned} \frac{\mathrm{d}}{\mathrm{d}t} p(\mathbf{x}_i, \boldsymbol{p}_i, t) &= \frac{\partial}{\partial t} p(\mathbf{x}_i, \boldsymbol{p}_i, t) + \left( \frac{\boldsymbol{p}_i}{m} \cdot \frac{\partial}{\partial \mathbf{x}_i} - \frac{\partial V}{\partial \mathbf{x}_i} \cdot \frac{\partial}{\partial \boldsymbol{p}_i} \right) p(\mathbf{x}_i, \boldsymbol{p}_i, t) \\ &= \Gamma(p). \end{aligned} \tag{5.14}$$

For a more general dynamic system, if the timescale is coarse-grained as well as the length scale, its stochastic process becomes a Markov process in the simplest case, and the master equation (5.7) is obtained.

### 5.2.3 Ergodicity

As a simple and important case, if $W(\boldsymbol{q}_1;\boldsymbol{q})$ does not depend on the time $t$, and if $W(\boldsymbol{q}_1;\boldsymbol{q})$ is a Hermitian operator, the master equation (5.7) can be solved as an eigenvalue problem.[1] From a simple argument, it is possible to prove that the eigenvalues $\lambda_n$ of $W(\boldsymbol{q}_1;\boldsymbol{q})$ have the following property:

$$\lambda_0 = 0, \qquad \mathrm{Re}\,\lambda_n < 0 \qquad (n \neq 0). \tag{5.15}$$

Then, if $\lambda_0$ is not degenerate, we have

$$p(\boldsymbol{q},t) \rightarrow \phi_0(\boldsymbol{q}) \equiv p^{\mathrm{eq}}(\boldsymbol{q}) \tag{5.16}$$

in the limit $t \rightarrow \infty$, starting from an arbitrary initial distribution. Therefore the distribution function is guaranteed to approach a unique equilibrium distribution. If there is a unique equilibrium state without any periodic motion, this Markov process is called ergodic and this property of approaching a unique final state from an arbitrary initial state is called "ergodicity". Thus we see that the highest eigenvector of the transition matrix $W$ is the equilibrium distribution $p^{\mathrm{eq}}(\boldsymbol{q})$. The second highest eigenvalue $\lambda_1$ gives the relaxation time necessary to reach the equilibrium state through $\tau_{\mathrm{relax}} = -1/\lambda_1$. On the other hand, if $W(\boldsymbol{q}_1;\boldsymbol{q})$ depends on time, some of the eigenvalues $\lambda_n$ may become pure imaginary, except for $\lambda_0 = 0$. If at least one such eigenvalue exists, the distribution in the limit $t \rightarrow \infty$ changes periodically in time and no unique equilibrium state can be determined.

Generally, either in a Markov process or in a non-Markov process, a system obeying ergodicity finally approaches a certain well-defined state. This final state depends on the environmental parameters such as the temperature but does not on the choice of the initial condition, the random numbers or other details of the time-evolution (relaxation) process. The fact that the final goal can be reached from any initial conditions means, considering the process in reverse, that any local configuration is reachable from any starting local configuration, although the probability might be very small.

For example, let us consider a system of many free particles confined in a box. Initially we separate this box into two areas (left and right) by placing a partition at the center (Fig. 5.6a). Now, starting from the initial condition that all the particles are

---

[1] If $W(\boldsymbol{q}_1;\boldsymbol{q})$ is Hermitian and we write its eigenvalues and eigenvectors as $-\lambda_n$ and $\phi_n$, respectively, we have

$$\int \phi_n(\boldsymbol{q}_1)d\boldsymbol{q}_1 W(\boldsymbol{q}_1;\boldsymbol{q}) = -\lambda_n\phi_n(\boldsymbol{q}).$$

Then, the solution of (5.7) can be expanded as

$$p(\boldsymbol{q},t) = \sum_n e^{\lambda_n t}\phi_n(\boldsymbol{q}) \int \phi_n(\boldsymbol{q}')d\boldsymbol{q}' p(\boldsymbol{q}',0)$$

owing to the completeness of the eigenfunctions.

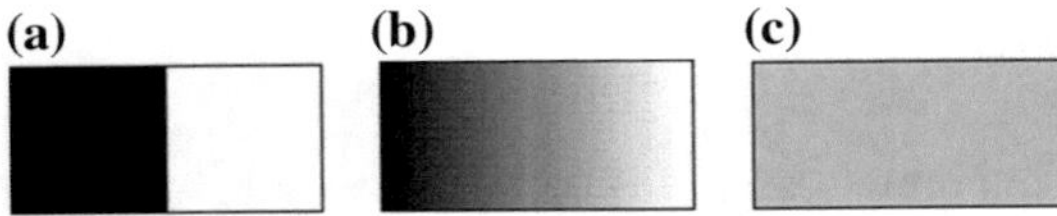


**Fig. 5.6** An example of ergodicity. The gas (**a**) confined to the left-hand side of a partition diffuses to the right-hand side as in (**b**) after the removal of the partition, and finally approaches the uniform distribution (**c**)

confined to the left-hand area, we remove the partition from the box. As the particles fly freely to all areas inside the box, they gradually diffuse into the right-hand part (Fig. 5.6b), and finally distribute themselves homogeneously throughout the whole box (Fig. 5.6c). This final state is the same if we start from the other initial state, where all the particles are confined to the right-hand area. It is clear that the final distribution of the particles in equilibrium is independent of the initial condition of the system.

In many thermodynamic systems, the final state after enough time has past is the thermal-equilibrium state. According to statistical mechanics [5] a system at thermal equilibrium obeys the Boltzmann distribution (also called the "canonical distribution")

$$p^{\mathrm{eq}}(\boldsymbol{q}) = \frac{1}{Z}\exp[-E(\boldsymbol{q})/k_{\mathrm{B}}T], \tag{5.17}$$

where $E(\boldsymbol{q})$ denotes the total energy of the whole system having the dynamic variable $\boldsymbol{q}$, and $Z$ is the partition function, given by

$$Z = \sum_{\boldsymbol{q}} \exp[-E(\boldsymbol{q})/k_{\mathrm{B}}T]. \tag{5.18}$$

Thermal equilibrium corresponds to the state with the lowest free energy in a given environment. If the partition function is calculated, the free energy $F$ and internal energy $E$ of the system can be obtained from

$$F = -k_{\mathrm{B}}T\log Z, \qquad E = -k_{\mathrm{B}}\frac{\partial}{\partial(1/T)}\log Z. \tag{5.19}$$

Then, from these values the specific heat, for example, is evaluated via $C = \partial E/\partial T$. However, in a Monte Carlo simulation, it is more convenient to evaluate the internal energy, for example, by taking the direct average

$$E = \sum_{\boldsymbol{q}} E(\boldsymbol{q})p(\boldsymbol{q}). \tag{5.20}$$

Many problems in statistical mechanics are associated with thermal equilibrium itself or the response of the system to a small deviation from thermal equilibrium.

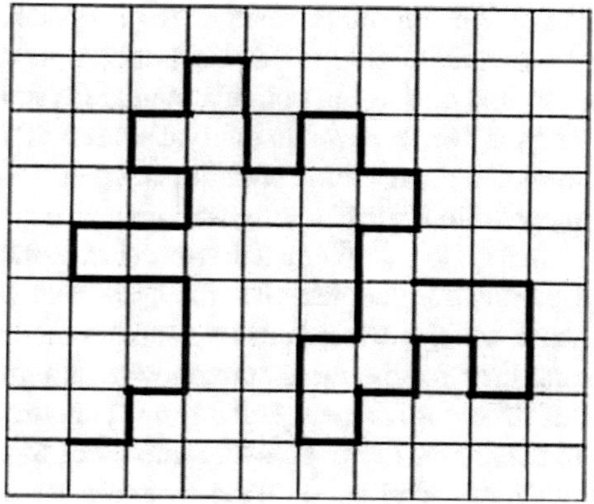

**Fig. 5.7** A self-avoiding walk, which represents a polymer chain in a good solvent. The segments constructing the chain are connected so as not to cross the other segments. The right-hand end is trapped in a blind alley

A glassy state which is obtained, for example, by quenching the system is not the true thermal-equilibrium state corresponding the minimum free energy, but a metastable state corresponding to a local minimum of the free energy. Before reaching the true thermal equilibrium, each degree of freedom freezes into a glassy state. Such freezing is not likely to occur in a system in which all the particles always interact with each other isotropically. But if there is an anisotropy in the interactions between molecules because of, for example, their special shape or their multipole moments, such freezing occurs more frequently. Even in finite-size systems, there may be a valley (or pocket) in phase space from which a particle cannot escape, once it is trapped. Examples are the motion of a particle that cannot go over an energy barrier at low temperatures, and a self-avoiding walk (a random walk with the condition that the trajectory cannot cross itself, which is often used to model a polymer chain in solution) when the walker is surrounded by preexisting trajectories (i.e. when it arrives at a blind alley) and cannot go further (Fig. 5.7). In real materials, even if such a freezing occurs, the movements of the surrounding particles often help the frozen particle to escape. Near the glass transition point, such a process takes a long time. This time is called the "ergodic time". Even if the system actually has ergodicity, the ergodicity seems to be broken when we observe the system over a time interval shorter than the ergodic time [3]. The ergodic time increases even in a finite system, when the system approaches the glassy state. Also, for general systems, if the number of particles $N$ goes to infinity, the ergodic time, in general, goes to infinity as well.

In thermal equilibrium, the physical quantities averaged over a very long time, comparable to the ergodic time, must be equivalent to the physical quantities averaged over samples ("ensemble average" ) obtained by starting from various initial states:

$$\bar{A} = \frac{1}{t_1 - t_0} \int_{t_0}^{t_1} A(\boldsymbol{q}, t)dt = \frac{1}{M} \sum_{l=1}^{M} A(\boldsymbol{q}_l, t). \tag{5.21}$$

This is an important relation for a system with ergodicity. In these averages, the glassy state might provide only negligible contributions compared to the crystalline state, and one would obtain a regular crystal from the average. However, if one takes the time average over an interval shorter than the ergodic time, the glassy state gives a dominant contribution which corresponds to the ensemble average of samples having many "blind-alley" configurations.

In a system showing a first-order transition, after the system is cooled down below the transition temperature, the system stays for some time in the high-temperature phase which corresponds to a supercooled state, and it sometimes requires rather, a long time for the more stable low-temperature phase to extend all over the system; this time, too, is the ergodic time. This time delay is mainly due to the existence of an energy barrier to be overcome (Sect. 5.4.6).

On the other hand, in the case of a second order phase transition, a spontaneous breaking of the symmetry of a dynamic variable takes place below a certain temperature (the transition temperature), and a clear ordering gradually appears from a high-symmetry disordered phase. In a ferromagnet, this is the magnetization; in a superconductor, this ordering is the persistent current. The spontaneous symmetry breaking should be regarded as a phenomenon occurring only in the thermodynamic limit of $N \to \infty$ (infinite number of particles) and $t \to \infty$ (infinite time). In an infinite system, the order parameter, which is the average of some dynamic variable and should be zero in the disordered phase, becomes nonzero only below the critical temperature. In some sense, the symmetry breaking can be regarded as an ergodicity-breaking. For example, in the ferromagnetic state, the direction of the total magnetic moment cannot change from upward to downward; or in the superconducting state, the direction of the persistent current cannot change from rightward to leftward. Since a second-order phase transition can occur in the limit of an infinite number of particles, the ergodic time becomes infinite only in the limit of $N \to \infty$.

In finite systems, the order parameter obtained by a short-time average is not always zero even above the critical temperature, because of the existence of thermal fluctuations. It becomes zero only when one takes an average over a long time comparable to the ergodic time. The ergodic time becomes infinite as $N \to \infty$. Similarly, even below the transition temperature, the order parameter in a finite system changes its sign (or rotates) from time to time, and hence the order parameter becomes zero in a rigorous sense. The approximate period for the change of the order parameter is a measure of the ergodic time. This ergodic time also goes to infinity as $N \to \infty$. If we interpret the above sentence literally, it means that we cannot treat second order phase transitions as long as we use finite systems. The usual strategy to prevent this problem is to introduce the average of the squared order parameter. The effect of finite size on a second-order phase transition has been discussed by using several scaling relations among the critical exponents [3].

## 5.3 Algorithms for Monte Carlo Simulation

In the Monte Carlo method, one performs trials of a stochastic process on the basis of a stochastic model. Except for special cases, one sample is insufficient, and it is necessary to provide many samples and estimate their average. One of the important reasons why Monte Carlo methods are used in many problems is that we can evaluate the correct average with rather a small amount of effort compared to directly counting all possible configurations of the system, even if we have to handle quite many samples.

### *5.3.1 Random Numbers*

In a Monte Carlo simulation, one treats a stochastic process by generating random numbers on a computer. The most basic algorithm of the Monte Carlo method is very simple as follows:

(1) Generate a random number.
(2) Take or do not take a new random step, depending on the generated random number.
(3) Repeat trial.

This algorithm offers a basis for solving more the complex problems which will appear later. Here, we consider uniform random numbers in the closed interval [0, 1]. Even if we confine ourselves to these random numbers, there are many ways to generate them on a computer; from pure software algorithms to hardware algorithms, e.g. using the clock on the computer. These random numbers called "pseudo-random numbers" and are not mathematically ideal random numbers. Pseudo-random numbers are, as their name implies, an imitation of ideal, uniform random numbers; the numbers may show a very high correlation with a long but finite period, e.g. of the order of $10^{10}$–$10^{20}$. That is, if one continues to generate the pseudorandom numbers more times than this period, their random character becomes worse. The longer the period, the better the random quality of the numbers. Moreover, within this period, the correlation between the values of the pseudorandom numbers should be as small as possible. Since the pseudorandom numbers are not completely uniform, it becomes impossible to get a good result when one has to generate random numbers many times in Monte Carlo simulations. In other words, in any calculation based on a Monte Carlo method, the quality of the random numbers becomes a key factor in getting good results.

As the simplest algorithm which generates random numbers purely by software, the "Lehmer method" is famous and popular. This method provides first an integer random number in the interval $[0, M)$ by the recursive equation $X_{k+1} = aX_k + c$ (mod $M$) and then gives a uniform random number in [0, 1] by the equation $x_k = X_k/M$. In particular, the method with $c = 0$ is called the "multiplicative congruential method".

The method with $c \neq 0$ is generally called the "mixed congruential method". In the multiplicative congruential method, $M$ is empirically chosen to be a prime number and $a$ satisfies $a^M = 1 \pmod M$ and $a^n \neq 1$ for arbitrary integers $n$ in the range $0 < n < M$. In this case, the random-number sequence repeats after $M - 1$ steps. That is, the periodicity of the random-number generator is $M - 1$. In the usual 32-bit integer used in computers, the uppermost bit represents the sign, while the lower 31 bits represent positive integer numbers. We may use $M = 2^{31} - 1 = 2147483647$ as the greatest number available on the computer. For the mixed congruential method, there is no such rule, so that one usually adopts the maximum integer, $M = 2^{31} = 2147483648$. The way of choosing $a$ and $c$ to get good random numbers has been studied. For example, according to D.E. Knuth, $a$ satisfying $(a \bmod 8) = 5$, $M$ satisfying $M/100 < a < M - \sqrt{M}$ and having no special pattern in its binary-number representation, and $c$ being an odd number satisfying $c/M \sim 1/2 - \sqrt{3}/6 \sim 0.21132$ give good random numbers.

These methods are widely used because of their simplicity. However, there is a problem due to the correlation in random numbers generated in these ways. It is known that there appears a significant correlation (or an unexpected regular pattern) in the multidimensional distribution if we use the Lehmer method or its slight modification which introduces shuffling, in which the order of the random-number sequence is randomly changed. Therefore, in order to perform a large-scale simulation using many random numbers, it is better to use another method which is not based on the Lehmer method and which uses logical algebra. Such a method is called a "non-Lehmer method". However, if we write down the algorithm of a non-Lehmer method in the FORTRAN language, the generation of each random number takes a longer time. In particular, if there is an "IF" statement, it prevents vectorization on a supercomputer and is not suitable in a high-speed Monte Carlo simulation. The Tausworthe method, which we will show below, is in principle vectorizable and simple enough since it uses only one "exclusive OR" (`IEOR`) among integers; see the program listing given in Fig. 5.8. It has been confirmed by Knuth [6] that distributions generated by these random numbers behave well in multidimensional space, and that their autocorrelation function coincides well with that of ideal, uniform random numbers. The following parameters are based on the hardware architecture of the 32-bit integer in which the uppermost bit represents the sign part and the lower 31 bits represent the natural-integer part. Therefore, if one wants to use a computer with a different integer architecture, one should change the parameter `ISIZE` from 31 to the number of bits for the natural-integer part for that machine. The subroutine is composed of the random generator `RAND` and the initialization part `INIR`, and it is necessary to call `INIR` in the main program before calling `RAND`. The values of the parameters `P`, `Q` and `IDELAY` used in this example can be changed so long as they are not too different from the original value.

In the random generator `RAND`, there is a DO-loop to generate the (input) desired number of random numbers simultaneously (this number should not be greater than `MAXIMUM`). Since this DO-loop has no "IF" statement inside it, this DO-loop is vectorizable, although the vector length has to be smaller than P.

```
C*********** RANDOM NUMBER GENERATOR *****************
      REAL X(100)
      INTEGER P,Q,M(100)
      COMMON /RPARAM/ M,P,Q,ISIZE,RMAX,IDELAY
      DATA P,Q,ISIZE,IDELAY/98,27,31,9800/
      RMAX=FLOAT((2**(ISIZE-1)-1)*2+1)
      READ(5,*) NUM
      IF(NUM.LE.0.OR.NUM.GE.P) GOTO 9
      CALL INIR()
      CALL RAND(NUM,X)
      DO 1 I=1,NUM
    1  WRITE(6,2) X(I)
    2  FORMAT(4X,' RANDOM NUMBER = ',E15.8)
    9  STOP
      END
C
      SUBROUTINE RAND(NUM,X)
      REAL X(100)
      INTEGER P,Q,M(100),W(100)
      COMMON /RPARAM/ M,P,Q,ISIZE,RMAX,IDELAY
      COMMON /RJ/ J0
      DO 1 I=1,NUM
      K=MOD(I+J0+Q-1,P)+1
    1  W(I)=M(K)
      DO 2 I=1,NUM
      J=MOD(I+J0-1,P)+1
      M(J)=IEOR(M(J),W(I))
    2  X(I)=FLOAT(M(J))/RMAX
      J0 = J0 + NUM
      RETURN
      END
C
      SUBROUTINE INIR()
      INTEGER P,Q,M(100)
      COMMON /RPARAM/ M,P,Q,ISIZE,RMAX,IDELAY
      COMMON /RJ/ J0
      J0=0
      L=2**(ISIZE-1)
      DO 1 I=1,P
    1  M(I)=L
      DO 4 K=1,ISIZE
      DO 2 I=1,IDELAY
      J=MOD(J,P)+1
      K=MOD(J+Q-1,P)+1
    2  M(J)=IEOR(M(J),M(K))
      NSTEP=0
      DO 3 I=1,P
      LTEMP=L/2**(K-1)
      LTEMP=(M(I)-M(I)/L*L)/LTEMP
      IF(LTEMP.EQ.1) NSTEP=NSTEP+1
      IF(K.EQ.ISIZE) GOTO 3
      M(I)=M(I)/2+L
    3  CONTINUE
    4  CONTINUE
      DO 6 I=1,5000*P
      J=MOD(J,P)+1
      K=MOD(J+Q-1,P)+1
    5  M(J)=IEOR(M(J),M(K))
    6  CONTINUE
      RETURN
      END
```

**Fig. 5.8** Random-number generator by the Tausworthe method

### 5.3.2 Simple Sampling Technique

As the most simple example of a Monte Carlo calculation, one can consider a mathematical "numerical integration". For example, in the case of one dimension, let us consider

$$\int_0^1 x dx = \frac{1}{2}. \tag{5.22}$$

Corresponding to the $x$ axis, we generate a sequence of random numbers in the interval [0, 1] and add all these values. If we repeat this process $M$ times with ideal, uniform random numbers, the sum of all these random numbers divided by $M$, i.e. the average value of the random numbers, would clearly approach the desired result of this integral, $\frac{1}{2}$, in the limit $M \to \infty$. However, in practice, since we use a finite number of pseudorandom numbers, the result is some value close to $\frac{1}{2}$, but not exactly $\frac{1}{2}$.

As a simple example in two dimensions, one can consider the problem of obtaining the area of a quarter-circle (Fig. 5.9). Successively generating random numbers in the interval [0, 1] corresponding to the $x$ and $y$ axes, we plot the position $(x, y)$ in the first quadrant of the $xy$ plane. For each point, we calculate $r^2 = x^2 + y^2$. If $r^2 \leq 1$, we increment the counter by unity. By plotting many points in this way, we may estimate the area by dividing the counter value by the total number of random numbers $(x, y)$ generated.

In the case of numerical integration, by simply repeating the trials using ideal random numbers, the result will get closer and closer to the exact value (or behavior), in principle. Namely, if we regard each point as a sample, the result approaches the correct expectation value when one increases the sampling number $M$, in principle. Owing to the central-limit theorem, the average of the $M$ random numbers in the interval [0, 1] approaches a Gaussian distribution with mean $\frac{1}{2}$ and variance $1/\sqrt{M}$, as $M$ increases. Therefore, in the first example of the one-dimensional integration,

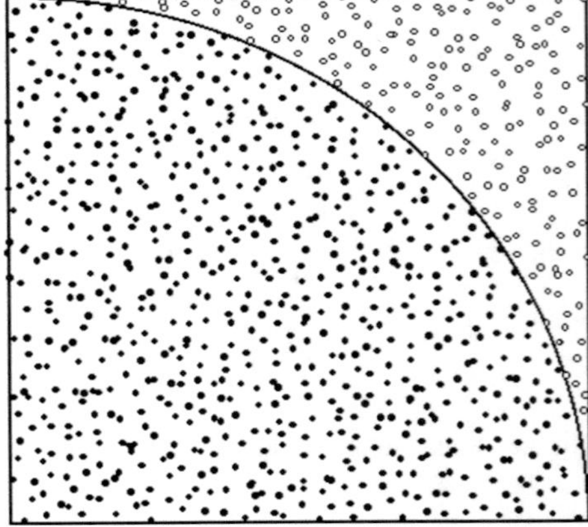

**Fig. 5.9** Evaluation of the circular constant $\pi$ by means of a Monte Carlo integral

the result approaches the correct value if the sample number $M$ increases, and its error decreases like $1/\sqrt{M}$.

As a different kind of numerical integration, the evaluation of an expectation value of a physical quantity $A(\boldsymbol{q})$ with respect to a dynamic variable $\boldsymbol{q}$, which obeys a distribution function $p(\boldsymbol{q})$, is an important theme in the field of Monte Carlo methods. Here we need to evaluate a multiple integral of the form

$$\bar{A} = \int d\boldsymbol{q} A(\boldsymbol{q}) p(\boldsymbol{q}). \tag{5.23}$$

To evaluate this integral, we may use a method similar to the previous examples; i.e. we choose one point $\boldsymbol{q}_l$ of the multidimensional variable $\boldsymbol{q}$ and calculate $p(\boldsymbol{q}_l)$ and $A(\boldsymbol{q}_l)p(\boldsymbol{q}_l)$. Repeating this procedure $M$ times, we may evaluate its average value from the following formula:

$$\bar{A} \simeq \frac{\sum_{l=1}^{M} A(\boldsymbol{q}_l) p(\boldsymbol{q}_l)}{\sum_{l=1}^{M} p(\boldsymbol{q}_l)}. \tag{5.24}$$

In the three examples described above, the Monte Carlo method is introduced to extract samples of the system in a completely random way; this method actually offers a well-defined stochastic process. Such a method of Monte Carlo sampling is called a "simple sampling technique". An example which frequently appears in statistical physics is the "random walk" (RW) which realizes Brownian motion on a lattice. Similarly to the RW, a slightly different model, which is commonly used to model a polymer chain in a good solvent, is the "self-avoiding walk" (SAW). For example, in the case of an RW on a two-dimensional square lattice, starting from one lattice point, the random walker moves to one of the four nearest-neighbor points (up, down, left, right) with equal probability. That is, generating one random number in the interval [0, 1), we select the movement direction among the four (up, down, left, right) possibilities, according to whether the random number is inside the region $[0, \frac{1}{4})$, $[\frac{1}{4}, \frac{1}{2})$, $[\frac{1}{2}, \frac{3}{4})$ or $[\frac{3}{4}, 1)$.

A thermodynamic ensemble based on a stochastic model could, in principle, be handled by the simple sampling technique. However, this does not give a practical method, for the following reason. Let $l$ be an index indicating the different samples, and $\boldsymbol{q}_l$ a set of dynamic variables corresponding to the $l$th sample. If we write $E(\boldsymbol{q}_l)$ and $A(\boldsymbol{q}_l)$ as the energy and a physical quantity of the state $\boldsymbol{q}_l$ of the sample $l$, the expectation value of the physical quantity $A$ is given by

$$\bar{A} = \frac{\sum_{l=1}^{M} \exp[-E(\boldsymbol{q}_l)/k_{\mathrm{B}}T] A(\boldsymbol{q}_l)}{\sum_{l=1}^{M} \exp[-E(\boldsymbol{q}_l)/k_{\mathrm{B}}T]}. \tag{5.25}$$

Here, we notice that all these samples are selected completely randomly, irrespective of the Boltzmann weight.

Now suppose that, in (5.25), the physical quantity $A(\boldsymbol{q}_l)$ is the energy $E(\boldsymbol{q}_l)$ of the system itself. Generally, the relative width of the energy fluctuation decreases with the number of particles, $N$, in the system:

$$\frac{\langle E^2 \rangle - \langle E \rangle^2}{\langle E \rangle^2} \sim \frac{1}{N}. \tag{5.26}$$

Thus the one-particle energy distribution function $p(\epsilon)$ shows a peak of height $\sqrt{N}$ and width $1/\sqrt{N}$ around its expectation value, $\langle E \rangle / N$. Far from the second-order phase transition point, where fluctuations become quite important, the energy distribution function $p(\boldsymbol{e})$ has generally a Gaussian form for large $N$:

$$p(\boldsymbol{e}) \sim \exp\left( -\frac{(\boldsymbol{e} - \langle E \rangle / N)^2}{2Ck_{\mathrm{B}}T^2} N \right), \tag{5.27}$$

where $C$ is the specific heat per particle. For this reason, the probability of generating the states with a one-particle energy $\boldsymbol{e}$ close to $\langle E \rangle / N$ by the simple sampling technique becomes exponentially small as the number of particles $N$ increases, and the simple sampling algorithm becomes very inefficient since much time will be spent in generating states with infinitesimally small probability.

### 5.3.3 Importance Sampling Technique

As is readily understood from the example of the numerical integration (5.23) for evaluating the average of a physical quantity $A(\boldsymbol{q})$, discussed above, the simple sampling technique offers an efficient algorithm only if the distribution function $p(\boldsymbol{q})$ resembles a more or less uniform distribution. In contrast, if the distribution function has a big value only at an isolated point, it becomes more efficient to choose the sample integration points with the same probability as that given by the distribution function $p(\boldsymbol{q})$. Such a biased sampling with a probabilistic weight is called "importance sampling".

If the sample integration points are chosen with the same probability as the distribution function $p(\boldsymbol{q})$, the expression for the expectation value (5.24) can be replaced simply by

$$\bar{A} = \frac{1}{M} \sum_{l=1}^{M} A(\boldsymbol{q}_l). \tag{5.28}$$

As a guiding principle for performing this importance sampling, Metropolis et al. [7] proposed an efficient algorithm. They considered a Markov process that generates a descendant sample $l'$ from the present sample $l$ according to a transition rate $W(\boldsymbol{q}_l; \boldsymbol{q}_{l'})$. They showed, by suitably choosing the transition rate $W(\boldsymbol{q}_l; \boldsymbol{q}_{l'})$, that it is possible to bring the distribution of the dynamic variable close to the expected (prescribed) distribution $p(\boldsymbol{q}_l)$ in the limit $M \to \infty$. This Markov process is based on a random walk in the space of the dynamic variable $\boldsymbol{q}$, called also a Markov chain. We consider one particle in the multidimensional space of the dynamic variable $\boldsymbol{q}$ and move it step by step according to random numbers. Here, the movement should obey the following rule. Calculate the value of the distribution function $p(\boldsymbol{q}')$ at the new position $\boldsymbol{q}'$ and compare it to the present value of the distribution function $p(\boldsymbol{q})$ at the position $\boldsymbol{q}$. If $p(\boldsymbol{q}') \geq p(\boldsymbol{q})$, then we move the particle to the new position. If not, i.e. if $p(\boldsymbol{q}') < p(\boldsymbol{q})$, generate one random number in the interval [0, 1]. If this random number is smaller than $x = p(\boldsymbol{q}')/p(\boldsymbol{q})$, we adopt the new position, and, if not, we leave the particle at the same position. That is, we update the particle position with the probability min[$x$,1]. We repeat this process successively. One can prove that, after repeating this process enough times, the distribution of this random walker approaches $p(\boldsymbol{q})$.

Now we turn our attention to the specific problem of thermodynamic ensembles which was posed in the last part of the previous section. We assume that the samples are chosen with the same probability as the distribution function $p(\boldsymbol{q}_l)$. Then, (5.25) for the expectation value becomes

$$\bar{A} = \frac{\displaystyle\sum_{l=1}^{M} \exp[-E(\boldsymbol{q}_l)/k_{\mathrm{B}}T] A(\boldsymbol{q}_l)/p(\boldsymbol{q}_l)}{\displaystyle\sum_{l=1}^{M} \exp[-E(\boldsymbol{q}_l)/k_{\mathrm{B}}T]/p(\boldsymbol{q}_l)}. \tag{5.29}$$

The simplest and most natural choice for the distribution function $p(\boldsymbol{q}_l)$ in (5.29) is the Boltzmann distribution, $p(\boldsymbol{q}_l) \sim \exp[-E(\boldsymbol{q}_l)/k_{\mathrm{B}}T]$. In this case, the samples themselves represent the Boltzmann distribution (or the "canonical distribution") and behave as if they are at the true thermal equilibrium. This is what we call "importance sampling", and because of the cancellation of the Boltzmann weight, (5.29) reduces to the simple arithmetic average (5.28). Using Metropolis's idea of importance sampling, one may bring the distribution function $p(\boldsymbol{q}_l)$ of the samples generated by the Markov process close to the thermal-equilibrium distribution

$$p^{\mathrm{eq}}(\boldsymbol{q}_l) = \frac{1}{Z} \exp\left(-\frac{E(\boldsymbol{q}_l)}{k_{\mathrm{B}}T}\right) \tag{5.30}$$

in the limit $M \to \infty$. A sufficient condition to achieve the equilibrium distribution is to impose the "principle of detailed balance"

$$p^{\mathrm{eq}}(\boldsymbol{q}_l) W(\boldsymbol{q}_l; \boldsymbol{q}_{l'}) = p^{\mathrm{eq}}(\boldsymbol{q}_{l'}) W(\boldsymbol{q}_{l'}; \boldsymbol{q}_l). \tag{5.31}$$

However, the principle of detailed balance does not uniquely determine the transition rate $W(\boldsymbol{q}_l, \boldsymbol{q}_{l'})$ but leaves an ambiguity in the choice of $W$. The two often used choices are given by

$$\begin{aligned} W(\boldsymbol{q}_l; \boldsymbol{q}_{l'}) &= \frac{1}{\tau}\left[1 - \tanh\left(\frac{\delta E}{2k_{\mathrm{B}}T}\right)\right] \\ &= \frac{1}{\tau}\frac{\exp(-\delta E/k_{\mathrm{B}}T)}{[1+\exp(-\delta E/k_{\mathrm{B}}T)]} \end{aligned} \tag{5.32}$$

and

$$W(\boldsymbol{q}_l; \boldsymbol{q}_{l'}) = \begin{cases} \frac{1}{\tau}\exp(-\delta E/k_{\mathrm{B}}T) & \delta E > 0, \\ \frac{1}{\tau} & \text{otherwise,} \end{cases} \tag{5.33}$$

where $\delta E = E(\boldsymbol{q}_{l'}) - E(\boldsymbol{q}_l)$. The method using (5.32) is called the "Glauber algorithm" [8] or the "heat bath method", while the method using (5.33) is called the "Metropolis algorithm" (Fig. 5.10). It is readily seen that these two choices certainly satisfy the detailed-balance which is required for importance sampling. A Monte Carlo simulation using this Metropolis algorithm is often referred to as a Metropolis Monte Carlo simulation.

By using random numbers, produce a new state from the present state.
↓
Calculate the energy difference $\delta E$ of the new state relative to the present state.
↓
If $\delta E \le 0$, then update to the new state.
If $\delta E > 0$, then generate a random number $\eta \in [0,1]$, and
if $\exp(-\delta E/k_{\mathrm{B}}T) > \eta$
update to the new state
else
leave the state as it is.

**Fig. 5.10** The Metropolis algorithm. This is a standard, important sampling technique to realize the canonical ensemble

### 5.3.4 General Comments on Dynamic Models

In some cases, the dynamic variables can be changed in a stochastic way except from some constraints, e.g. the connectivity of the polymer chain or the fixed magnitude of each spin. In such a case, we can introduce a stochastic model. We cannot say that it is always possible to derive a stochastic model since it depends on how we choose the dynamic variables. But, in many cases, by coarse-graining both the spatial and the temporal scales, we introduce stochastic models. By this coarse-graining process the energy quantum, for example, becomes a continuous variable in Einstein's model for lattice vibrations, and a quantum spin variable at finite temperatures becomes a classical spin variable. In these models, the original microscopic variables are replaced by continuous stochastic variables.

The important fact in this coarse-graining procedure is that the detailed degrees of freedom and interactions of each particle can be treated continuously in both space and time, and that, except for the more important variables (dynamic variables) $\boldsymbol{q}$, all other degrees of freedom and details of the system can be treated as a "thermal fluctuation" or a "random force" acting on the dynamic variables. This procedure may also be possible at the absolute zero of temperature. In this case, the random force acting stochastically on the dynamic variables is not a thermal fluctuation but a "quantum fluctuation".

As we saw in the previous section, if the updating of the dynamic variable is described as a Markov process, there is a clear correspondence with a stochastic process. Writing the distribution function for the dynamic variable $\boldsymbol{q}$ at a time $t$ as $p(\boldsymbol{q}, t)$, this function satisfies the following master equation for the Markov process (see (5.7) and (5.8)):

$$p(\boldsymbol{q}, t + \Delta t) - p(\boldsymbol{q}, t) = \sum_{\boldsymbol{q}_1} p(\boldsymbol{q}_1, t) W(\boldsymbol{q}_1; \boldsymbol{q}) - \sum_{\boldsymbol{q}_1} p(\boldsymbol{q}, t) W(\boldsymbol{q}; \boldsymbol{q}_1). \quad (5.34)$$

This is the expression describing the balance in (5.31) by means of the "rate equation". The first term on the right-hand side represents the incoming process to the state $\boldsymbol{q}$ (the existence probability of $\boldsymbol{q}$ increases), while the second term represents the outgoing process from the state $\boldsymbol{q}$ (the probability decreases); see Fig. 5.5. In the thermal equilibrium state, the principle of detailed balance (5.31) guarantees the cancellation of these two terms, and therefore for $p(\boldsymbol{q}, t) = p^{\text{eq}}(\boldsymbol{q}, t)$ we have $\partial p(\boldsymbol{q}, t)/\partial t = 0$. In fact, $p^{\text{eq}}(\boldsymbol{q}, t)$ is a steady-state solution of the above master equation.

In the "thermodynamic limit" $N \to \infty$, $t \to \infty$, regardless of the choice of updating algorithms, all ensembles lead to the same result, if they are in thermal equilibrium. In other words, if one is interested only in the static properties in thermal equilibrium, one can adopt any updating algorithm, e.g. one which is artificially devised for fast computation. This argument is true in the thermodynamic limit. However, because of the finite-size effect, different behaviors may happen even for $t \to \infty$ with different choices of the updating algorithm. Also, irrespective of the size of the system, the relaxation process of the approach to thermal equilibrium usually depends strongly on the updating algorithm for the dynamic variables. For this reason, it is very important to select the appropriate updating algorithm and statistical ensemble before performing the Monte Carlo simulation.

If there is no restriction (or conservation law) on the dynamic variables, we may use the steepest-descent method, which moves the dynamic variables toward the state characterized by the minimum of the free energy, to derive a time-evolution equation for the dynamic variables. That is, writing $\mathcal{F}$ as the free energy of the system, we obtain the "Langevin equation"

$$\frac{\partial \boldsymbol{q}}{\partial t} = -\Gamma \frac{\partial \mathcal{F}}{\partial \boldsymbol{q}} + \eta(\boldsymbol{q}, t). \tag{5.35}$$

Here, $\eta(\boldsymbol{q}, t)$ is a "random force" simulating thermal and/or quantum fluctuations. For some simple systems one can derive the master equation from the Langevin equation by using the "fluctuation–dissipation theorem", although the details are not presented here. In general, if there is no conservation law, one may update each of the dynamic variables independently.

On the other hand, for systems with some conservation law concerning the dynamic variables, the dynamic variables are not allowed to change their values independently. In such a case, we may often impose a rule on very local dynamic variables when one updates the dynamic variables. One example is an exchange rule among the dynamic variables at the adjacent positions. Such an exchange rule holds for a system in which the total value of the dynamic variables is conserved. That is, we may impose an exchange rule for updating a local quantity so as to preserve the total value. Such a local exchange rule is called "Kawasaki's (exchange) algorithm", and the dynamics described in this algorithm is called "Kawasaki dynamics".

If the free energy functional is given for a system with a conservation law on the total value of the dynamic variables, the Langevin equation is given by

$$\frac{\partial \boldsymbol{q}}{\partial t} = -\Gamma \nabla^2 \frac{\partial \mathcal{F}}{\partial \boldsymbol{q}} + \eta(\boldsymbol{q}, t) \tag{5.36}$$

in place of (5.35). Here $\nabla^2$ is the Laplacian with respect to the coordinate. The dynamics obeying (5.36) is also called "Kawasaki dynamics" [9].

In order to discuss the updating algorithm precisely, it is necessary to recognize also the physical background of the relaxation process. For example, in order to determine which relaxation process is dominant among processes such as lattice vibrations, thermal diffusion in a solid or a spin wave in a spin system, etc., it is necessary to compare the relaxation times. If the lattice relaxation time is much shorter than the spin relaxation time, we can treat the pure spin system at a fixed temperature, but in the opposite case, it becomes important to consider the coupling between the lattice vibrations and the spin degrees of freedom. In a model containing a coupling with the lattice vibrations, we treat the lattice vibration by the usual diffusion equation, add a term representing the coupling with this mode to the free energy and construct a Langevin equation such as (5.35) or (5.36). Such models have been studied in detail by Halperin and Hohenberg [10] using a continuous spin model called the "$\phi^4$ model".

## 5.4 Applications

Examples of the application of Monte Carlo simulations in the field of condensed-matter physics and materials science are quite widespread and varied, and it is impossible to introduce all of them in this small chapter. Here we separate the subjects into classical particles, percolation, polymers, classical spins, nucleation, crystal growth and fractals, and describe them briefly, including recent topics.

### 5.4.1 Systems of Classical Particles

A wide variety of many-atom systems, including disordered materials and fluids [11–14], defects [15, 16], molecules [17–19] and surfaces [20] can be simulated by Monte Carlo methods as well as molecular-dynamics methods. Consider a Monte Carlo simulation of a system of $N$ classical interacting particles obeying the canonical distribution. The Hamiltonian of this system is given by

$$\mathcal{H} = \sum_{i=1}^{N} \frac{\boldsymbol{p}_i^2}{2m} + U(\boldsymbol{r}_1, \boldsymbol{r}_2, \ldots, \boldsymbol{r}_N). \tag{5.37}$$

In general, denoting Planck's constant by $h$, we have

$$Z = \frac{1}{h^{3N} N!} \int \mathrm{e}^{-\mathcal{H}/k_\mathrm{B}T} \prod_{i=1}^{N} d\boldsymbol{r}_i d\boldsymbol{p}_i = \left(\frac{2\pi m k_\mathrm{B} T}{h^2}\right)^{3N/2} Q(T, V), \tag{5.38}$$

where

$$Q(T, V) = \frac{1}{N!} \int \mathrm{e}^{-U(\boldsymbol{r}_1,\ldots,\boldsymbol{r}_N)/k_\mathrm{B}T} \prod_{i=1}^{N} d\boldsymbol{r}_i. \tag{5.39}$$

as the partition function of the classical system. Thus the distribution function for the coordinates of the $N$ particles is given by

$$p(\boldsymbol{r}_1, \ldots, \boldsymbol{r}_N) = \frac{1}{N! Q(T, V)} \mathrm{e}^{-U(\boldsymbol{r}_1,\ldots,\boldsymbol{r}_N)/k_\mathrm{B}T}. \tag{5.40}$$

Construction of the distribution function $p^{\text{eq}}$ at thermal equilibrium which obeys (5.40) can be performed by introducing importance sampling (see (5.28)–(5.33)) and performing a Monte Carlo simulation.

We choose one particle among the $N$ particles randomly (or we may select them one by one from 1 to $N$), and move the position $\boldsymbol{r}_i$ of the selected particle $i$ to the new position $\boldsymbol{r}_i + \Delta\boldsymbol{r}$. Here $\Delta\boldsymbol{r}$ is a random vector. According to the Metropolis algorithm (5.33), if a new configuration is energetically more stable than the original configuration we allow this movement; otherwise (if it is less stable), we calculate the ratio of the distribution function before and after this movement,

$$\frac{p(\boldsymbol{r}_1, \ldots, \boldsymbol{r}_i + \Delta\boldsymbol{r}, \ldots, \boldsymbol{r}_N)}{p(\boldsymbol{r}_1, \ldots, \boldsymbol{r}_i, \ldots, \boldsymbol{r}_N)}, \tag{5.41}$$

and allow this movement only if this ratio is greater than a uniform random number generated between 0 and 1.

The random vector $\Delta\boldsymbol{r}$ is chosen so that the allowance rate of the movement is roughly several tens of per cent. The "Monte Carlo step" (MCS) is defined as the time required to repeating this process $N$ times. The Monte Carlo procedure is usually performed with a fixed particle number $N$, temperature $T$ and volume $V$; this ensemble is called the NTV ensemble.

In particular, the simulation of a system composed of many hard spheres (the "hard-sphere system") has an important historical significance. In 1954, Rosenbluth and Rosenbluth [21] examined the hard-sphere system by the Monte Carlo method but could not find a liquid–solid phase transition. This phase transition was first identified in 1957 by Alder and Wainwright [22] who used a molecular-dynamics simulation. However, almost at the same time, Wood and Jacobson [23] succeeded in observing this liquid–solid phase transition by means of the Monte Carlo method. The history is precisely described in the articles by D. Levesque and J. J. Weis in [2], and by D. Levesque, J. J. Weis and J. P. Hansen in [4].

Among soft-core potentials, a general pair potential with an inverse-power-law form,

$$V(r) = \frac{1}{\varepsilon}\left(\frac{\sigma}{r}\right)^n, \tag{5.42}$$

is theoretically interesting because of its simplicity. In particular, the homogeneity (i.e. unique power with respect to $r$) of the potential guarantees that the excess thermodynamic quantities and the scaled radial distribution function $g(r/\sigma)$ depend only on a unique parameter $x = \rho\sigma^3(1/\varepsilon k_{\mathrm{B}}T)^{3/n}$, with $\rho$ as the density. This scale invariance is very powerful since it allows one to determine all the thermodynamic properties everywhere on the $(\rho, T)$ plane by calculating the equation of state along only a single line at the isotherm. The hard-sphere model is the simplest example of (5.42), in the limit $n \to \infty$. The other special case is the Coulomb system, in which $n = 1$. This case will be mentioned later.

The Monte Carlo method was applied by Wood and Parker [24] to a system described by a Lennard–Jones potential (see Sect. 4.2.2) which is a slightly more complex short-range soft potential. In their paper, there is a very good description of the Monte Carlo method and the result for the equation of state and the internal energy at $\varepsilon k_B T = 2.74$.

When we treat a system of charged particles, since the Coulomb interaction (the potential with $n = 1$ in (5.42)) is long-ranged, we have to pay careful attention to its treatment. Usually, charge neutrality should be preserved. For simplicity, let us consider the system of equal numbers of particles having charges $\pm Ze$. If the unique parameter $x = \rho(\sigma\varepsilon/k_B T)^3$ (here, $\sigma = (Ze)^2$ and $\varepsilon$ is the dielectric constant of the medium) is less than unity, one is in the "weak coupling" regime. On the other hand, if it is greater than unity, the system is in the "strong coupling" regime. Looking at the averaged charge-density distribution around one particle (the pair distribution function), we find an almost uniform distribution outside a few atomic shells from the origin, in the case of weak coupling, while strong oscillations associated with the presence of atoms persist much farther from the origin in the case of strong coupling. Therefore, in the latter case, if we impose periodic boundary conditions, we have to take into account interactions with the particles in the periodic cells which are distant from the center cell. Thus, even if the number $N$ of particles is small, we have to evaluate an infinite sum of Coulomb interactions.

The Ewald sum serves as a convenient and relatively fast method to evaluate this Coulomb sum. This method first assumes, additionally, an extended charge distribution with a Gaussian distribution around each atomic center. (This additional charge distribution can be subtracted later, most conveniently in the Fourier-space summation.) Each Gaussian distribution is chosen to have the same integrated charge as the point charge at the center but with opposite sign. Therefore, it almost completely screens the point charge at the center, because of the charge neutrality and the spherical symmetry (recall Gauss's theorem in electrostatics). Then, the Coulomb potential from each point charge can be replaced by a quite rapidly decreasing potential which decreases like the tail of a Gaussian distribution, and we can sum the contributions from few neighboring cells rather easily. More details of this technique are described in Appendix C.

But even with the Ewald method, one has to evaluate a sum of order $10^2$ for each ion, and it is still heavy as a routine in the Monte Carlo simulation. However, there are several ways to improve on this naive Ewald method: one can store the calculated potential in an array at the outset, fit the calculated potential with some appropriate function or improve the method of summation further by changing the Gaussian distribution to some other function. It is also possible to use the acceleration methods described in Appendix A.

This treatment can be extended simply to a system of positively charged particles in a uniform negative charge background (a one-component plasma). One can also take account of the effect of electron screening rather than a uniform negative charge background.

### 5.4.2 Modified Monte Carlo Techniques

McDonald and Singer proposed a "histogram method" [25] which allows one to derive the thermodynamic properties at slightly different temperatures (for the system of the same volume) from the data determined at one temperature for a system with given volume. The method is to store the histogram $f_T(U)$ which gives the number of the configurations with a given energy $U$, as well as the sample configurations, with a population weight $\exp(-U/k_B T)$. If the volume is the same, putting $\beta = 1/k_B T$, we readily find that $f_{T'}(U) = \exp[-(\beta' - \beta)U]$ for a temperature $T'$ ($\beta' = 1/k_B T'$). The internal energy or the specific heat at constant volume is then calculated by averaging according to this histogram. This method was later extended by Ferrenberg and Swendsen [26] to the multiple-histogram method (sometimes called also the weighted-histogram method), which optimizes the combination of data from an arbitrary number of simulations to obtain information over a wide range of the parameter values, although its application was first restricted to the Ising model. It has been applied to classical particle systems by Kumer et al. [27] and Boczko and Brooks [28].

The extension of the Monte Carlo method to the "grand canonical ensemble", where the number of particles can change, has been performed by Adams [29] for a hard-sphere system and Rowley et al. [30] for a Lennard–Jones system.

For the purpose of simulating two-phase coexistence, Panagiotopoulos has proposed a "Gibbs-ensemble Monte Carlo simulation" [31]. In this method, the two systems are treated simultaneously, and the sum of the two volumes is held constant as well as the sum of the two numbers of particles; one may move particles from one system to the other or change the volume. By doing this, it becomes possible to identify thermal equilibrium between these two systems. With this method, one can also determine the coexistence curve rather accurately without simulating the interface in the system.

The chemical potential or the free energy is an essential quantity in determining the phase coexistence of a first-order transition. A simple and elegant method to compute the chemical potential $\mu$ of a fluid or a mixture is Widom's test article insertion method [220]. If we write the partition function of an $N$-particle system as $Z_N$, the chemical potential is given by

$$\begin{aligned} \mu &= -k_B T \log \frac{Z_{N+1}}{Z_N} \\ &= \mu_{\mathrm{ideal}} + \mu_{\mathrm{ex}}, \end{aligned} \tag{5.43}$$

where $\mu_{\mathrm{ideal}}$ means the ideal-gas part

$$\mu_{\mathrm{ideal}} = -k_B T \log \frac{\Omega (2\pi m k_B T/h^2)^{d/2}}{N+1} \tag{5.44}$$

($\Omega$ denotes the system volume), while $\mu_{ex}$ is the excess part given by

$$\mu_{ex} = -k_B T \log \frac{\int d\boldsymbol{r}_1 \dots d\boldsymbol{r}_{N+1} \exp[-\beta U(\boldsymbol{r}_1, \dots, \boldsymbol{r}_{N+1})]}{\int d\boldsymbol{r}_1 \dots d\boldsymbol{r}_N \exp[-\beta U(\boldsymbol{r}_1, \dots, \boldsymbol{r}_N)]}. \tag{5.45}$$

Now we separate the potential energy of the $(N+1)$-particle system into the potential energy of the $N$-particle system, $U(\boldsymbol{r}_1, \dots, \boldsymbol{r}_N)$, and the external potential caused by the introduction of the $(N+1)$th particle, $U_{ex} = U(\boldsymbol{r}_1, \dots, \boldsymbol{r}_{N+1}) - U(\boldsymbol{r}_1, \dots, \boldsymbol{r}_N)$. The excess chemical potential is then given by

$$\mu_{ex} = -k_B T \log \int d\boldsymbol{r}_{N+1} \langle \exp[-\beta U_{ex}(\boldsymbol{r}_1, \dots, \boldsymbol{r}_{N+1})] \rangle_N, \tag{5.46}$$

where the average $\langle \cdots \rangle_N$ is evaluated in the $N$-particle system. In Widom's test particle insertion method, (5.46) is evaluated by integrating the ensemble average with respect to the coordinate $\boldsymbol{r}_{N+1}$.

One way to calculate the difference in the Helmholtz free energy between the system of interest and a reference system is Kirkwood's method, described in Sect. 2.5.3, which integrates the difference in the total energy over the switching constant [13, 33, 34]; see (2.400) and (2.401).

An alternative way to determine the free energy difference between the system of interest and a reference system is to evaluate the ratio of the partition functions of these systems at the same temperature as follows:

$$\begin{aligned} F_1 - F_0 &= -k_B T \log \frac{Z_1}{Z_0} \\ \frac{Z_1}{Z_0} &= \frac{Z_1 \int d\boldsymbol{r}_1 \dots d\boldsymbol{r}_N w(\boldsymbol{r}_1, \cdot, \boldsymbol{r}_N) \exp[-\beta(U_0 + U_1)]}{Z_0 \int d\boldsymbol{r}_1 \dots d\boldsymbol{r}_N w(\boldsymbol{r}_1, \cdot, \boldsymbol{r}_N) \exp[-\beta(U_0 + U_1)]} \\ &= \frac{\langle w(\boldsymbol{r}_1, \dots, \boldsymbol{r}_N) \exp(-\beta U_1) \rangle_0}{\langle w(\boldsymbol{r}_1, \dots, \boldsymbol{r}_N) \exp(-\beta U_0) \rangle_1}, \end{aligned} \tag{5.47}$$

where $w(\boldsymbol{r}_1, \dots, \boldsymbol{r}_N)$ is an arbitrary function, and $\langle \cdots \rangle_i$ means the ensemble average evaluated for the system having a potential $U_i$. If we put here $w = \exp(\beta U_0)$, we obtain

$$F_1 - F_0 = -k_B T \log \langle \exp[-\beta(U_1 - U_0)] \rangle_0. \tag{5.48}$$

Here, it is instructive to derive from this formula an important thermodynamic inequality, called the Gibbs–Bogoliubov inequality,

$$F_1 \leq F_0 + \langle (U_1 - U_0) \rangle_0. \tag{5.49}$$

Equation (5.48) is useful for estimating the free-energy difference between two systems that resemble each other. However, in most cases of practical interest, the largest contributions $\exp[-\beta(U_1 - U_0)]$ in (5.48) come from the region where $\exp(-\beta U_0)$ is small.

In this case one has to choose the function $w$ in (5.47) more appropriately in order to obtain a better estimate of the free-energy difference. Bernett [35] found that the optimal choice is

$$w(\boldsymbol{r}_1, \ldots, \boldsymbol{r}_N) \propto \frac{1}{\exp(-\beta U_1) + \exp[-\beta(U_0 + C)]}, \tag{5.50}$$

where the parameter $C$ should be determined so as to satisfy

$$\begin{aligned} &\int d\boldsymbol{r}_1 \ldots d\boldsymbol{r}_N \frac{1}{1 + \exp[-\beta(U_0 - U_1 + C)]\exp(-\beta U_1)} \\ &= \int d\boldsymbol{r}_1 \ldots d\boldsymbol{r}_N \frac{1}{1 + \exp[-\beta(U_1 - U_0 - C)]\exp(-\beta U_0)}. \end{aligned} \tag{5.51}$$

This scheme is called the "acceptance ratio method".

As an alternative way to evaluate the free-energy difference, one may use the so-called "umbrella sampling" proposed by Torrie and Valleau [36]. In this method, the probability distribution realized in the simulation itself is changed from the original canonical distribution $p_0(\boldsymbol{r}_1, \ldots, \boldsymbol{r}_N) = \exp[-\beta U_0(\boldsymbol{r}_1, \ldots, \boldsymbol{r}_N)]$ to an artificial distribution

$$\pi(\boldsymbol{r}_1, \ldots, \boldsymbol{r}_N) = \exp\Big\{-\beta\big[U_0(\boldsymbol{r}_1, \ldots, \boldsymbol{r}_N) + W(\boldsymbol{r}_1, \ldots, \boldsymbol{r}_N)\big]\Big\}, \tag{5.52}$$

which is appropriate for the evaluation of the free energy difference. The potential $W(\boldsymbol{r}_1, \ldots, \boldsymbol{r}_N)$ is called the umbrella potential. For this purpose, the right-hand side of (5.48) is rewritten as

$$\begin{aligned} \langle \exp[-\beta(U_1 - U_0)]\rangle_0 &= \frac{\int d\boldsymbol{r}_1 \ldots d\boldsymbol{r}_N \pi(\boldsymbol{r}_1, \ldots, \boldsymbol{r}_N) \mathrm{e}^{-\beta U_1}/\pi(\boldsymbol{r}_1, \ldots, \boldsymbol{r}_N)}{\int d\boldsymbol{r}_1 \ldots d\boldsymbol{r}_N \pi(\boldsymbol{r}_1, \ldots, \boldsymbol{r}_N) \mathrm{e}^{-\beta U_0}/\pi(\boldsymbol{r}_1, \ldots, \boldsymbol{r}_N)} \\ &= \frac{\langle \exp(-\beta U_1)/\pi\rangle_\pi}{\langle \exp(-\beta U_0)/\pi\rangle_\pi}, \end{aligned} \tag{5.53}$$

where the notation $\langle \cdots \rangle_\pi$ is used to express an average over a probability distribution $\pi(\boldsymbol{r}_1, \ldots, \boldsymbol{r}_N)$. The umbrella potential should be chosen to make the configurations preferred in system 0 and the configurations preferred in system 1 equally probable. In the umbrella sampling (this name comes from this bridging nature of the distribution function), one modifies the Markov chain in the Monte Carlo algorithm to approach the distribution $\pi(\boldsymbol{r}_1, \ldots, \boldsymbol{r}_N)$ instead of $p_0(\boldsymbol{r}_1, \ldots, \boldsymbol{r}_N)$. Since the function $\pi(\boldsymbol{r}_1, \ldots, \boldsymbol{r}_N)$ is not known a priori, one has to search for an appropriate one.

For example, let us consider an example of a molecular liquid, in which each molecule has internal degrees of freedom such as a bending angle characterizing the trans, cis and gauche configurations, which are energetically close to each other. We assume that the reference system is an ideal gas, i.e. $U_0 = 0$, for simplicity. In such a case, if the total potential $U_1$ can be divided simply into the intramolecular potential $U_{\text{intra}}$ and the intermolecular potential $U_{\text{inter}}$, one optimal choice of the umbrella potential $W$ is $W = U_{\text{inter}}$, and (5.48) with (5.53) becomes

$$F_1 = -k_B T \log\Big[\langle \exp(-\beta U_{\text{intra}})\rangle_{\text{inter}}\Big]. \tag{5.54}$$

This choice of the umbrella potential gives a uniform distribution for the intramolecular degrees of freedom, so that it may offer a very good estimate for the free energy. In general, however, the search for the optimal umbrella potential is not so easy. Harvey and Prabhakaran [37] discussed useful ways to judge whether the choice of the umbrella potential is adequate or not.

One way to obtain a good estimate for the free-energy difference by using umbrella sampling is to divide the interval between 0 and $\Delta U = U_1 - U_0$ into $n$ "windows" and perform $n$ Monte Carlo sampling runs. One Monte Carlo trajectory can be regarded as a Markov process or a random walk in the potential energy space. We assume $n$ different umbrella potentials $W_i = (i/n)\Delta U$ $(i = 1, 2, \ldots, n)$, i.e. we divide the energy range into $n$ pieces of width $\Delta U/n$. The relaxation time associated with sampling one energy interval $\Delta U/n$ is given by $\tau_n = \tau_0 + (\Delta U/n)^2/D_U$. Here $\tau_0$ stands for a basic relaxation time required for a change in dynamic variables in a direction perpendicular to the energy, and $D_U$ stands for the diffusion constant in energy space which is characteristic of this Markov chain. The total time required to sample all $n$ windows is then given by $\tau_{\text{total}} = n\tau_n = n\tau_0 + \Delta U^2/nD_U$, which turns out to have a minimum at a certain value of $n$ when $\tau_n$ is comparable to $\tau_0$ [34, 38].

Another way is to adopt adaptive umbrella sampling [39–41], which is a systematic method to update and determine the umbrella potential $W_i(\boldsymbol{r}_1, \cdot, \boldsymbol{r}_N)$ such that uniform sampling can be achieved in a given energy interval.

Recently, the "multicanonical Monte Carlo method" [42–45] was proposed for studying "slow dynamics". In the multicanonical Monte Carlo method, one divides the energy into $N + 1$ intervals labeled by $k = 0, 1, \ldots, N$. For each interval, the distribution function is defined as $p_k(E) = \exp(\alpha_k - \beta_k E)$, $E_k < E \le E_{k+1}$. The idea is similar to umbrella sampling. This new distribution is chosen so as to have a nearly uniform distribution in the energy region among unstable, metastable and the most stable configurations. Physically, this situation can be achieved if the temperature is set high in the energy region around the metastable states and low in the energy region around the unstable states. With this ensemble, system that goes into a blind alley has more chance to leave it. Thus this method seems to be valid not only for classical particles but also for systems such as polymers and spin glasses. The idea of the multicanonical ensemble is not restricted only to the Monte Carlo method but is applicable also to the molecular-dynamics method [41, 46, 47].

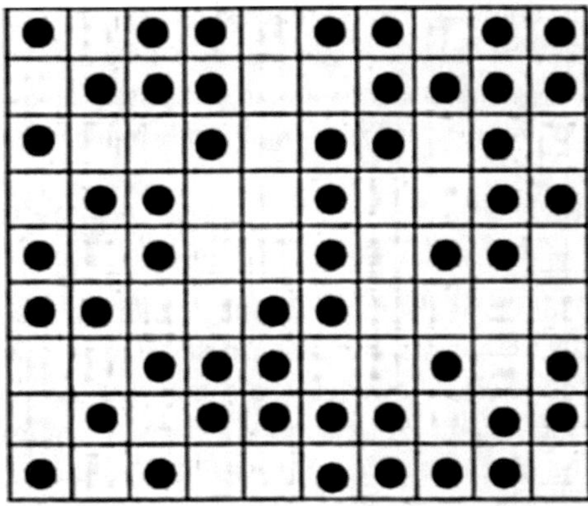

**Fig. 5.11** Site percolation model. Distribute particles (*black circles*) on the lattice sites randomly with a probability (occupancy ratio) $p$. If $p$ approaches a well-defined threshold $p_c$, there appears a very big cluster connecting one side to the other side of the system

## 5.4.3 Percolation

What is the electric conductivity of a material made by mixing conducting and insulating particles in some ratio randomly? This is the main theme of the percolation problem [49].

For simplicity, let us consider a finite two-dimensional square lattice. Assume first that the lattice sites are all covered by insulating particles. Then select one lattice site randomly, and replace the insulating particle by a conducting one, and repeat this procedure many times, without replacing a lattice site which is already occupied by a conducting particle. One can stop this process at an appropriate ratio of the number of conducting particles to the total lattice sites (Fig. 5.11). For small ratios, say 0.1, many isolated conducting clusters appear, while for ratios near unity almost all the lattice sites are occupied by conducting particles. This process resembles the percolation of particles in a medium, and this specific model is called the "site percolation model". Alternatively, considering bonds between adjacent lattice sites in place of the sites themselves, we could introduce a similar but different process which replaces insulating bonds by conducting bonds one by one. This alternative model is called the "bond percolation model".

In either case, when the conducting-element (site/bond) density $p$ approaches a well-defined threshold $p_c$, a very big conducting cluster connecting one side to the other side of the system appears. The percolation threshold $p_c$ is known to be 0.5 in the cases of a square lattice for bond percolation and a triangular lattice for site percolation, while it is generally different from 0.5 for other cases, e.g. $p_c = 0.592746$ in the case of a square lattice for site percolation [49]. Once the biggest cluster size becomes comparable to the system size, the electric conductivity emerges abruptly in this system. The conducting-element density at this point is called the "percolation threshold" $p_c$. This model might present the simplest classical picture of the "metal–insulator transition" when we consider impurity conduction. However, in real materials, the real metal–insulator transition point is different from the percolation threshold because of the quantum effect called "localization of wavefunctions". The phenomenon of localization due to the randomness of the atomic configuration is called "Anderson localization". On the other hand, in materials with strong electron correlation such as NiO, localization is also present, and the corresponding metal–insulator transition is called the "Mott transition", which is generally distinguished from Anderson localization. In addition to the problem of the metal–insulator transition, the percolation model is also used to describe the distribution of the magnetic and nonmagnetic atoms in dilute magnetic alloys.

Since the percolation process is completely random, the resulting clusters has no spatial correlation among them except for the self-avoiding effect. On the other hand, in a model called the "directed percolation model", which has a direction preference, the generated clusters show a clear spatial correlation. However, in either case, there is no clear difference in the symmetry of the configurations above and below the percolation threshold. It can be shown that this system is identical to the "$q$-state Potts model", which will be described in (**b**), in the limit $q \to 1$ [50–52]. Therefore, at the percolation threshold, there should occur some anomalous behavior which is characteristic of second-order phase transitions, although the relaxation time of this system is always zero [49]. The shape of the infinite cluster at the percolation threshold is known to be "scale invariant". Such a pattern without any specific scales is generally called "fractal" or a "fractal pattern" [53, 54]. At a density $p$ slightly above the percolation threshold $p_c$, the electric conductivity $\Sigma$ and the probability $P$ that one site (or bond) belongs to the infinite cluster behaves like

$$\Sigma \sim (p - p_c)^{\mu}, \qquad P \sim (p - p_c)^{\beta} \tag{5.55}$$

and the correlation length $\xi$ diverges according to $\xi \sim |p - p_c|^{-\nu}$. Generally, the exponents $\mu$, $\beta$, $\nu$ are the fractal number and are the same in both the site problem and the bond problem. Such fractal exponents are called "critical exponents" and are specific to systems showing a second-order phase transition or other fractal behaviors. Moreover, the number $n_s(p)$ of clusters composed of $s$ sites (or bonds) obeys the scaling form

$$n_s(p) = s^{-\tau} f[(p - p_c)s^{\sigma}], \tag{5.56}$$

and the "scaling relations"

$$\beta = \frac{\tau - 2}{\sigma}, \qquad d\nu = \frac{\tau - 1}{\sigma} \tag{5.57}$$

hold among the critical exponents. Here $d$ is the spatial dimensionality. From the relation with the Potts model, there is an exact solution based on the conformal invariance in two dimensions (the values of the exponents are given, in two dimensions, by $\beta = 5/36$, $\nu = 4/3$, $\sigma = 36/91$, $\tau = 187/91$, $\mu \sim 1.30$; in three dimensions, $\beta \sim 0.41$, $\nu \sim 0.88$, $\sigma \sim 0.45$, $\tau \sim 2.18$, $\mu \sim 2.0$ [49]). To determine the critical exponents, Monte Carlo renormalization group method, which will be explained briefly in (**b**), is known to be very efficient [55–60]; see also [49].

If one extracts one percolation cluster, this cluster is called a "lattice animal" [61, 62]. The values for the critical exponents characterizing the spatial pattern of the lattice animals in two and three dimensions are obtained exactly [63]. In three dimensions, its critical exponent $\nu$ is given exactly by 1/2 for three dimensions [63]. Lattice animals often appear as a model for describing the synthesis of the branched polymer network by the condensation polymerization reaction. In this case, the state having only finite-size clusters (below the threshold) corresponds to the high fluidity and is called "sol", while the state having an infinite cluster (above the threshold) corresponds to low fluidity and is called the "gel". As other types of model describing the condensation-polymerization reaction and colloidal aggregation, there are the diffusion-limited CCA model and the reaction-limited CCA model, explained in Sect. 5.4.8.

### 5.4.4 Polymer Systems

A single "polymer chain" dissolved in a solvent gives the simplest example of studying the spatial distribution and other statistical quantities of polymer solutions. Since the number of monomers in an actual polymer chain is of the order of 100 000, we have to use a coarse-grained model to simulate polymers on a computer [64, 65]. The model frequently used for a polymer chain in a good solvent is the self-avoiding-walk (SAW) model, which represents a single chain connected by bonds on a lattice (Fig. 5.7). In the SAW model, the chain cannot occupy the same lattice site more than once, because of the self-avoiding effect. There are typically two ways to realize SAWs on a computer; one is the direct enumeration algorithm which counts thoroughly all possible configurations, and the other is the simple sampling algorithm to collect the possible configurations randomly. The following gives some details of these two algorithms.

The direct enumeration algorithm, so to speak, searches everywhere from side to side in a maze. Let us describe it in the case of a two-dimensional square lattice. Take the starting point as X$(L{=}0) = 0$, Y$(L{=}0) = 0$, and set the segment number counter $L$ to $L = 1$. And then repeat the following sequence (A)–(C).

(A) After setting the $L$th element of the array NEL($L$), which stores the number of possible directions of adding the $L$th segment to the end of a chain, to zero (*initialization*), assume one direction for the $L$th segment. If the self-avoiding condition is fulfilled for this direction (*checking*), increment the array element NEL($L$) by unity. Simultaneously, record this direction of elongation (*recording*) in an array element IW($L$, NEL($L$)) as 0, 1, 2 or 3 corresponding, respectively, to up, down, left or right. Repeat this checking of the self-avoiding condition and recording the data, for all the directions of elongation of the $L$th segment.

(B) Actually add the $L$th segment to the chain (*elongation*) in the last direction IW($L$,NEL($L$)). Record the position of this new end as X($L$)$= 0$, Y($L$)$= 1$, and decrement NEL($L$) by unity: NEL($L$)=NEL($L$)$-1$.

(C) Increment $L$ by unity: $L = L + 1$ (*increment of the segment number*).

Repeating (A)–(C), one finally finds all the possible directions of the $N$th segment, which is actually the last end segment, but just for one realization among many chain configurations. Then go on to the following process (D)–(E). Note that the process (E) is repeatable by itself.

(D) Since the $N$th segment is the last end segment, the process of adding this segment to the chain can be omitted. However, it is necessary to decrement `NEL(L = N)` one by one as in the process of actually adding the $N$th segment to the chain, and eventually to set `NEL(L = N)` to 0.

(E) If `NEL(L)` is 0, then decrement $L$ by unity, i.e. set $L = L - 1$ (*decrement of the segment number*). If $L = 1$ and `NEL(L = 1) = 0`, stop all the processes (*goal*).

As soon as nonzero `NEL(L)` is found, go back to the process (B), and repeat again the process (A)–(C).

In contrast, the SAW Monte Carlo algorithm ("enrichment algorithm") based on the simple sampling technique is very simple:

- Extracting one sample of the $L$th generation, elongate the end by one segment. That is, choose one direction which would not fold backward on itself by means of a random number.
- Check if this elongation violates the self-avoiding condition. If not, store this realization as one sample of the $(L + 1)$th generation.

Thus, in order to generate $M_{L+1}$ samples at the $(L + 1)$th generation, it is necessary to perform $\alpha M_{L+1}/M_L$ trials on each of the $M_L$ samples at the $L$th generation. Here $\alpha$ is a value of the order of the inverse of the success ratio. Writing $z$ as the coordination number of the lattice and $\mu$ as the effective coordination number, which is known to be 2.6386 for two dimensions and 4.6838 for three dimensions, we find that $\alpha \sim (z-1)/\mu$ in the limit of long chains [66]. When the chain length $L$ is shorter, $\alpha$ becomes larger than the above value because the rate of violating the self-avoiding condition increases. Therefore, for extremely short chains such as $L = 1$ or 2, the enrichment algorithm can be slower than the direct enumeration method described above.

In the case of the SAW, the simple sampling (enrichment) algorithm has the following two advantages.

(i) One simulation provides information on all chain lengths from 1 to $N$.
(ii) Each SAW configuration is statistically independent and a simple error analysis can be applied.

The simple sampling method is generally considered to be useful in problems such as the random walk (RW), SAW and percolation. The SAW is different from the RW because the walker cannot come back to its path. The mean square distance $\langle R^2 \rangle$ in the RW increases linearly with the number of particles $N$. In the case of the SAW, it grows like

$$\langle R^2 \rangle \sim N^{2\nu}, \tag{5.58}$$

with the critical exponent $\nu \sim 0.59$ for three-dimensional systems and $\nu = 3/4$ (exactly) for two-dimensional systems [67]. To estimate such a quantity accurately, it is necessary to generate a huge number of samples and take the sample average as in (5.24), since the resulting patterns depend strongly on the initial conditions and the random numbers used in the sample generation. As a quantity related to the entropy, one may count the total number of possible walker's trajectories (configurations) on the lattice. This total number of configurations, $\mathcal{N}_L$, behaves as $\mathcal{N}_L = z^L$ in the case of the RW, obviously ($z$ is the coordination number). In the case of the SAW, it behaves as

$$\mathcal{N}_L \sim L^{\gamma-1}\mu^L, \tag{5.59}$$

with the critical exponent $\gamma = 43/32$ (exactly) for two dimensions and $\gamma \sim 1.17$ for three dimensions [67]. Using the enrichment algorithm, the ratio $\mathcal{N}_{L+1}/\mathcal{N}_L$ of the successive numbers of configurations is given just by the value of $\alpha$ discussed above.

It is also possible to consider a solution of a polymer network with a complex architecture in place of a linear chain. In this case also the SAW algorithm can be applied to generate the samples if we pay some attention to the network architecture. Figure 5.12 shows an example of a star polymer which was generated by the enrichment algorithm [66]. Star polymers resemble micelles with long "arm" chains. (Micelles are formed, for example, by AB block copolymers composed of short hydrophilic (A) and long hydrophobic (B) chain parts when dissolved in water.)

Mathematically, the SAW model is equivalent to the $n$-vector model, which will be described in Sect. 5.4.5, in the limit $n \to 0$ [68]. From this fact, we know that the configuration of a flexible polymer network which is made by connecting very long

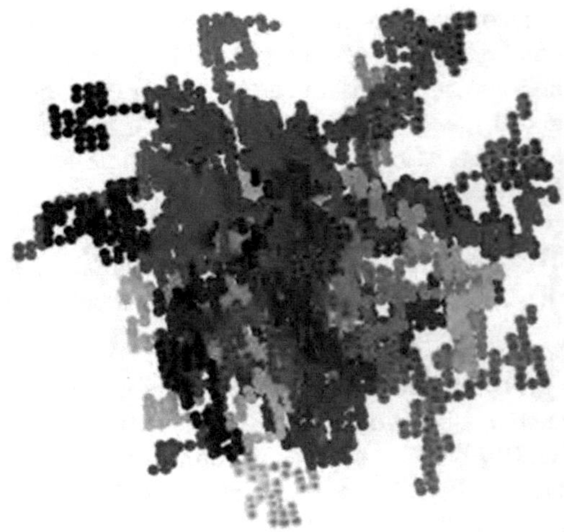

**Fig. 5.12** Star polymer with 32 arms generated by the SAW enrichment algorithm

chains is characterized by the correlation-length exponent $\nu$ which is the same [69] as that of a single self-avoiding-walk (SAW) in a dilute solution. The value of the exponent $\nu$ is related to the fractal dimension $D$ of the polymer conformation. The relation between $\nu$ and $D$ for SAW systems was obtained first by des Cloizeaux [70] and is given by $D = 1/\nu$. Therefore, in real polymer solutions, the fractal dimension $D$ of an arbitrary isolated polymer network is given by $D \sim 1.7$. In contrast, the configuration number exponent $\gamma$, of an isolated polymer network, defined in the same way as (5.59) depends strongly on the network topology. The value of $\gamma$ can, however, be obtained from that of star polymers via a scaling relation [69]. So far, many Monte Carlo simulations have been performed in order to investigate conformational properties of branched polymers as is reviewed recently by Freire [71].

In real polymer solutions, the SAW becomes a very good approximation in the good-solvent limit. However, when the solvent does not dissolve polymers so well, we call it a poor solvent. In a poor solvent, polymers have a tendency to segregate in order to decrease the interfacial region between the polymer and the solvent, and therefore they feel similar circumstances to a dense solution of a polymer. In such cases, polymers do not feel the excluded-volume effect, because of the screening effect, and behave as RWs. Anyway, in the case of a poor solvent, strong entanglement of the polymer often freezes the polymer flexibility and the simple sampling method using the SAW algorithm becomes useless. The boundary between a good solvent and a poor solvent emerges when the second virial coefficient becomes zero. This point is called the $\Theta$ point. At this point, the behavior of the polymer chains is somehow RW-like, but it is not an easy problem to simulate it very accurately. Approximately, it may be described by a nonreversal random walk (NRRW), which is essentially the RW but with no segment folding backward.

The SAW algorithm serves as a quite powerful tool to understand polymer conformations (geometrical configurations) in dilute solution in a good solvent. The free energy and the osmotic pressure are also fundamental quantities in polymer solutions, which give rise to a variety of complex phases. Flory and Huggins' mean-field theory [72, 73] has been widely accepted as a basic and simple theory for the free energy of polymer solutions. The deviation from this mean-field theory, which accounts for the effect of fluctuation of the segment density, has been analyzed by a cluster-expansion series; the first-order correction is the famous Miller [74] and Guggenheim [75] theory. However, these theories cannot explain des Cloizeaux's $c^{9/4}$ rule for the osmotic pressure in the semidilute regime [67, 76], for example, because mean-field theory breaks down in dilute and semidilute regimes, where fluctuation plays an important role. Recently, Shida et al. [77] extended the enrichment SAW algorithm to calculate the free energy and the osmotic pressure of a semidilute solution. They succeeded in observing des Cloizeaux's $c^{9/4}$ behavior. Similarly, the second virial coefficient can be estimated by considering just two polymers (see [78, 79] for star polymers).

In the case where polymer chains interact with each other by soft potentials, Kumar et al. [80, 81] estimated the chemical potential (the increase of the free energy on adding one polymer chain to the system) of monodisperse solutions of chains with length $L$ from Monte Carlo simulations, which were based on Widom's test particle insertion [82] and elongating one special chain step by step from 1 to $L$. But the estimation of the osmotic pressure by this method requires $L \times N$ simulations ($N$ is the number of chains in the system), and becomes quite computationally heavy. Madden et al. [83] have calculated the coexistence curve of a three-dimensional polymer–solvent system, but in their approach, the estimation of the free energy of the entire system is not straightforward and rather intractable, as is that in Kumar's algorithm. Recent progress has been made by Smit et al., who proposed a more efficient algorithm [84], but no precise analysis of the free energy has been reported yet in the dilute and semidilute regimes.

None of these methods can simulate the movement of polymers. There is a way to estimate the relaxation time from static simulations [85–87] (see also [88, 89, 92] for star polymers), but, in order to follow the relaxation process at all times, it is necessary to introduce an essentially dynamic algorithm. Besides the models using molecular-dynamics simulations, there are also several models suitable for lattice Monte Carlo simulations. One of the simplest models is a snake model which removes one segment from one side of a polymer chain and adds it to the other side of the chain, which describes the reptation process of polymer chains (this model is called also the reptation model) [90]. The bond-fluctuation model is a simple but elaborated lattice model with changeable bond lengths [91]. Figure 5.13 demonstrates a bond-fluctuation model of a star-polymer [92]. Each end of each segment (monomer) occupies a small cubic cell composed of eight lattice points. Monomers are the bonds between these cubic cells and their bond lengths can change from $l_{\min} = 2$ to $l_{\max} = \sqrt{10}$. But one must exclude the bond length $l = \sqrt{8}$. The prohibition of bond lengths of $l = \sqrt{8}$ and $l > \sqrt{10}$ automatically guarantees that two bonds do not overlap when one considers only the self-avoidance of the cubic cells. There are

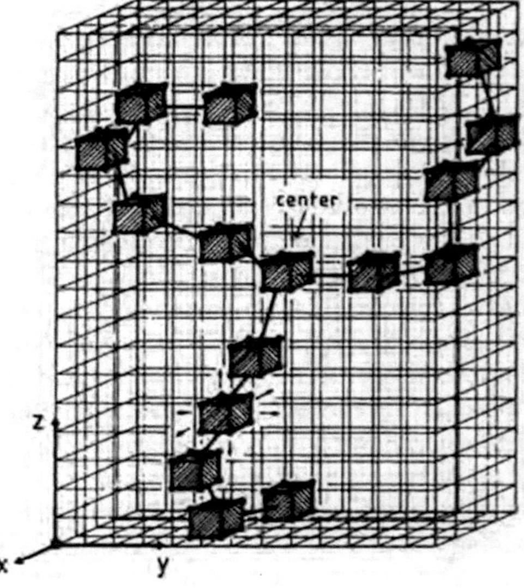


**Fig. 5.13** Bond-fluctuation model of a star polymer with three arms [92]

also many other models describing the movements of polymer chains. More recently, Urakami and Takasu [93] used the multicanonical Monte Carlo method and the bead jump algorithm devised originally by Murat and Witten [94] to simulate a polymer chain with stickers.

### 5.4.5 Classical Spin Systems

Systems with many degrees of freedom can be more easily handled with statistical physics. A typical problem in statistical physics is to evaluate the average of a macroscopic property of the system when the system Hamiltonian is given. One of the typical systems is an interacting spin system in which spins on a lattice interact with the adjacent spins. The behavior of spin systems is especially interesting when their size is macroscopic. Usually, a spin system approaches thermodynamic equilibrium, showing either a magnetically ordered or a disordered structure during annealing, but it sometimes goes to a glassy state after quenching.

#### (a) Ising Model

Let us consider first a ferromagnetic (F) system as an example. If a magnet has a very strong axial magnetic anisotropy, each spin can take only the values $\pm 1$ along the easy axis. These spins interact with each other by an exchange interaction and with the external field by the Zeeman energy, and fluctuate in both space and time. This system is none other than the Ising model illustrated in Fig. 5.14. The Hamiltonian of this system reads

$$\mathcal{H}_{\text{Ising}} = -\sum_{i>j} J_{ij} S_i S_j - \sum_i H_i S_i, \qquad S_i = \pm 1, \tag{5.60}$$

where the sum in the first term on the right-hand side is taken over all the pairs $i$ and $j$ of the lattice sites. $J_{ij}$ and $H_i$ represent, respectively, the magnitudes of the exchange constant and the external magnetic field along the easy axis. These magnitudes may change from point to point according to the composition or the shape of the magnet.

In real magnetic materials, there is a long-range magnetic dipole–dipole interaction (it has the same form as the electrostatic dipole–dipole interaction, (4.7)) in addition to the short-range exchange interaction. As a result, a variety of interesting magnetic domain structures including magnetic bubbles, lamellar phases or microdomain structures [95, 96] may appear. However, for simplicity, we will not consider the dipole–dipole interaction here.

In usual magnetic problems, where no conservation law exists, one may flip each spin independently during the process of updating spins. The kinetic Ising model is

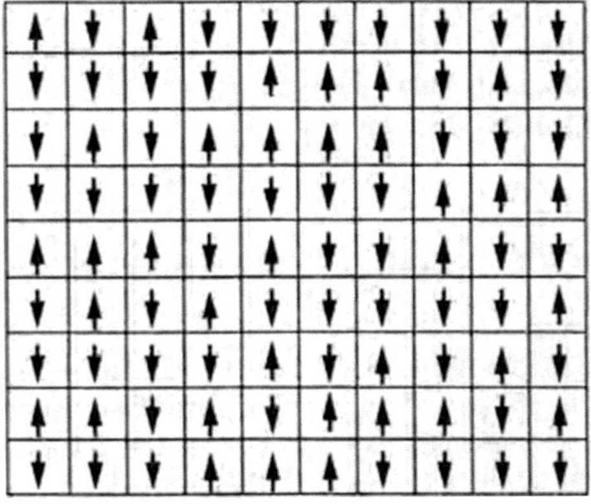

**Fig. 5.14** Ising model of a spin system. Each spin on a lattice point can point either upward or downward. Adjacent spins interact with each other according to their directions

a model describing the updation of each spin state in terms of the Glauber algorithm (5.32).

On the other hand, the Ising model describing particle occupancy according to a spin state is also called the lattice gas model. Regarding the Ising spin ($S_i = \pm 1$) as the particle occupancy or the particle species in a binary mixture (binary alloy), one can investigate with this model the following phenomena: two-phase coexistence, an order–disorder phase transition in solid solutions or binary mixtures, the liquid–vapor transition, particle diffusion in solids and on surfaces and, as will be mentioned in a later section, nucleation and solidification, phase separation and spinodal decomposition. In these systems, the external field generally corresponds to the chemical potential. In the lattice gas models, the conservation law on the particle number means the conservation of the total spin. Therefore, the exchange algorithm, where two adjacent spins can interchange their spin states is very useful for these systems; such an algorithm is also called Kawasaki's (exchange) algorithm (see Sect. 5.3.4).

For example, for the order–disorder phase transition of FCC $Cu_3Au$ and CuAu, Binder [97] performed a Metropolis Monte Carlo simulation of a nearest-neighbor FCC lattice gas model with Kawasaki's exchange algorithm and determined the phase diagram. This system corresponds to the magnetic Ising model showing a frustrated antiferromagnetic phase in the FCC lattice. There are several types of ordering; the most stable structure is $L1_2$ for $Cu_3Au$ and $L1_0$ for CuAu. His result can be compared with an analysis by Kikuchi's cluster variation method (CVM) [98–100]. Equation (5.60) has only the usual two-body interaction. It is, however, possible to introduce a more complicated many-body interaction. Matsumiya et al. [101] used ab initio results by Terakura et al. [102], and calculated the many-body interactions among Cu, Ag and Au atoms by means of the Connolly–Williams method, in order to perform a more realistic Monte Carlo simulation of the Cu–Au system. Their result, as well as the result obtained by Mohri [103], who used the CVM, for the coexistence curve is, however, about 300 K higher than the experimental results. In order to treat interactions between real atoms by means of lattice gas models, one may adopt the potential renormalization discussed in Sect. 4.4. Using the Finnis–Sinclair-type embedded-atom potential determined by Ackland and Vitek [104], we also find that the discrepancy between the experiments and theory is 300 K before renormalization, namely, if we use this potential directly on the discrete lattice space. But the two-step renormalization of this potential reduces this discrepancy to 50 K [105]. The agreement is still not perfect; but the reason may be attributed to the incompleteness of the use of the two-step renormalization and also to the MD potential assumed for the system.

For an Ising model on a one-dimensional lattice at thermal equilibrium, it is possible to evaluate the exact partition function

$$Z = \prod_{i=1}^{N} \sum_{S_i=(-1,1)} \exp\left[-\frac{J}{k_B T} S_i S_{i+1}\right]. \tag{5.61}$$

Here we set all the exchange interaction parameters between nearest neighbors the same (we write the value as $J$) and the external field $H^{\text{ext}}$ to zero for simplicity. To solve this problem, we adopt the transfer-matrix method. Let us choose just two spins interacting directly by the exchange interaction. Identifying the two spin states $(-1, 1)$ of the left spin as a column of a $2\times2$ matrix, and the spin states $(-1, 1)$ of the right spin as a row of the matrix, we may introduce a transfer matrix

$$T = \begin{pmatrix} \exp(-K) & \exp(K) \\ \exp(K) & \exp(-K) \end{pmatrix} \tag{5.62}$$

$(K = J/k_B T)$, whose elements are given by the corresponding Boltzmann weight. Then (5.61) can be rewritten as

$$Z = \text{Tr}T^N = \text{Tr}\begin{pmatrix} \lambda_+ & 0 \\ 0 & \lambda_- \end{pmatrix}^N = \lambda_+^N + \lambda_-^N \sim \lambda_+^N, \tag{5.63}$$

where $\lambda_+ = 2\cosh K$ and $\lambda_- = 2\sinh K$ are, respectively, the larger and smaller eigenvalues of the matrix $T$. The symbol $\sim$ indicates that the equation becomes accurate in the limit where the particle number $N$ goes infinity. Equation (5.63) can be readily extended to the case with an external magnetic field.

The exact solution of the two-dimensional Ising model with zero magnetic field and a constant nearest-neighbor exchange interaction was first solved by Onsager. In contrast to the one-dimensional model, the two-dimensional model undergoes a second-order phase transition at a temperature $T = T_c$, below which a spontaneous magnetization occurs. The phase transition temperature $T_c$ is called also the critical temperature or the critical point. In the case of a square lattice, it is given by

$$k_B T_c/J = 2/\ln(1+\sqrt{2}) = 2.26918531. \tag{5.64}$$

Around the critical point, there emerge the critical phenomena characteristic of second order phase transitions; for example, the specific heat $C$ or the magnetic susceptibility $\chi$ diverges at the critical point [1, 5, 52]. By introducing the reduced temperature $t = T - T_c$, one may characterize these anomalies as

$$C \sim t^{-\alpha}, \qquad \chi \sim t^{-\gamma}, \tag{5.65}$$

and then the critical exponents are given by $\alpha = 0$ (logarithmic divergence) and $\gamma = 7/4$. Such a critical anomaly is due to the fact that the correlation length becomes infinitely large at the critical point. (In this circumstance, the system has no special length scale, and the ordering pattern becomes fractal.) The correlation length $\xi$ diverges as

$$\xi \sim |t|^{-\nu} \tag{5.66}$$

and we have $\nu = 1$ in the two-dimensional Ising model. The critical exponents are usually "universal", i.e. they do not depend on the details of the system but depends only on the basic symmetry of the system such as the spatial dimensionality, the dimensionality of the spin space where the spin degrees of freedom exist, or the spatial homogeneity or inhomogeneity. In some two-dimensional Ising models with the nearest-neighbor exchange interaction parameter changing spatially in one direction, one may also obtain the exact solution of the system by the transfer-matrix method [52, 106].

Besides the exact solution, the high-temperature series expansion has been extensively applied to the two-dimensional Ising model by Domb and Sykes [107] and to a more general $n$-vector model, which will be introduced below, by Stanley [108]. The coherent anomaly method (CAM) has been applied to the two-dimensional Ising model by Suzuki [109] and co-authors [110] first to demonstrate its ability to determine accurately the critical exponents.

It is also important to investigate the dynamics exhibited by the spin models such as an Ising model. It is well-known that the dynamics becomes, in particular, slow when the temperature approaches the critical temperature. This behavior is called the "critical slowing down" and is investigated first by a Monte Carlo simulation by Binder for the two-dimensional Ising model [111]. The relaxation time of the system behaves as

$$\tau \sim t^{-z}, \tag{5.67}$$

where $t$ is the reduced temperature and the index $z$ is called the "dynamic critical exponent". The value of $z$ is close to its mean-field value, 2, regardless to the spatial dimensions of the system. The dynamic critical behavior of the two-dimensional Ising model has been investigated so far with the Monte Carlo method by many authors [112–114]. This study is related to the dynamic scaling of spinodal decomposition which will be explained in Sect. 5.4.6.

For the three-dimensional Ising model, although its exact solution is not known, its second-order phase transition in the case of a ferromagnetic exchange interaction has been well studied by means of Monte Carlo simulations based on the Metropolis algorithm or the Glauber algorithm, [2–4, 115, 116] as well as the high-temperature series expansion [107, 108, 117] or the field-theoretical renormalization-group calculations [118–123]. The three-dimensional Ising model with a constant nearest-neighbor exchange interaction without an external magnetic field has a critical temperature equal to $k_B T/J = 4.51337$ in the case of a simple cubic (SC) lattice, and the critical exponents are given by $\alpha \simeq 0.12$, $\gamma \simeq 1.25$ and $\nu \simeq 0.63$, independent of the lattice (SC, FCC, BCC, etc.) [1, 124]. The spontaneous magnetization $M$ and the spin-spin correlation function $g(r_{ij}) = \langle \sigma_i \sigma_j \rangle$ at the critical point ($t = 0$) behave as

$$M \sim (-t)^{\beta}, \qquad g(r) \sim r^{-(d-2+\eta)}. \tag{5.68}$$

There are several scaling relations among the critical exponents:

$$\beta = \frac{\nu}{2}(d - 2 + \eta), \qquad \alpha = 2 - \nu d, \qquad \gamma = \nu(2 - \eta), \tag{5.69}$$

where $d$ denotes the spatial dimensionality. Therefore there are only two independent critical exponents in the simple spin models. The first two equations of (5.69) depend explicitly on the spatial dimensionality $d$, are called the hyperscaling relations and are valid for $d \le 4$. On the other hand, for $d \ge 4$, fluctuations become less important and the mean-field theory becomes exact. That is, the equations (5.69) with $d = 4$, which are the results from mean-field theory, hold also for $d \ge 4$. The mean-field values for the critical exponents are $\alpha = 0$, $\nu = \frac{1}{2}$, $\gamma = 1$, $\beta = \frac{1}{2}$ and $\eta = 0$.

Besides the homogeneous Ising system, various more complicated systems have been investigated so far also. If the exchange constant $J$ becomes minus, the system has a tendency to exhibit an antiferromagnetic (AF) phase. This is certainly true in the case of simple cubic (SC) and BCC lattices that can be easily divided into two sublattices. The critical temperature and the critical behavior of such systems do not change at all from the ferromagnetic (F) case, if one considers the staggered magnetization instead of the homogeneous magnetization. However, in the case of closed-packed lattices that cannot be divided into sublattices, e.g. in the case of FCC and HCP lattices, it becomes nontrivial to determine the most stable phase even at the absolute-zero temperature. This is the simplest example of frustrated AF phases. Similar problem occurs also in the antiferromagnetic triangular lattice. Another type of frustrated systems can be constructed by setting an antiferromagnetic coupling $J_2$ ($J_2 < 0$) to the next nearest-neighbor interaction, though a ferromagnetic coupling $J_1$ ($J_1 > 0$) is set to nearest neighbors. Such a model is called the axial-next nearest-neighbor Ising (ANNNI) model [125]. Many theoretical approaches have been applied to this model, including the low-temperature series expansion [126, 127], Monte Carlo simulations [128–130] In this model, when $-J_2/J_1$ is relatively small, the usual ferromagnetic (F) phase appears. If $-J_2/J_1$ becomes larger, the two step phase transitions take place. At $T = T_{c_1}$, the system changes from the paramagnetic (P) phase to a modulated phase (M), and then, at $T = T_{c_2}$ ($T_{c_2} < T_{c_1}$), the system changes into the F phase. The point that $T_{c_1}$ coincides with $T_{c_2}$ is called the Lifshitz point. The Lifshitz point occurs at a certain value of $-J_2/J_1$ which

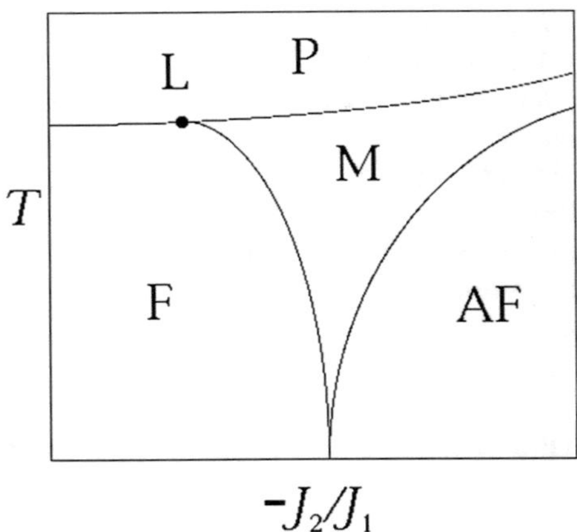


**Fig. 5.15** Typical phase diagram of the ANNNI model. The point indicated by L is the Lifshitz point, and P, F, AF and M stand for respectively, the paramagnetic, ferromagnetic, antiferromagnetic and modulated phases

is relatively small. As $-J_2/J_1$ increases further, many different modulated phases appear and multi-step phase transitions take place between the P and F (or AF) phases; see Fig. 5.15, for example. This model is closely related to the commensurate-incommensurate phase transition, although the introduction of the frustration in spin systems is not restricted to the Ising model but possible to other models which will be discussed below. (For example, ANNNXY model is studied by the Monte Carlo method [131]. The high-temperature series expansion for the ANNN $n$-vector model is treated in [132].) There is a good review article on this subject by Selke [133].

Another direction of extending the model is to consider a spin-glass problem, where the nearest-neighbor exchange constants $J_{ij}$ are randomly chosen from $\pm J$, according to the distribution function

$$p(J_{ij}) = p\delta(J_{ij} - J) + (1 - p)\delta(J_{ij} + J), \tag{5.70}$$

where $p$ is a parameter between (0,1); the two extremums, $p = 1$ and $p = 0$, correspond to the homogeneous ferromagnetic and antiferromagnetic systems, respectively. Many Monte Carlo studies using a standard method [134, 135], replica Monte Carlo method [136] and Monte Carlo renormalization group [181]. method [137], and the multicanonical Monte Carlo method [43, 138], and the related approach called "tempering" [139] have been devoted to this problem. The three-dimensional system with the value of $p$ around $1/2$ shows a spin glass phase, in which the time average of each spin has nonzero value but the ensemble average of each spin is zero:

$$\bar{S}_i \neq 0, \qquad \langle S_i \rangle = 0. \tag{5.71}$$

At the spin-glass transition temperature, the magnetic susceptibility shows, in general, a cusp-like anomaly while the magnetic nonlinear susceptibility diverges.

The effect of surfaces in semi-infinite systems have also been widely discussed by means of mean-field theory including wetting transition [140, 141] and uni-

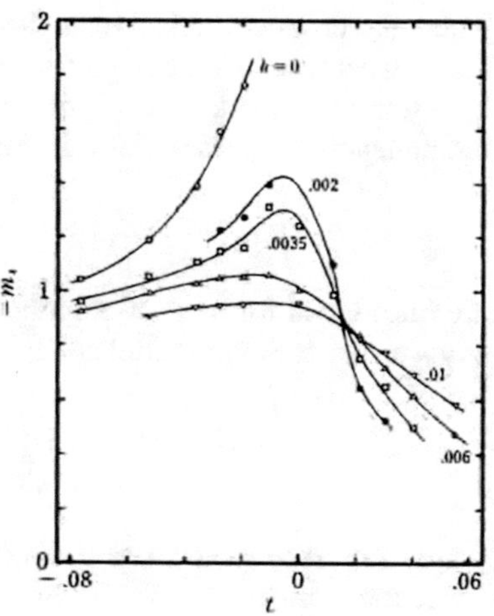


**Fig. 5.16** Temperature ($t = (T - T_c)/T_c$) dependence of the negative surface magnetization $-m_s$. This figure represents results with five different magnitudes of the homogeneous magnetic field $h = 0$–$0.01$. The exchange interaction parameter at the surface is chosen to be $J_s = 0.5J$ [146]

versal critical amplitude ratios [142], Monte Carlo simulations [124, 143–146], exact solutions in two-dimensional systems [106, 147], high-temperature series expansions [148, 149] and rigorous field-theoretical renormalization-group methods [150, 151]. Figure 5.16 represents the negative surface magnetization, defined as $-m_s = \int_0^\infty [m(\infty) - m(z)]dz$, obtained by a Monte Carlo simulation [146]. It diverges as $t \to -0$ for zero magnetic field $h = 0$, since $-m_s \sim (-t)^{\beta_s}$ with the surface exponent $\beta_s = \beta - \nu \sim -0.31 < 0$.

Let us briefly comment on various new Monte Carlo algorithms of Monte Carlo simulations for spin systems. As we mentioned in Sect. 5.4.2, the multiple-histogram method was introduced in spin systems first by Ferrenberg and Swendsen [26]. Since the state of each Ising spin is characterized by only one bit of information, i.e. spin up (1) or spin down (0), it could be very wasteful of memory to use a full 32-bit integer to store only this single piece of information. Thus, the multispin coding method is more suitable for this purpose [124, 152]. This method puts 32 different spins (or samples) to an integer and processes all the 32 spins (or samples) simultaneously by an integer operation which is the same for all the bits. In order to speed up the Monte Carlo simulations, a new algorithm called the multi-spin-flip method [153] and the cluster algorithm [154] has been proposed, where many spins flip simultaneously. The cluster algorithm proposed by Wolff is applicable also to the $O(n)$ models explained below [155]. Other new Monte Carlo algorithms were reviewed in the article by Swendsen et al. [156].

Many classical and quantum lattice models can be described by the tensor network [157]. For example, for the two-dimensional nearest-neighbor Ising model (5.60) on a triangular lattice, the partition function (5.61) can be rewritten as

$$Z = \mathrm{Tr}_S \mathrm{e}^{-\beta H} = \mathrm{Tr} \prod_{\Delta_{ijk}} \mathrm{e}^{\beta J(S_i S_j + S_j S_k + S_k S_i)/2}, \tag{5.72}$$

where the product is taken over all small triangles and the factor $1/2$ accounts for the fact that each bond is shared by two triangles. This partition function can be expressed by a tensor network in terms of bond variables through a duality transformation to the honeycomb lattice. For this purpose, the bond variables

$$\sigma_{ij} = S_i S_j, \tag{5.73}$$

are introduced for nearest-neighbor spin pairs. They can take either $+1$ for parallel spins or $-1$ for antiparallel spins. The partition function is the expressed as

$$Z = \mathrm{Tr}_S \mathrm{Tr}_\sigma \prod_{\langle ij \rangle} \delta_{\sigma_{ij},\, S_i S_j} \prod_{\Delta_{ijk}} \mathrm{e}^{\beta J (S_i S_j + S_j S_k + S_k S_i)/2}, \tag{5.74}$$

where the trace means the summation over all spin ($S_i$) and bond ($\sigma_{ij}$) variables. Since the number of bonds is equal to the sum of the number of sites and the number of triangles, $\mathrm{Tr}_S \mathrm{Tr}_\sigma \prod_{\langle ij \rangle}$ can be rewritten as a product of $\mathrm{Tr}_S \prod_i$ and $\mathrm{Tr}_\sigma \prod_{\Delta_{ijk}}$. On each triangle, we have $\sigma_{ij}\sigma_{jk}\sigma_{ki} = S_i S_j S_j S_k S_k S_i = 1$ irrespective to the values of spin variables. Therefore, we can take the trace with respect to spin variables $\mathrm{Tr}_S \prod_i$ first to get

$$Z = \mathrm{Tr}_\sigma \prod_{\Delta_{ijk}} \frac{1 + \sigma_{ij}\sigma_{jk}\sigma_{ki}}{2} \mathrm{e}^{\beta J (S_i S_j + S_j S_k + S_k S_i)/2}, \tag{5.75}$$

where we used the identity $\delta_{\sigma_{ij}\sigma_{jk}\sigma_{ki},1} = (1 + \sigma_{ij}\sigma_{jk}\sigma_{ki})/2$. If we use $i$ to denote the site position of the dual lattice, the partition function becomes that of a standard tensor network model

$$Z = \mathrm{Tr} \prod_i T_{x_i y_i z_i}, \tag{5.76}$$

where the trace means the summation over all indices and

$$T_{x_i y_i z_i} = \frac{1 + \sigma_{ij}\sigma_{jk}\sigma_{ki}}{2} \mathrm{e}^{\beta J (\sigma_{x_i} + \sigma_{y_i} + \sigma_{z_i})/2} \tag{5.77}$$

is a third-order tensor on the hexagonal lattice.

This can be easily extended to the square lattice, whose dual lattice is the square lattice. The partition function for the square lattice Ising model is given by

$$Z = \mathrm{Tr} \prod_i T_{x_i y_i x_i' y_j'}, \tag{5.78}$$

where the trace means the summation over all indices and

$$T_{x_i y_i x_i' y_j'} = \frac{1 + \sigma_{x_i}\sigma_{y_i}\sigma_{x_i'}\sigma_{y_i'}}{2} \mathrm{e}^{\beta J (\sigma_{x_i} + \sigma_{y_i} + \sigma_{x_i'} + \sigma_{y_i'})/2} \tag{5.79}$$

is a fourth-order tensor on the (dual) square lattice. Similarly, for the honeycomb lattice, its dual lattice is the triangular lattice, which has six nearest neighbors. Therefore, we have a sixth-order tensor in this case. This duality transformation can be extended to other models such as $q$-state Potts model (see next section) as well [157]. The tensor network can be numerically treated by the tensor renormalization group (TRG) [157, 158] based on the higher-order singular value decomposition (HOTRG) [159].

### (b) Classical Spin Models

When the magnet has planar magnetic anisotropy, one has the $XY$ model, whose Hamiltonian is given by

$$\mathcal{H}_{XY} = -J \sum_{i>j} (S_i^x S_j^x + S_i^y S_j^y) - H^x \sum_i S_i^x. \tag{5.80}$$

On the other hand, if the magnet does not have any magnetic anisotropy at all, one should consider the Heisenberg model, given by

$$\mathcal{H}_{\text{Heisenberg}} = -J \sum_{i>j} (S_i \cdot S_j) - H^z \sum_i S_i^z. \tag{5.81}$$

In these two models, however, each spin variable $S_i$ is a vector operator and should be regarded as a quantum mechanical variable. According to quantum mechanics, the magnitude of the spin angular momentum $S$ is restricted to integers or half-integers $(0, \frac{1}{2}, 1, \frac{3}{2}, 2, \frac{5}{2}, \ldots)$ in units of $\hbar$. If this magnitude becomes sufficiently large, each spin behaves as a simple classical variable. In the limit $S \to \infty$, the spins in the $XY$ model obey

$$(S_i^x)^2 + (S_i^y)^2 = S^2, \tag{5.82}$$

while those in the Heisenberg model obey

$$(S_i^x)^2 + (S_i^y)^2 + (S_i^z)^2 = S^2, \tag{5.83}$$

and they are classical vectors rotating under these constraints (see Fig. 5.17 for the classical $XY$ model). If we regard each classical Heisenberg spin as a crystalline orientation of a microzone partitioned regularly by a lattice with a lattice constant of order $100 \sim 1000\,\text{Å}$, one may model polycrystals or grain boundaries of polar crystals, and discuss their macroscopic energetics, statistics and also dynamics.

Extending these classical $XY$ and Heisenberg models to a more general spin space, one may introduce a model called the $n$-vector model or the $O(n)$ model (sometimes called the nonlinear $\sigma$ model), whose Hamiltonian is given by

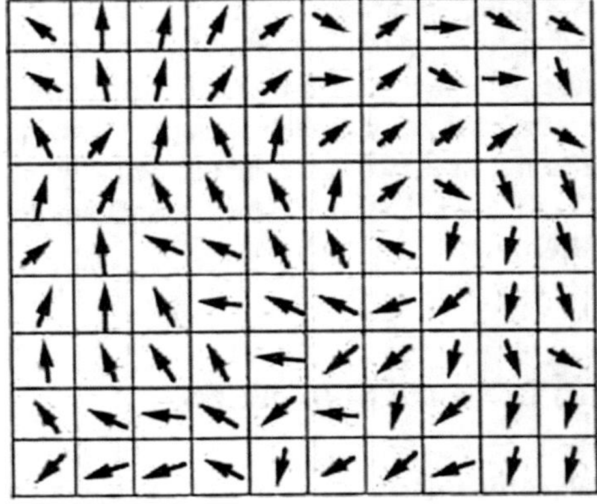

**Fig. 5.17** Classical $XY$ model. Each spin has a fixed length $S$ and can rotate only in the $XY$ plane. As in the Ising model, adjacent spins interact with each other according to the angle between the spins

$$\mathcal{H}_{n\text{-vector}} = -J\sum_{\langle ij\rangle}(S_i \cdot S_j) - H_z \sum S_i^z,$$

$$|S_i|^2 = \sum_{\mu=1}^{n} S_i^{(\mu)^2} = S^2. \tag{5.84}$$

This model reduces to the Ising model, the classical $XY$ model and the classical Heisenberg model, respectively, in the cases $n = 1, 2$ and 3. Moreover, it is known that this model becomes equivalent to a self-avoiding walk, which frequently appears in polymer physics, in the limit $n \to 0$ [67]. The correspondence between the $n$-vector model in the limit $n \to 0$ and the configurations of a polymer chain in a good solvent is quite a general theorem, and holds for many chains or polymer networks in solution with arbitrary architecture. On the other hand, the limit $n \to \infty$ can be treated exactly in the $n$-vector model, since, in this limit, the constraint (5.84) imposed on each spin quantity becomes equivalent to the constraint on the total quantity

$$\sum_i |S_i|^2 = \sum_i \sum_{\mu=1}^{n} S_i^{(\mu)^2} = NS^2, \tag{5.85}$$

where $N$ is the total number of spins in the system. The model described by the Hamiltonian (5.84) with the constraint (5.85) is called the spherical model and can be solved exactly [160]. This equivalence between the $n$-vector in the large $n$ limit and the spherical model was first pointed out by Stanley [161]. All these models are known to exhibit a second-order phase transition in three dimensions, while no order–disorder phase transition of the usual kind takes place for $n \geq 2$ in two dimensions. This can be shown by the Mermin–Wagner theorem [162].

In the case of a classical $XY$ model, the spin variable $S_i = (S_i^x, S_i^y)$ in (5.80) is the two-dimensional classical spin and can point in an arbitrary direction in the $XY$ plane with a fixed length $|S_i| = S_0$. In place of a spin variable which can point to an arbitrary angle in the interval [0, 360°], one may discretize the full angle interval [0, 360°] into the $m$ fixed angles 0, $360°/m$, $720°/m$, …, $360°(m-1)/m$. This discretized model should be distinguished from the classical $XY$ model and is called the $m$-clock model or simply the clock model.

The two-dimensional classical $XY$ model also describes superfluidity and shows the Kosterlitz–Thouless transition concerning vortex condensation and melting [163, 164]. This phase transition is not the usual order–disorder-type phase transition but a very weak one, and no other type of phase transition exists in this model according to the Mermin–Wagner theorem. Near the Kosterlitz–Thouless transition point $T_{\mathrm{K}}$, the correlation length $\xi$ of the fluctuations diverges like

$$\xi \sim \exp\left(\frac{\text{const.}}{|T-T_{\mathrm{K}}|^{1/2}}\right), \tag{5.86}$$

$\xi(T)$ does not obey a power law (5.66). This model has been investigated extensively in terms of Monte Carlo simulations by many authors [165–169]. The other systems to which the $XY$ model can be applied (in three dimensions) are the commensurate–incommensurate phase transition (the commensurate–incommensurate transition is generally first-order and cannot be dealt within the classical $XY$ model, but the transition between the random normal phase and the incommensurate phase is second-order and can be dealt within the classical $XY$ model), the $\lambda$ transition at the occurrence of superfluidity, and superconductivity.

It is known from Monte Carlo simulations [170–172] and high-temperature series analyses [108, 149, 173, 174] that the critical exponents of the three-dimensional classical $XY$ model are $\alpha \simeq -0.02$ and $\gamma \simeq 1.32$. On the other hand, the three-dimensional classical Heisenberg model has the critical exponents $\alpha \simeq -0.14$ and $\gamma \simeq 1.38$ [108, 149, 175, 176]. For Monte Carlo study of the three-dimensional classical Heisenberg model, see [177–180]

One possible procedure in the investigation of macroscopic properties is to repeatedly coarse-grain both the spatial and the temporal scales. The coarse-graining procedure itself is a very simple idea, but its repeated application is related to an important idea about the renormalization group which has led to very successful progress in the current theory of statistical physics. In the vicinity of a second-order phase transition, the coarse-graining procedure may also play an important role in the analysis of the behavior of spin systems by means of the Monte Carlo method. The coarse-graining procedure can be formulated in a mathematically rigorous fashion and one may call this the renormalization group transformation. When applying the renormalization group transformations repeatedly, the dynamic variables transforming into those which can be regarded as thermal agitations or random forces are called irrelevant variables, while those which become more and more important are called relevant variables. In fact, the dynamic variables introduced in Sect. 5.2.1 should contain all the relevant variables in some form. The first coarse-graining may amount to modeling stochastically the microscopic system as has been described so far. The second coarse-graining is to construct the next model by the well-defined renormalization group transformation from the model obtained at the first step. The third coarse-graining is to renormalize the model one step further to get the next model. One may repeat similar renormalizations many times. In some cases, during the successive renormalization transformations, a large number of irrelevant variables are removed and only a small number of (renormalized) relevant variables remain. In such cases, one may find that a universal part is extracted from the original system. In such circumstances, the universal part is changed no more by the renormalization transformation. This situation is called the fixed point. At the fixed point, there is no specific length scale in the system, so that all the spatial patterns of the system become fractal. The critical point in a second-order phase transition is one of its significant examples. This is why the critical exponents characterizing the critical anomalies are, in many cases, universal and do not depend on the details of the original system (there are also many exceptions, e.g. the two-dimensional $XY$ model previously mentioned). Now one may understand why many critical exponents measured in real magnets, critical binary mixtures and so forth coincide to high accuracy with the exponents of one of the models described above. This fact is called the universality. We say that all the models and real systems showing the same universal behavior belong to the same universality class. Therefore, systems belonging to different universality classes exhibit different critical behaviors.

In the analysis of a second-order phase transition, it is effective to combine the Monte Carlo simulation with the renormalization group transformation. This method is called the Monte Carlo renormalization group [181]. Using the Monte Carlo renormalization group helps a very accurate determination of both the critical temperature and the critical exponents of various spin models. For example, the three-dimensional Ising model was treated by Pawley et al. [182]. Typically, one uses a block spin transformation [183] in which one gathers several neighboring spins and regards them as a single spin. If the ordering pattern does not change under the block spin transformation (renormalization transformation), the system is just at the critical point. In practice, one may evaluate the change of the magnetization or the susceptibility instead of the ordering pattern, which is very abstract. Figure 5.18 represents an example of the block spin transformations in the three-dimensional 8-clock model on a simple cubic (SC) lattice.

Similar models which are described as spin models include the $q$-state Potts model, the dimer model, the ice model, the eight-vertex model and the $\phi^4$ model, for each of which exact solutions, high-temperature expansions and Monte Carlo simulations exist [52].

The $q$-state Potts model [184] is a lattice model on which Potts spins are allocated. Each Potts spin can have one of $q$ different states. The adjacent Potts spins interact only when they are in the same state. Its Hamiltonian reads

$$\mathcal{H}_{\text{Potts}} = -J \sum_{\langle ij \rangle} \delta_{\sigma_i,\sigma_j} . \tag{5.87}$$

where $\sigma_i$ and $\sigma_j$ represent the Potts spins. First we consider as usual the ferromagnetic case where $J$ is positive. The two-dimensional $q$-state Potts model shows a second-order phase transition for $q \leq 4$ and a first-order phase transition for $q > 4$. The

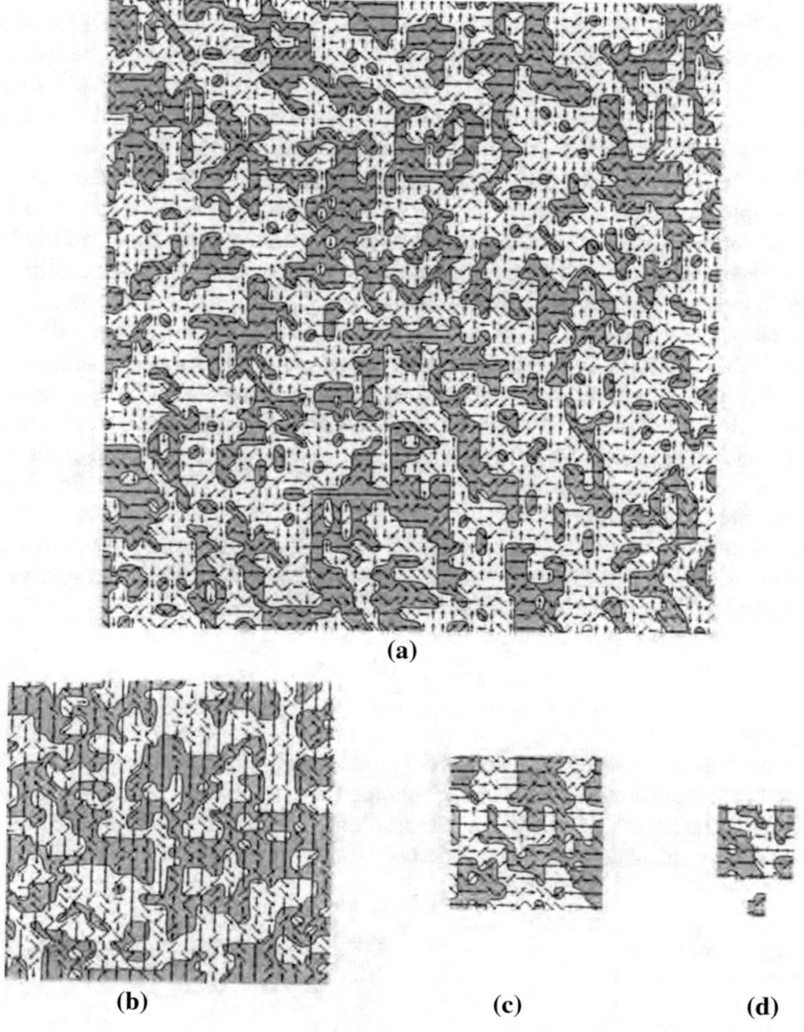


**Fig. 5.18** **a** Monte Carlo simulation of a three-dimensional 8-clock model. The figure describes the spins within one two-dimensional cross section. The *left* surface has a the fixed boundary condition, the *right* surface has a the free boundary condition, while the other two surfaces(top and bottom) have a the periodic boundary condition. **b** A spin configuration obtained from **a** by one block spin transformation. **c**, **d** The configurations obtained from **a** by two and three block spin transformations

transition temperature of a two-dimensional $q$-state Potts model is known exactly [52]. Monte Carlo simulations of two-dimensional $q$-state Potts models with various $q$ values have also been performed by many authors; see, for example, [185, 186]. On the other hand, the three-dimensional Potts model has this boundary between the second- and first-order transitions somewhere between $q = 2$ and 3. Monte Carlo simulations have been performed for the tree-dimensional Potts model by Jensen and Mouritsen [187], Herrmann [188] and Blöte and Swendsen [189].

The three-dimensional $q$-state Potts model with *negative* $J$ (i.e. the antiferromagnetic coupling), which is called the antiferromagnetic (AF) Potts model, has been attracted some interest, because this model shows a second-order phase transition and because Banavar et al. [190] suggested from their field-theoretical renormalization-group calculation and Monte Carlo simulations that this model belongs to the same universality class as that of the three-dimensional $O(q-1)$ model. Regarding several available results of Monte Carlo simulations [191–193] and a high-temperature series expansion [194], there is a good agreement in the critical exponents between the three-dimensional AF 3-state Potts model and the three-dimensional classical $XY$ model ($O(2)$ model). A random Potts model, where the nearest-neighbor interactions $J_{ij}$ are chosen randomly from F and AF as (5.70), has been investigated also by the Monte Carlo method [195].

In contrast to the $n$-vector model, there is a model describing the orientational phase transition of molecular crystals or liquid crystals, which is called the $n$-component quadrupolar model or the RP$^{n-1}$ model (Fig. 5.19):

$$\mathcal{H}_{\text{Quadrupole}} = -\sum_{i>j} J_{ij}(S_i \cdot S_j)^2. \tag{5.88}$$

This model is known to exhibit a first-order phase transition. There are exact solutions for the one-dimensional system, a high-temperature expansion, a low-temperature expansion, a mean-field theory and an exact solution in the limit $n \to \infty$ [196]. Monte Carlo simulations have been also carried out for this system with $n = 3$ [197]. The

**Fig. 5.19** Quadrupolar model. The crystal model is composed of oval-shaped molecules and describes the orientational phase transitions of the molecules. Although this model is a simple extension of the classical $XY$ model, it exhibits a first-order phase transition even though there is no external magnetic field orientational glass in which the coupling constant $J_{ij}$ is randomly distributed spatially has also attracted interest. The orientational glass was reviewed together with the Potts glass by Binder and Reger [198]. By extending the $O(n)$ model ($n$-vector model) and the $\mathrm{RP}^{n-1}$ model into a quantum mechanical gauge theory, one may construct the $SU(n)$ model and the $\mathrm{CP}^{n-1}$ model. On the other hand, if we regard again each quadrupole of the 3-component quadrupolar (or $\mathrm{RP}^2$) model as a crystalline orientation of a microzone partitioned regularly by a lattice with a lattice constant of order $100 \sim 1000$ Å, one may model polycrystals or grain boundaries of layered crystals, and discuss their macroscopic energetics, statistics and also dynamics. It is also possible to extend the quadrupolar model to multipolar models [199]. Then, one is allowed to treat more complicated interactions of the form assumed usually between multipoles.

### 5.4.6 *Nucleation*

Apart from quantum liquids like helium, liquids generally solidify when the pressure increases or the temperature decreases. Similarly, vapors liquefy as the pressure increases or the temperature decreases. The liquid–solid phase transition is always first-order and the liquid–vapor phase transition is usually first-order except at the critical point. Therefore, solid nuclei or liquid droplets appear at the initial stage of solidification or the liquefaction. (Since there is variety of solid phases, solidification here does not necessarily mean a unique crystallization. For example, depending on the cooling rate, liquids sometimes become glasses or amorphous.)

If the shape of a nucleus (or a droplet) is spherical and its radius is given by $R$, the surface energy is proportional to $R^2$ and the volume energy gain (the energy difference between the liquid and solid (or vapor) phases) is proportional to $R^3$, its formation energy $\varepsilon$ can be estimated as $aR^2 - bR^3$. Here the proportionality constants $a$ and $b$ are functions of the temperature. The condition that $d\varepsilon/dR$ is negative is required for the nucleus to grow, and it is satisfied only when the radius of the nucleus is larger than the critical radius $R_\mathrm{c} = 2a/3b$. Therefore, a nucleus whose size is greater than the critical size must exist in the liquid (or vapor) for solidification (or condensation) to occur. Otherwise, crystal growth does not start even below the melting temperature; the system stays in a supercooled state, which is metastable. Generally, the critical radius $R_\mathrm{c}$ becomes smaller as the temperature decreases. This fact is related to the Gibbs–Thomson effect. That is, if the radius $R$ of a nucleus decreases, the equilibrium pressure $p(R)$ between the nucleus and the surrounding liquid increases (The Gibbs–Thomson equation is given by $\log[p(R)/p(\infty)] = 2v\sigma/k_\mathrm{B}TR$, if the volume per atom in the solid is $v$ and the surface tension is $\sigma$), and in turn the effective melting temperature at a given pressure $p$ decreases; therefore the growth is suppressed. This is one of the important conclusions of the isothermal nucleation theory [200, 201]. Even in second-order phase transitions, in order to form a stable domain which has an order different from the remaining region, the domain size must be larger than the critical size. Such a nucleus may be formed by occasional thermal fluctuations or by

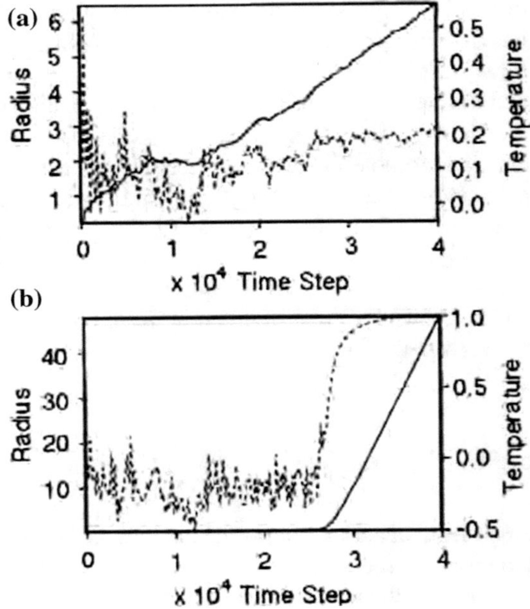


**Fig. 5.20** Simulations of isothermal and adiabatic nucleation [203]: **a** corresponds to fast thermal diffusion, and **b** to slow thermal diffusion. The *dotted lines* denote the surface temperature of the nucleus and the *solid line* is the radius of the nucleus

an impurity. More often, the nucleation starts on a surface. If the atomic constituents are the same in the substrate and the nuclei, the nucleation is called homogeneous nucleation, while if they are different in the substrate and the nuclei, the nucleation is called inhomogeneous.

The above argument about the nucleation assumed that the nucleation proceeds isothermally. Does the isothermal nucleation theory correctly describe the initial stage of a nucleation process? According to Mayer [202], there is a possibility that the speed of nucleation and growth exceeds the speed of the thermal diffusion by which the latent heat can diffuse into the surrounding liquid. He called such an anomalous nucleation adiabatic nucleation. In fact, in a computer simulation, it is possible to choose parameters so as to make the speed of the nucleation and growth faster than the speed of latent-heat diffusion. However, both the nucleation and growth process and thermal diffusion are mediated by atomic vibrations, so that these two processes should proceed roughly with the same speed. Moreover, in the case of metals, thermal diffusion mediated by electrons proceeds much faster than processes mediated by atomic vibrations. Therefore, there is a question as to whether adiabatic nucleation is possible. In order to simulate these processes, we may introduce a simple time-evolution equation. Figure 5.20 shows numerical results with two different parameters [203]. Figure 5.20a describes the case of rapid thermal diffusion and corresponds to an isothermal nucleation, while Fig. 5.20b describes the case of slow thermal diffusion and corresponds to adiabatic nucleation. It is still unclear whether such an anomalous condition can be realized in normal materials.

In the late stages of crystal growth and phase separation, mutually growing clusters collide with each other, and this phenomena should be taken into account. The kinetic theory of the nucleation and growth processes was first presented by Becker and Döring [204] in 1935. Later, Lifshitz and Slyozov [205] found that the size of the droplet increases with time as $R \sim t^{1/3}$ and its surface energy decreases as $u \sim t^{-1/3}$. Cluster surfaces always undergo repeated melting and solidification. When a large cluster is surrounded by small clusters, particles diffusing out of the small clusters can be incorporated into the bigger cluster. During such processes, small clusters are repeatedly created and annihilated, and larger clusters grow gradually and stabilize themselves. This is called Ostwald ripening. Binder and Stauffer [206] discussed the possibility that the exponents appearing in the growth process may change according to the mechanism of diffusion. Moreover, they performed extensive Monte Carlo simulations of the three-dimensional Ising model using both Kawasaki dynamics and Glauber dynamics, and found that the domain size increases according to $R \sim t^{1/3}$ in the case of Kawasaki dynamics and $R \sim t^{1/2}$ in the case of Glauber dynamics. In order to describe the dynamics of the first-order phase transitions realized in real materials, Monte Carlo simulations of Ising models with more realistic many-body interactions are desirable [207]. It should be noted that, in real situations, the value of the exponent changes because of fluid-dynamics effects.

Generally, when the supercooling (the temperature measured from the ideal melting temperature) becomes large, the size of the critical nucleus becomes small, and the energy barrier required for the nucleation process becomes lower. Therefore, for sufficiently large supercooling, spontaneous nucleation due to thermal fluctuations (or quantum fluctuations in special cases) takes place at various points. In the case of phase separation processes in binary mixtures or solid solutions, the temporal evolution of the density $\phi$ obeys, according to Kawasaki dynamics (5.36), the partial differential equation

$$\frac{\partial \phi}{\partial t} = M\left(a\nabla^2\phi + 2b\nabla^2\phi^3 - \epsilon\nabla^4\phi\right). \tag{5.89}$$

This equation, which is frequently used in the phase-field model for the conserved order parameter [208, 209], may be obtained as follows. First, the density $\phi$ satisfies the continuity relation

$$\frac{\partial \phi}{\partial t} = -\nabla \cdot \boldsymbol{J}, \tag{5.90}$$

where $\boldsymbol{J}$ is the flux density, which is related to the chemical potential $\mu$ and the particle mobility $M$ via

$$\boldsymbol{J} = -M\nabla\mu. \tag{5.91}$$

One may write the chemical potential as

$$\mu = \frac{\partial F}{\partial \phi} - \epsilon\nabla^2\phi, \tag{5.92}$$

where $F$ is the free energy and $\epsilon$ is the coefficient representing how much the density gradient contributes to the chemical potential. Combining all these equations, we derive the Cahn–Hilliard equation [210, 211]

$$\frac{\partial \phi}{\partial t} = M\nabla^2 \left( \frac{\partial F}{\partial \phi} - \epsilon \nabla^2 \phi \right). \tag{5.93}$$

Then, if we use the $\phi^4$ model for the free energy, $F = (a\phi^2 + b\phi^4/2)/2$, we will have (5.89). Since $a$ and $b$ are empirical parameters, there is unfortunately no connection between this model to the ab initio methods. Therefore, it is generally difficult to use this model to predict microstructures without knowing the experimental data. In order to overcome this difficulty, one may use the cluster expansion together with the potential renormalization described in Chap. 4 [212]. That is, dividing the space in fine grid and assuming that each grid point corresponds to, for example, a simple cubic unit cell including one tetrahedron, one may use the renormalized potential as the local free energy, which include the entropy for the length scale less than the lattice constant. In order to incorporate the entropy for the length scale larger than the lattice constant, one may add a random force in (5.93). The random force is usually assumed as a white noise and its strength is determined by the fluctuation-dissipation theorem [213]. One should note that, in this treatment, the chemical potential $\mu$ should be calculated not by the free energy derivative $\partial F/\partial \phi$ by the difference of the free energy $F(\phi) - F(\phi - 1)$ if $\phi$ is defined as the number of atoms occupying the tetrahedron. This procedure based on the ab initio renormalized potential including vacancies [214, 215] works well for FePt and AlNi alloys [212].

Equation (5.89) can be solved by linearization and Fourier transformation and has a solution of the form

$$\phi - \phi_0 = \sum_{\boldsymbol{k}} \mathrm{e}^{R(\boldsymbol{k})t} \Big[ A(\boldsymbol{k}) \cos(\boldsymbol{k} \cdot \boldsymbol{r}) + A(\boldsymbol{k}) \sin(\boldsymbol{k} \cdot \boldsymbol{r}) \Big]. \tag{5.94}$$

Since $R(\boldsymbol{k})$ has a sharp maximum at $|\boldsymbol{k}_\mathrm{m}| = (-a/\epsilon)^{1/2}$, the phase separation proceeds as a wave given by a superposition of many wave fluctuations with wavelength $\lambda_\mathrm{m} = 2\pi/|\boldsymbol{k}_\mathrm{m}|$. Figure 5.21 shows a snapshot obtained by a numerical analysis of (5.89), representing a pattern of randomly connected domains which have approximately the same thickness $\lambda_\mathrm{m}$ [216]. This process proceeds quite rapidly and can be observed in systems that separate into two phases and show slow dynamics, such as multicomponent glasses, alloys and polymer mixtures. This phenomenon is called spinodal decomposition. In spinodal decomposition, the dynamic structure factor $S(q, \omega)$ is well described by a special scaling function; this phenomenon is essentially different from the usual critical phenomena observed in the vicinity of a second-order phase transition and is called dynamic scaling [217].

A stability limit at which the supercooled liquid can exist in a metastable state is called the spinodal line (Fig. 5.22). According to the Monte Carlo study by Binder and Stauffer [218], the spinodal line is not a well-defined boundary line but depends on the length of the observation time, etc., in particular near the critical point where

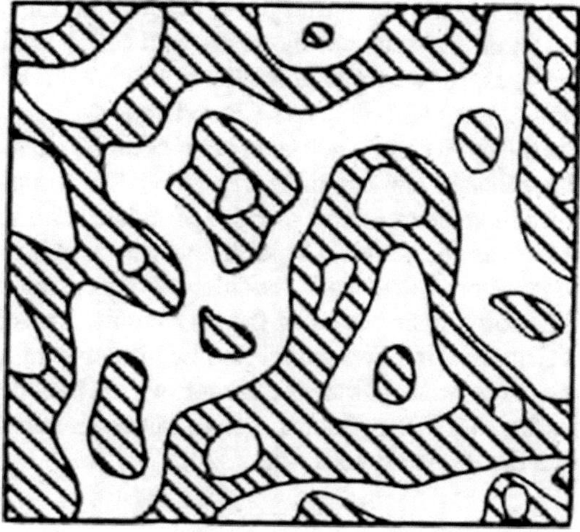

**Fig. 5.21** Schematic illustration of spinodal decomposition [216]

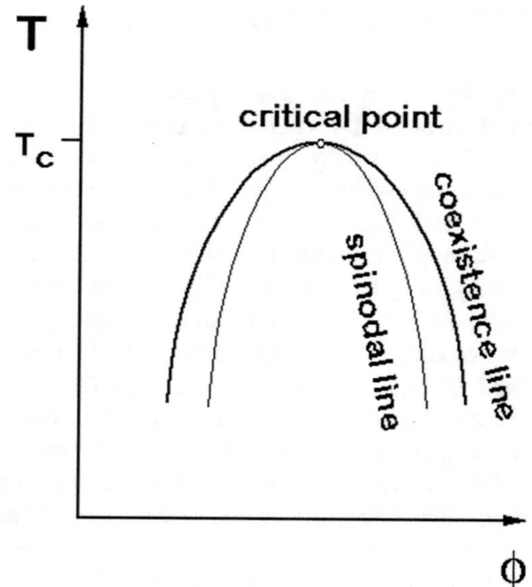


**Fig. 5.22** Coexistence line near the critical point and the spinodal line in a phase diagram. The symbol $\phi$ stands for the concentration

the critical slowing down becomes important. However, for many purposes, the idea of spinodal decomposition is still useful. On the other hand, however, when one quenches a solution of molecules with large molecular weight, the structure often freezes into an amorphous state because it cannot easily change into a more stable regular structure. Such a phenomenon is called a glass transition and should be distinguished from spinodal decomposition.

As another phenomenon closely related to phase separation, a three-component system composed of surfactants together with a mixture of water and oil forms a very interesting ordering pattern called the microphase separation. In this system, the interface between the water and oil is stabilized by surfactants, so that there emerges, for example, a lamellar phase, which has a multisandwich structure with water layers and oil layers aligned alternately, and a microemulsion phase, in which the oil and water invade each other microscopically and their interfaces fold in a very complicated and irregular way [219]. The simplest way to model this system is to introduce the one-order-parameter model. This model assumes a free energy which increases for a spatial modulation with a relatively long wavelength but decreases for a short-wavelength modulation. Since this model does not contain explicitly the spatial distribution of the surfactants, a possible extension is to introduce their distribution into the model (the two-order-parameter model). A further extension is the three-order-parameter model, which includes the molecular orientation of the surfactants [219]. Such a system can be simulated by an Ising model [95, 220–223] or a three-state Potts model [224].

### 5.4.7 Crystal Growth

There are many forms of crystals in nature. The crystal forms at equilibrium with the environment are called the equilibrium shapes of the crystals. The flat surfaces of crystal polyhedrons are called facets. The reason that crystals at low temperatures have various facets in the equilibrium shape is mainly due to the anisotropy in the surface tension caused by the discreteness of the crystal lattice. If there is no anisotropy in the surface tension, the crystal becomes spherical, like a liquids in zero gravity, in order to minimize its surface area. The equilibrium shape is determined mathematically by the variational problem (the problem of finding the minimum) of the surface (free) energy. Its solution is known as the Wulff theorem. On the other hand, owing to thermal agitation, crystals have fluctuating surfaces instead of facets at high temperatures. Therefore the edges and corners are rounded off and the crystal shape approaches a somewhat spherical shape. Generally, facets disappear above a well-defined temperature $T = T_{\mathrm{R}}$ called the roughening transition temperature. This roughening transition is one of the characteristics of a discrete lattice and has been well studied in statistical physics. This transition occurs at the point where the step free energy becomes zero. In order to discuss the roughening transition, it is possible to introduce the solid-on-solid (SOS) model, which does not allow overhangs as in Fig. 5.23, or the lattice Ising model with an appropriate surface tension.

While there are many beautiful regular crystals, some crystals have very rough, i.e. uneven, surfaces. Ice crystals have beautiful irregular patterns and some other crystals have a dendritic structure which is far from regular. Whether the crystal surface is flat on the atomic scale depends not only on the atomic alignment of the substrate and the binding morphology of the adatoms, but also strongly on the environment, e.g. whether the particle concentration in the environment is dilute (as in vapor growth or solution growth) or dense (as in melt growth), and also the temperature and pressure in the environment.

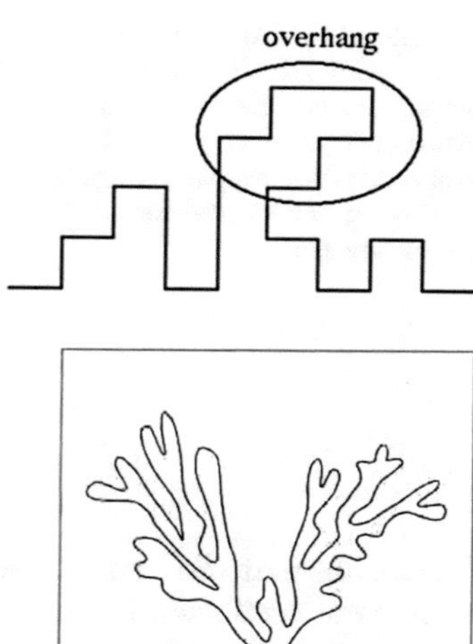


**Fig. 5.23** The SOS model does not allow the overhangs illustrated in this figure

**Fig. 5.24** Dendritic pattern caused by the Mullins–Sekerka instability

The instability of the crystal interface in crystal growth has been discussed by Mullins and Sekerka [225]. Generally, when the radius of interface curvature exceeds some value $R_0$ determined by the supercooling, smooth surfaces become unstable, and finally dendritic branching appears (Fig. 5.24). For example, if the radius of a spherical crystal exceeds this $R_0$, it cannot grow with a spherical shape any more. When the tip of a crystal grows selectively and branches as a dendrite, it is important to consider this instability, which is called the Mullins–Sekerka instability.

There are several growth modes at the atomistic level of heterogeneous crystal growth. Regular layer-by-layer growth on flat substrates is called the Frank–van der Merwe mode. Island growth on flat surfaces is called the Volmer–Weber mode. If the first layer covers the substrate completely and island growth takes place in the second layer, this growth mode is called the Stranski–Krastanov mode. A model for the regular layer-by-layer growth process for an ideal perfect crystal was first introduced by Kossel [226, 227] and studied by Stranski and Kaischew [228], Becker and Döring [204], and Volmer [229]. Later, in 1951, this model was extended to the step–kink–terrace mechanism by Burton, Cabrera and Frank [232] (Fig. 5.25). This is known as the BCF theory. Regular crystal growth in a dilute environment (e.g. vapor or solution) can be well described by the SOS model and be studied by Monte Carlo simulations [230, 231]. To be realistic, it is often required to treat all dynamic processes, such as three-dimensional diffusion in the environment, surface collisions, surface diffusion, surface adsorption and surface dissociation. During the surface diffusion of the particles, some atoms are adsorbed at kink positions. Then, the kink moves along the step edge and, finally, the step grows by one row. In a much longer time period, the step moves over the terrace and a new terrace appears on top of the old terrace. Sometimes, two-dimensional nucleation takes place on a terrace and an atomic island is created or annihilated; a big island may spread into a new terrace. This phenomenon manifests itself experimentally by RHEED oscillations.

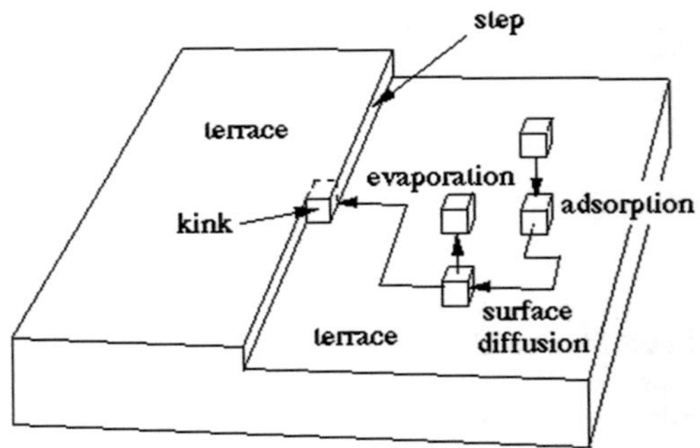


**Fig. 5.25** Step–kink–terrace model. Particles falls onto a terrace, diffuse on it and evaporate with some probability. When a particle is adsorbed at a kink or step, the surface grows

That is, the periodic change between nucleation on a terrace and step growth gives a temporal oscillation in the reflection high-energy electron diffraction (RHEED) pattern [200, 233–235].

In the Frank–van der Merwe mode of crystal growth, creation of a new atomic layer generally takes a longer time compared to the other modes, because the flat surface is unfavorable owing to the loss of the surface energy, and this fact makes two-dimensional nucleation on a terrace rarer. In this case, the speed of crystal growth is dominated by the creation rate of new atomic layers. This situation is similar to the traffic jam caused by road works; when one special process requires a much longer time compared to all other processes, the slowest process essentially determines the speed of the total process. Such a process is called the rate-determining process. In the case of the present example of crystal growth, the mode creating a new atomic layer is the rate determining process. Generally, in crystal growth, the process determining the rate is very important and strongly influences the resulting crystal morphology. One extreme example is dendritic growth in diffusion-limited growth, where the rate-determining process is the diffusion of particles in the environment.

Crystal growth proceeds so as to minimize the loss of the surface energy, so that layer growth or spiral growth often takes place at crystal surfaces. A crystal surface grown by spiral growth has at least one imperfection at the center of the spiral but it should also be categorized as a smooth surface on the atomic scale. The step–kink–terrace growth theory developed by Burton, Cabrera and Frank can also handle spiral growth. It is well known that there is an energy barrier, called the Schwoebel barrier, suppressing the migration of adatoms from one terrace to another terrace separated by a step [236, 237]. (An ab initio study of the Schwoebel barrier is described briefly in Sect. 2.5.3.) A simple computer simulation based on the BCF theory, including the Schwoebel effect, has been performed by Uwaha and Saito [238]. The problem has been reconsidered more recently with the addition of an external electric field by Sato et al. [239].

On the other hand, if the crystal surface is very rough and uneven on the atomic scale, one may say that there are many kink positions. Therefore, the atoms falling onto the growing surface are almost all adsorbed in the vicinity of the falling position without much diffusion. That is, if the crystal mainly grows at the step positions on flat surfaces, the crystal growth has the form of step growth, while for a rough surface, the crystal growth can proceed anywhere on the surface and has the form of adsorption growth.

The thin-film growth of gold, copper, silicon, GaAs and other materials from the vapor phase can be understood by the step–kink–terrace growth mechanism. (If one wants to consider the surface reconstruction or the order of the layer-by-layer stacking, a more detailed picture at the atomic or electronic level is required.) On the other hand, the thin-film growth of diamond is possible only when radical molecules from methane, ethane or oxides are present in the vapor; this helps to etch unstable adatoms. This growth process is closer to adsorption growth on rough surfaces. In the next section, we describe fractal systems as an extreme example of crystal growth on rough surfaces.

### 5.4.8 *Fractal Systems*

Growth in the vapor phase or the liquid phase can take place in several ways: one is the case where the crystal growth takes place with a regular atomic alignment, and the other is growth with an irregular or random morphology, or with growth morphology called whiskers or dendrites. In colloidal solutions or electrolytic solutions, aggregations of colloids or metals can sometimes form fascinating dendritic patterns. Sometimes these patterns appear in the three-dimensional solution and sometimes on two-dimensional surfaces. There are several models (one may refer to several earlier reviews [53, 54, 240–242]) for simulating these dendritic patterns on computers; the DLA model, the ballistic deposition model and the CCA model.

The diffusion-limited aggregation (DLA) model [54, 240, 243, 244] is one of the simplest dendritic growth models in computer physics. Initially, a seed particle is located at the center of a sphere (or circle). Then one diffusing particle starts walking randomly from one point randomly selected on this spherical shell. If this particle goes out of the sphere, remove it and start a new random walk again from the spherical shell. If the diffusing particle touches the seed at the center, attach it at that position and regard it as part of the cluster. By repeating this process many times, the seed grows to a fascinating dendritic cluster as shown in Fig. 5.26. Why such a dendritic structure appears in this simple simulation seems somewhat tricky to explain, but there is a well-established physical background. Intuitively one may understand the mechanism as follows. Assume that there is a bay area surrounded by an already grown part of the cluster. Then most of the diffusing particles are adsorbed on the cluster before entering this bay. This is closely related to the fact that the particle diffusion is treated mathematically by a diffusion equation just like the equation in the electrostatic problem of determining the electric potential for a given conducting boundary condition. In the electrostatic theory, the electric potential is almost zero everywhere inside an area surrounded by a conducting boundary with $\phi = 0$. The same thing happens for the DLA, i.e. the probability of the particle existing inside

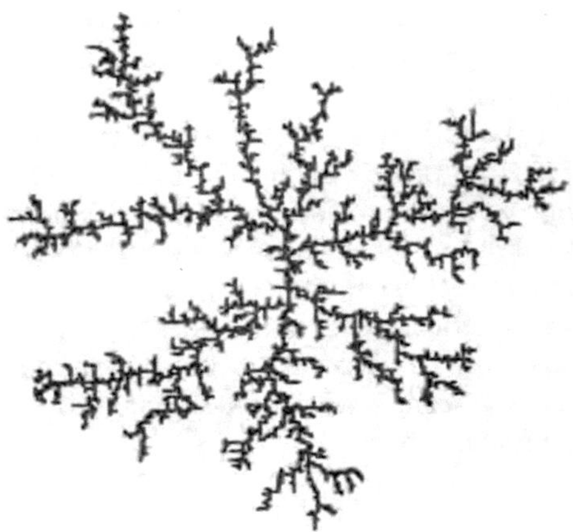

**Fig. 5.26** Pattern obtained from the DLA model [243]. The number of particles in this dendrite is 3600

the bay is almost zero. This effect is called the screening effect. Because of this effect, many bays remain in the growing morphology.

The DLA model in a $d$-dimensional hypercubic lattice is mathematically described by a set of four equations [245]:

$$\hat{P}(\boldsymbol{r},t) = [1-\hat{\rho}(\boldsymbol{r},t)]\left\{1-\prod_{i=1}^{2d}\left[1-\hat{\rho}(\boldsymbol{r}+a\boldsymbol{e}_i,t)\right]\right\} \tag{5.95}$$

defines a function that is unity at the perimeter sites of the cluster and vanishes otherwise. Here, $\hat{\rho}(\boldsymbol{r},t)$ is the density function which is unity at the cluster site and vanishes otherwise. Next,

$$\hat{\rho}(\boldsymbol{r},t+\delta t) = \hat{\rho}(\boldsymbol{r},t) + \hat{P}(\boldsymbol{r},t)\hat{v}(\boldsymbol{r},t) \tag{5.96}$$

specifies how the dendrite grows. Moreover,

$$\hat{u}(\boldsymbol{r},t+\delta t) = [1-\hat{P}(\boldsymbol{r},t)]\hat{v}(\boldsymbol{r},t) \tag{5.97}$$

and

$$\hat{v}(\boldsymbol{r},t) = \sum_{i=1}^{2d}\hat{w}_i(t)\hat{u}(\boldsymbol{r}+a\boldsymbol{e}_i,t) \tag{5.98}$$

describe the movement of a diffusive particle. In these equations, $\boldsymbol{r}$, $t$ and $\delta t$ denote, respectively, the spatial coordinate of the lattice, the real time and the basic time step. Each variable, $\hat{u}$, $\hat{v}$ and $\hat{w}_i$ can have only two values, one or zero; this also applies to $\hat{\rho}$ and $\hat{P}$. The variables have the following meaning: $\hat{u}$ characterizes the site occupancy of the diffusive particle outside the cluster; obviously, there is only one site with $\hat{u} = 0$ at any time. The variable $v$ characterizes the incoming flux of the

**Fig. 5.27** Density profile $\rho$ (solid line) and the diffusive particle density $u$ of the DLA under the continuous mean-field theory [245]. Five curves correspond, respectively, to $t = 200, 400, 600, 800, 1000$ steps

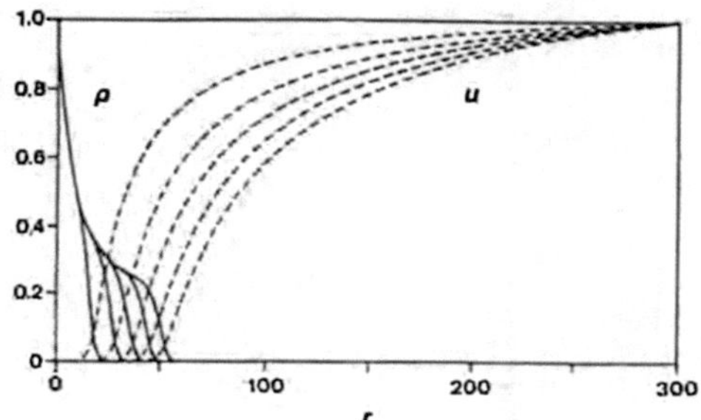


diffusive particle. Finally, $\hat{w}_i$ $(i = 1, 2, 3, \ldots, 2d)$ is a random variable introduced for particle diffusion. Only one $\hat{w}_j$ is unity and all the other values are zero $(\hat{w}_{k(\neq j)} = 0)$; this means that the diffusive particle moves toward the $j$th direction. Therefore the averaged value of $\hat{w}_j$ is $\langle \hat{w}_i \rangle = 1/(2d)$. As in the original simulation algorithm due to Witten and Sander [243], there is only one seed particle at the center and only one diffusive particle on the spherical shell of radius $A$ at the initial time. At the next time step, the diffusive particle moves by one lattice bond.

Equations (5.95)–(5.98) can be solved within the mean-field theory and in the continuum limit [245]. Figure 5.27 shows the temporal evolution obtained from such an analysis. The cluster density profile in the radial direction shows a power-law decay in some interval and decreases exponentially outside this interval. To study this power, we replot the density on a log–log scale in Fig. 5.28. Clearly, ln $\rho$ decreases linearly for $r \leq R$ and more rapidly for $r \geq R$. Here $R$ is the radius of gyration. Writing the density profile as $\rho(r) \sim r^{D-d}$, we call $D$ the fractal dimension. The estimated results for the fractal dimension in this continuous mean-field (MF) analysis are listed in Table 5.1. These results are consistent with the results obtained by the original simulation for $d = 3, 4$. The analytic result $D = (d^2 + 1)/(d + 1)$, based on Flory's mean-field-like theory given by Muthukumer, and Tokuyama and Kawasaki [246] is also listed in Table 5.1.

As an extension of the DLA model, we may assume that the diffusive particle attaches to the cluster with some probability $p$ and continues to diffuse with the probability $1 - p$. If this sticking probability becomes very small, the resulting pattern changes qualitatively. In the limit $p \to 0$, the reaction-limited model is defined. In the reaction-limited model, the screening effect is weakened and there appears a complex branching pattern inside the bays. Generally, the reaction-limited model has a higher fractal dimension than the diffusion-limited model. However, in this case of this particle-cluster model, the reaction-limited model results in a dense morphology corresponding to $D = d$. It is also possible to modify the model so as to assume that only after several hits at the same point does the particle become attached at that position. With this modification, the resulting pattern also becomes denser than in the original DLA model and is called the dense-branching morphology [247–249]. On the other hand, it is possible to construct a model without any diffusion. In this case, we select one perimeter site of the growing cluster and just attach a new particle

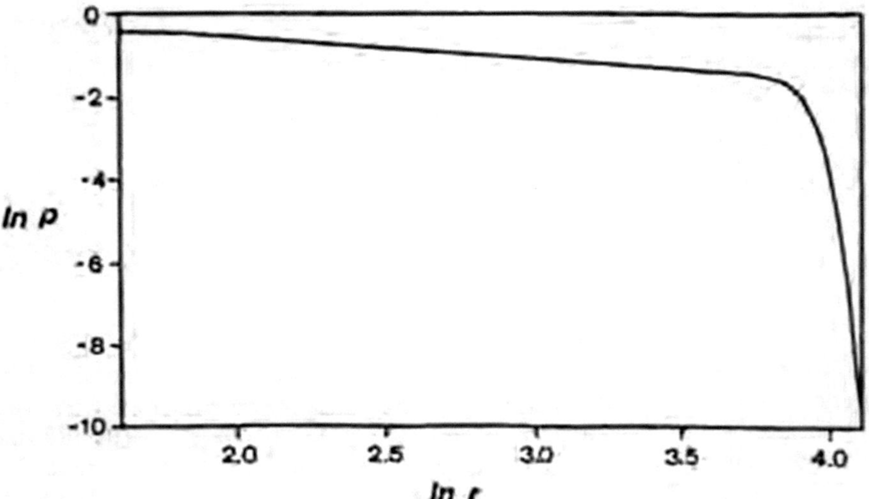


**Fig. 5.28** The log-log plot of the density $\rho(r)$ at $t = 1000$. From the slope of this linear-dependence, we get the estimation $D = 2.50 \pm 0.05$ [245]

**Table 5.1** Fractal dimension for the DLA model obtained by various methods

| Spatial dimensionality | 2 | 3 | 4 | 6 |
|---|---|---|---|---|
| Computer simulation [244] | 1.67 | $2.53 \pm 0.06$ | $3.31 \pm 0.10$ | ~5.35 |
| Continuous MF theory [245] | – | $2.49 \pm 0.05$ | $3.43 \pm 0.05$ | $5.39 \pm 0.05$ |
| $D = (d^2 + 1)/(d + 1)$ [246] | 1.67 | 2.50 | 3.40 | 5.29 |

at that position. This model is called the Eden model [250] and the resulting cluster is completely dense.

Another type of dendritic growth pattern can be created by the ballistic deposition model [54, 251–256]. In this model, one selects one point far above the crystal surface, vertically drops a particle without any diffusion and attaches the particle to the substrate at the position where it first touches the substrate. The resulting pattern is fractal only in one direction and is called self-affine. Figure 5.29a shows a pattern generated by the simplest ballistic deposition model [257]. In growing surface models such as the ballistic deposition model, the height $h_i$ at the position $i$ can be regarded as the dynamic variable. The temporal evolution of $h_i$ can be described as a Markov process, and if there is no characteristic length scale, the resulting rough surface becomes fractal. In order to describe the surface fractal, we introduce the surface width $\sigma$ and the average height $h$ defined by

$$\sigma(L, h) = \left[\frac{1}{N_s}\sum_{i=1}^{N_s}(h_i - h)^2\right]^{1/2}, \quad h = \frac{1}{N_s}\sum_{n=1}^{N_s} h_i, \tag{5.99}$$

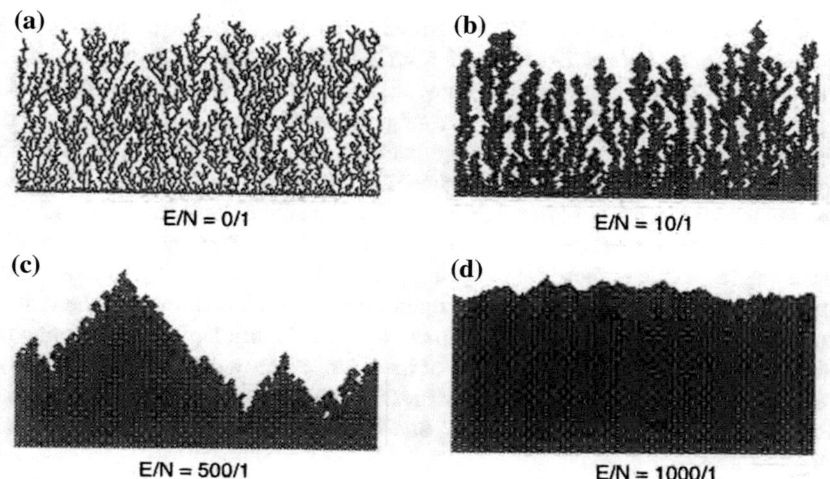


**Fig. 5.29** Pattern in the EBM. $E/N$ is the ratio of etchant to normal particles. **a** is the case of $E/N = 0$ and corresponds to the usual ballistic deposition model. From [257]

where $L$ is the width of the periodic boundary condition. In this kind of film growth, the width $\sigma$ of the surface fluctuation increases with $h$ for $h \ll L$, while it increases with $L$ for $h \gg L$; in both cases, it obeys a power law. Therefore the following scaling law holds:

$$\sigma(L, h) \sim L^{\alpha} f\left(\frac{h}{L^z}\right). \tag{5.100}$$

Here the scaling function $f(x)$ behaves as

$$f(x) \sim \begin{cases} x^{\beta}, & x \ll 1 \\ \text{const.} & x \gg 1. \end{cases} \tag{5.101}$$

This is equivalent to

$$\sigma(L, h) \sim \begin{cases} h^{\beta}, & h/L^z \ll 1 \\ L^{\alpha}, & h/L^z \gg 1, \end{cases} \tag{5.102}$$

with $z = \alpha/\beta$. According to an analysis based on the renormalization group theory by Kardar Parisi and Zhang [258] (the KPZ theory), the critical exponents of the ballistic deposition model are exactly $\alpha = 1/2$ and $\beta = 1/3$, which are consistent with the results obtained in the simulation of Meakin et al. [255].

As was briefly mentioned at the end of Sect. 5.4.7, the thin-film growth of diamond can be described by the ballistic deposition model, extended to include the etchants in addition to the usual deposition particles [257, 259]. We will call this model the etchant ballistic model (EBM). The role played by etchants is to remove adatoms attached at unstable positions. To be explicit, one can consider a square-shaped particle on a square lattice. The topmost normal particle, which is supported only at a lower-left or lower-right corner, is removed, whenever an etchant falls on top of it or on an upper corners. The topmost normal particle, which is located on top of the second-layer particle and has no other contacts, is also removed by an etchant. The processes are illustrated in Fig. 5.30. Figure 5.30a shows the possible places at which normal particles can stick. Figure 5.30b shows the unstable particles (gray squares), which are removed by the etchant particles, while Fig. 5.30c shows the particles that are stable against etching. The etching process makes the growth process much slower, and the time required for the computer simulation increases with the ratio of the number of etchant particles to the number of normal particles $E/N$. Although the EBM reproduces typical characteristics of the diamond growth process from the gas phase, it has the following two disadvantages: (i) the process is complicated because of the introduction of etchant particles, and (ii) the simulation requires very long computer times.

We show the patterns resulting from the EBM for different values of $E/N$ in Fig. 5.29. We can classify the growth process into three stages.

- *First stage*: an almost flat surface is maintained and the resulting pattern is completely dense. The sites are almost all stable.
- *Second stage*: inclined facets with angles of $\pm 45°$ appear. The surface is covered almost everywhere with these inclined facets at the end of this stage. Since any particles attached to the facets are unstable particles, the regular crystal growth stops when these facets are created. There then remains almost no stable site for new incident particles.
- *Third stage*: the growth morphology spontaneously changes to columnar growth. At this stage, the growth of the structure becomes two orders of magnitude slower compared to the first two stages. It is worth noticing that the patterns of columnar growth in the third stage are not sensitive to the value of $E/N$ when $E/N$ is somewhat larger than unity, i.e. $E/N > 3$.

Because of the etching process, the height of the surface may decrease but its decrement depends strongly on the history. Therefore, this model is a non-Markov process. The characteristics of this EBM are the creation of the $\pm 45°$ facets and the spontaneous change of the growth morphology to columnar growth, but these are not specific to non-Markov processes. In fact, in place of the EBM, it is possible to introduce the probabilistic ballistic model (PBM) [257], in which one introduces a sticking probability $p$ without introducing etchant particles (see Fig. 5.31). When $p$ is smaller than unity, the pattern resulting from the PBM resembles that from the EBM. Strictly speaking, the two models can give equivalent patterns when the relation $p = 1/[3(1 + E/N)]$ is fulfilled [257]. Another merit of introducing the PBM is that it reduces the computational time dramatically, because the ratio of etchants to normal particles, $E/N$, is sometimes very high in the EBM.

There is another model of dendritic aggregation called the cluster–cluster aggregation (CCA) model [54, 240, 260, 261], in which the cluster seeds themselves

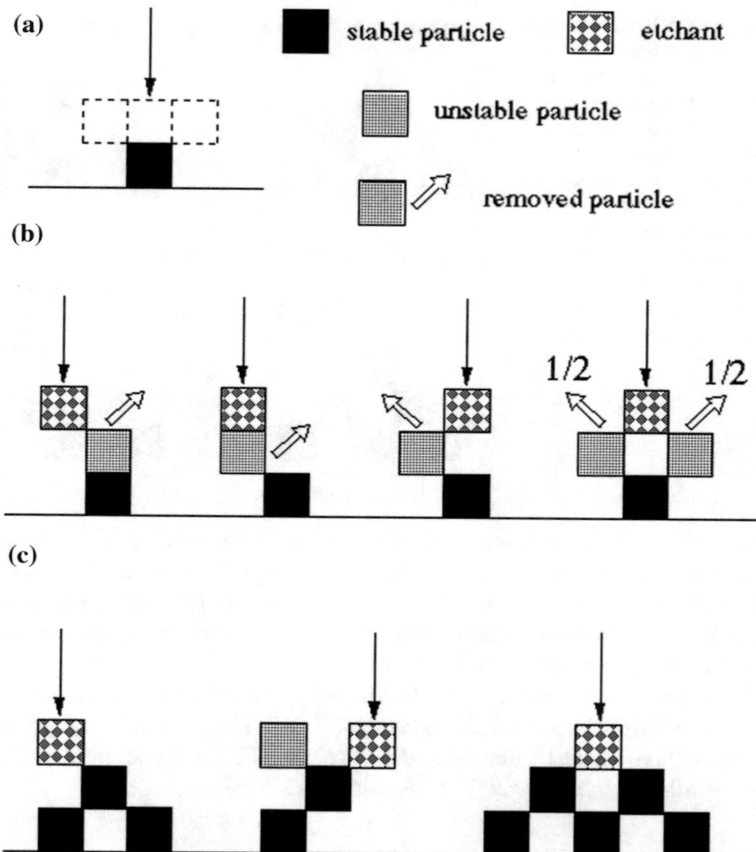


**Fig. 5.30** Etchant ballistic model: **a** shows possible places at which normal particles can stick, **b** shows unstable particles (gray squares), which are removed by the etchant particles, and **c** shows particles that are stable against etching

can diffuse. For example, colloidal aggregation can be described by this model. In the CCA model, there are two cases according to the sticking probability: one is the diffusion-limited case [260, 261], where the sticking probability is equal to unity, and the other is the reaction-limited case [262–264], where the sticking probability is very small. The resulting fractal dimensions differ in these two cases. Here, we consider a simulation of a silica-glass formation process by the sol–gel method. Generally, gelation occurs as a result of a condensation polymerization by complex chemical reactions. In producing the final fractal gel structure, the amount of the 3- or 4-functional monomers relative to the amount of the usual 2-functional monomers has a crucial role. From naive intuition, one would expect to obtain a higher fractal dimension when the number of 3- or 4-functional units is increased relative to the number of 2-functional units, because this might make the resulting gel denser.

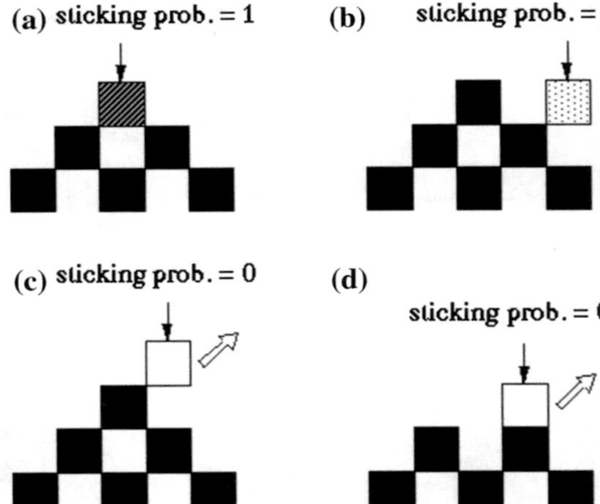


**Fig. 5.31** Probabilistic ballistic model. A normal particle in the $i$th row sticks with a probability $p$ if the new height $h'_i$ is the same as the height of a neighbor, $h_{i-2}$ or $h_{i+2}$, and if only one of the lower corners is attached to an upper corner of a surface particle. If both of the lower corners are attached simultaneously, the normal particle sticks with a probability unity

However, this is not simply the case in a situation such as silica gel. In contrast to the above intuition, there is a wide region where the greater the number of the 3- or 4-functional units, the lower the fractal dimension of the silica gel. This phenomenon has been explained theoretically by using a computer simulation of a new cluster–cluster aggregation (CCA) model [265].

To be explicit, we consider the following chemical reaction between water molecules and $Si(OR)_4$, with acids as catalyst [266]. Here R is an alkyl group such as methyl or ethyl. At the beginning of the reaction, $Si(OR)_4$ reacts successively with water molecules to produce silicic acids, $Si(OR)_{4-n}(OH)_n$:

$$\begin{aligned}
Si(OR)_4 &+ H_2O \rightarrow Si(OR)_3(OH) + ROH \\
&+ H_2O \rightarrow Si(OR)_2(OH)_2 + ROH \\
&+ H_2O \rightarrow Si(OR)(OH)_3 + ROH \\
&+ H_2O \rightarrow Si(OH)_4 + ROH.
\end{aligned}$$

Silicic acid molecules have a tendency to join together by a condensation-polymerization reaction with the elimination of a water molecule:

1 2 3 4 5 6 7

**Fig. 5.32** The seven monomer types appearing in the CCA sol–gel model in two dimensions [265]

$$\equiv\text{Si-OH} + \text{HO-Si}\equiv \rightarrow \equiv\text{Si-O-Si}\equiv + \text{H}_2\text{O}$$

The amount of $Si(OR)(OH)_3$ and $Si(OH)_4$ relative to $Si(OR)_2(OH)_2$ is very roughly proportional to the amount of water added initially. If we regard $Si(OR)_{4-n}(OH)_n$ as an $n$-functional monomer, we can model the condensation-polymerization reaction process by introducing an extended CCA model. In a more realistic situation, depending on pH, temperature and how much water is included. The condensation-polymerization reaction proceeds also by elimination of alcohol molecules:

$$\equiv\text{Si-OH} + \text{RO-Si}\equiv \rightarrow \equiv\text{Si-O-Si}\equiv + \text{ROH},$$

and the process becomes more complex. However, here we simply assume that the reaction with the elimination of water occurs more frequently than the reaction with the elimination of alcohol.

The model proposed by Kudoh et al. [265] assumes only two types of monomers in the system, i.e. monomers with 2- and 4-functionalities. In the case of two-dimensional square lattice, there are seven types of monomers shown in Fig. 5.32 in the system. (Here we have neglected 1- and 3-functional units for simplicity. The point here is to have as few parameters as possible.) The central circle of each monomer represents a Si atom, and the bars attached correspond to active reaction sites. We assume the reaction to occur only when two arms meet in opposite directions. The last monomer, labeled 7, corresponds to a fully active Si atom, which may yield a cross-link. The other six monomers model Si atoms which have two inactive and two activate sites, which allow only a linear bond. More general case in three dimensions is modeled by [267]. Since the condensation-polymerization process occurs according to the available OH groups, the number of cross monomers roughly corresponds to the amount of water added. The sol–gel transition process of $SiO_2$ is simulated in the same way as in the usual CCA model except for the active-site constraint.

First, we randomly place a large number of monomers on a lattice. These monomers move randomly on the lattice, for example with a mobility proportional to the inverse square root of the mass, aggregating into large clusters. To simulate infinite space, periodic boundary conditions are used.

The numerical simulation of this model was performed in both two and three dimensions. The only parameter was the relative concentration of the 4-functional monomers, $p$. The relative concentration of the 2-functional monomers is given by $1 - p$.

The number of mesh points was varied according to the number of monomers handled. The number of monomers $N$ that we distributed on the lattice was from 30 to 1000. The size of the lattice was determined so as to achieve the same density for the different numbers of monomers. We confirmed the density independence of the fractal structure for small values of the occupancy rate of monomers on the lattice, and set this occupancy rate in the present simulation to a very small value of 0.004 for 2D and 0.0007 for 3D.

The time evolution of the aggregation process can be simulated by moving each monomer. Because of the nature of the model, it is easy to achieve the final single gel in the case of a large fraction of the cross monomers ($p > 0.5$). On the other hand, for cases with low $p$ values, the aggregation process often takes an extraordinarily long time (more than a million time steps) to reach the final stage. This stage sometimes cannot be reached, because of the initial condition or intermediate stages. It would be very difficult to obtain a the gel state for a cluster having few available arms. In Figs. 5.33 and 5.34, we show typical gel patterns obtained for two-dimensional and three-dimensional models, respectively.

The scaling analysis enables one to reduce the value of the error in the effective fractal dimension. The result for the fractal dimension is shown in Fig. 5.35 as a function of $p$.

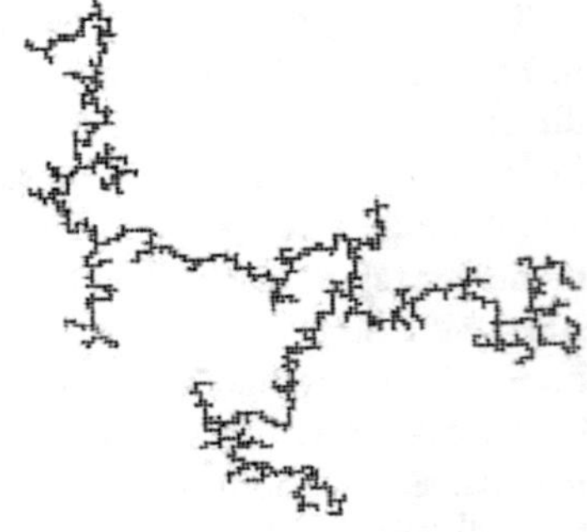


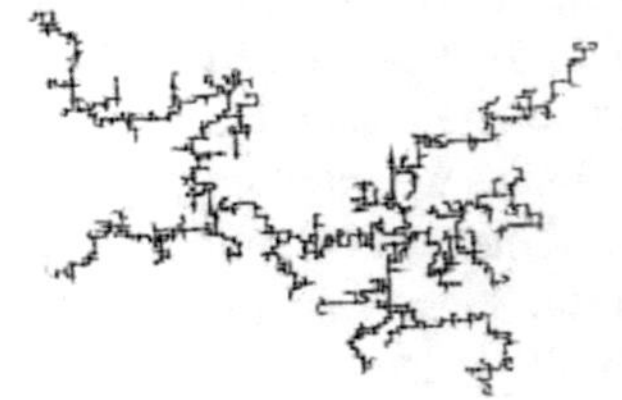


**Fig. 5.33** Typical patterns of the gel structure obtained in the two-dimensional CCA sol–gel model [265]

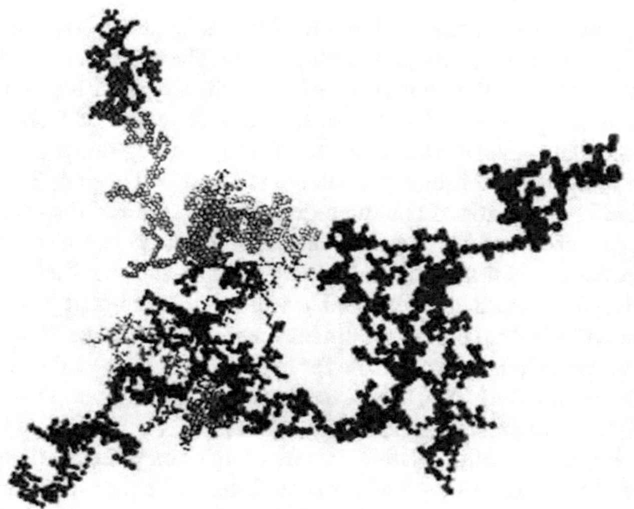

**Fig. 5.34** Typical patterns of the gel structure obtained in the three-dimensional CCA sol–gel model [267]

**Fig. 5.35** Obtained fractal dimension as a function of the ratio $p$ of cross monomers, for **a** a two-dimensional system [265] and **b** a three-dimensional system [267]

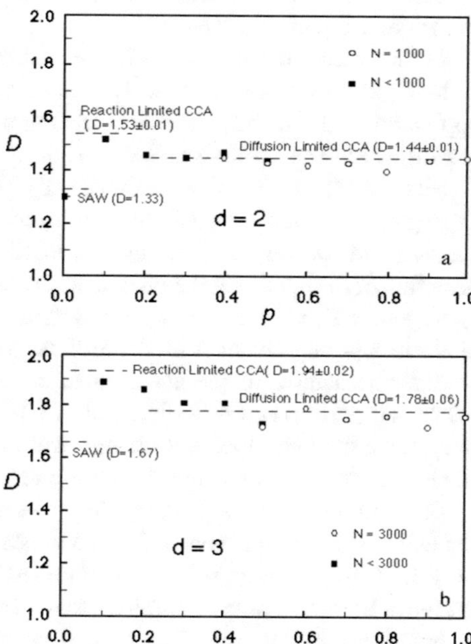

The $p$ dependence of the values of the effective fractal dimension can be explained as follows. First, consider the two-dimensional case. The values for $p = 0.4$ to $p = 1$ are around 1.42, which is the same as the value for the simple diffusion-limited CCA model. The values for $p = 0.1$–$0.4$ correspond to that expected from the reaction-limited CCA model, which is equal to 1.53. The fractal dimension at $p = 0$ estimated here is close to the value for the self-avoiding walk $D = 1/\nu = 4/3$.

For the fractal dimension of real three-dimensional systems, the overall behavior of the $p$ dependence can be explained in the same way. For large $p$ values the expected mechanism is that of diffusion-limited CCA and the fractal dimension is $D = 1.75 \pm 0.05$. For small to medium $p$ values, the dominant process is that of reaction-limited CCA, and the fractal dimension is expected to be $D = 2.00 \pm 0.05$. The same comments that we have made for 2D also hold for $p = 0$ in 3D space, and the fractal dimension is expected to become that of the 3D SAW, i.e. $D = 1/\nu \sim 1.69$, although it is very difficult to simulate this region.

When $p$ becomes small, gelation requires large computation times, since the intermediate clusters have very few arms available for aggregation, to realize the final gel state. The reason why the fractal dimension increases with decreasing $p$ can be explained as follows: for lower ratios of the 4-functional monomers, it takes a longer time to aggregate into a single gel. This yields more complex structures; for example, after the formation of an open loop, some of the small clusters enter to fill in the central part of the loop.

In the present model, just as in the usual CCA model, all aggregates are assumed to be rigid and non-rotatable. This is of course an assumption, but this model is applicable, at least to some extent, also to the case of flexible aggregates, due to the reason which has been discussed in [267].

First of all, in the $p = 0$ limit, only possible way of elongating the polymer aggregate is to add a two-functional monomer (or a short linear chain) at one of the two ends of the polymer aggregate. The monomer (short chain) is added in the direction of its first touch which is arbitrary and random. Thus, even when the growing aggregate is rigid, it should have configurations similar to those of a flexible linear chain in a good solvent limit. Therefore, a grown polymer aggregate in the $p = 0$ limit is characterized by a usual linear chain in a good solvent, and can be modeled accurately with a self-avoiding-walk (SAW). That is, the fractal dimension of the polymer aggregate does not change from the SAW value, even if polymers move flexibly as in a good solvent during the aggregation process.

In the dilute limit, each aggregate grows nearly *independently* at different places in the existence of monomers and small clusters. In such a situation, one may consider only one big aggregating cluster surrounded by monomers and small clusters. Assume that each branch of polymer aggregates can move freely under a topological condition of branching. A rotation of a polymer aggregate as a whole can be considered as a special case of such a free motion of polymers. Here we note that the aggregation process is Markovian and ergodic. Then, the configuration of the flexible polymer aggregate at one time can be identified as one of various configurations of rigid aggregates.

There are two ways of the touch among clusters, i.e., the touch through (A) a diffusive motion of the clusters as a whole, and (B) a flexible motion of an arm of the clusters. The (A) is the common process in both cases of rigid and flexible aggregates. If the touch occurs only through the process (A), the resulting conformation should not change in between the rigid and flexible cases. Then the problem to be considered is the effect of the process (B). When a small cluster touches a bigger cluster via a flexible motion (process (B)) of the small cluster, it is possible to neglect the detail at length scale of the size of this small cluster by a suitable coarse-graining of the space. In such a case, one may regard the process (B) as if it is the process (A). Hence, one should pay attention to the process (B) when the touch occurs by a local movement of a flexible arm chain of a large cluster.

In the case of relatively low $p$ values, however, even if the process (B) occurs, the new connection cannot be made easily because of the limitation of the number of active positions along the arms of the polymer aggregates. Therefore, in this case, one may neglect the aggregation process via the process (B), and we expect that the fractal dimensionality obtained by the present simulation at small $p$ values does not change even if we take the flexibility of polymer aggregates into account.

On the other hand, the process (B) in flexible aggregates would become important if new connections can be made anywhere on the chains as in the usual diffusion-limited CCA model. This might be the case also in the present model, when the value of $p$ is relatively high. However, in such a case also, the diffusion time needed for a monomer (or a small cluster) to touch finally the aggregating cluster is relatively short, and the process (B) becomes again negligible when such a diffusion time is shorter than the relaxation time on local conformations of arms of polymer aggregates.

There are two typical time scales which are relevant in the present discussion: One is the relaxation time on polymer conformations in a solvent and the other is the gelation time of the system. The effect of flexibility of growing polymer arms during the aggregation process becomes important if polymer aggregates have plenty of active positions along arms and if the length of each arm is long enough and the relaxation time on polymer conformations is faster than the gelation time. However, this may usually be not the case when the size of polymer aggregates is of order of 100Å which is the typical length scale observed by small-angle X-ray scattering measurements of $SiO_2$. A small-angle X-ray scattering experiment has recently been performed by Mikami [268] for the silica aggregates of size of order $\sim$100Å. This experiment shows the clear dependence of the fractal dimension $D$ on the amount of added water, which corresponds to that observed in the present CCA sol–gel model. This model has an advantage compared to the other existing models, since it can show the single parameter dependence of the fractal dimension characterizing the resulting gel structure, and therefore can be directly applied to the gelation process of, for example, $SiO_2$. Another type of gelation process is experimentally studied for $TiO_2$ [268]. This is a good candidate for the present model to be applied to. Similar dependence of the fractal dimension have been found also in the colloidal aggregates of silica particles by Aubert and Cannel [269].

## References


1. H.E. Stanley, *Introduction to Phase Transitions and Critical Phenomena* (Oxford University Press, Oxford, 1971)
2. K. Binder (ed.), *The Monte Carlo Method in Condensed Matter Physics* (Springer, Berlin, 1992)
3. K. Binder, D.W. Heermann, *Monte Carlo Simulation in Statistical Physics* (Springer, Berlin, 1988)
4. K. Binder (ed.), *Monte Carlo Methods in Statistical Physics* (Springer, Berlin, 1979)
5. L.D. Landau, E.M. Lifshitz, *Statistical Physics* (Pergamon, Oxford, 1959, 1980)
6. D.E. Knuth, *The Art of Computer Programming*, vol. 2, ed. by M.A. Harrison (Addison-Wesley, Reading, Massachusetts, 1981). Chapter 3
7. N. Metropolis, A.W. Rosenbluth, M.N. Rosenbluth, A.M. Teller, E. Teller, J. Chem. Phys. **21**, 1087 (1953)
8. R.J. Glauber, J. Math. Phys. **4**, 294 (1963)
9. K. Kawasaki, in *Phase Transitions and Critical Phenomena*, vol. 2, ed. by C. Domb, M.S Green (Academic Press, London, 1972). Chapter 11
10. B.I. Halperin, P.C. Hohenberg, S.-K. Ma, Phys. Rev. Lett. 29, 1548 (1972); P.C. Hohenberg, B.I. Halperin. Rev. Mod. Phys. **49**, 435 (1977)
11. J.M. Ziman, *Models of Disorder* (Cambridge University Press, Cambridge, 1979)
12. R. Evans, in *Liquids at Interfaces*, ed. by J. Charvolin, J.F. Joanny, J. Zinn-Justin (North-Holland, Amsterdam, 1990)
13. D. Frenkel, B. Smit, *Understanding Molecular Simulation: From Algorithms to Applications* (Academic Press, San Diego, 1996)
14. F. Yonezawa, in *Solid State Physics*, vol. 45, ed. by H. Ehrenreich, D. Turnbull (Academic Press, San Diego, 1991), p. 179
15. R.H. Jones, W.W. Gerberich (eds.), *Modeling Environmental Effects on Crack Growth Processes* (Metallurgical Society of the AIME, Warrendale, Pennsylvania, 1986)
16. W. Eckstein, *Computer Simulation of Ion-Solid Interactions*. Materials Science Series, vol. 10 (Springer, Berlin, 1991). Chapter 4
17. A.R. Leach, *Molecular Modeling* (Longman, Harlow, 1996)
18. L. Monnerie, U.W. Suter, *Atomistic Modeling of Physical Properties*, vol. 116 (Advances in Polymer Science (Springer, Berlin, 1994)
19. A. Gavezzotti, *Theoretical Aspects and Computer Modeling of the Molecular Solid State* (Wiley, New York, 1997)
20. M.-C. Desjonquères, D. Spanjaard, *Concepts in Surface Physics*, 2nd edn. (Springer, Berlin, 1995)
21. M.N. Rosenbluth, A.W. Rosenbluth, J. Chem. Phys. **22**, 881 (1954)
22. B.J. Alder, T.E. Wainwright, J. Chem. Phys. **27**, 1208 (1957)
23. W.W. Wood, J.D. Jacobson, J. Chem. Phys. **27**, 1207 (1957)
24. W.W. Wood, F.R. Parker, J. Chem. Phys. **27**, 720 (1957)
25. T.R. McDonald, K. Singer, Faraday Disc. **43**, 40 (1967)
26. A.M. Ferrenberg, R.H. Swendsen, Phys. Rev. Lett. **63**, 1195 (1989)
27. S. Kumer, D. Bouzida, R.H. Swendsen, P.A. Kollman, J.M. Rosenberg, J. Comput. Chem. **13**, 1011 (1992)
28. E.M. Boczko, C.L. Brooks, III, J. Phys. Chem. **97**, 4509 (1993); Science **269**, 393 (1995)
29. D.J. Adams, Mol. Phys. **28**, 1241 (1974)
30. L.A. Rowley, D. Nicholson, N.G. Parsonage, J. Comput. Phys. **17**, 401 (1975)
31. A.Z. Panagiotopoulos, Mol. Phys. **61**, 813 (1987); **62**, 701 (1987)
32. B. Widom, J. Chem. Phys. **39**, 2802 (1963)
33. J.G. Kirkwood, J. Chem. Phys. **3**, 300 (1935)
34. D. Frenkel, A.J.C. Ladd, J. Chem. Phys. **81**, 3188 (1984)
35. C.H. Bernett, J. Comput. Phys. **22**, 245 (1976)

36. G.M. Torrie, J.P. Valleau, J. Comput. Phys. **23**, 187 (1977)
37. S.C. Harvey, M. Prabhakaran, J. Phys. Chem. **91**, 4799 (1987)
38. D. Chandler, *An Introduction to Modern Statistical Mechanics* (Oxford Universion Press, New York, 1987)
39. J. Mezei, J. Comput. Phys. **68**, 237 (1987)
40. R.W.W. Hooft, J. Chem. Phys. **97**, 6690 (1992)
41. C. Bartels, M. Karplus, J. Comput. Chem. **18**, 1450 (1997); J. Phys. Chem. B **102**, 865 (1998)
42. B.A. Berg, T. Neuhaus, Phys. Lett. B **267**, 249 (1991)
43. B.A. Berg, T. Celik, Phys. Rev. Lett. **69**, 2292 (1992)
44. U.H.E. Hansmann, Y. Okamoto, J. Comput. Chem. **14**, 1333 (1993)
45. Y. Okamaoto, U.H.E. Hansmann, J. Phys. Chem. **99**, 11276 (1995)
46. U.H.E. Hansmann, Y. Okamoto, F. Eisenmenger, Chem. Phys. Lett. **259**, 321 (1996)
47. N. Nakajima, H. Nakamura, A. Kidera, J. Phys. Chem. B **101**, 817 (1997)
48. C. Bartels, M. Karplus, J. Phys. Chem. **102**, 865 (1998)
49. D. Stauffer, A. Aharony, *Introduction to Percolation Theory*, 2nd edn (Taylor and Francis, London, 1992, 1994)
50. P.W. Kasteleyn, P.W. Fortuin, J. Phys. Soc. Jpn. Suppl. **26**, 11 (1969); P.W. Fortuin, P.W. Kasteleyn, Physica (Utrecht) **57**, 536 (1972)
51. R.J. Baxter, J. Stat. Phys. **15**, 485 (1976)
52. R.J. Baxter, *6 Exactly Solved Models in Statistical Mechanics* (Academic Press, London, 1982)
53. F. Family, T. Vicsek, *Dynamics of Fractal Surfaces* (World Scientific, Singapore, 1991)
54. T. Vicsek, *Fractal Growth Phenomena*, 2nd edn. (World Scientific, Singapore, 1992)
55. R.J. Reynolds, H.E. Stanley, W. Klein, J. Phys. C: Condens. Matt. **10**, L167 (1977); Phys. Rev. B **21**, 1223 (1980)
56. J. Bernasconi, Phys. Rev. B **18**, 2185 (1978)
57. H. Nakanishi, R.J. Reynolds, Phys. Lett. A **71**, 252 (1979)
58. P.D. Eschbach, D. Stauffer, H.J. Herrmann, Phys. Rev. B **23**, 422 (1981)
59. A. Conglio, Phys. Rev. Lett. **46**, 250 (1981)
60. R.M. Ziff, Phys. Rev. Lett. **69**, 2670 (1992)
61. T.C. Lubensky, J. Issacson, Phys. Rev. Lett. 41, 829 (1978). Phys. Rev. A **20**, 2130 (1979)
62. J. Issacson, T.C. Lubensky, J. Physique **41**, L469 (1980)
63. G. Parisi, N. Sourlas, Phys. Rev. Lett. **46**, 871 (1981)
64. S. Kumer, in *Computer Simulation of Polymers*, ed. by E.A. Colbourn (Longman Scientific & Technical, London, 1994). Chapter 7; K. Binder, ibid., Chapter 3
65. K. Binder (ed.), *Monte Carlo and Molecular Dynamics Simulations in Polymer Science* (Oxford University Press, New York, 1995)
66. K. Ohno, K. Binder, J. Stat. Phys. **64**, 781 (1991); K. Ohno, Macromolecular symposia **81**, 121 (1994)
67. P.G. de Gennes, *Scaling Concepts in Polymer Physics* (Cornell University Press, Ithaca, New York, 1979)
68. P.G. de Gennes, Phys. Lett. A **38**, 339 (1972)
69. K. Ohno, K. Binder, J. Physique **49**, 1329 (1988)
70. J. des Cloizeaux, J. Physique **42**, 635 (1981)
71. J.J. Freire, *in Advances in Polymer Science*, vol. 143 (Springer, Berlin, 1999), p. 37
72. P.J. Flory, J. Chem. Phys. **9**, 660 (1941); P.J. Flory, J. Chem. Phys. **10**, 51 (1942)
73. M.L. Huggins, J. Chem. Phys. **9**, 440 (1941)
74. A.R. Miller, Proc. Camb. Philos. Soc. **39**, 54 (1942)
75. E.A. Guggenheim, Proc. R. Soc. Lond. A **183**, 213 (1944)
76. J. des Cloizeaux, J. Physique **36**, 281 (1975)
77. K. Shida, K. Ohno, M. Kimura, Y. Kawazoe, Comput. Theor. Polymer Sci., in submission
78. K. Ohno, K. Shida, M. Kimura, Y. Kawazoe, Macromolecules **29**, 2269 (1996)
79. A. Rubio, J. Freire, Macromolecules **29**, 6946 (1996)
80. S. Kumer, I. Szleifer, A. Panagiotopoulos, Phys. Rev. Lett. **66**, 2935 (1991)

81. S. Kumer, in *Computer Simulation of Polymers*, ed. by E.A. Colbourn (Longman Scientific & Technical, London, 1994). Chapter 7
82. B. Widom, J. Chem. Phys. **39**, 11 (1963); J. Chem. Phys. **39**, 2808 (1963)
83. W.G. Madden, A.I. Pesci, K.F. Freed, Macromolecules **23**, 1181 (1990)
84. B. Smit, S. Karaborni, J.I. Siepmann, Macromol. Symp. **81**, 343 (1994)
85. H.A. Kramers, J. Chem. Phys. **14**, 415 (1946)
86. H.L. Frisch, N. Pistoor, A. Sariban, K. Binder, S. Fesjian, J. Chem. Phys. **89**, 5194 (1988)
87. B.H. Zimm, Macromolecues 13, 592 (1980); Macromolecules 17, 795 (1984). Macromolecules **17**, 2441 (1984)
88. M.K. Wilkinson, D.S. Gaunt, L.E.G. Lipson, S.G. Wittingon, Macromolecules **21**, 1818 (1988)
89. K. Shida, K. Ohno, M. Kimura, Y. Kawazoe, Y. Nakamura, Macromolecules **31**, 2343 (1998)
90. A.K. Kron, Vysokomol. Soedin. 7, 1228 (1965) [Polym. Sci. USSR 7, 1361 (1965)]; F.T. Wall, F. Mandel. J. Chem. Phys. **63**, 4592 (1975)
91. I. Carmesin, K. Kremer, J. Physique **51**, 915 (1990)
92. K. Ohno, M. Schulz, K. Binder, H.L. Frisch, J. Chem. Phys. **101**, 4452 (1994)
93. N. Urakami, M. Takasu, J. Phys. Soc. Jpn. **65**, 2694 (1996)
94. M. Murat, T.A. Witten, Macromolecules **23**, 520 (1990)
95. M. Seul, D. Andelman, Science **267**, 476 (1995)
96. M. Ishihara, H. Mizuseki, K. Ohno, Y. Kawazoe, in *Proc. Fourth Int. Symp. Phys. Magn. Mater.* (1998), to appear
97. K. Binder, Phys. Rev. Lett **45**, 811 (1980)
98. R. Kikuchi, Phys. Rev. **81**, 988 (1951)
99. N.S. Golsov, L.E. Popov, L. Ya. Pudan, J. Phys. Chem. Solids **34**, 1149 (1973); J. Phys. Chem. Solids **34**, 1157 (1973)
100. R. Kikuchi, H. Sato, Acta Metall. **22**, 1099 (1974)
101. T. Matsumiya, H. Sawada, W. Yamada, A. Nogami, in *Computer Aided Innovation of New Materials*, ed. by M. Doyama, et al. (Elsevier, Amsterdam, 1991), p. 779
102. K. Terakura, T. Oguchi, T. Mohri, K. Watanabe, Phys. Rev. B **35**, 2169 (1987)
103. T. Mohri, Bull. JIM **28**, 268 (1989)
104. G.J. Ackland, V. Vitek, Phys. Rev. B **41**, 10324 (1990)
105. R. Sahara, H. Ichikawa, Master Thesis, Tohoku University (1999). R. Sahara, H. Ichikawa, H. Mizuseki, K. Ohno, H. Kubo, Y. Kawazoe. J. Chem. Phys. **120**, 9297 (2004)
106. B.M. McCoy, T.T. Wu, *The Two-Dimensional Ising Models* (Harbard University Press, Cambridge, 1973)
107. C. Domb, M.F. Sykes, Proc. R. Soc. A 240, 214 (1957). J. Math. Phys. **2**, 63 (1961)
108. H.E. Stanley, in *Phase Transitions and Critical Phenomena*, vol. 3, ed. by C. Domb, M.S. Green (Academic Press, New York, 1974), p. 485
109. M. Suzuki, J. Phys. Soc. Jpn. **55**, 4205 (1986); Prog. Theor. Phys. Suppl. **87**, 1 (1986)
110. M. Suzuki, M. Katori, X. Hu, J. Phys. Soc. Jpn. **56**, 3092 (1987); X. Hu, M. Katori, M. Suzuki, J. Phys. Soc. Jpn. **56**, 3865 (1987)
111. E. Stoll, K. Binder, T. Schneider, Phys. Rev. B **8**, 3266 (1973)
112. J. Tobochnik, S. Sarker, R. Cordery, Phys. Rev. Lett **46**, 1417 (1981)
113. M.C. Yalabik, J.D. Gunton, Phys. Rev. B **25**, 534 (1082)
114. S.L. Katz, J.D. Gunton, C.P. Liu, Phys. Rev. B **25**, 6008 (1982)
115. M.N. Barber, R.B. Pearson, J.L. Richardson, D. Touissaint, Phys. Rev. B **32**, 1720 (1985)
116. A.M. Ferrenberg, D.P. Landau, Phys. Rev. B **44**, 5081 (1991)
117. D.S. Gaunt, M.F. Sykes, J. Phys. A Math. Nucl. Gen. **6**, 1517 (1973)
118. K.G. Wilson, Phys. Rev. B 4, 3174 (1971); Phys. Rev. B 4, 3184 (1971). Phys. Rev. Lett. **28**, 548 (1972)
119. K.G. Wilson, M.E. Fisher, Phys. Rev. Lett. **28**, 240 (1972)
120. E. Brézin, J.C. Le Guillou, J. Zinn-Justin, B. Nickel, Phys. Lett. A **44**, 227 (1973)
121. G.A. Baker Jr., B.G. Nickel, M.S. Green, P.I. Meiron, Phys. Rev. Lett. **36**, 1351 (1976)
122. J.C. Le Guillou, J. Zinn-Justin, Phys. Rev. Lett. **39**, 95 (1977)

123. S.K. Ma, *Modern Theory of Critical Phenomena, Frontiers in Physics*, vol. 46 Lecture Note Series (Benjamin, Reading, 1976)
124. D.P. Landau, in *The Monte Carlo Method in Condensed Matter Physics*, ed. by K. Binder (Springer, Berlin, 1992, 1995). Chapter 2
125. R.J. Elliot, Phys. Rev. **124**, 346 (1961)
126. E.I. Dinaburg, A.E. Mazel, Ya G. Sinai, Sov. Sci. Rev. C **6**, 113 (1987)
127. M.E. Fisher, W. Selke, Phys. Rev. Lett. **44**, 1502 (1980)
128. W. Selke, Z. Phys, B 29, 133 (1978); Z. Phys. B **43**, 335 (1981)
129. W. Selke, M.E. Fisher, Phys. Rev. B 20, 257 (1979). Z. Phys. B **40**, 71 (1980)
130. E.B. Rasmussen, S.J. Knak, Jensen. Phys. Rev. **24**, 2744 (1989)
131. W. Selke, J. Magn. Mag. Mat. **9**, 7 (1978); Sol. State Commun. **27**, 1417 (1978); Z. Phys. C: Condens. Matt. **13**, L261 (1980)
132. S. Redner, H.E. Stanley, J. Phys. C: Condens. Matt. **10**, 4765 (1978); Phys. Rev. B **16**, 4901 (1977a, b)
133. W. Selke, in *Phase Transitions and Critical Phenomena*, vol. 15, ed. by C. Domb, J.L. Lebowitz (Academic Press, London, 1992), p. 1
134. K. Binder, A.P. Young, Rev. Mod. Phys. **58**, 801 (1986)
135. M. Mezard, G. Parisi, M.A. Virasoro, *Spin Glass Theory and Beyond*. Lecture Notes in Physics, vol. 9 (World Scientific, Singapore, 1987)
136. R.H. Swendsen, J.-S. Wang, Phys. Rev. Lett. **57**, 2607 (1986)
137. J.S. Wang, R.H. Swendsen, Phys. Rev. B **37**, 7745 (1988)
138. B.A. Berg, U.E. Hansmann, T. Celik, Nucl. Phys. B **42**, 905 (1995)
139. E. Marinari, G. Parisi, Europhys. Lett. **19**, 451 (1992)
140. H. Nakanishi, M.E. Fisher, Phys. Rev. Lett. **49**, 1565 (1982)
141. S. Dietrich, in *Phase Transitions and Critical Phenomena*, vol. 12, ed. by C. Domb, J.L. Lebowitz (Academic Press, London, 1988), p. 1
142. Y. Okabe, K. Ohno, Phys. Rev. **30**, 6573 (1984)
143. K. Binder, in *Phase Transitions and Critical Phenomena*, vol. 8, ed. by C. Domb, J.L. Lebowitz (Academic Press, London, 1983) pp. 1, 467
144. K. Binder, P.C. Hohenberg, Phys. Rev. B 6, 3461 (1972). Phys. Rev. B **9**, 2194 (1974)
145. M. Kikuchi, Y. Okabe, Prog. Theor. Phys. **74**, 458 (1985)
146. Y. Okabe, M. Kikuchi, K. Ohno, Prog. Theor. Phys. **75**, 496 (1986)
147. K. Ohno, Y. Okabe, J. Phys. Soc. Jpn. **55**, 2627 (1986)
148. S.G. Whittington, G.M. Torrie, A.J. Guttmann, J. Phys. A. Math. Gen. **12**, 2449 (1979)
149. K. Ohno, Y. Okabe, A. Morita, Prog. Theor. Phys. **71**, 714 (1984)
150. H.W. Diehl, in *Phase Transitions and Critical Phenomena*, vol. 10, ed. by C. Domb, J.L. Lebowitz (Academic Press, London, 1986), p. 75
151. K. Ohno, *in Trends in Statistical Physics* (Research Trends, Poojapura, Trivandrum, 1999). (in press)
152. R. Zorn, H.J. Herrmann, C. Rebbi, Comput. Phys. Commun. **23**, 337 (1981)
153. R.H. Swendsen, J.-S. Wang, Phys. Rev. Lett. **58**, 86 (1987)
154. U. Wolff, Phys. Rev. Lett. **62**, 361 (1989)
155. U. Wolff, Phys. Lett. B 228, 379 (1989). Nucl. Phys. B **322**, 759 (1989)
156. R.H. Swendsen, J.-S. Wang, A.M. Ferrenberg, in *The Monte Carlo Method in Condensed Matter Physics*, ed. by K. Binder (Springer, Berlin, 1992, 1995). Chapter 4
157. H.H. Zhao, Z.Y. Xie, Q.N. Chen, Z.C. Wei, J.W. Cai, T. Xiang, Phys. Rev. B **81**, 174411 (2010)
158. M. Levin, C.P. Nave, Phys. Rev. Lett. **99**, 120601 (2007)
159. Z.Y. Xie, J. Chen, M.P. Qin, J.W. Zhu, L.P. Yang, T. Xiang, Phys. Rev. B **86**, 045139 (2012)
160. T.H. Berlin, M. Kac, Phys. Rev. **86**, 821 (1952)
161. H.E. Stanley, Phys. Rev. **176**, 718 (1968)
162. N.D. Mermin, H. Wagner, Phys. Rev. Lett. **17**, 1133 (1966)
163. J.M. Kosterlitz, D.J. Thouless, J. Phys. **C6**, 1181 (1973)
164. D.R. Nelson, B.I. Halperin, Phys. Rev. B **19**, 2457 (1979)

165. S. Miyashita, H. Nishimori, A. Kuroda, M. Suzuki, Prog. Theor. Phys. **60**, 1669 (1978)
166. J. Tobochnik, G.V. Chester, Phys. Rev. B **20**, 3761 (1979)
167. J.E. Van Himbergen, S. Chakravarty, Phys. Rev. B **23**, 359 (1981)
168. H. Betsuyaku, Physica A **106**, 311 (1981)
169. D.P. Landau, J. Mag. Mag. Mat. **31–34**, 1115 (1983)
170. M. Hasenbusch, S. Meyer, Phys. Lett. B **241**, 238 (1990)
171. W. Janke, Phys. Lett. A **148**, 306 (1990)
172. A.P. Gottlob, M. Hasenbusch, Physica A **201**, 593 (1993)
173. R.G. Bowers, G.S. Joyce, Phys. Rev. Lett. **19**, 630 (1967)
174. M. Ferer, M.A. Moore, M. Wortis, Phys. Rev. B **8**, 5205 (1973)
175. P.J. Wood, G.S. Rushbrooke, Phys. Rev. Lett. **17**, 307 (1966)
176. M. Ferer, M.A. Moore, M. Wortis, Phys. Rev. B **4**, 3954 (1971)
177. R.E. Watson, M. Blume, G.H. Vineyard, Phys. Rev. **181**, 811 (1969)
178. K. Binder, H. Rauch, Z. Phys. **219**, 201 (1969)
179. K. Binder, H. Rauch, V. Wildpaner, J. Phys. Chem. Solids **31**, 391 (1970)
180. P. Peczak, A.M. Ferrenberg, D.P. Landau, Phys. Rev. B **43**, 6087 (1991)
181. K. Binder, Phys. Rev. Lett. 47, 693 (1981). Z. Phys. B **43**, 119 (1981)
182. G.S. Pawley, R.H. Swendsen, D.J. Wallece, K.G. Wilson, Phys. Rev. B **29**, 4030 (1984)
183. L.P. Kadanoff, Physics (New York) **2**, 263 (1966)
184. P. Martin, *Potts Models and Related Problems in Statistical Mechanics* (World Scientific, Singapore, 1991)
185. C. Rabbi, R.H. Swendsen, Phys. Rev. B **21**, 4094 (1980)
186. S.Y. Tang, D.P. Landau, J. Appl. Phys. **61**, 4409 (1987); Phys. Rev. B **43**, 6006 (1991)
187. S.J.K. Jensen, O.G. Mouritsen, Phys. Rev. Lett. **43**, 1736 (1979)
188. H.J. Herrmann, Z. Phys. **35**, 171 (1979)
189. H.W.J. Blöte and R. H. Swendsen, J. Appl. Phys. **50**, 7382 (1979); J. Magn. Mag. Mat. **15-18**, 399 (1980)
190. J.R. Banavar, G.S. Grest, D. Jasnow, Phys. Rev. Lett. 45, 1424 (1980). Phys. Rev. B **25**, 4639 (1982)
191. B. Hoppe, L.L. Hirst, J. Phys. A Math. Gen. **18**, 3375 (1986); Phys. Rev. B **34**, 6589 (1986)
192. J.-S. Wang, R.H. Swendsen, R. Kotecký, Phys. Rev. Lett. 63, 109 (1989). Phys. Rev. B **42**, 2465 (1990)
193. A.P. Gottlob, Hasenbusch. Physica A **210**, 217 (1994)
194. K. Yasumura, T. Oguchi, J. Phys. Soc. Jpn. **53**, 515 (1984)
195. Y. Okabe, M. Kikuchi, in *Computational Approaches in Condensed-Matter Physics, Proceedings in Physics*, vol. 70, ed. by S. Miyashita, M. Imada, H. Takayama (Springer, Berlin, 1992), p. 193
196. K. Ohno, H.-O. Carmesin, H. Kawamura, Y. Okabe, Phys. Rev. B **42**, 10360 (1990)
197. P.A. Lebwohl, G. Lasher, Phys. Rev. A **6**, 426 (1972)
198. K. Binder, J.D. Reger, Adv. Phys. **41**, 547 (1992)
199. H.O. Carmesin, K. Ohno, J. Magn. Magn. Mater. **104–107**, 264 (1992)
200. I.V. Markov, *Crystal Growth for Beginners* (World Scientific, Singapore, 1979)
201. Y. Saito, *Statistical Physics of Crystal Growth* (World Scientific, Singapore, 1996)
202. E. Mayer, J. Cryst. Growth **74**, 425 (1986)
203. K. Ohno, H. Trinkaus, H. Müller-Krumbhaar, J. Cryst. Growth **99**, 68 (1990)
204. R. Becker, W. Döring, Ann. Phys. **24**, 719 (1935)
205. I.M. Lifshitz, V.V. Slyozov, J. Phys. Chem. Solids **19**, 35 (1961)
206. K. Binder, P. Stauffer, Phys. Rev. Lett. **33**, 1006 (1974)
207. R. Sahara, H. Mizuseki, K. Ohno, S. Uda, T. Fukuda, Y. Kawazoe, Sci. Rep. Res. Inst. Tohoku Univ. A **43**, 23 (1997)
208. I. Steinbach, Annual Rev. Mater. Res. **43**, 89 (2013)
209. N. Provatas, K. Elder, *Phase-Field Methods in Materials Science and Engineering* (Wiley-VCH, Weinheim, 2010)
210. W. Cahn, J.E. Hilliard, J. Chem. Phys. **28**, 258 (1958)

211. D. Lee, J.-Y. Huh, D. Jeong, J. Shin, A. Yun, J. Kim, Comput. Mater. Sci. **81**, 216 (2014)
212. S. Bhattacharyya, R. Sahara, K. Ohno, to appear
213. R. Kubo, Rep. Prog. Phys. **29**, Part I, 255 (1966)
214. S. Masatsuji, Y. Misumi, S. Ishii, K. Ohno **49**, 2424 (2008)
215. Y. Misumi, S. Masatsuji, R. Sahara, S. Ishii, K. Ohno, J. Chem. Phys. **128**, 234702 (2008)
216. I.M. Lifshitz, Sov. Phys. JETP **15**, 939 (1962)
217. J.D. Gunton, in *Phase Transitions and Critical Phenomena*, vol. 8, ed. by C. Domb, J.L. Lebowitz (Academic Press, London, 1983) pp. 267, 479
218. K. Binder, P. Stauffer, Adv. Phys. **25**, 343 (1976)
219. G. Gompper, M. Schick, Self-Assembling Amphiphilic Systems. Phase Transitions and Critical Phenomena, vol. 16, ed. by C. Domb, J.L. Lebowitz (Academic Press, London, 1994); K. Kawasaki, T. Kawakatsu. Physica A **164**, 549 (1990)
220. B. Widom, J. Chem. Phys. **84**, 6943 (1986)
221. S.T. Milner, T.A. Witten, J. Physique **49**, 1951 (1988)
222. P. Chandra, S.A. Safran, Europhys. Lett. **17**, 691 (1992)
223. D. Wu, D. Chandler, B. Smit, J. Phys. Chem. **96**, 4077 (1992)
224. Y. Okabe, Private Communication
225. W.W. Mullins, R.F. Sekerka, J. Appl. Phys. **34**, 323 (1963)
226. W. Kossel, Nach. Ges. Wiss. Göttingen 135 (1927)
227. W. Kossel, Naturwissenschaften **18**, 901 (1930)
228. I. Stranski, R. Kaischew, Z. Phys, Chem. B **26**, 31 (1934)
229. M. Volmer, *Kinetik der Phasenbildung* (Leipzig, Leipzig, 1939)
230. R.H. Swendsen, Phys. Rev. B 15, 5421 (1977). Phys. Rev. B **18**, 492 (1978)
231. G. Gompper, D.M. Kroll, R. Lipowsky, Phys. Rev. B **42**, 961 (1990)
232. W.K. Burton, N. Cabrera, F.C. Frank, Phil. Trans. R. Soc. Lond. A **243**, 299 (1950–1)
233. J.J. Harris, B.A. Joyce, P.J. Dobson, Surf. Sci. **103**, L90 (1981)
234. C.E.C. Wood, Surf. Sci. **108**, L441 (1981)
235. W. Braun, L. Däweritz, K.H. Ploog, Phys. Rev. Lett. **80**, 4935 (1998)
236. R.L. Schwoebel, E.J. Shipsey, J. Appl. Phys. **37**, 3682 (1966)
237. R.L. Schwoebel, J. Appl. Phys. **40**, 614 (1969)
238. M. Uwaha, Y. Saito, Phys. Rev. Lett. **68**, 224 (1992)
239. M. Sato, M. Uwaha, Y. Saito, Phys. Rev. Lett. **80**, 4233 (1998)
240. P. Meakin, in *Phase Transitions and Critical Phenomena*, vol. 12, ed. by C. Domb, J.L. Lebowitz (Academic Press, London, 1988), p. 335
241. H.E. Stanley, N. Ostrowsky, *Random Fluctuations and Pattern Growth: Experiments and Models* (Kluwer, Amsterdam, 1988)
242. B.H. Kaye, *A Random Walk Trhough Fractal Dimensions* (VCH, Weinheim, 1989)
243. A. Witten, L.M. Sander, Phys. Rev. Lett. **47**, 1400 (1981)
244. P. Meakin, Phys. Rev. A 27, 604 (1983). Phys. Rev. A **27**, 1495 (1983)
245. K. Ohno, K. Kikuchi, H. Yasuhara, Phys. Rev. A **46**, 3400 (1992)
246. M. Muthukumer, Phys. Rev. Lett. 50, 839 (1983); M. Tokuyama, K. Kawasaki. Phys. Lett. A **100**, 337 (1984)
247. H. Fujikawa, M. Matsushita, J. Phys. Soc. Jpn. **60**, 88 (1991)
248. H.C. Fogedby, J. Phys. Soc. Jpn. **60**, 704 (1991)
249. M. Ohgiwari, M. Matsushita, T. Matsuyama, J. Phys. Soc. Jpn. **61**, 816 (1992)
250. M. Eden, in *Proceedings of the Fourth Berkeley Symposium on Mathematical Statistics and Probability*, vol. IV, ed. by J. Neyman (University of California Press, Berkeley, 1961), p. 223
251. P. Ramanlal, L.M. Sander, Phys. Rev. Lett. **54**, 1828 (1985)
252. F. Family, T. Vicsek, J. Phys. A **18**, L75 (1985)
253. P. Meakin, J. Phys. A **18**, L661 (1985)
254. S. Liang, L.P. Kadanoff, Phys. Rev. A **31**, 2628 (1985)
255. P. Meakin, P. Ramanlal, L.M. Sander, R.C. Ball, Phys. Rev. A **34**, 5091 (1986)
256. P. Meakin, Phys. Rev. A **38**, 994 (1988)
257. M. Itoh, R. Sahara, M. Takahashi, X. Hu, K. Ohno, Y. Kawazoe, Phys. Rev. E **53**, 148 (1996)

258. M. Kardar, G. Parisi, Y. Zhang, Phys. Rev. Lett. **56**, 889 (1986)
259. C.T. Capraro, Y. Bar-Yam, Comput. Mater. Sci. **1**, 169 (1993)
260. P. Meakin, Phys. Rev. Lett. **51**, 1119 (1983)
261. M. Kolb, R. Botet, R. Jullien, Phys. Rev. Lett. **51**, 1123 (1983)
262. R. Jullien, M. Kolb, J. Phys. A Math. Gen. **17**, L639 (1984)
263. Brown, Ball (1985); W.D. Brown, R.C. Ball, J. Phys. A Math. Gen. **18**, L517 (1985)
264. F. Family, P. Meakin, T. Vicsek, J. Chem. Phys. **83**, 4141 (1985)
265. M. Kudoh, X. Hu, K. Ohno, Y. Kawazoe, J. Cryst. Growth **128**, 1162 (1993)
266. R.K. Iler, *The Chemistry of Silica* (Wiley, New York, 1979)
267. K. Ohno, Y. Kawazoe, Comput. Theor. Polymer Sci., in submission
268. M. Mikami, Master Thesis, Tohoku University (1991)
269. C. Aubert, D.S. Cannel, Phys. Rev. Lett. **56**, 738 (1986)

# Chapter 6
# Quantum Monte Carlo (QMC) Methods

## 6.1 Introduction

Since Metropolis's work , the Monte Carlo method has been applied not only to various statistical problems of classical systems, but also to many quantum mechanical systems. In this chapter, we briefly describe the extension of the Monte Carlo method to quantum mechanical systems.

The thermal average of a physical quantity $A$ with respect to the quantum mechanical Hamiltonian $\hat{H}$ is given, in general, by

$$\langle A \rangle_T = \frac{\mathrm{Tr}\, \hat{A} \exp(-\hat{\mathcal{H}}/k_{\mathrm{B}}T)}{\mathrm{Tr} \exp(-\hat{\mathcal{H}}/k_{\mathrm{B}}T)} = \frac{\sum_m \langle m|\hat{A}|m\rangle \exp(-E_m/k_{\mathrm{B}}T)}{\sum_m \exp(-E_m/k_{\mathrm{B}}T)}, \tag{6.1}$$

where the trace Tr denotes the sum of all the diagonal matrix elements (the expectation value) of $A$ sandwiched between $\langle m|$ and $|m\rangle$, where $|m\rangle$ are the eigenstates of the Hamiltonian operator and form a complete basis set. If all the eigenvalues $E_m$ and the eigenstates $|m\rangle$ are already known as is the case for a noninteracting particle system, the Ising model of magnets, or the Einstein model for lattice vibrations, (6.1) can be evaluated exactly in the same way as in the Monte Carlo method for classical systems (or, in simple cases, one may even obtain exact solutions without using the Monte Carlo method).

However, in general cases, these eigenvalues and eigenstates are not known, and (6.1) cannot be handled directly. The quantum Monte Carlo method can allow us to sample the phase space with a given distribution $\exp(-\hat{\mathcal{H}}/k_{\mathrm{B}}T)$. This is one of the current topics in the field of Monte Carlo methods that has kept many researchers interested in them.


K. Ohno et al., *Computational Materials Science*,
https://doi.org/10.1007/978-3-662-56542-1_6

## 6.2 Variational Monte Carlo (VMC) Method

The variational Monte Carlo method is based on the variational principle valid for the quantum mechanical zero-temperature ground state. One starts with a trial wavefunction and evaluates the expectation value of the Hamiltonian in this state as the variational ground-state energy,

$$E_{\mathrm{T}} = \frac{\langle \psi_{\mathrm{T}} | \mathcal{H} | \psi_{\mathrm{T}} \rangle}{\langle \psi_{\mathrm{T}} | \psi_{\mathrm{T}} \rangle} \geq E_0. \tag{6.2}$$

The variational Monte Carlo simulation for a quantum many-particle system can be performed by an importance sampling with respect to the square of the normalized trial wavefunction $|\psi_{\mathrm{T}}(R)|^2 / \int |\psi_{\mathrm{T}}(R)|^2 \, \mathrm{d}R$, which can be interpreted as a probability distribution function. Since $\mathcal{H}$ is an operator, the quantity that is averaged will be the *local energy* $E_{\mathrm{T}}(R) = \mathcal{H}(R)\psi_{\mathrm{T}}(R)/\psi_{\mathrm{T}}(R)$. Ideally, if $\psi_{\mathrm{T}}$ is the exact ground-state wavefunction, the local energy $E_{\mathrm{T}}(R)$ should be a constant. The trial wavefunction should therefore be chosen to make this number as constant as possible. Once the trial energy has been computed for a set of parameters used in the trial wavefunction, it is then possible to perform a minimization of $E_{\mathrm{T}}$ with respect to the latter, by doing the same calculation for other values of the parameters. One could also minimize the variance in $E_{\mathrm{T}}$.

Regarding the importance sampling (Sect. 5.3.3), the trial ground-state energy is given by

$$E_{\mathrm{T}} = \frac{1}{N_P} \sum_{i=1}^{N_P} E_{\mathrm{T}}(R_i) \tag{6.3}$$

where $N_P$ points $R_i$ are sampled according to the square of the trial wavefunction. One should keep in mind that $R_i$ is a point in a $3N$-dimensional space. The importance sampling can be performed by the Metropolis algorithm. One chooses an initial point $R$ in the phase space and computes the trial local energy. $R$ is then moved to $R'$ by adding a random vector to it (one can of course consider an arbitrary function $T(R, R')$ or a Gaussian probability function as well). If $|\psi_{\mathrm{T}}(R')|^2 > |\psi_{\mathrm{T}}(R)|^2$ then the move is accepted, otherwise, one compares a random number in [0, 1] to the ratio $|\psi_{\mathrm{T}}(R')|^2/|\psi_{\mathrm{T}}(R)|^2$. If the ratio is larger then the move is accepted, otherwise it is rejected and one chooses another point in the phase space by adding another random vector to $R$. One repeats the same procedure, but keeping the energy of the unmoved points in the statistics, as is usually done when using the Metropolis algorithm. In this way, provided enough points are taken to sample the distribution function, the resulting total energy converges to the exact trial energy with a standard deviation proportional to $1/\sqrt{N_{\mathrm{P}}}$. This way of calculating an integral requires a large number of points and is certainly not economic for one particle in one dimension, for example, as much better integration methods are available in this case. But in a high-dimensional space (or, in other words, when a large number of particles are involved), it becomes very effective; in fact it is the only way one can calculate such integrals.

Indeed, the usual quadrature techniques would require $p^{3N}$ function evaluations, $p$ being the number of points typically needed in one dimension, and $N$ the number of particles in the system. Clearly, this number becomes very large when more than even five particles are involved, not to mention the fact that at higher dimensions, by choosing a regular grid, one mostly samples the *surface* of the system!

The difficulty in the VMC method is the appropriate choice of the trial function $\psi_T(R)$. It must have the proper symmetry, include correlations as accurately as possible, have the correct nodes, etc. Fermion wavefunctions must be antisymmetric under exchange, whereas for bosons, they must be invariant under exchange. The most general form would be a symmetric function of the coordinates for bosons, and a product of a symmetric function with an antisymmetric function (such as a Slater determinant of $N$ orbitals) for fermions. More generally, the wavefunction must be an eigenstate of the symmetry group of the Hamiltonian, i.e. an eigenstate of all operators that commute with $\mathcal{H}$. The shape of the orbitals used fixes also the nodes of the wavefunction.

To include correlations, one multiplies the wavefunction by the so-called Jastrow factor (see below), which is a symmetric function of the coordinates. This usually includes pair correlations, but a more sophisticated form could also include three-body correlations, which turn out to be important in predicting the ground-state energy of liquid $^3$He, for example. Other corrections, such as the introduction of a back-flow term, can also improve the ground-state energy of liquids. For more technical details of these corrections, we refer the reader to [1, 2].

The problem of constructing a good trial function by means of the variational Monte Carlo method was first considered for a hard-sphere Bose liquid at the absolute zero of temperature by Bijl [3]. For hard-sphere systems, the wavefunction should become zero if two particles interpenetrate. The simplest way to construct a trial function obeying this condition is to use a Bijl-type function, i.e. to write the trial function as the product of the two-body correlation functions $f(|\mathbf{r}_i - \mathbf{r}_j|)$:

$$\psi_T = \prod_{i<j} f(r_{ij}) \equiv \exp\left[-\frac{1}{2}\sum_{i<j} u(r_{ij})\right], \tag{6.4}$$

where the pseudopotential $u(r)$ depends only on the radial distance for isotropic liquids, and can be taken as $u(r) = A/r^5$ for Lennard–Jones systems, for example. The square of this trial function is similar to the usual Boltzmann distribution function of classical systems if one replaces $u(r)$ by (interatomic potential)/$k_B T$, hence the name "pseudopotential".

Jastrow et al. extended a similar idea to the case of a Fermi liquid [4]. They adopted

$$\psi_T = D(\boldsymbol{R}) \exp\left[-\frac{1}{2}\sum_{i<j} u(r_{ij})\right] \tag{6.5}$$

as a trial function for the Fermi liquid. Here $D(\boldsymbol{R})$ denotes the wavefunction in an ideal Fermi gas, i.e. a Slater determinant of plane waves. This type of wavefunction is called the Jastrow function, and its prefactor $D(\boldsymbol{R})$ is called the Jastrow factor.

A possible form for the trial function for quantum solids can be chosen as,

$$\psi_{\mathrm{T}} = \exp\left[-\frac{1}{2}\sum_{i<j} u(r_{ij})\right] \sum_{P} (\pm 1)^P \prod_i \phi(\boldsymbol{r}_i - \boldsymbol{s}_{P_i}) n_{\sigma_i \sigma_{P_i}}, \tag{6.6}$$

where the $\phi(\boldsymbol{r})$ is the atomic orbital centered at a site $\boldsymbol{s}$ and characterizes the one-particle trajectory around that site. The shape of the orbital is usually taken to be a Gaussian. $P$ denotes a permutation and relates each particle to a lattice site $\boldsymbol{s}$, $n$ is the spinor operator and $\pm$ denotes Bose $(+)$ or Fermi $(-)$ statistics.

The effect of exchange in quantum solids is sometimes small. In this case, one can adopt Gaussian orbitals with small overlap, and there is no need to impose any spin symmetry on the wavefunction of the solid, since the particles are distinguishable. This is valid far from the melting point, i.e. where the mean square lattice displacement is less than 25%. In cases where exchange is important, however, one needs, to take the proper symmetry of the wavefunction into account. This involves either taking sums of many-body wavefunctions in which particle coordinates are permuted, or using just a simple Slater determinant. The number of permutations being $N!$ and therefore extremely large, a Monte Carlo sampling of the permutations is also usually performed, just as in the PIMC (discussed later). In the case of a Slater determinant, one must first calculate the inverse of the Slater matrices and then use the property that if one particle is moved, only one row of the matrix is changed, but its cofactors remain the same. Then the task of updating the wavefunction or rather computing the ratio $|\psi_{\mathrm{T}}(R')|^2/|\psi_{\mathrm{T}}(R)|^2$ is much lighter than the computation of the full determinants at each step. In general, owing to the form of the trial function, fermion calculations take much more CPU time and memory than boson calculations.

## 6.3 Diffusion Monte Carlo (DMC) Method

One can consider the imaginary time evolution equation satisfied by the many-body wavefunction.

$$\frac{d\psi}{dt} = -(\mathcal{H} - E)\psi. \tag{6.7}$$

This equation looks like a diffusion equation, and its effect is to converge the initial wavefunction to the ground state. This can be seen by expanding $\psi$ in the set of eigenstates of $\mathcal{H}$. The formal solution of this equation is

$$\psi(R, t) = \sum_{\lambda} c_{\lambda} e^{-(E_{\lambda} - E)t} \psi_{\lambda}(R). \tag{6.8}$$

One can easily check that the above equation satisfies (6.7), and that, furthermore, as $t \to \infty$, the solution tends to the ground state with the lowest $E_\lambda$ provided that the constant $E$ is the ground-state energy. To keep proper normalization during the dynamics, the arbitrary parameter $E$ can be adjusted so that the amplitude of the wavefunction remains constant at each time step. In this way, while evolving $\psi$ to the ground state, one also obtains simultaneously the ground-state energy as the final value of $E$. In principle, therefore, the diffusion equation yields the exact ground state wave function at large times. A more quantitative formula for $E$ is

$$E = \frac{\int \psi(R,0)\mathcal{H}\psi(R,t)\mathrm{d}R}{\int \psi(R,0)\psi(R,t)\mathrm{d}R} \tag{6.9}$$

since the wavefunction $\psi(R,t)$ converges to the lowest eigenstate of $\mathcal{H}$ for $t \to \infty$. The above formula should really be a time average for times $t$ larger than the relaxation time of the system, i.e. after the convergence to the ground state has been achieved. The initial wavefunction is usually known analytically, and the final one is represented by an ensemble of delta functions or walkers (see below) peaked at the positions of the particles, so performing the integrals is a trivial task. In fact, one would just get an average of the previously introduced local energy over the walker trajectories at large times.

Starting with a variational wavefunction, one can use (6.7) to "diffuse" it further towards the ground state. For bosons, since the wavefunction has no nodes, one can consider it positive without any loss of generality. In this case, $\psi$ (and not $\psi^2$) can be considered as a density or population of a set of diffusing (unphysical) objects, called "walkers" in this context. A walker is represented by a point in the $3N$-dimensional space. To represent the wavefunction, one samples it by considering an *ensemble* of walkers which will move or evolve in time to reach the optimum equilibrium distribution corresponding to the ground state. The algorithm to update the position of the walkers that we will adopt is called the short-time approximation to the Green's Function Monte Carlo (GFMC) method. The Green's function is the evolution operator (in imaginary time), also called the propagator, and takes the initial wavefunction $|R_\mathrm{i}(t=0)\rangle$ at time $t=0$ to its final value $|R_\mathrm{f}(t)\rangle$ at time $t$:

$$G(R_\mathrm{f}, R_\mathrm{i}; t) = \langle R_\mathrm{f}| e^{-(\mathcal{H}-E)t} |R_\mathrm{i}\rangle. \tag{6.10}$$

The final wavefunction can be obtained from the convolution of $G$ with the initial wavefunction:

$$\psi(R_\mathrm{f}, t) = \int G(R_\mathrm{f}, R_\mathrm{i}; t)\psi(R_\mathrm{i}, 0)\,\mathrm{d}R_\mathrm{i}.$$

One can start with a set of walkers, distributed according to $\psi(R_\mathrm{i}, 0)$ obtained from a variational calculation, and move them according to some Metropolis algorithm which simulates their time evolution $\mathrm{e}^{-(\mathcal{H}-E)t}$. For short times $\delta t$, we have

$$\langle R|\mathrm{e}^{-(\mathcal{H}-E)\delta t}|R'\rangle = \frac{1}{(2\pi\,\delta t)^{3N/2}}\mathrm{e}^{-m(R-R')^2/(2\delta t)}\,\mathrm{e}^{-[V(R)-E]\delta t}. \tag{6.11}$$

From a walker located at $R$, one goes to $R'$ by using the importance sampling technique in two steps. The first is with respect to the choice of $R'$ (choosing a Gaussian distribution because of the kinetic-energy term or a random move of amplitude $\sqrt{\delta t/m}$) and the second with respect to the potential-energy term $e^{-[V(R)-E]\delta t}$. One can continue this process for larger times until the final state is reached. Alternatively, one can make this faster by generating or eliminating walkers in the regions where the potential energy is low or high, respectively: $m$ copies of the walker are generated at the same position $R$, where $m = \mathrm{int}(w + r)$, $r$ is a random number between 0 and 1, and $w = \mathrm{e}^{-[V(R)-E]\delta t}$. For $m = 0$, the walker is deleted. This operation of creating or deleting walkers is called branching. This can then be interpreted as a diffusion process of free particles (kinetic-energy term) with source and sink terms (potential-energy part). This is more efficient than the previous algorithm since walkers will accumulate in regions of lower potential energy much faster than if they had to diffuse there from a distant point. The arbitrary value of $E$ can be adjusted so as to keep a constant population of walkers. The time for converging to the ground state is quite short; most of the CPU time used is to obtain a good sampling of the ground-state wavefunction (or to reduce noise) by continuing the run for longer times.

Using this diffusion equation method to update the walkers is called the diffusion Monte Carlo (DMC) method in which one samples the time-dependent Green's function $G(R_\mathrm{f}, R_\mathrm{i}; t)$.

Another way of treating the diffusion problem which was introduced more recently, is to consider the diffusion equation (6.7), and introduce importance sampling by multiplying it by a trial wavefunction $\psi_\mathrm{T}$ with energy $E_\mathrm{T}$, obtained from a variational calculation [5]. Defining $f(R,t) = \psi_\mathrm{T}(R)\psi(R,t)$, one can write the equation satisfied by $f$ as

$$\frac{\partial f}{\partial t} = \frac{\nabla^2 f}{2m} - \nabla\cdot\frac{f\boldsymbol{F}(R)}{m} - [E_\mathrm{T}(R) - E_\mathrm{T}]f(R,t), \tag{6.12}$$

where $\boldsymbol{F}(R) = \nabla\ln\psi_\mathrm{T}(R)$. Using this equation has the advantage that fluctuations in $V(R)$ are reduced by the use of the local energy $E_\mathrm{T}(R)$ which could have very small fluctuations if the trial function is well chosen. There would be much less branching in this case. Furthermore, a new *drift* term $\nabla\cdot[f\boldsymbol{F}(R)/m]$ has been introduced, and its effect, as can be seen from the short-time approximation of the propagator below, is to displace each walker by an additional amount of $\delta t\boldsymbol{F}(R)/m$ The short-time approximation of the propagator can be written in this case as

$$\begin{aligned} G(R, R'; \delta t) &= \frac{1}{(2\pi\,\delta t)^{3N/2}}\mathrm{e}^{-m[R-R'-F(R)\delta t]^2/(2m\delta t)} \\ &\times\ \mathrm{e}^{-[E_\mathrm{T}(R)/2+E_\mathrm{T}(R')/2-E_\mathrm{T}]\delta t}. \end{aligned} \tag{6.13}$$

Similarly to the previous algorithm, a walker at $R$ is moved to $R'$ sampled from the kinetic part of the exponent and its weight is changed to $w' = we^{-[E_T(R)/2+E_T(R')/2-E_T]\delta t}$. Branching could occur if $w'$ becomes too large or too small compared to one. The ground state energy is given by

$$E = \frac{\sum_i w_i E_T(R_i)}{\sum_i w_i}, \tag{6.14}$$

where the local energy $E_T(R) = \mathcal{H}\psi_T(R)/\psi_T(R)$ is defined as previously, and $w_i$ is the weight of each walker $R_i$.

For bosons, the ground-state wavefunction has no nodes and does not change sign; it can therefore be interpreted as the distribution function of a set of walkers. For fermions, however, the wavefunction will have nodes. This makes such an interpretation impossible. But within each of the nodal volumes where the wavefunction has a constant sign, one can still use such a trick and sample $\psi$ by walkers. The only constraint imposed would be that the nodes remain fixed at those of the trial wavefunction $\psi_T$. Within each nodal volume, $\psi$ would converge to the best possible wavefunction (i.e. with lowest energy). This is also called the "fixed-node" approximation. The divergence of the function $F(R) = \nabla \ln|\psi_T(R)|$ in (6.12) acts as a repulsive potential which keeps the walkers away from the nodal surfaces and makes QMC calculations for fermions stable. The total energy would then, unlike the case for bosons, still be an upper bound to the true ground-state energy.

It is also possible to use the integrated form of $G$ and obtain the ground-state wavefunction by iterating the following:

$$\psi^{[n+1]}(R) = \int G(R, R')\psi^{[n]}(R')\mathrm{d}R', \tag{6.15}$$

where, this time, $G$ is integrated over time:

$$G(R, R') = \langle R|(E_T + E_c)\int_0^\infty \mathrm{e}^{-(\mathcal{H}+E_c)t}\,\mathrm{d}t\,|R'\rangle = \langle R|\frac{E_T + E_c}{\mathcal{H} + E_c}|R'\rangle. \tag{6.16}$$

This method is called the GFMC method. $E_c$ is chosen such that all eigenvalues of the denominator are positive. In this case, one must sample the inverse of $\mathcal{H}$. This can be done by using the identity $(1-x)^{-1} = \sum_k x^k$, $x$ being the properly rescaled and shifted Hamiltonian operator. The Neumann–Ulam method [6] provides us with a way to sample $\mathcal{H}^k$. $\mathcal{H}_{ij}$ can be decomposed into a product of a transition probability and a residual weight: $\mathcal{H}_{ij} = P_{ij}W_{ij}$, where $\sum_j P_{ij} = 1$; $P_{ij} \geq 0$. One can then define a random walk of $k$ steps defined by the transition probabilities $P_{ij}$ with a weight of $W_{ij}$ for each step. If the final step is $j$, the final score is defined as $S_{ij} = W_{ii_1}W_{i_1i_2}\dots W_{i_{k-1}j}$, and zero otherwise. The ensemble average of $S_{ij}$ is then equal to $(\mathcal{H}^k)_{ij}$. More details about this method can be found in [7]. Both the GFMC and the DMC methods not using the short-time approximation, require similar tasks.

Whereas the DMC method can in principle give the exact ground-state energy for bosons, it can only give an upper bound for fermions because of the difficulty in determining the correct position of the nodes.

It must be added that up to now, most accurate ground state energies have been obtained with QMC methods. Recently, some work has been done in this field to treat real materials by using accurate pseudopotentials [8, 9]. One application has been to calculate the total energy of three isomers of $C_{20}$ [10] where ab initio LDA-based methods (including the GGA) and Hartree–Fock-based methods could not agree on the relative energies of these isomers. Even the VMC is capable of computing about 60% or more of the correlation energy provided good trial functions are used. In the case of the GFMC method, this accuracy could reach 95%.

Although we did not concern ourselves here with the application of the DMC method to the solid state, we remark that the special $k$-point method has been applied to calculate the electronic properties of solids [11]. For a review of the many-body theory and methods, we refer the reader to the books by Fetter and Walecka [12], and Negele and Orland [13].

## 6.4 Path-Integral Monte Carlo (PIMC) Method

As the most commonly used formula in the quantum Monte Carlo method, the Trotter formula[14] is given by

$$e^{\hat{\mathcal{H}}_1+\hat{\mathcal{H}}_2} = \lim_{M\to\infty}\left(e^{\hat{\mathcal{H}}_1/M}e^{\hat{\mathcal{H}}_2/M}\right)^M . \tag{6.17}$$

This equation holds exactly only in the limit where the Trotter number $M$ goes to infinity. In practice, one may use this equation approximately with a suitably large but finite $M$. Generally, one may use a relatively small $M$ in high-temperature regions, while one should increase $M$ when the temperature decreases in order to maintain the same accuracy. Therefore, how big the Trotter number should be depends on how low the temperatures one is investigating are. Using the Trotter formula, the partition function is rewritten as

$$\begin{aligned} Z^{(m)} &= \mathrm{Tr}\left(e^{-\beta\hat{\mathcal{H}}_1/m}e^{-\beta\hat{\mathcal{H}}_2/m}\right)^m \\ &= \sum_{\alpha_1,\ldots,\alpha_{2m}} \langle\alpha_1|e^{-\beta\hat{\mathcal{H}}_1/m}|\alpha_2\rangle\langle\alpha_2|e^{-\beta\hat{\mathcal{H}}_2/m}|\alpha_3\rangle\langle\alpha_3|e^{-\beta\hat{\mathcal{H}}_1/m}|\alpha_4\rangle \\ &\times\cdots\langle\alpha_{2m-2}|e^{-\beta\hat{\mathcal{H}}_2/m}|\alpha_{2m-1}\rangle\langle\alpha_{2m-1}|e^{-\beta\hat{\mathcal{H}}_1/m}|\alpha_{2m}\rangle\langle\alpha_{2m}|e^{-\beta\hat{\mathcal{H}}_2/m}|\alpha_1\rangle, \end{aligned} \tag{6.18}$$

where $\beta = 1/k_B T$ and $\langle\alpha_i|e^{-\beta\hat{\mathcal{H}}_j/m}|\alpha_{i+1}\rangle$ is called the propagator.

The path-integral Monte Carlo method [15] offers a powerful method applicable to many interesting quantum systems describing, for example, the melting transition of a Wigner lattice or the Bose–Einstein condensation, to which lattice models cannot be applied directly. Feynman's path-integral formula can be retrieved from the Trotter formula when $\hat{\mathcal{H}}_1$ and $\hat{\mathcal{H}}_2$ are regarded, respectively, as the kinetic-and potential-energy terms. The path-integral Monte Carlo method is also based on this method of decomposition. Here we put

$$\hat{\mathcal{H}}_1 = -\frac{\hbar^2}{2m}\sum_{i=1}^{N}\nabla_i^2, \qquad \hat{\mathcal{H}}_2 = V(\boldsymbol{r}_1, \ldots, \boldsymbol{r}_N). \tag{6.19}$$

Let us consider first one particle confined in a $d$-dimensional hypercubic box of length $L$. We can assume periodic boundary conditions. The propagator in the coordinate representation (we may regard $\alpha_i$ as the particle coordinate $\mathbf{r}$)

$$f(\boldsymbol{r}, \boldsymbol{r}') \equiv \langle \boldsymbol{r}|\mathrm{e}^{-(\beta\hat{\mathcal{H}}_1/M)}|\boldsymbol{r}'\rangle \tag{6.20}$$

can be evaluated in the limit $L \to \infty$, and is given by a Gaussian distribution function with respect to the distance between $\boldsymbol{r}$ and $\boldsymbol{r}'$:

$$f(\boldsymbol{r}, \boldsymbol{r}') = \left(\frac{Mm}{2\pi\beta\hbar}\right)^{d/2} \exp\left(-\frac{Mm}{2\beta\hbar^2}(\boldsymbol{r} - \boldsymbol{r}')^2\right). \tag{6.21}$$

Substituting this into (6.18), one may derive Feynman's path-integral formula for the quantum one-body problem in statistical physics.

Next, we turn our attention to Bose systems. The propagator of a system of $N$ bosons should describe the propagation from $[\boldsymbol{r}_1(j), \boldsymbol{r}_2(j), \ldots, \boldsymbol{r}_N(j)]$ at the time slice $j$ to $[\boldsymbol{r}_1(j+1), \boldsymbol{r}_2(j+1), \ldots, \boldsymbol{r}_N(j+1)]$ at the next time step, and is given by

$$\frac{1}{N!}\sum_{\mathrm{P}}\prod_{i=1}^{N} f\big[\boldsymbol{r}_{\mathrm{P}_i}(j+1), \boldsymbol{r}_i(j)\big], \tag{6.22}$$

where P represents an arbitrary permutation of the numbers from 1 to $N$, and $f$ is the same as the single-particle propagator given by (6.21). Since the boson wavefunction is symmetric under permutations, the partition function (6.18) is given by

$$Z = \sum_{\mathrm{P}} \int \prod_j \prod_i d\boldsymbol{r}_i(j) p[\mathrm{P}, \boldsymbol{r}_i(j)]; \tag{6.23}$$

$N!$ being the number of permutations P. The sum in $Z$ contains a very large number of terms. Hence, if we regard the pair [P, $\boldsymbol{r}_i(j)$] as the dynamical variable, the distribution function is

$$
\begin{aligned}
p[\mathrm{P}, \boldsymbol{r}_i(j)] &= \frac{1}{N!} \prod_{i=1}^{N} \left\{ f[\boldsymbol{r}_i(1), \boldsymbol{r}_{\mathrm{P}_i}(M)] \prod_{j=1}^{M-1} f[\boldsymbol{r}_i(j+1), \boldsymbol{r}_i(j)] \right\} \\
&\quad \times \exp\left\{ -\frac{\beta}{M} \sum_{j=1}^{M} V[\boldsymbol{r}_1(j), \ldots, \boldsymbol{r}_N(j)] \right\}. \qquad (6.24)
\end{aligned}
$$

Thus the average of an arbitrary physical quantity $A[\mathrm{P}, \boldsymbol{r}_i(j)]$ can be evaluated from

$$
\langle A[\mathrm{P}, \boldsymbol{r}_i(j)] \rangle = \frac{1}{Z} \sum_{\mathrm{P}} \int \prod_j \prod_i d\boldsymbol{r}_i(j) A[\mathrm{P}, \boldsymbol{r}_i(j)] p[\mathrm{P}, \boldsymbol{r}_i(j)]. \qquad (6.25)
$$

Now it is possible to introduce importance sampling *with respect to the permutations* to evaluate (6.25). Let us consider a change from $\mathrm{P} = \binom{1,\ldots,N}{\mathrm{P}_1,\ldots,\mathrm{P}_N}$ to $\mathrm{P}' = \binom{1,\ldots,i,\ldots,j,\ldots,N}{\mathrm{P}_1,\ldots,\mathrm{P}_j,\ldots,\mathrm{P}_i,\ldots,\mathrm{P}_N}$. Note that we have exchanged the $i$th particle with the $j$th particle. In this process, the partition function changes by just the following factor $W$:

$$
W = \frac{f[\boldsymbol{r}_i(1), \boldsymbol{r}_{\mathrm{P}_j}(M)] f[\boldsymbol{r}_j(1), \boldsymbol{r}_{\mathrm{P}_i}(M)]}{f[\boldsymbol{r}_i(1), \boldsymbol{r}_{\mathrm{P}_i}(M)] f[\boldsymbol{r}_j(1), \boldsymbol{r}_{\mathrm{P}_j}(M)]}. \qquad (6.26)
$$

Applying Metropolis's importance sampling (see Sect. 5.3.3) to this problem, $W$ can be regarded as a transition probability. Thus, by generating a pseudorandom number between 0 and 1, one adopts the permutation P′ if $W$ is greater than this random number, and leaves the system with the previous P otherwise. One should update the particle positions simultaneously. For the particle movement, one should update all the coordinates with different Trotter numbers ($j$) independently:

$$
\boldsymbol{r}(j) \to \boldsymbol{r}(j) + \Delta \boldsymbol{r}_1, \qquad (6.27)
$$

where $\Delta \boldsymbol{r}_1$ denotes a random vector. This update is done according to Metropolis's algorithm (5.33), i.e. by calculating the ratio of weights ($\mathrm{e}^{-\beta \Delta V}$) before and after a movement. The movement is adopted if this ratio is greater than a generated random number between 0 and 1 (if the ratio is greater than unity, it is unnecessary to generate a random number).

We show in Fig. 6.1 the result for the ideal Bose system due to Takahashi and Imada, who used importance sampling as described above [15] as well as the exact values at $N = \infty$. Of course, the result approaches the exact behavior when both the number of particles $N$ and the Trotter number $M$ increase. However, when $M$ increases, the particle movements become slower and more computational time is required. Therefore, for practical use, one should choose a moderate value for $\beta/M$.

Next, consider a fermion system as an example of the application of the path-integral Monte Carlo method. In fermion systems, owing to the antisymmetry of the wavefunction under particle permutations, an additional factor $\varepsilon(\mathrm{P}) = \pm 1$ must be included in the propagator; $\varepsilon(\mathrm{P}) = 1$ for even permutations (even number of interchanges), and $\varepsilon(\mathrm{P}) = -1$ for odd permutations.

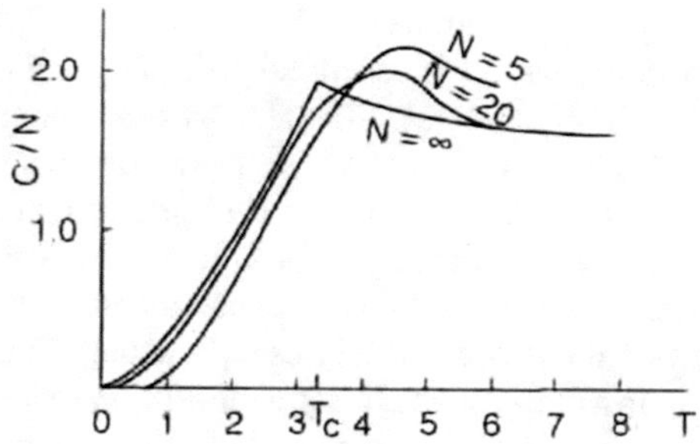


**Fig. 6.1** Specific heat of an ideal Bose gas (in the cases of particle number $N = 5$ and $N = 20$) calculated by the PIMC method [15]. The curve indicated by $N = \infty$ corresponds to the exact solution

Thus, the propagator has the form of a determinant with $f[\boldsymbol{r}_k(j+1), \boldsymbol{r}_l(j)]$ as the generic $k, l$ element. Then, one may ask: is it possible to consider a permutation P and a coordinate $\boldsymbol{r}_i(j)$ as a pair describing one state and introduce an importance sampling with respect to the permutations, similarly to the boson system? The answer is, unfortunately, no, since the transition probability $W$ and therefore the distribution function $p[\mathrm{P}, \boldsymbol{r}_i(j)]$ contain $\varepsilon(\mathrm{P})$ as a factor and may become negative. Thus we encounter a serious problem, called the sign problem. The probability that the distribution function has a negative value becomes larger as the temperature decreases [16, 17].

## 6.5 Quantum Spin Models

As was discussed already in Chap. 4, it is generally impossible to treat everything from ab initio calculations in particular for large-scale phenomena. For this reason, many quantum theories have also very often been simplified by introducing simple models. If such simplification does not lose any essential feature of the systems, the models can be used very successfully. One typical example is an interaction between electron spins when one discusses magnetic orderings of various magnetic materials. In this case, one can successfully introduces various quantum spin Hamiltonians.

One class of quantum Monte Carlo simulations is based on lattice models like a Heisenberg model or a $XY$ model. Here we do not enter the details of the derivation of these quantum spin models, besides mentioning that Heitler–London theory for weakly interacting atoms (or Wannier states) can lead to them (the spin anisotropy can be discussed on the basis of ab initio theories by considering the spin–orbit coupling).

The introduction of lattice models helps us to understand more simply or clearly the behavior of realistic continuous models, and may save computational time significantly. For some lattice systems, by introducing a suitable sublattice decomposition, Suzuki et al. [18] and Hirsch et al. [16] showed that the total Hamiltonian can be divided into two sublattice Hamiltonians $\hat{\mathcal{H}}_1$ and $\hat{\mathcal{H}}_2$, each of which is given by a sum of independent (mutually commuting) local Hamiltonians. In this case, one may easily evaluate $e^{-\beta\hat{\mathcal{H}}_1/M}$ and $e^{-\beta\hat{\mathcal{H}}_2/M}$, by using these $\hat{\mathcal{H}}_1$ and $\hat{\mathcal{H}}_2$ in (6.18). This decomposition is essentially a real-space decomposition, and (6.17) in such cases is called the Suzuki–Trotter formula. Suzuki showed that, if one regards the Trotter axis (the direction of $M$) as an additional spatial direction, the problem of evaluating the partition function of a $d$-dimensional quantum spin system becomes equivalent to the similar problem of a $(d+1)$-dimensional classical spin system. This mapping procedure of the $d$-dimensional quantum system to the $(d+1)$-dimensional classical system has been widely used in Monte Carlo simulations [19].

In a one-dimensional quantum spin system with nearest-neighbor interaction only, one may include the 1–2, 3–4, 5–6, ... spin pairs in $\mathcal{H}_1$ and include the 2–3, 4–5, 6–7, ... spin pairs in $\mathcal{H}_2$. Then, all the terms in $\mathcal{H}_1$ (and $\mathcal{H}_2$) commute with each other, and therefore the exponent $e^{-\hat{\mathcal{H}}_j/m}$ $(j = 1, 2)$ in (6.18) can be rewritten as a simple product of exponentials associated with each spin pair. It is also convenient in this problem to use a transfer-matrix method which was mentioned previously in connection with the one-dimensional Ising model in (5.62). Since there are four states, $(++)$, $(+-)$, $(-+)$, $(--)$, for a spin-pair, one may introduce a $4\times 4$ transfer matrix [20]. Since there are both the real axis and the Trotter axis (two dimensions), its trace cannot be evaluated exactly, but one can map this model to a two-dimensional Ising model with interactions between four-component spins in addition to the usual nearest-neighbor interaction. This algorithm works well for a regular lattice where $\mathcal{H}_1$ (and $\mathcal{H}_2$) is given by a sum of terms which commute. Okabe and Kikuchi carried out such MC simulations successfully both for the $XY$ chain [21] and for the Heisenberg chain [22]. However, for an irregular lattice with frustration, the commutation relation is violated and one encounters again the sign problem when one formulates the system in this way.

Here again, we briefly comment on the tensor network. Various quantum spin models can be converted to the tensor network models. Then tensor network models can be treated by the tensor renormalization group (TRG) [26] based on the higher-order singular value decomposition (HOTRG) [27].

## 6.6 Other Quantum Monte Carlo Methods

As another example of quantum Monte Carlo methods, one can mention the projection Monte Carlo method.

The projection Monte Carlo method was first introduced by Parisi and Wu [28] in lattice gauge theory. After applying the Trotter decomposition, one uses the Hubbard–Stratonovich transformation ($\exp(B^2) = \int_{-\infty}^{\infty} \exp(-\phi^2 + 2\phi B)d\phi/\sqrt{\pi}$) to linearize the fermion system ($B$). Then one may evaluate the trace with respect to the fermion operators first, and treat the boson field ($\phi$) as an auxiliary field. Hirsch [29] first introduced the method into the field of solid-state physics. Even in this method, one cannot avoid the sign problem.

On the other hand, in the variational and Green's function Monte Carlo methods, there is no sign problem, but these methods treat essentially the ground state at the absolute zero of temperature.

For more details, one may refer to the review articles by Ceperley and Kalos [1, 2], the recently published book on QMC methods by Hammond, Lester and Reynolds [9], and the book edited by Lester [10].

## References

1. D.M. Ceperley, M.H. Kalos, in *Monte Carlo Methods in Statistical Physics*, ed. by K. Binder (Springer, Berlin, 1979). Chapter 4
2. K.E. Schmidt, D.M. Ceperley, in *The Monte Carlo Method in Condensed Matter Physics*, 2nd edn., ed. by K. Binder (Springer, Berlin, 1995). Chapter 7
3. A. Bijl, Physica **7**, 869 (1940)
4. R.B. Dingle, Philos. Mag. 40, 573 (1949); R. Jastrow, Phys. Rev. 98, 1479 (1955); N.F. Mott. Philos. Mag. **40**, 61 (1949)
5. C.W. Greeff, W.A. Lester, B.L. Hammond, *Recent Advances in Quantum Monte Carlo Methods* (World Scientific, Singapore, 1997), p. 117
6. G.E. Forsythe, R.A. Leibler, MTAC **4**, 127 (1950)
7. J. Kuti, J. Polonyi, in *Proceedings of the XXI International Conference on High Energy Physics*, Paris (1982)
8. C.W. Greeff, W.A. Lester, B.L. Hammond, J. Chem. Phys. **104**, 1973 (1996)
9. B.L. Hammond, W.A. Lester, P.J. Reynolds, *Monte Carlo Methods in ab-initio Quantum Chemistry* (World Scientific, Singapore, 1994)
10. J.C. Grossman, L. Mitas, K. Raghavachari, Phys. Rev. Lett. **75**, 3870 (1995); W.A. Lester, in *Recent Advances in Quantum Monte Carlo Methods*, ed. by W.A. Lester, Jr. (World Scientific, Singapore, 1997), p. 133
11. G. Rajagopal, R.J. Needs, S. Kenny, W.M.C. Foulkes, A. James, Phys. Rev. Lett. **73**, 1959 (1994)
12. A.L. Fetter, J.D. Walecka, *Quantum Theory of Many-Particle Systems* (McGraw-Hill, Boston, 1971)
13. J.W. Negele, H. Orland, *Quantum Many-Particle Systems* (Addison-Wesley, Redwood City, 1988)
14. H.F. Trotter, Proc. Am. Math. Soc. **10**, 545 (1959)
15. M. Takahashi, M. Imada, J. Phys. Soc. Jpn. **53**, 963,3765,3770 (1984); F.F. Abraham, *Computer Simulation Studies in Condensed-Matter Physics*, ed. by D.P. Landau, K.K. Mon, H.-B. Schüttler (Springer, Berlin, 1988) p. 134
16. J. Hirsch, D.J. Scalapino, R.L. Sugar, R. Blankenbecker, Phys. Rev. Lett. 47, 1628 (1981). Phys. Rev. **26**, 5033 (1982)
17. R.P. Feynman, *Statistical Mechanics* (W. A. Benjamin, New York, 1972)
18. M. Suzuki, Prog. Theor. Phys. 56, 1454 (1976); M. Suzuki, S. Miyashita, A. Kuroda. Prog. Theor. Phys. **58**, 1377 (1977)
19. M. Suzuki, *Quantum Monte Carlo Methods in Equilibrium and Nonequilibrium Systems*, vol. 74 (Solid-State Sciences (Springer, Berlin, 1987)
20. Y. Okabe, M. Kikuchi, Phys. Rev. B **34**, 7896 (1986)
21. Y. Okabe, M. Kikuchi, J. Phys. Soc. Jpn. **56**, 1963 (1987)
22. Y. Okabe, M. Kikuchi, J. Phys. Soc. Jpn. **57**, 4351 (1988)
23. G.M. Crosswhite, D. Bacon, Phys. Rev. A **78**, 012356 (2008)
24. G. Evenbly, G. Vidal, J. Stat. Phys. **145**, 891 (2011)
25. U. Schollwöck, Ann. Phys. **326**, 96 (2011)

26. M. Levin, C.P. Nave, Phys. Rev. Lett. **99**, 120601 (2007)
27. Z.Y. Xie, J. Chen, M.P. Qin, J.W. Zhu, L.P. Yang, T. Xiang, Phys. Rev. B **86**, 045139 (2012)
28. G. Parisi, Y. Wu, Sci. Sin. **24**, 483 (1981)
29. J.E. Hirsch, Phys. Rev. Lett. 51, 1960 (1983). Phys. Rev. B **31**, 4403 (1985)

# Appendix A
# Fourier Interpolation in 1 and 3 Dimensions

In perfect crystals, where the wavevector $k$ is a good quantum number, many functions of $k$ are used for calculations. They are periodic in the first Brillouin zone, and therefore can be represented by their Fourier series. Examples of such functions are electron and phonon band dispersions. In case they are smooth functions, they will have very few Fourier components, and can therefore be efficiently represented with only a few Fourier coefficients. These coefficients are functions in the real space, and would physically represent some kind of "interactions" between neighboring atoms. So the idea of Fourier interpolation is to compute the function (typically a band dispersion) at a few k-points placed on a regular mesh, Fourier transform them to get their Fourier coefficients in real space, and use them to interpolate the dispersion at any arbitrary k-point with minimal cost.

## A.1 Definitions in One Dimension

Fourier expansion is used for periodic functions. The basis for such an expansion is exponential functions $e^{ikx}$, and the expansion for a function of period $L$ takes the form:

$$f(x) = \sum_{n=1}^{\infty} F(n) e^{i2\pi nx/L} \tag{A.1}$$

where $n$ is an integer ($k = 2\pi n/L$) and the Fourier coefficients are:

$$F(n) = \frac{1}{L} \int_0^L dx\, f(x) e^{-i2\pi nx/L} \tag{A.2}$$

In a numerical computation however, the function is represented by a discrete set of points $x_j = j \times \Delta x, j = 1, \ldots, N$, with $N\Delta x = L$ and $x_i = x_{i+N}$ for any arbitrary integer $i$. In this case, the Fourier expansion is also finite and of the same size as the number of points $N$. For a given set of $N$ function values $f_j, j = 1, \ldots, N$



K. Ohno et al., *Computational Materials Science*,
https://doi.org/10.1007/978-3-662-56542-1

the expansion is defined by:

$$f(x_j) = f_j = \sum_{n=1}^{N} F_n e^{i2\pi n x_j/L} = \sum_{n=1}^{N} F_n e^{i2\pi nj/N} \tag{A.3}$$

with the Fourier coefficients being defined by:

$$F_n = \frac{1}{N} \sum_{j=1}^{N} f_j \, e^{-i2\pi n x_j/L} = \frac{1}{N} \sum_{j=1}^{N} f_j \, e^{-i2\pi nj/N} \tag{A.4}$$

Note we have the periodicity relation $F_n = F_{n+N}$ since $j$ in the above relations is integer. We have therefore the freedom to choose in which interval $n$ resides: $n = 1, \ldots, N$ or $n = -N/2, \ldots, N/2 - 1$ etc. As far as reproducing the original set $f_j$ is concerned all choices will lead to the exact same $f_j$. This is called *aliasing*. For intermediate points which are not in the form $x = j\Delta x$, however, every choice of the n-mesh will lead to a different result. Which is the "correct" choice? The correctness of the choice comes from our requirement on the interpolating function: If we require it to be as smooth as possible, then we should have only long wavelength components, meaning $n$ has to be as small as possible, which considering symmetry, leads to $n$ values centered around 0:

$$n = -p, \ldots, p - 1 \ \ (\text{if } N = 2p) \qquad\qquad n = -p, \ldots, p \ \ (\text{if } N = 2p + 1) \tag{A.5}$$

Note that for even values of $N$, the result will not be symmetric. We can add the two end points $-N/2$ and $N/2$ each with weights 1/2, when we do interpolation. The interpolation formula is then:

$$f(x) = \sum_{n=-p+1}^{p-1} F_n \, e^{i2\pi nx/L} + F_{N/2} \cos\left(\frac{\pi N x}{L}\right) \ \ (\text{if } N = 2p) \tag{A.6}$$

$$f(x) = \sum_{n=-p}^{p} F_n \, e^{i2\pi nx/L} \ \ (\text{if } N = 2p + 1) \tag{A.7}$$

Finally, going back to the result $F_n = F_{n+N}$: in practice if the function $f(x)$ is smooth, its Fourier components will decay to zero as $n$ is increased from 0, before going back up when $n = N$. So when we restrict $n$ to be in $-N/2, N/2 - 1$, it is hoped that if $N$ is large enough and the function $f$ smooth enough, at these end points the coefficients $F_{N/2}$ have decayed to a negligible value.

## A.2 Extension to 3 Dimensions

Given a primitive lattice defined by three vectors $(\boldsymbol{a}_1, \boldsymbol{a}_2, \boldsymbol{a}_3)$ one defines the reciprocal lattice by $(\boldsymbol{b}_1, \boldsymbol{b}_2, \boldsymbol{b}_3)$ such that $\boldsymbol{a}_i \cdot \boldsymbol{b}_j = 2\pi\delta_{ij}$

Since it is customary to use the Fourier interpolation to calculate functions in the reciprocal space, we will consider our function in the k-space defined by $\boldsymbol{b}_i$, and have its Fourier coefficients be in the real space defined by $\boldsymbol{a}_i$. These coefficients will be indexed by translation vectors $\boldsymbol{R} = \sum_{i=1}^{3} n_i \boldsymbol{a}_i$, and will be limited to few neighbors, as physically the range of "interactions" is usually short due to screening effects. A widely-used example of such interpolation is for the dynamical matrix when we wish to compute phonon dispersions at many arbitrary kpoints. The computation of the dynamical matrix from DFPT for instance being very costly, we perform these calculations on a finite and coarse mesh. Then Fourier transform to obtain the force constants (in real space) and use the interpolation formula to obtain the dynamical matrix at an arbitrary k-point. The important point to note here is that the force constants obtained from direct Fourier transformation will be periodic, while physically they should decay to zero beyond some range. So, similar to the 1D case, we have to provide a rule to get such physical result when "Fourier transforming", since having $N$ large would be costly.

So, let us assume that we have a periodic function $g(\boldsymbol{k})$ periodic within the first Brillouin zone, i.e. $g(\boldsymbol{k}) = g(\boldsymbol{k} + \boldsymbol{b}_i)\ \forall i = 1, 2, 3$, and let us assume we have calculated its values on a mesh within the parallelepiped defined by $(\boldsymbol{b}_1, \boldsymbol{b}_2, \boldsymbol{b}_3)$: $g_{h_1h_2h_3} = g(\frac{h_1}{N_1}\boldsymbol{b}_1 + \frac{h_2}{N_2}\boldsymbol{b}_2 + \frac{h_3}{N_3}\boldsymbol{b}_3)$, where $h_i = 1, \ldots, N_i$, and we want to interpolate to any arbitrary $\boldsymbol{k}$ point which is not on this mesh. We will first write its three-dimensional Fourier expansion in the real space as an extension of the 1D formula:

$$g_{h_1h_2h_3} = \sum_{n_1n_2n_3} G_{n_1n_2n_3} \exp\left[i2\pi\left(\frac{h_1n_1}{N_1} + \frac{h_2n_2}{N_2} + \frac{h_3n_3}{N_3}\right)\right] \tag{A.8}$$

We generally write this in short vector form as:

$$g(\boldsymbol{k}) = \sum_{\boldsymbol{R}} G(\boldsymbol{R})\, e^{i\boldsymbol{k}\cdot\boldsymbol{R}} \tag{A.9}$$

in which $\boldsymbol{R} = n_1\boldsymbol{a}_1 + n_2\boldsymbol{a}_2 + n_3\boldsymbol{a}_3$ and $\boldsymbol{k} = \frac{h_1}{N_1}\boldsymbol{b}_1 + \frac{h_2}{N_2}\boldsymbol{b}_2 + \frac{h_3}{N_3}\boldsymbol{b}_3$, so that the dot product $\boldsymbol{k} \cdot \boldsymbol{R}$ produces the stated exponent in (A.8). The Fourier components $G_{n_1n_2n_3} = G(\boldsymbol{R})$ are given by the inverse formula:

$$G_{n_1n_2n_3} = \frac{1}{N_1N_2N_3} \sum_{h_1h_2h_3} g_{h_1h_2h_3} \exp\left[-i2\pi\left(\frac{h_1n_1}{N_1} + \frac{h_2n_2}{N_2} + \frac{h_3n_3}{N_3}\right)\right] \tag{A.10}$$

or equivalently

$$G(\boldsymbol{R}) = \frac{1}{N_1 N_2 N_3} \sum_{\boldsymbol{k}} g(\boldsymbol{k})\, e^{-i\boldsymbol{k}\cdot\boldsymbol{R}} \tag{A.11}$$

Similar to the one-dimensional case, to avoid aliasing and keep the longest wavelength modes, the integers $n_1, n_2, n_3$ are chosen to be the smallest possible, i.e. between $-N_i/2$ and $N_i/2$. We see therefore that this expansion is over translation vectors of range $N_i\boldsymbol{a}_i/2$. In the language of solid-state physicists, we restrict the R-mesh to be in the Wigner–Seitz (real-space version of the first Brillouin zone) supercell (WSSC) defined by the three vectors $(N_1\boldsymbol{a}_1, N_2\boldsymbol{a}_2, N_3\boldsymbol{a}_3)$.

As mentioned at the end of the previous section, there might be a few $\boldsymbol{R}$ points at the WSSC boundary where $n_i = N_i/2$. Let us call this number $N_B$ and define new Fourier coefficients $G'$ as:

$$G'(\boldsymbol{R}) = G(\boldsymbol{R}) \text{ if } \boldsymbol{R} \text{ is inside WSSC;}$$

$$G'(\boldsymbol{R}) = G(\boldsymbol{R})/N_B \text{ if } \boldsymbol{R} \text{ is on the WSSC boundary}$$

The interpolation formula in 3D for an arbitrary $\boldsymbol{k}$ vector will then be:

$$g(\boldsymbol{k}) = \sum_{\boldsymbol{R}\in \text{WSSC}} G'(\boldsymbol{R})\, e^{i\boldsymbol{k}\cdot\boldsymbol{R}} \tag{A.12}$$

# Appendix B
# Molecular Dynamics and Mechanical Properties

The method of classical molecular dynamics was first proposed by Alder and Wainwright [1] in 1957. In order to perform molecular dynamics, one must first assume an interatomic or intermolecular force function. Alder and Wainwright applied the method to the simplest system, i.e. hard spheres, whose interparticle potential $V(r)$ is given by $V = \infty$ for $r < a$ and $V = 0$ for $r > a$, and found a liquid–solid phase transition in this hard-sphere system. This was conjectured by Kirkwood [2] in 1939 theoretically, by invoking statistical mechanics. This phase transition is now called the Alder transition, and is important because a system without an attractive force may have a cohesive phase.

Later, simulation techniques were also applied to various systems with soft potentials [3] such as the Lennard–Jones potential, and have played an important role in the study of the structure and dynamics of simple liquids.

What is the basic equation of motion for atoms and molecules? If we assume the NVE ensemble (also called the microcanonical ensemble), in which the number of particles $N$, the volume $V$ and the total energy $E$ are constant, the equation of motion for the atoms is the usual Newtonian equation

$$m_i \frac{d^2 \boldsymbol{r}_i}{dt^2} = \boldsymbol{F}_i, \quad \boldsymbol{F}_i = -\nabla_i \sum_{j=1}^{N} V_{ij}, \tag{B.1}$$

where $m_i$ is the mass of the $i$th particle, $\boldsymbol{r}_i$ its coordinate and $\boldsymbol{F}_i$ the force acting on it. Equation (B.1) can be solved by numerical integration. The most popular methods are the Verlet [4], leapfrog [5] and velocity Verlet [6] methods. In order to integrate the temporal evolution equation for longer times with more accuracy, the predictor–corrector method, e.g. the Gear method [7], is superior. Recently, Tuckerman and Berne [8] developed the multi-time-step method, which is expected to improve the accuracy and reduce the required computational time.

The effects of finite system size and surfaces are always a severe problem in simulations. In order to reduce these effects, one usually uses periodic boundary conditions. All the particles are put inside a box called the unit cell, and if the



K. Ohno et al., *Computational Materials Science*,
https://doi.org/10.1007/978-3-662-56542-1

particle goes outside the cell boundary, it is brought back in from the opposite side of the cell.

In order to evaluate the force acting on, say, the $i$th particle, it is necessary to sum all forces exerted by all other particles inside a certain sphere of a cutoff radius $R_c$ centered at the $i$th particle. Of course, this summation must be performed even if the particles are not in the same unit cell but in the image cells [9]. One should note that this method is not exact but, rather, depends on the size and shape of the unit cell. Nevertheless, this is an effective way to deal with the finiteness of the simulated system.

In 1972, Lees and Edwards [10] proposed a method to "directly" treat nonequilibrium systems with gradients in the spatial velocity distribution. In contrast to the popular method of deriving transport properties by using the linear-response theory [11], they realized this nonequilibrium state on computers and discussed transport properties directly from the result. This method was later extended by Gillan and Dixon [7] to systems with thermal gradients to discuss the thermal conductivity and then was widely used in the field of nonequilibrium molecular-dynamics simulations [13]. If one is interested in equilibrium properties, some extensive or intensive parameter of the simulation must be fixed. In statistical thermodynamics, there are several possible ensembles, called NVE, NTV, NHP, NTP, $\mu$TV and $\mu$TP, where N, E, V, T, P, H and $\mu$ denote, respectively, the number of particles, the total energy, the volume, the temperature, the pressure, the enthalpy and the chemical potential. For example, the NVE ensemble is the so-called microcanonical ensemble, where the quantities $N$, $V$ and $E$ are held constant. (In the NVE ensemble the basic equation of motion is of course the Newtonian equation, but in other ensembles this is not necessarily the case.) The NTV ensemble is the canonical ensemble, where $N$, $T$ and $V$ are held constant, while the $\mu$TV ensemble is the grand canonical ensemble, where $\mu$, $T$ and $V$ are held constant. An elaborate molecular-dynamics technique for constant pressure was developed by Andersen [14] and Parrinello and Rahman [15]. A similar technique to simulate constant temperature was developed by Nosé [16]; see also Appendix B. More recently, in 1991, Çaǧin and Pettitt [18] proposed a molecular-dynamics method to treat the grand canonical ensemble. The basic equations for the evolution of particle positions are selected according to the different ensembles one wishes to simulate. For example, by solving the equations of motion given by the Parrinello–Rahman and Nosé techniques, after a long time period one may obtain an equilibrium state which is very close to the true thermal equilibrium at the given temperature and pressure.

From such a simulation, one collects a times series of coordinates (trajectories) and kinetic and potential energies that one can use for performing time averages. These averages replace the ensemble average required to calculate the thermodynamic properties of a system. Therefore, the simulation time should be large enough, that one reaches thermodynamic equilibrium, and has enough data after this equilibrium point to compute averages. Other than information on the thermodynamic properties of a system, MD provides us with structural quantities such as the pair correlation function, which gives the distribution of distances between pairs of atoms, and the structure factor $S(k) = 1/N\langle\rho(k)\rho(-k)\rangle$, which can be measured in neutron-scattering experiments, for example, and, finally, a way to calculate *response functions*. The latter are obtained from time correlation functions between two different quantities: $C_{AB}(t) = \langle \delta A(t) \delta B(0) \rangle$. Among useful response functions, one can mention the diffusion coefficient, viscosity, compressibility, electrical and heat conductivities, and dynamic structure factor. The fluctuation–dissipation theorem relates such response functions, characterizing a *dissipation* in the system, to the time correlation functions above, which include *fluctuations* of the system. The latter can be obtained from an MD simulation or a model calculation by using Green's-function techniques (see for example the book by Doniach and Sondheimer [19]).

In a solid, it is also possible to obtain from the velocity autocorrelation function the phonon frequencies of the system. Actually, by Fourier transforming the velocity autocorrelation function, one gets the power spectrum, which has peaks at the phonon frequencies. Anharmonic effects and phonon lifetimes are also included in such a spectrum.

More descriptions of these methods and of the general techniques used in molecular dynamics can be found in [9].

## B.1 Time Evolution of Atomic Positions

Using the Hellmann–Feynman theorem, we have shown that it is possible to calculate the forces acting on each atom. It is then possible to perform molecular-dynamics simulations by integrating Newton's equations of motion using finite-difference methods. One such method is due to Verlet [4]. Expanding the atomic positions at two times, $t + \Delta t$ and $t - \Delta t$, as Taylor series, we obtain

$$\boldsymbol{r}_i(t + \Delta t) = \boldsymbol{r}_i(t) + \Delta t\, \boldsymbol{v}_i(t) + \frac{(\Delta t)^2}{2} \boldsymbol{a}_i(t) + \cdots , \tag{B.2}$$

$$\boldsymbol{r}_i(t - \Delta t) = \boldsymbol{r}_i(t) - \Delta t\, \boldsymbol{v}_i(t) + \frac{(\Delta t)^2}{2} \boldsymbol{a}_i(t) - \cdots . \tag{B.3}$$

Adding these two equations, we obtain

$$\boldsymbol{r}_i(t + \Delta t) = 2\boldsymbol{r}_i(t) - \boldsymbol{r}_i(t - \Delta t) + (\Delta t)^2 \frac{\mathbf{F}_i(t)}{m_i}, \tag{B.4}$$

where we have used $\boldsymbol{a}_i(t) = \boldsymbol{F}_i(t)/m_i$. The error in this difference equation is of the order of $(\Delta t)^4$, since the odd terms in $\Delta t$ disappear. Subtracting (B.3) from (B.2), we obtain the velocity,

$$\boldsymbol{v}_i(t) = \frac{\boldsymbol{r}_i(t + \Delta t) - \boldsymbol{r}_i(t - \Delta t)}{2\Delta t}. \tag{B.5}$$

The error in the velocity is of the order of $(\Delta t)^2$. The first and second terms on the right-hand side of (B.4) are of the order of $(\Delta t)^0$, while the third term is of the order of $(\Delta t)^2$. The addition of very large and very small values introduces a numerical error even though the number of significant figures of the two values is the same, because some of the last figures in the smaller value will be discarded. The error will accumulate with the time steps and may lead to a serious error in the final result. One way to avoid this accumulation of errors is the leapfrog method by Hockney [5].

First, introducing the velocities, we rewrite the second-order differential equation as two first-order differential equations:

$$m_i \frac{d\boldsymbol{v}_i}{dt} = \boldsymbol{F}_i, \tag{B.6}$$

$$\frac{d\boldsymbol{r}_i}{dt} = \boldsymbol{v}_i. \tag{B.7}$$

Then we replace the derivative of the velocity by the difference between the velocities at $t + \Delta t/2$ and $t - \Delta t/2$:

$$\frac{\boldsymbol{v}_i(t + \Delta t/2) - \boldsymbol{v}_i(t - \Delta t/2)}{\Delta t} = \frac{\boldsymbol{F}_i(t)}{m_i}. \tag{B.8}$$

Thus, the velocity at time $t + \Delta t/2$ is given by

$$\boldsymbol{v}_i\left(t + \frac{\Delta t}{2}\right) = \boldsymbol{v}_i\left(t - \frac{\Delta t}{2}\right) + \frac{\Delta t}{m_i}\boldsymbol{F}_i(t). \tag{B.9}$$

On the other hand, the derivative of the atomic position is

$$\frac{\boldsymbol{r}_i(t + \Delta t) - \boldsymbol{r}_i(t)}{\Delta t} = \boldsymbol{v}_i\left(t + \frac{\Delta t}{2}\right). \tag{B.10}$$

From this equation, the atomic position at time $t$ is given by

$$\boldsymbol{r}_i(t + \Delta t) = \boldsymbol{r}_i(t) + \Delta t\, \boldsymbol{v}_i\left(t + \frac{\Delta t}{2}\right). \tag{B.11}$$

Note that in (B.9) and (B.11) the ratio of the first and second terms on the right-hand side is only of order $\Delta t$. In the leapfrog method, the updates of the velocity and position are shifted by $\Delta t/2$. First, one calculates the force from the coordinate data at time $t$. Then, using this force, one estimates the velocity at time $t + \Delta t/2$. Next, using this velocity, one obtains the position at $t + \Delta t$. The time integration of one basic step is completed in this way. The leapfrog method is more tractable than the Verlet method when one introduces velocity scaling in a system with periodic boundary conditions. This is because, even if the new position is outside the cell boundary, the velocity is not affected, even though the particle will jump to the other side of the cell at the next time step.

Because the leapfrog method cannot handle the velocities properly, a third method, called *the velocity Verlet method* is usually adopted. It uses the following equations for updating positions and velocities:

$$\boldsymbol{r}_i(t+\Delta t) = \boldsymbol{r}_i(t) + \Delta t\, \boldsymbol{v}_i(t) + \frac{1}{2m}\boldsymbol{F}_i(t)\Delta t^2, \tag{B.12}$$

$$\boldsymbol{v}_i(t+\Delta t) = \boldsymbol{v}_i(t) + \frac{1}{2m}[\boldsymbol{F}_i(t+\Delta t) + \boldsymbol{F}_i(t)]\,\Delta t^2. \tag{B.13}$$

The algorithm requires the storage of coordinates, velocities and forces at every time step, but it minimizes round-off errors. It involves two stages since the forces must be evaluated after the updating of the coordinates and before the updating of the velocities. In terms of stability and accuracy, it is the best of the three proposed methods. If memory is no major problem, and force calculation is expensive, then higher-order methods that allow the use of larger time steps are preferable. The time step $\Delta t$ must be chosen as large as possible, with the constraint that during an (NVE) microcanonical simulation, the total energy is conserved. In order to run the simulation with still higher values of $\Delta t$, or to obtain more accurate trajectories with the same time step, one can store and use the forces at previous time steps and update the positions and velocities by the *Gear predictor–corrector* algorithm. Since the trajectory of the atomic motions is a continuous function of time, the position, velocity and acceleration at time $t + \Delta t$ can be expanded as Taylor series as follows:

$$\begin{aligned}
\boldsymbol{r}_i^{\mathrm{P}}(t+\Delta t) &= \boldsymbol{r}_i(t) + \Delta t\, \boldsymbol{v}_i(t) + \frac{(\Delta t)^2}{2}\boldsymbol{a}_i + \frac{(\Delta t)^3}{6}\boldsymbol{b}_i(t) + \cdots, \\
\boldsymbol{v}_i^{\mathrm{P}}(t+\Delta t) &= \boldsymbol{v}_i(t) + \Delta t\, \boldsymbol{a}_i(t) + \frac{(\Delta t)^2}{2}\boldsymbol{b}_i(t) + \cdots, \\
\boldsymbol{a}_i^{\mathrm{P}}(t+\Delta t) &= \boldsymbol{a}_i(t) + \Delta t\, \boldsymbol{b}_i(t) + \cdots, \\
\boldsymbol{b}_i^{\mathrm{P}}(t+\Delta t) &= \boldsymbol{b}_i(t) + \cdots.
\end{aligned} \tag{B.14}$$

Using the new position $\boldsymbol{r}_i^{\mathrm{P}}(t+\Delta t)$, one can calculate the force $\boldsymbol{F}(t+\Delta t)$ at time $t+\Delta t$, and in turn the acceleration $\boldsymbol{a}_i^{\mathrm{C}}(t+\Delta t)$, which can be used as a corrector. By means of this corrector, it is possible to estimate the error in the predicted acceleration, $\boldsymbol{a}_i^{\mathrm{P}}(t+\Delta t)$, as follows:

$$\Delta \boldsymbol{a}_i(t+\Delta t) = \boldsymbol{a}_i^{\mathrm{C}}(t+\Delta t) - \boldsymbol{a}_i^{\mathrm{P}}(t+\Delta t). \tag{B.15}$$

Assuming the difference between the predicted and corrected quantities is small, one can say that they are all proportional to each other and, therefore, to the $\Delta \boldsymbol{a}$. Adding this difference to the predicted values, one obtains the corrected quantities:

**Table B.1** The coefficients for the correctors in the Gear method for a second-order ordinary differential equation

| Order | $c_0$ | $c_1$ | $c_2$ | $c_3$ | $c_4$ | $c_5$ |
|---|---|---|---|---|---|---|
| 3 | 0 | 1 | 1 | | | |
| 4 | 1/6 | 5/6 | 1 | 1/3 | | |
| 5 | 19/120 | 3/4 | 1 | 1/2 | 1/12 | |
| 6 | 3/20 | 251/360 | 1 | 11/18 | 1/6 | 1/60 |

$$\begin{aligned} \boldsymbol{r}_i^{\mathrm{C}}(t+\Delta t) &= \boldsymbol{r}_i^{\mathrm{P}}(t+\Delta t) + c_0 \Delta \boldsymbol{a}_i(t+\Delta t) \\ \boldsymbol{v}_i^{\mathrm{C}}(t+\Delta t) &= \boldsymbol{v}_i^{\mathrm{P}}(t+\Delta t) + c_1 \Delta \boldsymbol{a}(t+\Delta t) \\ \boldsymbol{a}_i^{\mathrm{C}}(t+\Delta t) &= \boldsymbol{a}_i^{\mathrm{P}}(t+\Delta t) + c_2 \Delta \boldsymbol{a}_i(t+\Delta t) \\ \boldsymbol{b}_i^{\mathrm{C}}(t+\Delta t) &= \boldsymbol{b}_i^{\mathrm{P}}(t+\Delta t) + c_3 \Delta \boldsymbol{a}_i(t+\Delta t) \end{aligned} \tag{B.16}$$

Gear [7] has determined the coefficients $c_0, c_1, c_2, c_3$ so as to achieve the best accuracy and stability. These values are listed in Table B.1. The predictor–corrector method is one of the superior methods for solving nonlinear ordinary differential equations. It can also be used, as an alternative to the Runge–Kutta method, to obtain accurate solutions of eigenvalue problems.

The above descriptions are used for a microcanonical simulation (i.e. in the NVE ensemble), where the total energy is a conserved quantity. To perform simulations comparable to the experimental situation, one might wish to keep the temperature or the pressure constant (i.e. use the NVT or NTP ensembles). In such cases, integrating Newton's equations of motion is not enough, and one needs to add the effect of a thermostat interacting with the system.

For canonical simulations (NVT ensemble), the total energy is allowed to fluctuate, but its conjugate intensive parameter, the temperature, is held fixed. The easiest way to achieve this is to use the equipartition theorem and equate the kinetic energy to $\frac{3}{2}Nk_{\mathrm{B}}T$. This constraint fixes the value of the kinetic energy. One only needs to rescale the velocities at each time step or every few time steps so that the total kinetic energy satisfies $\sum_i \frac{1}{2}m_i v_i^2 = \frac{3}{2}Nk_{\mathrm{B}}T$. For simulation of polyatomic molecules, one should also take into account their rotational, vibrational and translational kinetic energies separately. From these kinetic energies, one can deduce the respective temperatures, and perform the velocity scaling in such a way as to equilibrate the three temperatures.

Other more sophisticated thermostats can be used, among which we can mention Nosé's thermostat [16] or Langevin's algorithm, which is stochastic in nature.

Nose's original thermostat was further modified by Hoover [17] to what we now call the Nose–Hoover thermostat. He has shown that the new equations of motion lead to the canonical distribution. Basically an additional degree of freedom $\xi(t)$ is introduced to model the "dynamics" of the thermostat. Its dynamics is such that it should keep the kinetic energy of the system, in accordance with the equipartition theorem, equal to $gk_BT/2$, where $g$ is the number of degrees of freedom of the system (typically $g = 3N - 3$ is the simulation is done in box with periodic boundary conditions where rotations are not allowed). The coupling of the system to the thermostat is taken to be linear. This leads to the following equations of motion:

$$\frac{d\boldsymbol{p}_i}{dt} = \boldsymbol{F}_i - \xi \boldsymbol{p}_i \tag{B.17}$$

$$\frac{d\xi}{dt} = \frac{1}{\tau^2}\left(\frac{\sum_i m_i v_i^2}{g k_B T} - 1\right) \tag{B.18}$$

As one can see, the coupling with the thermostat acts as a damping force in case $\xi > 0$. If there is excess kinetic energy, the second equation causes $\xi$ and therefore the damping force to increase. If on the other hand the kinetic energy is below the target value, $\xi$ will decrease until it becomes negative, causing an increase in the particle momenta and therefore total kinetic energy, so as to keep the kinetic energy fluctuating around a target value of $gk_BT/2$. The only parameter characterizing this thermostat is the relaxation time $\tau$, the inverse of which models the strength of the coupling to the thermostat: large $\tau$ values cause slow changes in $\xi$. To implement this thermostat, the equations used to update particle positions and velocities need to be modified. The total energy of the system is not conserved as it exchanges energy with the thermostat. In this case, one can show that the conserved quantity is $G = E_{total} + (\xi^2\tau^2/2 + s) \times gk_BT$ where the parameter $s$ is also related to the thermostat through $ds/dt = \xi$. So if the latter can also be updated during the simulation, one can monitor the constancy of $G$ to insure of the validity of the simulation. Again the MD time step should be chosen so that the total energy of the system plus the thermostat is conserved. The value of the relaxation time $\tau$ is arbitrary and can be fixed by trial and error. It must not be so large that there is no energy flow between the thermostat and the system. And it must not be so small as to cause large fluctuations in the energy of the system. A similar scheme of coupling can be used to simulate an NTP ensemble by adding another "pressure thermostat" to the composite system and introducing a coupling to the system's degrees of freedom.

## B.2 Acceleration of Force Calculations

For short-range potentials, there is no problem in the potential or force calculation if one cuts it off at a certain range and uses a neighbor list. This type of calculation in coordinate space is called a particle–particle method. The amount of calculation in this case is of $O(N)$ for an $N$-particle system. On the other hand, for Coulomb systems, one has to evaluate the sum of long-range potentials acting on each atom. This becomes quite heavy, especially when the number of particles increases, because the particle–particle method requires a computation of $O(N^2)$. This becomes a problem when one treats more than one million atoms simultaneously in simulations such as theory of micelle formation, the (protein-)folding problem, visco-elastic behavior, phase transitions, etc., of polymers, liquid crystals or biological macromolecules.

In order to decrease the amount of calculation in Coulomb systems, one may use the Ewald sum described in Appendix D. But even in the Ewald method, one has to evaluate a sum of order $10^2$ for each ion, and the task still becomes heavy in large systems. In such a case, one may use several new methods, which have been developed mainly in the field of cosmological physics. (In cosmological physics, the gravitational force is always long-ranged because it cannot be screened.)

### B.2.1 Particle–Mesh Method

The particle–mesh method, which discretizes the whole space (unit cell) into regular meshes, uses fast Fourier transformation (FFT) to evaluate the potential or force acting on each mesh point [20]. The advantage of using the FFT is that it requires a computational amount of $O(M \log M)$ instead of $O(M^2)$, if we write the total number of mesh points as $M$. The calculation is very simple. First, the charge density $\rho(\boldsymbol{R})$ is evaluated at each mesh point. Then, by means of the FFT, it is transferred into Fourier space:

$$\bar{\rho}(\boldsymbol{G}) = \frac{1}{M} \sum_{\boldsymbol{R}} \mathrm{e}^{-i\boldsymbol{G}\cdot\boldsymbol{R}} \rho(\boldsymbol{R}). \tag{B.19}$$

Finally, the potential and the force acting on each mesh point is calculated simultaneously by the backward transformation using the FFT:

$$V(\boldsymbol{R}) = \sum_{\boldsymbol{G}} \mathrm{e}^{i\boldsymbol{G}\cdot\boldsymbol{R}} \frac{\bar{\rho}(\boldsymbol{G})}{G^2}, \tag{B.20}$$

$$F(\boldsymbol{R}) = -i \sum_{\boldsymbol{G}} \boldsymbol{G} e^{i\boldsymbol{G}\cdot\boldsymbol{R}} \frac{\bar{\rho}(\boldsymbol{G})}{G^2}. \tag{B.21}$$

In this scheme, one has to make the approximation of replacing the potential acting on a particle with that acting on the mesh point which is closest to the particle position. Therefore, the resolution of the force is not high, although the speed is fast.

In order to increase the resolution, one can combine the particle–particle method with the particle–mesh method [21]. This strategy is somewhat similar to that used in the Ewald method described in Appendix D, but the difference is the use of the FFT. That is, we first separate the short-range part from the long-range part of the potential and treat only the long-range part in Fourier space. To calculate Fourier series, one first discretizes the real space into meshes and utilizes the FFT to speed the process up. On the other hand, the short-range part can be rapidly evaluated by the usual particle–particle method. For example, one such simulation has been done recently by Ihara [22].

There are several other modifications of the method described here, such as the particle–multiple-mesh method [23] and the tree code method [24], but we shall not go into details here.

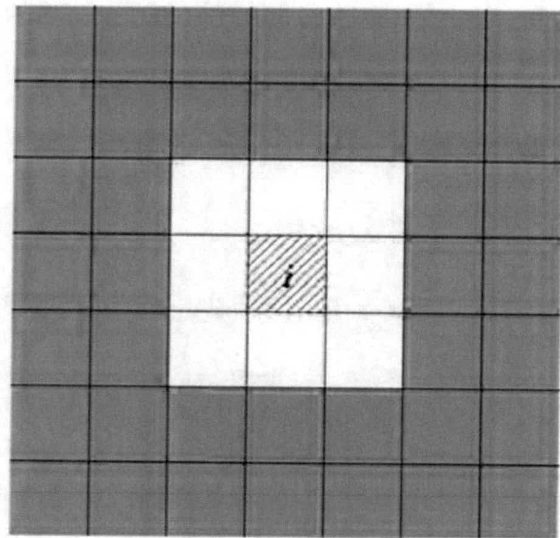


**Fig. B.1** Discretization of the unit cell

Instead of accelerating the force calculation, it is also possible to integrate the Hamiltonian as rigorously as possible to allow one to make the basic time step $\Delta t$ larger. This method is called the symplectic method [25–27].

### B.2.2 The Greengard–Rockhlin Method

The Greengard–Rockhlin method [28] serves a good basis for the calculation of a large system. This method is also called the fast multipole method. It is usually expected to reduce significantly the computational task from $O(N^2)$ to $O(N)$, although there is a criticism [29].

In the usual multipole method, one discretizes the unit cell into many subcells as is shown in Fig. B.1. Consider one atom $i$ in the hatched cell in this figure. The potential acting on this atom can be decomposed into two parts. One is the direct interaction caused by the white adjacent cells and the other is the multipole interactions caused by the atoms in the gray cells:

$$V(\boldsymbol{r}_i) = \sum_{j \in \text{near}} q_j |\boldsymbol{r}_i - \boldsymbol{r}_j|^{-p} + \sum_{A \in \text{far}} V_A^{\text{pole}}(\boldsymbol{r}_i - \boldsymbol{r}_A), \tag{B.22}$$

where $\boldsymbol{r}_i$ is the position of the $i$th atom and $\boldsymbol{r}_A$ is the coordinate of the center of the $A$ cell. The Coulomb potential corresponds to $p = 1$, while the van der Waals potential corresponds to $p = 6$. The multipole field $V_A^{\text{pole}}(\boldsymbol{r}_i - \boldsymbol{r}_A)$ is given by

$$V_A^{\text{pole}}(\boldsymbol{r}_i - \boldsymbol{r}_A) = \frac{Z_A}{R^p} + \frac{\mu_\alpha R_\alpha}{R^{p+2}} + \frac{Q_{\alpha\beta} R_\alpha R_\beta}{R^{p+4}} + \frac{O_{\alpha\beta\gamma} R_\alpha R_\beta R_\gamma}{R^{p+6}} + \cdots, \tag{B.23}$$

where $\alpha, \beta, \gamma$ denote the $x, y, z$ components; $Z$, $\mu_\alpha$, $Q_{\alpha\beta}$ and $O_{\alpha\beta\gamma}$ are the multipole moments given by

$$Z_A = \sum_i q_i \tag{B.24}$$

$$\mu_\alpha = p \sum_i q_i r_{i\alpha} \tag{B.25}$$

$$Q_{\alpha\beta} = p \sum_i q_i [(p+2) r_{i\alpha} r_{i\beta} - \delta_{\alpha\beta} r_i^2]/2 \tag{B.26}$$

$$O_{\alpha\beta\gamma} = p(p+2) \sum_i q_i [(p+4) r_{i\alpha} r_{i\beta} r_{i\gamma} - (r_{i\alpha}\delta_{\beta\gamma} + r_{i\beta}\delta_{\alpha\gamma} + r_{i\gamma}\delta_{\alpha\beta}) r_i^2]/6, \tag{B.27}$$

respectively, where $r_i$ is the coordinate of the center of each smallest-size cell.

This multipole algorithm accelerates the speed of computation by a factor of several tens but not more, when one discretizes the whole space into cells of the same size. Greengard and Rockhlin [28] showed that the computational task can be decreased considerably by grouping small cells to make bigger cells hierarchically for farther cells. Ding et al. [30] showed by applying this method to a macromolecular system composed of 1.2 million atoms that the computation becomes faster by a factor of 2400.

The calculation is performed hierarchically. The form of this hierarchical decomposition is shown in Fig. B.2. The smallest cells are chosen so as to contain four atoms on average. The number of the smallest cells is given by $M = 8^L$. Here $L$ denotes the number of levels of the hierarchy. Since $M = N/4$, we have

$$L = \log_8(N/4). \tag{B.28}$$

One may suppose that the multipole field varies only slowly inside the cell at each level. It can be then expressed in terms of a Taylor series:

$$\sum V_A^{\text{pole}}(r_i - r_A) = V^{(0)} + V_\alpha^{(1)} r_{i\alpha} + V_{\alpha\beta}^{(2)} r_{i\alpha} r_{i\beta} + V_{\alpha\beta\gamma}^{(3)} r_{i\alpha} r_{i\beta} r_{i\gamma} + \cdots . \tag{B.29}$$

In order to obtain the coefficients of this Taylor expansion, i.e. $V^{(0)}$, $V_\alpha^{(1)}$, $V_{\alpha\beta}^{(2)}$, $V_{\alpha\beta\gamma}^{(3)}$, one expands each term of $V_A^{\text{pole}}$ given by (B.23) with respect to $r_{i\alpha}$ to obtain

$$\begin{aligned} &\frac{Z_A}{|\boldsymbol{r}_i - \boldsymbol{r}_A|^p} \\ &= \frac{Z_A}{r_A^p} + \frac{p Z_A \boldsymbol{r}_A}{r_A^{p+2}} \cdot \boldsymbol{r}_i \\ &+ \frac{p Z_A}{r_A^{p+4}} + [(p+2)(\boldsymbol{r}_A \cdot \boldsymbol{r}_i)^2 - r_A^2 r_i^2] + \cdots , \end{aligned} \tag{B.30}$$

**Fig. B.2** Hierarchical cell decomposition of the unit cell

$$\frac{\boldsymbol{\mu}\cdot(\boldsymbol{r}_i-\boldsymbol{r}_A)}{|\boldsymbol{r}_i-\boldsymbol{r}_A|^{p+2}}=\frac{-\boldsymbol{\mu}\cdot\boldsymbol{r}_A}{r_A^{p+2}}+\left(\frac{\boldsymbol{\mu}}{r_A^{p+2}}-\frac{(p+2)(\boldsymbol{\mu}\cdot\boldsymbol{r}_A)\boldsymbol{r}_A}{r_A^{p+4}}\right)\cdot\boldsymbol{r}_i+\cdots. \tag{B.31}$$

Summing all the terms with the same power of $r$, we find

$$V^{(0)}=\sum\left(\frac{Z_A}{r_A^p}-\frac{\boldsymbol{\mu}\cdot\boldsymbol{r}_A}{r_A^{p+2}}+\cdots\right), \tag{B.32}$$

$$V_\alpha^{(1)}=\sum\left[\frac{p\,Z_A\,r_{A\alpha}}{r_A^{p+2}}+\left(\frac{\mu_\alpha}{r_A^{p+2}}-\frac{(p+2)(\boldsymbol{\mu}\cdot\boldsymbol{r}_A)r_{A\alpha}}{r_A^{p+4}}\right)+\cdots\right], \tag{B.33}$$

and so on.

If we assume that the number of levels is three (hereafter we will distinguish the cells at different levels by $A$, $B$ and $C$) and that the atom $i$ is in one of the smallest cells $C_0$, we first calculate the potential (i.e. the Taylor coefficients $V^{(0)}$, $V_\alpha^{(1)}$, ...) at the center of the largest cell $A_0$ in which $C_0$ exists:

$$V_{A_0}(\boldsymbol{r})=\sum_A V_A^{\text{pole}}(\boldsymbol{r}-\boldsymbol{r}_A), \tag{B.34}$$

where the summation runs over all the $A$ cells which are not adjacent to $A_0$. Next, for the second biggest cell $B_0$ in which $C_0$ exists, we have

$$V_{B_0}(\boldsymbol{r})=V_{A_0}(\boldsymbol{r}+\boldsymbol{r}_{B_0})+\sum_B V_B^{\text{pole}}(\boldsymbol{r}-\boldsymbol{r}_B). \tag{B.35}$$

Furthermore, we evaluate the potential at the center of the $C_0$ cell:

$$V_{C_0}(\boldsymbol{r}) = V_{B_0}(\boldsymbol{r} + \boldsymbol{r}_{C_0}) + \sum_C V_C^{\text{pole}}(\boldsymbol{r} - \boldsymbol{r}_C). \tag{B.36}$$

Finally, we evaluate the potential acting on the atom $i$ as follows:

$$V_i(\boldsymbol{r}_i) = V_{C_0}(\boldsymbol{r}_i) + \sum_j \frac{q_j}{|\boldsymbol{r}_i - \boldsymbol{r}_j|^p}, \tag{B.37}$$

where the second term on the right-hand side is the direct interaction between the $i$th atom and the atoms $j$ in the $C$ cells which are adjacent to $C_0$.

In the algorithm described so far, one needs the multipole moments at each level of the cell. Since eight smaller cells form one bigger cell, one may note that the multipole moments of this bigger cell are given by

$$Z^{l-1} = \sum_{k=1}^{8} \sum_i q_{ik} = \sum_{k=1}^{8} Z_k^{(l)}, \tag{B.38}$$

$$\mu_\alpha^{(l-1)} = \sum_{k=1}^{8} \sum_i q_{ik}(r_{ik\alpha} + R_{k\alpha}) = \sum_{k=1}^{8} \left(\mu_{k\alpha}^{l} + Z_k^{(l)} R_{k\alpha}\right) \tag{B.39}$$

$$Q_{\alpha\beta}^{(l-1)} = \sum_{k=1}^{8} \left(Q_{k\alpha\beta}^{(l)} - \mu_{k\alpha}^{(l)} R_{k\beta} + Z_k^{(l)} R_{k\alpha} R_{k\beta}\right). \tag{B.40}$$

The fast multipole method described above can be accelerated further by combining it with the FFT [31].

## References


1. B.J. Alder, T.E. Wainwright, J. Chem. Phys. **27**, 1208 (1957)
2. J.G. Kirkwood, J. Chem. Phys. **7**, 919 (1939)
3. A. Rahman, Phys. Rev. A **136**, 405 (1963)
4. L. Verlet, Phys. Rev. **159**, 98 (1967)
5. R.W. Hockney, Methods Comput. Phys. **9**, 136 (1970)
6. W.C. Swope, H.C. Andersen, P.H. Berens, K.R. Wilson, J. Chem. Phys. **76**, 637 (1982)
7. C.W. Gear, *Numerical Initial Value Problems in Ordinary Differential Equations* (Prentice-Hall, Englewood Cliffs, 1971)
8. M. Tuckerman, B.J. Berne, J. Chem. Phys. **97**, 1990 (1992)
9. M.P. Allen, D.J. Tildesley, *Computer Simulation of Liquids* (Clarendon, Oxford, 1989)
10. A.W. Lees, S.F. Edwards, J. Phys. C **5**, 1921 (1972)
11. E. Helfand, Phys. Rev. **119**, 1 (1960)
12. M.J. Gillan, M. Dixon, J. Phys. C **15**, 869 (1983)
13. D.J. Evans, G.P. Morriss, *Statistical Mechanics of Nonequilibrium Liquids* (Academic Press, London, 1990)

14. H.C. Andersen, J. Chem. Phys. **72**, 2384 (1980)
15. M. Parrinello, A. Rahman, J. Appl. Phys. **52**, 7182 (1981)
16. S. Nosé, J. Chem. Phys. **81**, 511 (1984)
17. W.G. Hoover, Phys. Rev. A **31**, 1695 (1985)
18. T. Çagĭn, B.M. Pettitt, Mol. Simul. **6**, 5 (1991)
19. S. Doniach, E.H. Sondheimer, *Green's Functions for Solid State Physicists* (W. A. Benjamin, Reading, 1974)
20. E. Bertschinger, J.M. Gelb, Comput. Phys. **5**, 164 (1991)
21. B.A. Luty, I.G. Tironi, W.F. van Gunsteren, J. Chem. Phys. **103**, 3014 (1995); B.A. Luty, W.F. van Gunsteren, J. Phys. Chem. **100**, 2581 (1996)
22. S. Ihara, Supercomputing (1999), in press
23. C. Jessop, M. Duncan, W.Y. Chau, J. Comput. Phys. **115**, 339 (1994)
24. J. Barnes, P. Hut, Nature **324**, 446 (1986)
25. P.J. Channell, C. Scovel, Nonlinearity **3**, 231 (1990)
26. J. Candy, W. Rozmus, J. Comput. Phys. **92**, 230 (1991)
27. J. Wisdom, M. Holman, Astron. J. **102**, 1528 (1991)
28. L. Greengard, V.I. Rokhlin, J. Comput. Phys. **73**, 325 (1987)
29. S. Aluru, SIAM, J. Sci. Comput. **17**, 773 (1996)
30. H. Ding, N. Karasawa, W.A. Goddard, J. Chem. Phys. **97**, 4309 (1992)
31. W.D. Elliott, J.A. Board Jr., SIAM, J. Sci. Comput. **17**, 398 (1996)

# Appendix C
# Vibrational Properties

In addition to the method described in Sect. 2.3.5, this appendix discusses the theory of lattice dynamics.

A few words regarding the actual calculation of the force-constants matrix $\Phi$ are in order. In the cluster case, an element of the force-constants matrix can be found by displacing an atom by a small amount and calculating the force on all other atoms. The displacement should be small enough that the force–displacement relationship is a linear one. In the periodic case, one actually calculates the force on an atom when another atom and all its images in the other cells have moved. In this case, one must make sure that the effect of the displaced atoms in the other cells is negligible. For this purpose, one must use a supercell large enough that the range of the force constants is less than the size of the cell. This so-called "direct method" maybe compared to the linear response method of calculating the dynamical matrix which was mentioned in the chapter on density-functional methods. The "direct method" is conceptually simple, but requires a heavy calculation in a supercell with a large number of atoms. The linear response method may be used in a small unit cell and is preferable for metals, but is more complicated in terms of its formalism since the numerically calculated finite difference is computed analytically by using perturbation theory. For references on the direct method, the reader may consult [1] where some more technical details regarding how to treat the dynamical matrix so as to conserve basic symmetries of the elastic tensor are included. The paper by Giannozzi et al. [2] treats the linear-response formalism. Finally, we recommend Born and Huang's [3] book, which is a standard textbook on lattice vibrations.

## C.1 Vibration Modes of a Cluster

Consider a cluster or a molecule formed of an assembly of atoms bound due to a specific potential. First, the structure must be relaxed to its ground state, or at least a local minimum. This can be done numerically by several local optimization methods. The most widely used methods are steepest descent (SD), conjugate gradients (CG)



K. Ohno et al., *Computational Materials Science*,
https://doi.org/10.1007/978-3-662-56542-1

and quasi-Newton (QN). Once the local minimum is reached, one can expand the potential energy about this minimum in terms of the powers of atomic displacements. Since all forces on all atoms are zero, the Taylor expansion does not have linear terms. The Harmonic approximation (HA) consists in neglecting all the powers of displacements larger or equal to 3. Being at a local minimum, the matrix of second derivatives must be positive definite, and thus will have only positive eigenvalues. The potential energy thus becomes:

$$E(r_1, \ldots, r_N) = E_0 + \frac{1}{2} \sum_{\tau,\tau'} \Phi_{\tau\tau'}\, \boldsymbol{u}_\tau\, \boldsymbol{u}_{\tau'} \tag{C.1}$$

where $\boldsymbol{u}$ is the "small" displacement vector from the equilibrium position: $\boldsymbol{r}_i(t) = \boldsymbol{R}_i^0 + \boldsymbol{u}_\tau(t)$, and the label $\tau = (i, \alpha)$ refers to an atom $i$ and the cartesian component $\alpha$ of its displacement. For an $N$ atom cluster in 3D, $\tau$ varies from 1 to $3N$, and $\Phi$ is a $3N \times 3N$ matrix. The latter is the second derivative of the potential energy evaluated at the equilibrium position and is called the **force constants matrix**.

$$\Phi_{\tau\tau'} = \frac{\partial^2 E}{\partial \boldsymbol{u}_\tau \partial \boldsymbol{u}_{\tau'}} = -\frac{\partial \boldsymbol{F}_\tau}{\partial \boldsymbol{u}_{\tau'}} = -\frac{\partial \boldsymbol{F}_{\tau'}}{\partial \boldsymbol{u}_\tau} = \Phi_{\tau'\tau}$$

### *C.1.1 Symmetries of the Force Constants*

Apart from the invariance under permutation of atoms expressed above, which comes from the fact that the total energy is an analytic function of the atomic coordinates, there are other relations between different elements of this tensor due to symmetries of the system. An important relation satisfied by the force constants comes from the translational invariance of the potential energy: under any arbitrary translation, the potential energy and the forces should remain the same: $E(\boldsymbol{u} + \boldsymbol{c}) = E(\boldsymbol{u})$; $\boldsymbol{F}(\boldsymbol{u} + \boldsymbol{c}) = \boldsymbol{F}(\boldsymbol{u})$. Substituting for the forces its harmonic expression, we find:

$$\sum_{\tau'} \Phi_{\tau,\tau'} \boldsymbol{u}_{\tau'} = \sum_{\tau'} \Phi_{\tau,\tau'} (\boldsymbol{u}_{\tau'} + \boldsymbol{c})$$

implying, since the displacements are arbitrary:

$$\sum_{\tau'} \Phi_{\tau\tau'} = 0 \tag{C.2}$$

This relation defines the diagonal element of the force constants matrix as a function of its non-diagonal elements: $\Phi_{\tau\tau} = -\sum_{\tau' \neq \tau} \Phi_{\tau\tau'}$ meaning that, effectively, the atom $\tau$ is bound by a spring to its equilibrium position. Other relations come from symmetry operations, such a rotations or mirror symmetries, elements of its point group, which leave the molecule invariant. If such a symmetry operation is denoted by $\boldsymbol{S}$, and as $\Phi$ is a second rank tensor, we must have:

$$\Phi^{\alpha\beta}_{S\tau,\,S\tau'} = \sum_{\alpha',\,\beta'} \Phi^{\alpha'\beta'}_{\tau\tau'}\; S_{\alpha\alpha'}\; S_{\beta\beta'}$$

where $\boldsymbol{S}_{\alpha,\alpha'}$ are the $3 \times 3$ matrix elements of the operation $\boldsymbol{S}$. This means that the spring tensor between rotated atoms $\boldsymbol{S}\tau - \boldsymbol{S}\tau'$ is the rotated spring tensor between the original pair $\tau - \tau'$.

Another symmetry relation *in crystals* expresses he fact that the spring tensor between a pair is the same as that between any other pair obtained from a translation of the original pair by a translation vector of the lattice:

$$\Phi_{\boldsymbol{R}\tau,\boldsymbol{R}'\tau'} = \Phi_{0\tau,(\boldsymbol{R}'-\boldsymbol{R})\tau'}$$

where $\boldsymbol{R}$, $\boldsymbol{R}'$ are translation vectors of the lattice and $\tau$, $\tau'$ are labeling atoms within the unit cell.

### C.1.2 Classical Theory of Vibrations

Given the expression for the potential energy as a function of displacements, it is an easy task to derive the Newtonian equations of motion:

$$m_\tau \frac{\mathrm{d}^2 \boldsymbol{r}_\tau(t)}{\mathrm{d}t^2} = \boldsymbol{F}_\tau \simeq -\sum_{\tau'} \Phi_{\tau\tau'}\;[\boldsymbol{r}_{\tau'}(t) - \boldsymbol{R}^0_{\tau'}] = -\sum_{\tau'} \Phi_{\tau\tau'}\;\boldsymbol{u}_{\tau'}(t)\;(\forall\,\tau) \quad \text{(C.3)}$$

This formula is linear in the atomic coordinates, and can be interpreted, within the harmonic approximation, as the particles being connected by "springs". A (harmonic) solution of the form $A\cos(\omega t + \phi)$ can be substituted in (C.3). The resulting set of linear equations in the amplitudes and frequencies $\omega$ define the vibrational modes of the system. First to make the system of equations symmetric, one needs to make a change of variable by setting $\sqrt{m_\tau}\boldsymbol{u}_\tau = \boldsymbol{v}_\tau$. The resulting equations on $\boldsymbol{e}_\tau$ become:

$$\frac{\mathrm{d}^2 \boldsymbol{v}_\tau(t)}{\mathrm{d}t^2} = -\sum_{\tau'} \frac{\Phi_{\tau\tau'}}{\sqrt{m_\tau m_{\tau'}}}\;\boldsymbol{v}_{\tau'}(t)\;(\forall\,\tau \in \{1,\ldots,3N\}), \quad \text{(C.4)}$$

Using the harmonic form proposed above, we obtain:

$$\sum_{\tau'} \left( \frac{\Phi_{\tau\tau'}}{\sqrt{m_\tau m_{\tau'}}} - \omega^2 \delta_{\tau\tau'} \right) \boldsymbol{v}_{\tau'} = 0\;(\forall\,\tau \in \{1,\ldots,3N\}). \quad \text{(C.5)}$$

This linear system has a nonzero solution for $\boldsymbol{v}$ if the determinant of the matrix $\Phi_{\tau\tau'}/\sqrt{m_\tau m_{\tau'}} - \omega^2$ is equal to zero. Consequently, the square of the vibrational frequencies of the cluster are the eigenvalues of the matrix $\Phi_{\tau\tau'}/\sqrt{m_\tau m_{\tau'}}$. The system is $3N \times 3N$, in three dimensions, and has $3N$ eigenvalues, six of which will be due to pure translations (3) and rotations (3), and therefore equal to zero. All the rest are positive, reflecting the fact that the total energy was a minimum and any deviation of the atoms from their equilibrium position results in an increase in $E$ (in other words, $\Phi$ is positive definite). For each eigenvalue $\omega_\lambda$, there is an eigenvector of $3N$ components, defined by $\boldsymbol{e}_\tau^\lambda$, which is also called the **normal mode**. For this mode $\lambda$, the polarization vector $\boldsymbol{e}_\tau^\lambda$ can be represented by $N$ three-dimensional vectors associated with the $N$ atoms in the cluster, and showing the amplitude and the direction along which the atom $\tau$ oscillates in that mode $\lambda$.

$$\sum_{\tau'} \frac{\Phi_{\tau\tau'}}{\sqrt{m_\tau m_{\tau'}}} \boldsymbol{e}_{\tau'}^\lambda = \omega_\lambda^2 \boldsymbol{e}_\tau^\lambda \tag{C.6}$$

Since the matrix is positive-definite, its eigenvectors are real. Furthermore the eigenvectors $\boldsymbol{e}$ being a complete orthonormal set satisfy two relations of orthonormality and completeness:

$$\sum_{\tau} \boldsymbol{e}_\tau^\lambda \cdot \boldsymbol{e}_\tau^{\lambda'} = \delta_{\lambda\lambda'} \text{ orthonormality} \tag{C.7}$$

$$\sum_{\lambda} \boldsymbol{e}_\tau^\lambda \cdot \boldsymbol{e}_{\tau'}^{\lambda} = \delta_{\tau\tau'} \text{ completeness} \tag{C.8}$$

The normal mode solution is then:

$$v_\lambda = A_\lambda \cos(\omega_\lambda t + \phi_\lambda)$$

A general displacement of the atoms defined by the two initial conditions on positions and velocities and the equations of motion, can be expanded on the set of eigenvectors which form a complete orthonormal basis of the 3N-dimensional space:

$$\boldsymbol{u}_\tau(t) = \frac{1}{\sqrt{m_\tau}} \sum_{\lambda=1}^{3N} A_\lambda\, \boldsymbol{e}_\tau^\lambda \cos(\omega_\lambda t + \phi_\lambda)$$

where the $6N$ coefficients $(A_\lambda, \phi_\lambda)$ must be determined from the $6N$ initial conditions on positions and velocities of the particles. The eigenvectors not only give information about the polarization of the mode, but also allow one to calculate mechanical properties of the system under study. They can also be used to calculate infrared (IR) or Raman active spectra of molecules if information on the induced charge distribution (dipole moment in the case of IR, and polarizability in case of Raman) under that mode are available. From the frequencies, one can deduce a criterion for mechanical stability: if the lowest eigenvalue is small, this means the period of the oscillations for that mode is large, and the corresponding increase of the energy is small (since it depends on $m\omega^2 \boldsymbol{e}^2/2$), or, in other words, the mode is *soft*. The softening of a mode

($\omega \rightarrow 0$) is a signature of its becoming mechanically unstable, and thus leading possibly to a phase transition.

### C.1.3 Quantum Theory of Vibrations of a Cluster

It is also possible to obtain the vibrational frequencies of a cluster by using the quantum formalism and start from the Hamiltonian of the system. For the sake of completeness, we will derive phonon[1] frequencies from both methods. Again, using the Harmonic approximation, we keep up to second order terms in the potential energy of the particles, and write the Hamiltonian as follows:

$$\mathcal{H} = \sum_{\tau} \frac{\boldsymbol{P}_{\tau}^2}{2m_{\tau}} + \frac{1}{2} \sum_{\tau\tau'} \Phi_{\tau\tau'}\, \boldsymbol{u}_{\tau} \boldsymbol{u}_{\tau'} \tag{C.9}$$

The index $\tau$, as before, labels the atom and its cartesian component and goes from 1 to 3N. This is a system of $3N$ coupled harmonic oscillators. To solve it, we can first make the change of variable $\sqrt{m_{\tau}}\boldsymbol{u}_{\tau} = \boldsymbol{v}_{\tau}$. The corresponding conjugate momenta become $\boldsymbol{Q}_{\tau} = -i\partial/\partial \boldsymbol{v}_{\tau} = \boldsymbol{P}_{\tau}/\sqrt{m_{\tau}}$, and the Hamiltonian, in terms of $\boldsymbol{v}_{\tau}$, $\boldsymbol{Q}_{\tau}$ becomes:

$$\mathcal{H} = \frac{1}{2} \sum_{\tau} \boldsymbol{Q}_{\tau}^2 + \frac{1}{2} \sum_{\tau\tau'} \frac{\Phi_{\tau\tau'}}{\sqrt{m_{\tau} m_{\tau'}}}\, \boldsymbol{v}_{\tau} \boldsymbol{v}_{\tau'} \tag{C.10}$$

As before, one can diagonalize the matrix $D_{\tau\tau'} = \Phi_{\tau\tau'}/\sqrt{m_{\tau} m_{\tau'}}$ and write the Hamiltonian in the basis of its eigenvectors, which are real since $D$ is symmetric:

$$\sum_{\tau'} D_{\tau\tau'}\, \boldsymbol{e}_{\tau'}^{\lambda} = \omega_{\lambda}^2\, \boldsymbol{e}_{\tau}^{\lambda}$$

Using $\boldsymbol{v}_{\tau} = \sum_{\lambda=1}^{3N} v_{\lambda}\, \boldsymbol{e}_{\tau}^{\lambda}$, with $v_{\lambda} = \sum_{\tau} \boldsymbol{v}_{\tau} \cdot \boldsymbol{e}_{\tau}^{\lambda}$, and taking $v_{\lambda}$ as the new dynamical variables having as conjugate momenta $\pi_{\lambda} = -i\partial/\partial v_{\lambda} = \sum_{\tau} \boldsymbol{Q}_{\tau} \cdot \boldsymbol{e}_{\tau}^{\lambda}$, then the Hamiltonian can be rewritten in terms of $(v_{\lambda}, \pi_{\lambda})$. To write $\boldsymbol{Q}_{\tau}$ in terms of $\pi_{\lambda}$ one needs to invert $\boldsymbol{e}_{\tau}^{\lambda}$ which can be thought of as a matrix of row index $\tau$ and column index $\lambda$. It is the matrix of the eigenvectors which is also the orthogonal matrix that takes one from the "$\tau$" basis to the "$\lambda$" basis. Its inverse is its transpose as one can also see from the completeness relation (C.8). We have therefore $\boldsymbol{Q}_{\tau} = \sum_{\lambda} \boldsymbol{e}_{\tau}^{\lambda}\, \pi_{\lambda}$, and using the orthogonality of the eigenvectors $\sum_{\tau} \boldsymbol{e}_{\tau}^{\lambda} \cdot \boldsymbol{e}_{\tau}^{\lambda'} = \delta_{\lambda\lambda'}$, the Hamiltonian becomes:

$$\mathcal{H} = \sum_{\lambda} \frac{\pi_{\lambda}^2}{2} + \frac{1}{2} \sum_{\lambda} \omega_{\lambda}^2\, v_{\lambda}^2 \tag{C.11}$$

[1]The word phonon is actually mostly used for quanta of vibrations in crystals.

which is now a set of $3N$ **decoupled** oscillators each oscillating independently with the frequency $\omega_\lambda$. The eigenvalues of $\mathcal{H}$ are well-known[2]:

$$\sum_{\lambda=7}^{3N} \hbar\omega_\lambda \left(n_\lambda + \frac{1}{2}\right)$$

and from it, one can easily extract thermodynamic properties within the harmonic approximation. Going back to the cartesian representation, the expression for the displacement operator becomes now:

$$\boldsymbol{u}_\tau(t) = \sum_{\lambda=7}^{3N} \sqrt{\frac{\hbar}{2m_\tau\omega_\lambda}}\, \boldsymbol{e}_\tau^\lambda \,(a_\lambda\, e^{-i\omega_\lambda t} + a_\lambda^\dagger\, e^{i\omega_\lambda t}) \tag{C.12}$$

where $a_\lambda$ and $a_\lambda^\dagger$ are respectively annihilation and creation operators of a phonon in the state $\lambda$.

## C.2 Vibration Modes of a Crystal

The treatment of a crystal is very similar to that of a cluster except that the number of atoms, or degrees of freedom becomes infinite. In this case, use has to be made of symmetry properties of the crystal, namely translational invariance, to simplify the decoupling problem. As mentioned in the case of a cluster, in the harmonic approximation, the Hamiltonian is quadratic, and therefore exactly solvable. This is called the **non-interacting** problem, because in principle, one can reduce it to a set of uncoupled harmonic oscillators. The interacting problem of lattice dynamics will consist in including higher anharmonic terms in the potential energy of the lattice. They define the thermal conductivity of the crystal. Perturbation theory is the most widely used method to treat this problem, but we will not discuss these methods here. The results concerning the thermal conductivity are in the main text.

### *C.2.1 Classical Theory of Lattice Vibrations*

As in the case of a cluster, one can get rid of unequal masses by making a change of variables from the displacements $\boldsymbol{u}$ to $\boldsymbol{v} = \boldsymbol{u}\sqrt{m}$, at the cost of modifying the force constant matrix from $\Phi$ to $\Phi_{\tau,\tau'}/\sqrt{m_\tau m_{\tau'}}$: Formally one can write this change of variable as:

[2]To consider purely vibrational states for a cluster, we need to exclude the 6 rotational and translational modes with frequency zero, and strictly sum $\lambda$ from 7 to $3N$.

$$M\frac{d^2\boldsymbol{u}}{dt^2} = -\Phi\boldsymbol{u} \;\;<=>\;\; \frac{d^2(\sqrt{M}\boldsymbol{u})}{dt^2} = -\sqrt{M^{-1}}\Phi\sqrt{M^{-1}}\,(\sqrt{M}\boldsymbol{u})$$

where $\sqrt{M}$ is the matrix with atomic masses on its diagonal. Next, using invariance of the crystal under discrete translation by vectors $\boldsymbol{R}$, we have $\Phi_{\boldsymbol{R}\tau,\,\boldsymbol{R}'\tau'} = \Phi_{0\tau,\,(\boldsymbol{R}'-\boldsymbol{R})\tau'}$, where each atom is identified by the unit cell it is in (denoted by the translation vector $\boldsymbol{R}$), and its label within the unit cell (denoted by $\tau$, including, as in the case of clusters, the cartesian component of the displacement, so that $\tau$ runs from 1 to 3p, p being the number of atoms per unit cell).

Now we can use Bloch's theorem and go to new variables. The displacements $\boldsymbol{u}$ about the equilibrium position for an atom $\tau$ in the cell defined by the translation vector $\boldsymbol{R}$ can be written as

$$\boldsymbol{u}_{\tau\boldsymbol{R}}(t) = \sum_{\boldsymbol{k}} \frac{1}{\sqrt{Nm_\tau}}\,\boldsymbol{u}_{\tau\boldsymbol{k}}(t)\,\mathrm{e}^{i\boldsymbol{k}\cdot\boldsymbol{R}}. \tag{C.13}$$

the sum over $\boldsymbol{k}$ being restricted to the first Brillouin zone (FBZ). The inverse relation is:

$$\boldsymbol{u}_{\tau\boldsymbol{k}}(t) = \sum_{\boldsymbol{R}} \sqrt{Nm_\tau}\,\boldsymbol{u}_{\tau\boldsymbol{R}}(t)\,\mathrm{e}^{-i\boldsymbol{k}\cdot\boldsymbol{R}}. \tag{C.14}$$

Note that the above definition of $\boldsymbol{u}_{\tau\boldsymbol{k}}$ has the periodicity of the reciprocal space: $\boldsymbol{u}_{\tau\boldsymbol{k}} = \boldsymbol{u}_{\tau\boldsymbol{k}+\boldsymbol{G}}$. Substituting this into the expansion of the potential energy in powers of the displacements about the equilibrium positions, and truncating the sum at the second powers of $\boldsymbol{u}$ (harmonic approximation), we find

$$\begin{aligned}
E - E_0 &= \frac{1}{2}\sum_{\boldsymbol{R}\boldsymbol{R}',\tau\tau'} \boldsymbol{u}_{\tau\boldsymbol{R}}\,\boldsymbol{u}_{\tau'\boldsymbol{R}'}\,\Phi_{\boldsymbol{R}\tau,\boldsymbol{R}'\tau'}\\
&= \frac{1}{2N}\sum_{\boldsymbol{R}\boldsymbol{R}'\boldsymbol{k}\boldsymbol{k}'\tau\tau'} \boldsymbol{u}_{\tau\boldsymbol{k}}\,\boldsymbol{u}_{\tau'\boldsymbol{k}'}\,\frac{\Phi_{0\tau,(\boldsymbol{R}'-\boldsymbol{R})\tau'}}{\sqrt{m_\tau m_{\tau'}}}\,\mathrm{e}^{i\boldsymbol{k}\cdot\boldsymbol{R}}\,\mathrm{e}^{i\boldsymbol{k}'\cdot\boldsymbol{R}'}\\
&= \frac{1}{2N}\sum_{\boldsymbol{k}\boldsymbol{k}'\tau\tau'} \boldsymbol{u}_{\tau\boldsymbol{k}}\,\boldsymbol{u}_{\tau'\boldsymbol{k}'}\left(\sum_{\boldsymbol{R}}\frac{\Phi_{0\tau,\,\boldsymbol{R}\tau'}}{\sqrt{m_\tau m_{\tau'}}}\,\mathrm{e}^{-i\boldsymbol{k}\cdot\boldsymbol{R}}\right)\left(\sum_{\boldsymbol{R}'}\mathrm{e}^{i(\boldsymbol{k}+\boldsymbol{k}')\cdot\boldsymbol{R}'}\right)\\
&= \frac{1}{2}\sum_{\boldsymbol{k}\tau\tau'} \boldsymbol{u}_{\tau\boldsymbol{k}}\,\boldsymbol{u}_{\tau'-\boldsymbol{k}}\left(\sum_{\boldsymbol{R}}\frac{\Phi_{0\tau,\,\boldsymbol{R}\tau'}}{\sqrt{m_\tau m_{\tau'}}}\,\mathrm{e}^{-i\boldsymbol{k}\cdot\boldsymbol{R}}\right)\\
&= \sum_{\boldsymbol{k}}\left[\frac{1}{2}\sum_{\tau\tau'} \boldsymbol{u}_{\tau\boldsymbol{k}}\,\boldsymbol{u}_{\tau'-\boldsymbol{k}}\,D_{\tau\tau'}(\boldsymbol{k})\right].
\end{aligned} \tag{C.15}$$

The second line was obtained using the translational invariance of the force-constants matrix $\Phi$, implying that it depends only on the distance between the two cells, and the last line maybe taken as the definition of the dynamical matrix $D$. Note that this definition of $D$ is very similar to the definition of the Hamiltonian matrix in the tight-binding formalism, which contained a sum of the neighboring cells of a short-ranged matrix. In the case of TB, the short-ranged matrix was that of the Hamiltonian, and in the case of lattice dynamics, it is that of the force constants which also extends to neighboring atoms. Note that similar to the TB case, $D(-\boldsymbol{k}) = D(\boldsymbol{k})^*$ implying, since both have the same set of eigenvalues, that the eigenvalues $\omega_{\lambda \boldsymbol{k}}$ are even functions of $\boldsymbol{k}$. Note that here again, using the "Bloch" transformation, one was able to do the decoupling in the $\boldsymbol{k}$ space and change the force constants matrix to the dynamical matrix which is block-diagonal in the $\boldsymbol{k}$ space, meaning that it does not couple $\boldsymbol{k}$ to any other vector $\boldsymbol{k}'$ within the first Brillouin zone. A last remark concerns the correspondence between the eigenvalues: $\epsilon_\lambda \leftrightarrow \omega_\lambda^2$ where both have a quadratic dispersion near $\boldsymbol{k} = 0$, implying that the phonon dispersion at the $\Gamma$ point is linear in $\boldsymbol{k}$ for acoustic modes. This will be discussed in more detail in the next section where an example will illustrate better this theory.

Now that modes of different $\boldsymbol{k}$ are separated, and one can diagonalize the hermitian matrix $D_{\tau\tau'}(\boldsymbol{k})$ for each $\boldsymbol{k}$ in the FBZ[3]:

$$\sum_{\tau'} D_{\tau\tau'}(\boldsymbol{k})\, \boldsymbol{e}_{\tau'}^{\lambda}(\boldsymbol{k}) = \omega_{\lambda \boldsymbol{k}}^2\, \boldsymbol{e}_{\tau}^{\lambda}(\boldsymbol{k}) \;\rightarrow\; D_{\tau\tau'}(\boldsymbol{k}) = \sum_{\lambda} \omega_{\lambda \boldsymbol{k}}^2\, \boldsymbol{e}_{\tau}^{\lambda}(\boldsymbol{k})\, \boldsymbol{e}_{\tau'}^{\lambda}(\boldsymbol{k})^\dagger \tag{C.16}$$

where, similar to the case of a cluster, we have a completeness and orthonormality constraint on the eigenvectors of $D$, which are now complex. They form therefore a unitary matrix, as opposed to an orthogonal one in the case of a cluster. The two constraints can be written as:

$$\begin{aligned} \sum_{\tau} \boldsymbol{e}_{\tau}^{\lambda}(\boldsymbol{k})^\dagger \cdot \boldsymbol{e}_{\tau}^{\lambda'}(\boldsymbol{k}) &= \delta_{\lambda\lambda'} \quad \text{orthonormality} \\ \sum_{\lambda} \boldsymbol{e}_{\tau}^{\lambda}(\boldsymbol{k}) \cdot \boldsymbol{e}_{\tau'}^{\lambda}(\boldsymbol{k})^\dagger &= \delta_{\tau\tau'} \quad \text{completeness} \end{aligned} \tag{C.17}$$

Recall that the vector $\boldsymbol{e}^{\lambda}(\boldsymbol{k})^\dagger$ is the transpose and conjugate of $\boldsymbol{e}^{\lambda}(\boldsymbol{k})$, and that changing $\boldsymbol{k}$ to $-\boldsymbol{k}$ transforms $\boldsymbol{e}$ to $\boldsymbol{e}^*$. One can finally write the potential energy in the new basis as:

$$E - E_0 = \frac{1}{2} \sum_{\boldsymbol{k}\lambda} u_{\lambda \boldsymbol{k}}\, u_{\lambda -\boldsymbol{k}}\, \omega_{\lambda \boldsymbol{k}}^2, \tag{C.18}$$

where $u_{\lambda \boldsymbol{k}} = \sum_\tau \boldsymbol{e}_{\tau}^{\lambda}(\boldsymbol{k})^\dagger \cdot \boldsymbol{u}_{\tau \boldsymbol{k}}$. The potential energy is now diagonal in $\boldsymbol{u}_{\lambda \boldsymbol{k}}$ and can be combined with the transformed kinetic energy to yield a system of $3N$ uncoupled one-dimensional harmonic oscillators. Alternatively, one may write down the equation of motion for $\boldsymbol{u}_{\tau \boldsymbol{k}}(t)$ from the Newtonian equations for $\boldsymbol{u}_{\tau \boldsymbol{R}}(t)$. It is easy to verify that this equation is:

[3] In effect, the power of Bloch's theorem has allowed us to reduce the infinite crystal problem to a set of "cluster" problems for each k-point.

$$\frac{\mathrm{d}^2 \boldsymbol{u}_{\tau \boldsymbol{k}}(t)}{\mathrm{d}t^2} = -\sum_{\tau'} D_{\tau\tau'}(\boldsymbol{k})\, \boldsymbol{u}_{\tau' \boldsymbol{k}}(t) \;\; (\forall\, \tau \in \{1, \ldots, 3p\}). \tag{C.19}$$

After multiplying both sides by $\boldsymbol{e}_\tau^\lambda(\boldsymbol{k})$ and summing over $\tau$, we recover the uncoupled set of 1D harmonic oscillator equations for each $u_{\lambda \boldsymbol{k}}$ with frequency $\omega_{\lambda \boldsymbol{k}}$: $-\omega^2 u_{\lambda \boldsymbol{k}} = -\omega_{\lambda \boldsymbol{k}}^2 u_{\lambda \boldsymbol{k}}$.

The dynamical matrix $D$ for a crystal is the analog of the force-constants matrix $\Phi$ for a cluster (within a mass factor in the denominator). Their eigenvalues are the squares of the vibrational frequencies of the system. As we mentioned this whole treatment is, very analogous to the tight-binding formalism. The hopping matrix is replaced by the force constants and the electronic energy levels by the square of the phonon frequencies. They would follow the same symmetry rules if the orbitals are of p ($L = 1$) symmetry, otherwise the TB Hamiltonian can have different symmetry transformation properties if ($L \neq 1$). The general solution to the equations of motion in the normal mode representation is:

$$u_{\lambda \boldsymbol{k}}(t) = A_{\lambda \boldsymbol{k}} \cos(\omega_{\lambda \boldsymbol{k}} t + \phi_{\lambda \boldsymbol{k}})$$

where $(A_{\lambda \boldsymbol{k}}, \phi_{\lambda \boldsymbol{k}})$ are the two constant of integration to be determined from the 2 initial conditions on displacement and velocity. Going back to real space and cartesian representation, the general displacement of an atom can be written in terms of normal modes as:

$$\boldsymbol{u}_{\tau \boldsymbol{R}}(t) = \frac{1}{\sqrt{N m_\tau}} \sum_{\boldsymbol{k}\lambda} \boldsymbol{e}_\tau^\lambda(\boldsymbol{k})\, \mathrm{e}^{i\boldsymbol{k}\cdot\boldsymbol{R}}\, A_{\lambda \boldsymbol{k}} \cos(\omega_{\lambda \boldsymbol{k}} t + \phi_{\lambda \boldsymbol{k}}) \tag{C.20}$$

where the $6N$ integration constants $(A, \phi)$ are found from the initial conditions.

### C.2.2 *Quantum Theory of Phonons in a Crystal*

Now we can also quantize the theory, and consider the displacement $\boldsymbol{u}$ as an operator. In the uncoupled case, we have:

$$\boldsymbol{u}_{\lambda \boldsymbol{k}}(t) = \sqrt{\frac{\hbar}{2\omega_{\lambda \boldsymbol{k}}}}\, (a_{\lambda \boldsymbol{k}}\, \mathrm{e}^{-i\omega_{\lambda \boldsymbol{k}} t} + a_{\lambda -\boldsymbol{k}}^\dagger\, e^{i\omega_{\lambda -\boldsymbol{k}} t})$$

leading to

$$\boldsymbol{u}_{\tau \boldsymbol{R}}(t) = \sum_{\lambda \boldsymbol{k}} \sqrt{\frac{\hbar}{2N\, m_\tau\, \omega_{\lambda \boldsymbol{k}}}}\, \boldsymbol{e}_\tau^\lambda(\boldsymbol{k})\, e^{i\boldsymbol{k}\cdot\boldsymbol{R}}\, (a_{\lambda \boldsymbol{k}}\, \mathrm{e}^{-i\omega_{\lambda \boldsymbol{k}} t} + a_{\lambda -\boldsymbol{k}}^\dagger\, \mathrm{e}^{i\omega_{\lambda \boldsymbol{k}} t}) \tag{C.21}$$

This expression can be used when computing thermal or ground state averages of quantities such as $< \boldsymbol{u}_{\tau \boldsymbol{R}}(t) \boldsymbol{u}_{\tau' \boldsymbol{R}'}(t') >$ which appear in the calculation of structure factors and response functions.

## C.3 Phonon Density of States

The density of states (DOS) is a function of the energy and basically $DOS(E)dE$ counts the number of energy states within an energy window $dE$. In other words, it is the distribution function of the energy levels in a system. Its mathematical definition is therefore:

$$DOS(E) = \sum_{\lambda} \delta(E - E_{\lambda}) \tag{C.22}$$

In practice, given a set of eigenvalues $E_{\lambda}$, we substitute the Dirac delta function by either a Gaussian or a Lorentzian function of appropriate width:

$$\delta(x) = lim_{\sigma \to 0} \frac{1}{\sigma \sqrt{2\pi}} e^{-x^2/2\sigma^2} = lim_{\sigma \to 0} \frac{1}{\pi} \frac{\sigma}{x^2 + \sigma^2}$$

For the calculation of DOS in a crystal where the eigenvalue spectrum is defined on the first Brillouin zone, one should use the tetrahedron [4–6] method for an accurate evaluation of the DOS, because it will provide the same accuracy with fewer k-points.

If on the other hand, we do not have access to the spectrum but have generated atomic trajectories from a molecular dynamics simulation, one can compute the DOS from the velocity autocorrelation function. We can either adopt the classical or the quantum description. For simplicity, we will use the classical formula (C.20), from which by taking one time derivative, we obtain the velocity in terms of normal modes:

$$\boldsymbol{v}_{\tau \boldsymbol{R}}(t) = -\frac{1}{\sqrt{N m_{\tau}}} \sum_{\boldsymbol{k}\lambda} \boldsymbol{e}_{\tau}^{\lambda}(\boldsymbol{k})\, \mathrm{e}^{i\boldsymbol{k}\cdot\boldsymbol{R}}\, A_{\lambda \boldsymbol{k}}\, \omega_{\lambda \boldsymbol{k}} \sin(\omega_{\lambda \boldsymbol{k}} t + \phi_{\lambda \boldsymbol{k}}) \tag{C.23}$$

We define the frequency-dependent diffusivity as:

$$\mathcal{D}(\omega) = \int_{-\infty}^{\infty} dt\, e^{i\omega t} \sum_{\boldsymbol{R}\tau} m_{\tau} < \boldsymbol{v}_{\tau \boldsymbol{R}}(t) \cdot \boldsymbol{v}_{\tau \boldsymbol{R}}(0) >$$

Upon substitution, we obtain:

$$\mathcal{D}(\omega) = \frac{1}{N} \int_{-\infty}^{\infty} dt\, \mathrm{e}^{i\omega t} \sum_{\boldsymbol{R}\tau \boldsymbol{k}\boldsymbol{k}'\lambda\lambda'} \boldsymbol{e}_{\tau}^{\lambda}(\boldsymbol{k}) \cdot \boldsymbol{e}_{\tau}^{\lambda'}(\boldsymbol{k}')\, \mathrm{e}^{i(\boldsymbol{k}+\boldsymbol{k}')\cdot\boldsymbol{R}} < A_{\lambda \boldsymbol{k}} A_{\lambda' \boldsymbol{k}'} > \tag{C.24}$$

$$\times < \sin\phi_{\lambda \boldsymbol{k}} \sin\phi_{\lambda' \boldsymbol{k}'} > \omega_{\lambda \boldsymbol{k}}\, \omega_{\lambda' \boldsymbol{k}'} \cos(\omega_{\lambda \boldsymbol{k}} t) \tag{C.25}$$

The sum over $\boldsymbol{R}$ yields $\boldsymbol{k}' = -\boldsymbol{k}$ and cancels the factor $N$. Using the orthogonality relation, the sum over $\tau$ yields $\lambda = \lambda'$. The equipartition along with the virial theorem imply that at temperature $T$ each mode has on the average $k_B T/2$ potential energy, leading to $< A_{\lambda\boldsymbol{k}} A_{\lambda-\boldsymbol{k}} > \omega^2_{\lambda\boldsymbol{k}} = 2k_B T$. The phase factors average to one half and the square of frequencies cancel out.[4] So the final result becomes:

$$\mathcal{D}(\omega) = k_B T \sum_{\boldsymbol{k}\lambda} \int_{-\infty}^{\infty} dt\, e^{i\omega t} \cos(\omega_{\lambda\boldsymbol{k}} t) \tag{C.26}$$

$$= \frac{1}{2} k_B T \sum_{\boldsymbol{k}\lambda} \delta(\omega - \omega_{\lambda\boldsymbol{k}}) = k_B T \times DOS(\omega) \tag{C.27}$$

So we see that in the harmonic case the diffusivity reproduces the DOS. If anharmonicity is added the delta functions in the DOS acquire a broadening with an eventual small shift in the peak frequencies. These effects will be reflected in the velocity autocorrelation function. So in practice one may obtain the phonon frequency spectrum in a solid from the velocity autocorrelation function. An efficient way to compute it is to use the Wiener–Khinchin theorem and square the Fourier transform of the velocities instead of computing the autocorrelation function.

## C.4 An Example

To illustrate the theory previously outlined, let us consider a linear chain with two atoms A and B per unit cell and two different spring constants $\chi$ and $\xi$ between nearest neighbors only. The hard part in the lattice dynamics calculations is the correct set up of the force constants and then the dynamical matrix. The rest is a matter of a matrix diagonalization. There are two degrees of freedom in this problem, and the dynamical matrix is therefore $2 \times 2$, and can therefore be analytically diagonalized, and modes can easily be characterized. We can start by writing down the equations of motion for the two atoms. But this has been done before (see (C.19)), and we know that we first need to form the force constant matrix, renormalized by the mass factors, and Fourier transform it to get the dynamical matrix as in (C.15). Since atom A is second neighbor to itself, there is only FCs of the type AB (let us use $\chi$ for the hard spring or short AB bond, and $\xi$ for the soft spring or the long AB bond).

Due to the translational invariance constraint on $\Phi$, we have $\Phi_{AA} = -\chi - \xi = \Phi_{BB}$, whereas $\Phi_{AB} = \chi$ or $\Phi_{AB} = \xi$ depending on our choice for the unit cell. Let us choose the unit cell as the one containing the strong bond. In this case $\Phi_{AB} = \chi$.

Now, we can form the dynamical matrix. For the diagonal terms, since only the $R = 0$ term contributes to the sum, we have:

$$D_{AA}(k) = -\Phi_{AA}/m_A = (\chi + \xi)/m_A$$

[4] This is why we chose velocity autocorrelation as opposed to displacement autocorrelation.

In the off-diagonal terms, there are the contributions of $R = 0$ and $R = \pm 2a$ depending on which off-diagonal term we are considering:

$$D_{AB}(k) = -\frac{1}{\sqrt{m_A m_B}}(\Phi_{AB} + \Phi_{A+R,B}e^{2ika}) = -\frac{\chi + \xi e^{2ika}}{\sqrt{m_A m_B}}$$

Likewise, we can calculate $D_{BA}$ but we know that $D$ is hermitian. Finally $D_{BB}(k) = (\chi + \xi)/m_B$.

$$D(k) = \begin{bmatrix} \frac{\chi+\xi}{m_A} & -\frac{\chi+\xi e^{2ika}}{\sqrt{m_A m_B}} \\ -\frac{\chi+\xi e^{-2ika}}{\sqrt{m_A m_B}} & \frac{\chi+\xi}{m_B} \end{bmatrix}$$

Note that the lattice constant is $2a$, and $a$ is the interatomic distance. The eigenvalues can in principle be calculated analytically from $\det(D(k) - \omega^2 \boldsymbol{I}) = 0$.

Solving this equation for $\omega^2$ assuming for simplicity that the masses are equal, we obtain:

$$\omega^2 = \frac{\chi + \xi}{m} \pm \frac{1}{m}\sqrt{\chi^2 + \xi^2 + 2\chi\xi \cos 2ka} \tag{C.28}$$

Here we are mostly interested in their behavior near the zone center at $k = 0$. In this case, we can see that the determinant is zero implying, as expected, that one of the eigenvalues is zero. This corresponds to the translation of the whole lattice which does not cost any energy, and all crystals have such modes in d dimensions. These are called the **acoustic modes** since the motion of the atoms within the unit cell are in phase. For small $k$ one can simplify the dispersion relation:

$$\omega^2 \to \frac{\chi + \xi}{m} \pm \frac{\chi + \xi}{m}\left(1 - \frac{\chi\xi\, 4k^2a^2}{2(\chi + \xi)^2}\right)$$

The acoustic branch, obtained with the (−) sign has therefore a linear dispersion:

$$\omega_-(k) \to ka\sqrt{\frac{\chi\xi}{2m(\chi + \xi)}}$$

The fact that this mode is linear in $k$ near $k = 0$ is not always true, but it occurs in most cases where there are no long-range forces present in the system, and no peculiar interactions among the particles.

The slope at the origin is the group velocity of that mode and is called the **speed of sound**, as it is also the speed of propagation of elastic (long wavelength limit) waves. The latter can be obtained from the "wave equation" derived within elasticity theory, which is the long wavelength limit of the phonon theory.

The other mode is called the **optical mode**, and as can be deduced from the analysis of the eigenvectors of $D$, the two atoms in the unit cell are vibrating with a phase difference of $\pi$. This usually creates a dipole moment, and therefore couples to electromagnetic waves, hence the name of optical for these modes. The frequency

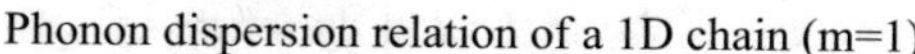


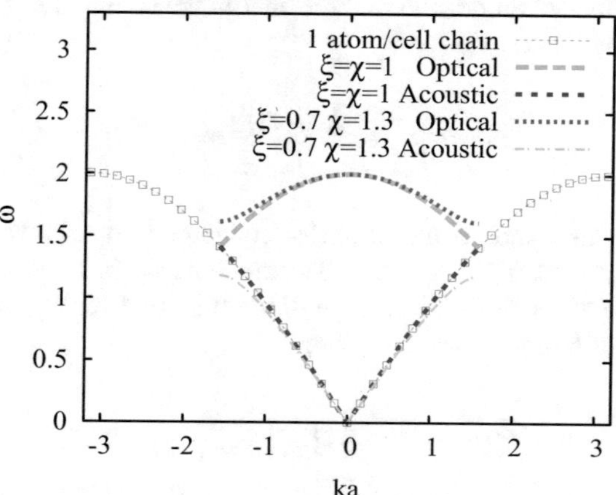


**Fig. C.1** Dispersion relations of the 1D chain with two atoms per unit cell for two different couplings, compared to the uniform chain (unfolded dispersion)

of optical modes at the zone center are particle like i.e. quadratic in $k$.

$$\omega_+(k) = \sqrt{\frac{\chi+\xi}{m}}\left(1 - \frac{\chi\xi\, 4k^2a^2}{4(\chi+\xi)^2}\right)$$

Note that in the case where both masses and spring constants are equal ($m_A = m_B$; $\chi = \xi$), then we have a simple linear chain with one atom per unit cell. The representation that we have just used has a twice larger unit cell and hence a twice smaller first Brillouin zone. The eigenvalues in both representations should however be the same. What we have found (C.28) becomes in this limit:

$$\omega^2 = \frac{2\chi}{m} \pm \frac{2\chi \cos ka}{m} = \frac{2\chi}{m}(1 \pm \cos ka) \tag{C.29}$$

with $a$ being the distance between two nearest neighbor atoms, also the lattice constant of the uniform chain for which it is straightforward to derive that the dispersion relation is $\omega = 2\sqrt{\chi/m}\,\sin ka/2$. From Fig. C.1, one can see that this is the unfolded representation of the previous dispersion with two branches.

## C.5 Thermodynamic Properties

In general, it is difficult to calculate thermodynamic properties such as entropy or free energy of a cluster or a solid. At very low temperatures, however, it is possible to use the harmonic approximation, and calculate exactly these properties. As the system of coupled oscillators was reduced to $3N - 6$ uncoupled ones, the partition function of the system is therefore the product of the partition functions for each normal mode, and the free energy the sum of the free energies. For a single harmonic oscillator, we have:

$$Z = \sum_{n=0}^{\infty} e^{-\beta\hbar\omega(n+1/2)} = e^{-\beta\hbar\omega/2}\,\frac{1}{1-e^{-\beta\hbar\omega}} = \frac{1}{2\sinh\left(\beta\hbar\omega/2\right)}$$

where the geometric series converged due to positivity of the frequency $\omega$. We put $\beta = 1/k_{\mathrm{B}}T$ as usual. The free energy is $F = -k_B T\ln Z = \hbar\omega/2 + k_B T\ln(1 - e^{-\beta\hbar\omega}) = k_B T\ln 2\sinh(\beta\hbar\omega/2)$ so that for a general molecule with many degrees of freedom, we can write

$$Z_{vib} = \Pi_{\lambda}^{vib}\,\frac{1}{2\sinh\left(\beta\hbar\omega_\lambda/2\right)} \tag{C.30}$$

$$F = k_B T\sum_{\lambda}\ln 2\sinh\left(\beta\hbar\omega_\lambda/2\right) \tag{C.31}$$

$$E = \sum_{\lambda}\frac{1}{2}\coth(\beta\hbar\omega_\lambda/2)\,\hbar\omega_\lambda = \sum_{\lambda}\left(n_\lambda+\frac{1}{2}\right)\hbar\omega_\lambda \tag{C.32}$$

$$\frac{S}{k_B} = \frac{E-F}{k_B T} = \sum_{\lambda}\beta\hbar\omega_\lambda/2\coth(\beta\hbar\omega_\lambda/2) - \ln 2\sinh\left(\beta\hbar\omega_\lambda/2\right) \tag{C.33}$$

Note that the above formulas are only the contributions of the vibrational degrees of freedom, and one must also add to them the contributions of rotational and translational degrees of freedom in order to obtain the total free/total energy.

$$Z = Z_{vib}Z_{rot}Z_{trans};\;\; F = F_{vib} + F_{rot} + F_{trans};\ldots$$

The vibrational heat capacity is defined as

$$C_v = (\partial E_{\mathrm{vib}}/\partial T)_V = k_B\sum_{\lambda}(x_\lambda/\mathrm{sinh}x_\lambda)^2$$

where $x_\lambda = \beta\hbar\omega_\lambda/2$. We can clearly see that at high temperatures $x_\lambda \to 0$, and therefore $C_v = k_B\sum_\lambda 1 = N_\lambda\, k_B$ with $N_\lambda = 3N-6$ being the number of vibrational modes in a $N$ atom cluster in 3 dimensions. The total energy also has a simple expression and goes to $E = k_B T(3N-6)$.

The equation of state relating the pressure $P$, volume $\Omega$ and temperature $T$ can be obatined from the definition of $P = -(\partial F/\partial\Omega)_T$. We find

$$P = -\frac{dE_0}{d\Omega} - \sum_{\lambda}\left(\frac{1}{2} + n_{\mathrm{BE}}(\beta\omega_\lambda)\right)\frac{d\hbar\omega_\lambda}{d\Omega} \tag{C.34}$$

If we introduce the **Gruneisen parameters**,[5] $\gamma_\lambda = -\mathrm{d}\ln\omega_\lambda/\mathrm{d}\ln\Omega$ we can write the equation of state in terms of these parameters (supposed to be known) as follows:

$$P = -\frac{dE_0}{d\Omega} - \sum_\lambda \left(\frac{1}{2} + n_{\mathrm{BE}}(\beta\omega_\lambda)\right) \frac{\hbar\omega_\lambda\gamma_\lambda}{\Omega} \tag{C.35}$$

which looks like a $\gamma$ times the total energy divided by the volume, if $\gamma_\lambda$ were constant.

In all above formulas the sum over modes can be replaced by a frequency integral of the frequency distribution function (or $g(\omega) = \mathrm{DOS}$). For instance

$$E = E_0 + \int d\omega\, g(\omega) \left(n_{\mathrm{BE}}(\beta\omega) + \frac{1}{2}\right) \hbar\omega$$

$$C_v = k_B \int d\omega\, g(\omega) \left(\frac{\beta\hbar\omega/2}{\sinh\beta\hbar\omega/2}\right)^2 \tag{C.36}$$

## References

1. M.H.F. Sluiter, M. Weinert, Y. Kawazoe, Europhys. Lett. **43**, 183 (1998)
2. P. Giannozzi, W. Andreoni, Phys. Rev. Lett. **76**, 4915 (1996); S. Baroni, P. Gianozzi, A. Testa, Phys. Rev. Lett. **59**, 2662 (1987)
3. M. Born, K. Huang, *Dynamical Theory of Crystal Lattices* (Oxford University Press, London, 1954)
4. G. Lehmann, M. Taut, Phys. Status Solidi (B) **54**, 469 (1972)
5. Ph. Lambin, J. Vigneron, Phys. Rev. B **29**, 3430 (1984)
6. P.E. Blöchl, O. Jepsen, O.K. Andersen, Phys. Rev. B **49**, 16223 (1994)

[5] Gruneisen parameters for each mode show by how many percents its frequency changes if there is a one percent change in the volume.

# Appendix D
# Calculation of the Ewald Sum

For long-range interactions, when simulating systems in a box with periodic boundary conditions, the direct summation of the interaction of particles in different boxes is slowly convergent. It is, in fact, divergent for the Coulomb interaction! To remedy this problem, one usually uses the trick called the "Ewald sum" method. This summation technique is needed to compute the ion–ion interaction (energy as well as force) in periodic systems. This interaction is Coulombic, but the ionic charge is an effective charge including the core electrons as well.

Below, we give a derivation of the Ewald sum for $S$, the interaction potential of one particle with another and all its images:

$$S(\boldsymbol{x}, d, n) = \sum_{\boldsymbol{R}} \frac{1}{|\boldsymbol{x} + \boldsymbol{R}|^n}.$$

The space dimension is denoted by $d$ and the translation vectors of the lattice by $\boldsymbol{R}$; $\boldsymbol{x}$ is the $d$ dimensional vector separating the two particles. We will first use the following identity:

$$\frac{1}{x} = \frac{1}{(x^2)^{(1/2)}} = \frac{1}{\Gamma(\frac{1}{2})} \int_0^\infty e^{-tx^2}\, t^{-1/2}\, dt$$

from which, after $p$ differentiations with respect to $x^2$, one gets

$$\frac{1}{(x^2)^{(p+1/2)}} = \frac{1}{\Gamma(p+\frac{1}{2})} \int_0^\infty e^{-tx^2}\, t^{(p-1/2)}\, dt$$

or after substituting $2p+1$ by $n$;

$$\frac{1}{x^n} = \frac{1}{\Gamma(\frac{n}{2})} \int_0^\infty e^{-tx^2}\, t^{(n/2-1)}\, dt.$$


K. Ohno et al., *Computational Materials Science*,
https://doi.org/10.1007/978-3-662-56542-1

We can then write the sum $S(\boldsymbol{x}, d, n)$ as follows:

$$\sum_{\boldsymbol{R}} \frac{1}{|\boldsymbol{x}+\boldsymbol{R}|^n} = \sum_{\boldsymbol{R}} \frac{1}{\Gamma(n/2)} \int_0^\infty e^{-t|\boldsymbol{x}+\boldsymbol{s}|^2}\, t^{(n/2-1)}\, dt. \tag{D.1}$$

Now we notice that if the integral is taken from some nonzero number $\epsilon$ to infinity, then the summation over $\boldsymbol{R}$ is very rapidly convergent. The rest of the integral, considered as a function of $\boldsymbol{x}$, is almost flat and very slowly convergent. Its Fourier transform, however, is rapidly convergent. We use the Poisson summation formula, which transforms sums on the real lattice to sums on the reciprocal lattice:

$$\sum_{\boldsymbol{R}} f(\boldsymbol{R}+\boldsymbol{x}) = \frac{1}{a_{\mathrm{c}}} \sum_{\boldsymbol{G}} \hat{f}(\boldsymbol{G}) \mathrm{e}^{i\boldsymbol{G}\cdot\boldsymbol{x}} \qquad \forall f. \tag{D.2}$$

Here $a_{\mathrm{c}}$ is the volume of the unit cell, the $\boldsymbol{G}$s are the reciprocal-lattice vectors with respect to the $\boldsymbol{R}$s, and $\hat{f}$ is the Fourier transform of the function $f$ defined by

$$\hat{f}(\boldsymbol{k}) = \int d^d\boldsymbol{r}\, e^{-i\boldsymbol{k}\cdot\boldsymbol{r}} f(\boldsymbol{r}).$$

So, after breaking up the integral into two parts, and applying the above summation formula to the slowly convergent part, we find

$$\begin{aligned} \sum_{\boldsymbol{R}} \frac{1}{|\boldsymbol{x}+\boldsymbol{R}|^n} &= \frac{1}{\Gamma(n/2)} \int_\epsilon^\infty \sum_{\boldsymbol{R}} e^{-t|\boldsymbol{x}+\boldsymbol{R}|^2}\, t^{(n/2-1)}\, dt \\ &\quad + \frac{1}{a_{\mathrm{c}}} \sum_{\boldsymbol{G}} \mathrm{e}^{i\boldsymbol{G}\cdot\boldsymbol{x}}\, \hat{f}(\boldsymbol{G}), \end{aligned} \tag{D.3}$$

where

$$\begin{aligned} \hat{f}(\boldsymbol{G}) &= \frac{1}{\Gamma(n/2)} \int d^d\boldsymbol{r}\, e^{-i\boldsymbol{G}\cdot\boldsymbol{r}} \int_0^\epsilon e^{-tr^2}\, t^{(n/2-1)}\, dt \\ &= \frac{1}{\Gamma(n/2)} \int_0^\varepsilon e^{-G^2/4t}\, (\pi/t)^{(d/2)}\, t^{(n/2-1)}\, dt \\ &= \frac{\pi^{d/2}}{\Gamma(n/2)} \left(\frac{G^2}{4}\right)^{(n-d)/2} \Gamma\left(\frac{d-n}{2}, \frac{G^2}{4\varepsilon}\right). \end{aligned} \tag{D.4}$$

The rapidly convergent part is expressed as an incomplete $\Gamma$ function.

This expression is divergent for $(n-d+1) \leq 1$. To remedy this problem, one must add a uniformly charged background of the same charge density with the opposite sign, and consider instead

$$\sum_{R} \frac{1}{|x + R|^n} - \frac{1}{a_c} \int d^d r \, \frac{1}{|x + r|^n}.$$

The effect of the background in the above is to cancel the contribution of large $\boldsymbol{R}$s, or equivalently, the infinity in the $\boldsymbol{G} = 0$ term. This cancellation is not complete, however, and one is left with a constant term equal to $-2\pi^{d/2}\,\varepsilon^{(n-d)/2}/[a_c\,(d-n)\,\Gamma(n/2)]$.

Finally, the Ewald sum can be written in the following form:

$$\sum_{R} \frac{1}{|x + R|^n} - \frac{1}{a_c} \int d^d r \, \frac{1}{r^n} = \frac{1}{\Gamma(n/2)} \sum_{R} \frac{\Gamma(n/2, \varepsilon |x + R|^2)}{|x + R|^n}$$
$$+ \frac{\pi^{d/2}}{a_c\,\Gamma(n/2)} \left\{ \sum_{G \neq 0} \left(\frac{G}{2}\right)^{(n-d)} \Gamma\left(\frac{d-n}{2}, \frac{G^2}{4\varepsilon}\right) e^{iG\cdot x} + \frac{2\,\varepsilon^{(n-d)/2}}{n-d} \right\}. \quad \text{(D.5)}$$

Moreover, if one is interested in the Madelung energy of a point charge $E_{\mathrm{M}} = \sum_{R \neq 0} R^{-n} -$ background term, an additional number arises from the $\boldsymbol{R} = 0$ term:

$$E_{\mathrm{M}} = \lim_{x \to 0} \left( \sum_{R} \frac{1}{|x + R|^n} - \frac{1}{x^n} \right) - \frac{1}{a_c} \int d^d r \, \frac{1}{r^n}$$
$$= \frac{1}{\Gamma(n/2)} \left\{ \sum_{R \neq 0} \frac{\Gamma(n/2, \varepsilon |R|^2)}{R^n} - \frac{2}{n} \varepsilon^{n/2} \right\}$$
$$+ \frac{\pi^{d/2}}{a_c\,\Gamma(n/2)} \left\{ \sum_{G \neq 0} \left(\frac{G}{2}\right)^{(n-d)} \Gamma\left(\frac{d-n}{2}, \frac{G^2}{4\varepsilon}\right) + \frac{2\,\varepsilon^{(n-d)/2}}{n-d} \right\}. \quad \text{(D.6)}$$

In all of this calculation, the choice of $\varepsilon$ was arbitrary. To obtain sums that are equally rapidly convergent in both spaces, the best choice is $\varepsilon = \pi/a_c^{2/d}$.

In the special case of $d = 2$ and $n = 1$, for example, the incomplete $\Gamma$ function is reduced to a complementary error function, and after renaming $\sqrt{\varepsilon}$ as $\varepsilon$, the Ewald sum becomes

$$\sum_{R} \frac{1}{|x + R|} - \frac{1}{a_c} \int d^2 r \, \frac{1}{|x + r|} = \sum_{R} \frac{\mathrm{Erfc}(\varepsilon |x + R|)}{|x + R|}$$
$$+ \frac{\pi}{a_c} \sum_{G \neq 0} \left(\frac{2}{G}\right) \mathrm{Erfc}\left(\frac{G}{2\varepsilon}\right) \mathrm{e}^{iG\cdot x} - \frac{2}{a_c} \frac{\sqrt{\pi}}{\varepsilon}. \quad \text{(D.7)}$$

Likewise, the Madelung energy becomes:

$$E_{\rm M} = \sum_{\boldsymbol{R}\neq 0} \frac{1}{R} - \frac{1}{a_{\rm c}} \int d^2\boldsymbol{r}\, \frac{1}{r} = \sum_{\boldsymbol{R}\neq 0} \frac{\mathrm{Erfc}(\varepsilon R)}{R} - \frac{2\varepsilon}{\sqrt{\pi}} + \frac{\pi}{a_{\rm c}} \sum_{\boldsymbol{G}\neq 0} \left(\frac{2}{G}\right) \mathrm{Erfc}\left(\frac{G}{2\varepsilon}\right) - \frac{2}{a_{\rm c}} \frac{\sqrt{\pi}}{\varepsilon}. \tag{D.8}$$

In three dimensions, the formula reduces to

$$\sum_{\boldsymbol{R}} \frac{1}{|\boldsymbol{x}+\boldsymbol{R}|} - \frac{1}{a_{\rm c}} \int d^3\boldsymbol{r}\, \frac{1}{|\boldsymbol{x}+\boldsymbol{r}|} = \sum_{\boldsymbol{R}} \frac{\mathrm{Erfc}(\varepsilon|\boldsymbol{x}+\boldsymbol{R}|)}{|\boldsymbol{x}+\boldsymbol{R}|} + \frac{\pi}{a_{\rm c}} \sum_{\boldsymbol{G}\neq 0} \left(\frac{4}{G^2}\right) \mathrm{e}^{-(G/2\varepsilon)^2}\, \mathrm{e}^{i\boldsymbol{G}\cdot\boldsymbol{x}} - \frac{\pi}{a_{\rm c}\,\varepsilon^2} \tag{D.9}$$

and the Madelung energy is

$$E_{\rm M} = \sum_{\boldsymbol{R}\neq 0} \frac{1}{R} - \frac{1}{a_{\rm c}} \int d^3\boldsymbol{r}\, \frac{1}{r} = \sum_{\boldsymbol{R}\neq 0} \frac{\mathrm{Erfc}(\varepsilon R)}{R} - \frac{2\varepsilon}{\sqrt{\pi}} + \frac{\pi}{a_{\rm c}} \sum_{\boldsymbol{G}\neq 0} \left(\frac{4}{G^2}\right) \mathrm{e}^{-(G/2\varepsilon)^2} - \frac{\pi}{a_{\rm c}\,\varepsilon^2}. \tag{D.10}$$

Higher-order sums can be deduced from direct differentiation, but keeping in mind the definition and the origin of the additional constants. The background charge is not always necessary and depends on the interaction.

Another general method has also been proposed [1] on the basis of similar ideas. The original sum can be written as a convolution of a charge density composed of a series of delta functions with a potential function ($1/R^n$ in this case). One adds and subtracts a Gaussian from each delta function (although it is also possible to chose any other short-ranged but smooth function $\sigma$):

$$\begin{aligned} \rho(\boldsymbol{r}) &= \rho_1(\boldsymbol{r}) + \rho_2(\boldsymbol{r}), \\ \rho_1(\boldsymbol{r}) &= \sum_{\boldsymbol{R}} \sigma(|\boldsymbol{r}-\boldsymbol{R}|), \\ \rho_2(\boldsymbol{r}) &= \sum_{\boldsymbol{R}} [\delta(\boldsymbol{r}-\boldsymbol{R}) - \sigma(|\boldsymbol{r}-\boldsymbol{R}|)], \end{aligned} \tag{D.11}$$

where $\sigma(r) = (1/\pi^{d/2}\alpha^d)\exp\left[-(r/\alpha)^2\right]$. The contribution of $\rho_1(\boldsymbol{r})$ is a smooth function, and therefore its Fourier series is rapidly convergent. By using the Poisson summation formula (which is the same as using the Fourier expansion), one can express $\rho_1(\boldsymbol{r})$ in reciprocal space. This corresponds to the $\boldsymbol{G}$ sum in the formulae previously derived.

The contribution of $\rho_2(\boldsymbol{r})$, however, can be made rapidly convergent in real space since the integral of the charge density about each point amounts to zero. Lattice points that are located far away will be "screened" and hence do not contribute to the sum. So this term too, will be short-ranged in real space and one has to compute only a few terms in each of the two sums. In three dimensions, with the choice of a Gaussian as the smoothing function $\sigma$, one also recovers the formula (D.9).

## Reference

1. E.F. Bertaut, J. Phys. Radiat. **13**(499), 633 (1952)

# Appendix E
# Optimization Methods Used in Materials Science

## E.1 Conjugate-Gradient Minimization

The conjugate-gradient method for the minimization of a given function $F(\boldsymbol{x})$ [1] is known as one of the most powerful and fast methods [2, 3]. It can be used to solve the matrix eigenvalue problem appearing in first-principles calculations if one uses the algorithm [4, 5] described in Sect. 2.4.1. Explicit methods for applying the conjugate-gradient method or the preconditioned conjugate-gradient method have been proposed and discussed by many authors [5–16].

Let us consider first a variational requirement that a bilinear (biconjugate) function of the form

$$F(\boldsymbol{x}) = \frac{1}{2}\boldsymbol{x}^{\dagger} \cdot \boldsymbol{A} \cdot \boldsymbol{x}, \tag{E.1}$$

is minimized with respect to $\boldsymbol{x}^{\dagger}$. Here $\boldsymbol{A}$ is an $N \times N$ Hermitian matrix and $\boldsymbol{x}$ an $N$-column vector. For this simple case, we know that this function is minimized when its gradient becomes zero:

$$\nabla F(\boldsymbol{x}) = \boldsymbol{A} \cdot \boldsymbol{x} = 0. \tag{E.2}$$

That is, the original variational requirement is identical to the problem of solving the linear equation (E.2).

Now, suppose that we know the functional form of $F(\boldsymbol{x})$ and $\nabla F(\boldsymbol{x})$ but not $\boldsymbol{A}$. In order to find the solution of (E.2), we consider a trial with $\boldsymbol{x} = \boldsymbol{x}_i$. A "residual vector" (or "steepest-descent vector") is defined as $\boldsymbol{s}_i = -\nabla F(\boldsymbol{x}_i)$ $(= -\boldsymbol{A} \cdot \boldsymbol{x}_i)$, which is the same as the vector appearing in the steepest-descent method. Next we search for the minimum point $\boldsymbol{x} = \boldsymbol{x}_{i+1}$ along a direction $\boldsymbol{h}_i$ starting from the first point $\boldsymbol{x}_i$, i.e. we minimize $F(\boldsymbol{x}_{i+1})$ at a point

$$\boldsymbol{x}_{i+1} = \boldsymbol{x}_i + \alpha_i \boldsymbol{h}_i \tag{E.3}$$


K. Ohno et al., *Computational Materials Science*,
https://doi.org/10.1007/978-3-662-56542-1

with respect to $\alpha_i$ ("line minimization"). It is easy to understand that this minimum occurs generally at the point where the steepest direction at the point $\boldsymbol{x}_{i+1}$ becomes orthogonal to $\boldsymbol{h}_i$, i.e. where the orthogonality relation

$$\boldsymbol{h}_i^\dagger \cdot \nabla F(\boldsymbol{x}_{i+1}) = -\boldsymbol{h}_i^\dagger \cdot \boldsymbol{s}_{i+1} = 0 \tag{E.4}$$

holds. Here, $\boldsymbol{s}_{i+1}$ is the steepest-descent vector at $\boldsymbol{x}_{i+1}$ (i.e. the residual vector for the next step), given by

$$\boldsymbol{s}_{i+1} = -\nabla F(\boldsymbol{x}_{i+1})(= -\boldsymbol{A} \cdot (\boldsymbol{x}_i + \alpha_i \boldsymbol{h}_i) = \boldsymbol{s}_i - \alpha_i \boldsymbol{A} \cdot \boldsymbol{h}_i). \tag{E.5}$$

From the relation inside the parentheses, we see that $\alpha_i$ satisfies

$$\alpha_1 = \frac{\boldsymbol{s}_i^\dagger \cdot \boldsymbol{h}_i}{\boldsymbol{h}_i^\dagger \cdot \boldsymbol{A} \cdot \boldsymbol{h}_i} = \frac{\boldsymbol{s}_i^\dagger \cdot \boldsymbol{s}_i}{\boldsymbol{h}_i^\dagger \cdot \boldsymbol{A} \cdot \boldsymbol{h}_i}. \tag{E.6}$$

Up to now, we have not yet prescribed the direction $\boldsymbol{h}_i$. How should one determine this direction? At the $i$th step, we have from (E.4) and (E.5),

$$(\boldsymbol{x}_i + \alpha_i \boldsymbol{h}_i)^\dagger \cdot \boldsymbol{A} \cdot \boldsymbol{h}_i = 0 \tag{E.7}$$

for $\alpha_i$. If we proceed one step further, we have

$$(\boldsymbol{x}_{i+1} + \alpha_{i+1} \boldsymbol{h}_{i+1})^\dagger \cdot \boldsymbol{A} \cdot \boldsymbol{h}_{i+1} (\boldsymbol{x}_i + \alpha_i \boldsymbol{h}_i + \alpha_{i+1} \boldsymbol{h}_{i+1})^\dagger \cdot \boldsymbol{A} \cdot \boldsymbol{h}_{i+1} = 0. \tag{E.8}$$

This step-by-step minimization does not determine the directions of the $\boldsymbol{h}$s. A way to select the most suitable directions can be found from a global search of the $\alpha$s. If we minimize $F(\boldsymbol{x}_{i+1})$ with respect not only to $\alpha_{i+1}$ but also to $\alpha_k$ with $k = 1, 2, \ldots, i$, we obtain

$$(\boldsymbol{x}_i + \alpha_i \boldsymbol{h}_i + \alpha_{i+1} \boldsymbol{h}_{i+1})^\dagger \cdot \boldsymbol{A} \cdot \boldsymbol{h}_k = 0. \tag{E.9}$$

In order for (E.7)–(E.9) to be consistent, the minimizations along all $\boldsymbol{h}_k$ $(k = 1, 2, \ldots)$ should be independent. Thus, one is led to the "conjugacy relations"

$$\boldsymbol{h}_i^\dagger \cdot \boldsymbol{A} \cdot \boldsymbol{h}_k = 0, \quad \text{for } i \neq k, \tag{E.10}$$

as well as the orthogonality relations

$$\boldsymbol{h}_i^\dagger \cdot \boldsymbol{s}_k = 0, \quad \text{for } i \neq k. \tag{E.11}$$

The conjugate (biconjugate)-gradient algorithm which will be described below presents a systematic way to determine the conjugate-gradient directions $\{\boldsymbol{h}_i\}$ which satisfy (E.10) and (E.11) automatically. The method is to successively determine the conjugate-gradient directions according to the recursion relation

$$\boldsymbol{h}_{i+1} = \boldsymbol{s}_{i+1} + \beta_i \boldsymbol{h}_i, \tag{E.12}$$

$$\beta_i = \frac{\boldsymbol{s}_{i+1}^{\dagger} \cdot \boldsymbol{s}_{i+1}}{\boldsymbol{s}_i^{\dagger} \cdot \boldsymbol{s}_i}, \tag{E.13}$$

together with the line minimization described in (E.3)–(E.6), to determine $\alpha_i$ at each step. One may show by straightforward induction [1] that this algorithm generates the conjugate-gradient directions ($\boldsymbol{h}$s) and the estimates for the minimum point ($\boldsymbol{x}$s) which satisfy (E.10) and (E.11). Multiplying (E.12) by $\boldsymbol{s}_k^{\dagger}$ from the left and using (E.10) and (E.11), we find that the following orthogonality relations also hold in the conjugate-gradient algorithm:

$$\boldsymbol{s}_i^{\dagger} \cdot \boldsymbol{s}_k = 0, \quad \text{for } i \neq k. \tag{E.14}$$

The ordinary conjugate-gradient method described above works well for well-conditioned matrices which are not very far from the identity matrix. In view of this fact, one may use

$$(\boldsymbol{K} \cdot \boldsymbol{A}) \cdot \boldsymbol{x} = 0 \tag{E.15}$$

instead of (E.2). The matrix $\boldsymbol{K}$ is called a "preconditioner", and can be chosen arbitrary in order to make $(\boldsymbol{K} \cdot \boldsymbol{A})$ close to the identity matrix. The preconditioned (bi)conjugate-gradient method introduces a preconditioned vector

$$z_i = \boldsymbol{K}\boldsymbol{s}_i. \tag{E.16}$$

Its formulation modifies some of the equations presented above. That is, (E.6), (E.12) and (E.13) are replaced with

$$\alpha_1 = \frac{\boldsymbol{s}_i^{\dagger} \cdot z_i}{\boldsymbol{h}_i^{\dagger} \cdot \boldsymbol{A} \cdot \boldsymbol{h}_i}, \tag{E.17}$$

$$\boldsymbol{h}_{i+1} = z_{i+1} + \gamma_i \boldsymbol{h}_i, \tag{E.18}$$

and

$$\gamma_i = \frac{\boldsymbol{s}_{i+1}^{\dagger} \cdot z_{i+1}}{\boldsymbol{s}_i^{\dagger} \cdot z_i}, \tag{E.19}$$

respectively.

In the actual implementation of this algorithm in the case of electronic-structure calculations, one should pay attention to the orthonormality of the wavefunctions which is generally needed at each step of the conjugate-gradient iterations. One way of doing this is a band-by-band minimization technique discussed by Teter et al. [5, 6]. In their algorithm, one can reduce the number of arrays (i.e. the computer memory) used in the conjugate-gradient calculation. Such a technique is elegant and used very widely. However, if the number of bands increases, even this method slows down considerably. To compensate such problems, one can use an alternative approach which updates all the bands simultaneously [7, 8]. Many other related approaches have also been proposed [9–15].

## E.2 Block Davidson Method

In the plane wave (PW) expansion, for example, the matrix size to be diagonalized is usually quite large, typically, 10,000 × 10,000 ∼ 100,000 × 100,000. If one tries to directly diagonalize such a large matrix, a large computational time is required even if one uses supercomputer. In DFT calculations, however, one needs the lowest $N$ eigenvalues and eigenfunctions only, where $N$ is typically 100 ∼ 2000. In such a case, block Davidson method [17–19] is useful for the diagonalization. This method needs $N'(\geq N)$ orthonormal guess vectors $| b_i \rangle$ satisfying $\langle b_i | b_j \rangle = \delta_{ij}$, which may be provided by the solution of the Kohn–Sham equation at one step before in the SCF loop. (Only first step needs to provide an initial guess by using, for example, random numbers.) Then, construct the Hamiltonian matrix in the subspace spanned by $\{ | b_i \rangle \}$. After diagonalizing this Hamiltonian, one calculates the residual vectors $| r_\lambda \rangle$ for its lowest $N$ eigenvalues ($h_\lambda$) in the original Fock space spanned by the (PW) basis functions $\{\xi\}$. For the residual vectors, whose norm $|| r_\lambda \rangle$ is larger than $\epsilon \sim 10^{-3}$–$10^{-6}$, form the preconditioned correction vectors $| \delta_{\lambda\xi} \rangle$ by dividing the residual vectors $| r_{\lambda\xi} \rangle$ by $\hat{H}_{\xi\xi} - h_\lambda$, where $\hat{H}_{\xi\xi}$ means the $\xi$th diagonal element of the Hamiltonian in the original Fock space. Orthogonalizing these preconditioned correction vectors to all $| b_i \rangle$ ($i = 1, 2, \ldots, N$), and normalizing the results $| \delta'_\lambda \rangle$ ($\lambda = 1, 2, \ldots, N$), one checks if the ratio of the norms $|| \delta'_\lambda \rangle | / || \delta_\lambda \rangle$ is smaller than $\tau \sim 10^{-3}$ or not. If this inequality is not satisfied, normalize $| \delta'_\lambda \rangle$ and add it to the subspace $\{ | b_i \rangle \}$. This procedure is repeated until convergence is achieved for all $N$ residuals.

The algorithm is conveniently summarized as follows [18, 19]:

1. Provide $N'(\geq N)$ orthonormal guess vectors $| b_i \rangle$ satisfying $\langle b_i | b_j \rangle = \delta_{ij}$.
2. Form the subspace Hamiltonian $H_{ij} = \langle b_i | \hat{H} | b_j \rangle$
3. Solve the subspace eigenvalue equation $\sum_j H_{ij} C_{j\lambda} = h_\lambda C_{i\lambda}$.
4. Obtain the lowest $N$ eigenvalues $h_\lambda$ and eigenvectors $| \Psi_\lambda \rangle = \sum_j | b_j \rangle C_{j\lambda}$ satisfying $\sum_j C^*_{j\lambda} C_{j\lambda'} = \delta_{\lambda\lambda'}$.
5. Construct $N$ residual vectors $| r_\lambda \rangle = \hat{H} | \Psi_\lambda \rangle - h_\lambda | \Psi_\lambda \rangle$.
6. **for** $\lambda = 1, N$, **do** followings:
7.     **if** $| | r_\lambda \rangle | > \epsilon$ **then**
8.         Form the preconditioned correction vector $| \delta_{\lambda\xi} \rangle = -(\hat{H}_{\xi\xi} - h_\lambda)^{-1} = | r_{\lambda\xi} \rangle$.
9.         Orthogonalize to the current subspace as $| \delta'_\lambda \rangle = | \delta_\lambda \rangle - \sum_i^N | b_i \rangle \langle b_i | \delta_\lambda \rangle$.
10.         **if** $| | \delta'_\lambda \rangle | / | | \delta_\lambda \rangle | > \tau$ **then**
11.             Normalize $| \delta'_\lambda \rangle$.
12.             Add $| \delta'_\lambda \rangle$ to the subspace $\{ | b_i \rangle \}$.
13.         **end if**

14. **end if**
15. **end for**
16. **goto 2** unless convergence is achieved for all $N$ residuals.

## E.3 Broyden's Method

Broyden's method [20] is an algorithm which searches for the nearest zero of an $N$-dimensional function, given a starting vector $X_0$. It can be used both for geometry optimization where one searches for the zero of the force vector, and to achieve quick convergence in self-consistent calculations. In the latter case, one is searching for the zero of the difference between the input and output functions, which are to converge to a self-consistent value. For example, in the case of the charge density, $\rho_{\text{out}}(r) - \rho_{\text{in}}(r)$ is the equivalent of the force vector, and it should be minimized with respect to the electronic degrees of freedom, i.e. the projection of the eigenstates on the basis functions, or the density matrix [21].

This algorithm uses information on all previous iterations to construct iteratively an approximation to the gradient of the function $F$, and then uses a quasi-Newton method to update the "position" $X$ which makes $F = 0$. Briefly, one starts with a first guess for the gradient (or Jacobian) $J$. The easiest one is just a constant matrix. Newton's method updates the variable $X$ according to

$$X^{\text{new}} = X^{\text{old}} - J^{-1} F. \tag{E.20}$$

For a constant matrix $J = -\alpha^{-1}\mathbf{I}$, this reduces to $X_{\text{in}}^{\text{new}} = (1-\alpha)X_{\text{in}}^{\text{old}} + \alpha\, X_{\text{out}}^{\text{old}}$, which is the often-used linear mixing scheme. This scheme converges rather slowly, and usually small mixing parameters of the order of $\alpha = 0.1, 0.05$ or less are needed for a monotonic convergence.

In Broyden's method, $G = J^{-1}$ is updated by using information on $X$ and $F$ of all previous iterations, even though one does not need to actually store $G$ in a matrix. Clearly, it requires more memory and more CPU time per iteration than a linear mixing scheme, but it converges much faster than the latter. If an approximation to the true $G$ is known after the $m$th iteration, the variable $X_{\text{in}}$ is updated for the next iteration as follows:

$$X_{\text{in}}^{[m+1]} = X_{\text{in}}^{[m]} - G^{[m]} F^{[m]}. \tag{E.21}$$

One must remember that in the above equation, $X$ and $F$ are one-dimensional arrays and $G$ is a matrix; $m$ is the iteration number. Since $G$ is not the exact inverse Jacobian for the new position $X_{\text{in}}^{[m+1]}$, the force $F^{[m+1]}$ is not zero. Therefore, one updates $G$ to correct for this error:

$$X_{\text{in}}^{[m+1]} - X_{\text{in}}^{[m]} = G^{[m+1]}(F^{[m+1]} - F^{[m]}). \tag{E.22}$$

Remember that in general, $F$ is not linear in $X$ and even if $G$ were exact, the updated force would not be zero. The above equation imposes $N$ constraints on the matrix elements of $G^{[m+1]}$. The other $N(N-1)$ elements are found by minimizing the change in $G$ after each iteration. This ensures the stability of the iteration process. The minimization of the change can be performed by using Lagrange multipliers. One updating procedure for $G$, called Broyden's second method is

$$G^{[m+1]} = G^{[m]} + \frac{|\Delta X_{in}^{[m]} - G^{[m]}\Delta F^{[m]}\rangle\langle\Delta F^{[m]}|}{\langle\Delta F^{[m]}|\Delta F^{[m]}\rangle} \tag{E.23}$$

where $|\Delta F^{[m]}\rangle = |F^{[m+1]}\rangle - |F^{[m]}\rangle$, and the usual ket–bra notation has been used to represent vectors and their transposes. It can be easily verified that $G^{[m+1]}$ defined above also satisfies (E.22). Using the above, without having to store $G$ at each iteration, one can update $X$ according to [22]:

$$|X_{\text{in}}^{[m+1]}\rangle = |X_{\text{in}}^{[m]}\rangle - G^{[1]}|F^{[m]}\rangle - \sum_{j=2}^{m} |U^{[j]}\rangle\langle V^{[j]}|F^{[m]}\rangle, \tag{E.24}$$

where

$$|U^{[i]}\rangle = -G^{[1]}|\Delta F^{[i-1]}\rangle + |\Delta X_{\text{in}}^{[i-1]}\rangle - \sum_{j=2}^{i-1} |U^{[j]}\rangle\langle V^{[j]}|\Delta F^{[i-1]}\rangle \tag{E.25}$$

and

$$|V^{[j]}\rangle = \frac{|\Delta F^{[j-1]}\rangle}{\langle\Delta F^{[j-1]}|\Delta F^{[j-1]}\rangle}.$$

## E.4 SA and GA as Global Optimization Methods

Traditional optimization methods based on Newton's method or steepest descent will always find the nearest local minimum to the starting point. These methods are ineffective when the phase space over which one wants to optimize some cost function $E(X)$ is of high dimensionality. Of course, having ideas about where one could eventually find minima would help, but it would not be a systematic search. Up to now, no foolproof algorithm has been developed to find the global minimum of a continuous functional, in a continuous space, within a finite time. Powerful methods, however, have recently been developed that will at least find a "good minimum" for the cost function. The simulated Annealing (SA) method [23] and Genetic Algorithm (GA) [24] are two of these.

### E.4.1 Simulated Annealing (SA)

The idea in SA comes from the annealing of metals, i.e. one starts from a high-temperature phase, and lowers the temperature slowly, so that the system has time to find its ground state (or at least a metastable state). When the temperature is high, the system can easily overcome energy barriers because of its large (thermal) kinetic energy. As one can see from the Boltzmann factor $e^{-E/kT}$, temperature provides a means of scaling down the energies. Therefore, provided that the annealing rate is slow enough, the system can move in a wide region of the phase space to find the deepest valley, overcoming the main problem with traditional optimization techniques, that one is stuck near the starting point. This method was originally proposed by Kirkpatrick et al. [23] in 1983, and since then has found many applications in all fields requiring optimization, from device design to thermodynamics. The art of performing the simulated annealing resides in controlling the cooling rate. It can be shown theoretically that if this rate is low enough, one ends up at the absolute minimum. But for practical purposes, this rate should be large enough that one reaches the final step within a reasonable time. Simple ways of cooling consist of choosing a stepwise, linear or exponential cooling rate. A better algorithm, developed recently, is called "adaptive simulated annealing" (ASA) [25]. In the ASA algorithm, one starts the molecular-dynamics simulation at a high temperature well above melting, and lowers it after some $M$ molecular-dynamics steps, according to the following equation:

$$\frac{dT}{dt} = -\nu T/\epsilon\sqrt{C}. \tag{E.26}$$

Here $\nu$ is a velocity scale taken as a constant (of the order of 1) to control the cooling rate, $C$ is the heat capacity of the system, indicating the fluctuations in the average configurational (or potential) energy, and $\epsilon$ represents the relaxation time; it is related to the second-highest eigenvalue of the transition matrix by the following relation: $\epsilon = -\text{Log}\ (q_2)$ (for more details see Sect. 5.2.3 and [25]). The thermodynamic properties of the system ($C$ and $\epsilon$) are computed from the information gathered during these $M$ steps, where the temperature is held constant. The number of steps necessary to obtain the averages depends of course on the system, and should be large enough that one has reached an equilibrium state and can obtain meaningful statistics.

Regarding the computation of $\epsilon$, one must calculate the transition matrix during the $M$ steps. This is a nonequilibrium property of the system. The $i, j$ element of this matrix is incremented by one every time there is a transition from state $i$ to state $j$. The number of states in practice being infinite, one needs to group them together, in order to have a finite transition matrix of manageable size. One way to do this is to include in the state $i$, all the states having an energy $E$ satisfying $E_i < E < E_i + \Delta E$, where $\Delta E$ is a window defining the size of the $T$ matrix and $E_i = E_{\min} + i\Delta E$ defines the subdivisions of the energy axis, starting from the lower bound $E_{\min}$. After the $M$ steps, one finds the transition matrix for the temperature considered. If the columns of this matrix are normalized to one, then $q_1 = 1$ is its largest eigenvalue having as its eigenvector the equilibrium state of the system, i.e. the probabilities of occupation of the different states $i$. The second-largest eigenvalue gives the relaxation time through $\epsilon = -\operatorname{Log}(q_2)$ which will be used in (E.26). The heat capacity $C$ is found from the fluctuations in the total energy: $C = \sqrt{(\langle E^2\rangle - \langle E\rangle^2)/T}$. The free parameter $v$ can be adjusted to control the cooling rate. Larger $v$s lead to faster annealing, with the danger of missing a good minimum.

### *E.4.2 Genetic Algorithm (GA)*

Similarly to SA, the GA provides us with a way to move almost randomly in the phase space and probe the properties of the energy surface. The algorithm is even more nonlocal than SA, i.e. one can move in a next step to a point very far from the starting one, whereas in SA, one is following a Markov process, and therefore follows a continuous trajectory. The idea of the GA, as its name indicates, is based on the theory of evolution and the survival of the fittest. In the algorithm, one starts with a set of individuals each having a "gene". In the case of structural optimization, the individuals are the structures, and their genes, are the atomic coordinates. One wishes to find individuals with the "best genes", i.e. the structures with lowest energy. One passes from one generation of individuals to the next by *mating* and *mutating* the genes of the individuals. A simple example is to map all the $3N$ coordinates of an individual to a binary string containing this information. Mating between two individuals consists of combining a piece (half) of the gene of one individual with half of the gene of the other. A mutation consists of taking one or a few bits in the gene sequence randomly, and invert them (i.e. change 0 to 1 and 1 to 0). After such operations, one ends up with the next *generation* of individuals, some of them being more fit (having lower energy), and some others with bad genes. Among the new population, one then takes the good individuals and performs mating and mutations among them to get the next generation. This way, one expects to get individuals with better genes after each generation. There is no criterion for, nor is there any guarantee of convergence. But one expects to reach a steady regime with at least a "good" minimum. The general algorithm outlined above is certainly too general, and one needs to adapt the mating (also called *crossover*) and mutations to the specific problem at hand. Figure E.1 summarizes schematically the passage from one generation to the next. For more details about this technique, one can refer to the book by Goldberg [24].

In the case of structural optimization, one way of performing the crossover is to cut each of the parents' structures in half, and connect the obtained halves. A mutation could be to move one or more atoms randomly by a small amount. Finally, a relaxation to the nearest local minimum is performed by the conjugate-gradient method or molecular dynamics, for example. Then the newly obtained children are selected for making the next generation, according to the probability $P_i = e^{-\beta E_i}$. Using the tight-binding molecular-dynamics method, it has been possible to obtain the soccer-ball fullerene $C_{60}$ as the lowest-energy configuration of 60 carbon atoms

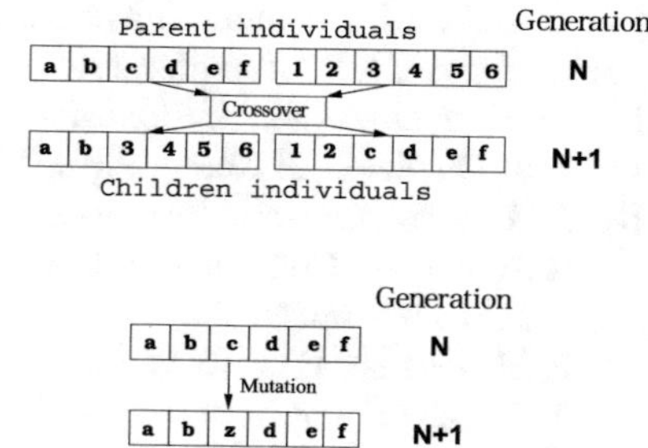


**Fig. E.1** Crossover and mutation take individuals from one generation to the next

[26]. Obtaining this structure by the usual MD or even SA techniques would take a very large number of time steps, and would not be as efficient as the GA method. One reason is that in the high temperature regime, the true ground state (soccer ball), despite its low energy, has a higher free energy than chains, for example. So, at such temperatures, the SA method finds chains or other high-entropy structures as the minimum. But after the temperature is lowered, one might get stuck in such a minimum and not have enough kinetic energy to cross the high barrier needed to reach the real ground state.

This method has also been applied to investigate the ground-state structure of the silicon [100] surface [27]. In general, the GA is more advantageous than SA if the dimensionality of the search space is large and if one does not have a good idea of the location of the minimum.

## References

1. R. Fletcher, in *Numerical Analysis Dundee*, vol. 506. Lecture Notes in Mathematics, ed. by A. Dold, B. Eckmann (Springer, Berlin, 1975), p. 73
2. P.E. Gill, W. Murray, M.H. Wright, *Practical Optimization* (Academic Press, London, 1981)
3. W.H. Press, S.A. Teukolsky, W.T. Vettering, B.R. Fiannery, *Numerical Recipes in FORTRAN: The Art of Scientific Computing*, 2nd edn. (Cambridge University Press, New York, 1992)
4. M.C. Payne, M. Needels, J.D. Joannopoulos, Phys. Rev. B **37**, 8137 (1988)
5. M.P. Teter, M.C. Payne, D.C. Allan, Phys. Rev. B **40**, 12255 (1989)
6. M.C. Payne, M.P. Teter, C. Allan, T.A. Arias, J.D. Joannopoulos, Rev. Mod. Phys. **64**, 1045 (1992)
7. M.J. Gillan, J. Phys. Condens. Matter **1**, 689 (1989)
8. R.D. King-Smith, D. Vanderbilt, Phys. Rev. B **49**, 5828 (1994)
9. P. Bendt, A. Zunger, Bull. Am. Phys. Soc. **27**, 248, (1982); Phys. Rev. B **26**, 3114 (1982)
10. D. Singh, Phys. Rev. B **40**, 5428 (1989)
11. I. Štich, R. Car, M. Parrinello, S. Baroni, Phys. Rev. B **39**, 4997 (1989)
12. D.M. Bylander, L. Kleinman, S. Lee, Phys. Rev. B **42**, 1394 (1990)
13. T.A. Arias, M.C. Payne, J.D. Joannopoulos, Phys. Rev. Lett. **69**, 1077 (1992)

14. G. Galli, M. Parrinello, Phys. Rev. Lett. **69**, 3547 (1992)
15. G. Kresse, J. Furthmüller, Phys. Rev. B **54**, 11169 (1996)
16. M. Kohyama, Model. Sim. Mater. Sci. Eng. **4**, 397 (1996)
17. E.R. Davidson, J. Comp. Phys. **17**, 87 (1975)
18. B. Liu, in *Numerical Algorithms in Chemistry: Algebraic Methods*. National Resource for Computation in Chemistry (CONF-780878) (Lawrence Berkeley Lab., 1978), pp. 49–52
19. R.M. Parrish, E.G. Hohenstein, T.J. Martıanez, J. Chem. Theory Comput. **12**, 3003 (2013)
20. C.G. Broyden, Math. Comput. **19**, 577 (1965)
21. D. Vanderbilt, S.G. Louie, Phys. Rev. B **30**, 6118 (1984)
22. D.J. Singh, H. Krakauer, C.S. Wang, Phys. Rev. B **34**, 8391 (1986)
23. K.S. Kirkpatrick, C.D. Gelatt, M.P. Vecchi, Science **220**, 671 (1983); K.S. Kirkpatrick, J. Stat. Phys. **34**, 975 (1984)
24. D.E. Goldberg, *Genetic Algorithms in Search, Optimization, and Machine Learning* (Addison-Wesley, Reading, 1989)
25. B. Andersen, J.M. Gordon, Phys. Rev. E **50**, 4346 (1994). References therein
26. D.M. Deaven, K.M. Ho, Phys. Rev. Lett. **75**, 288 (1995)
27. R.T. Fu, K. Esfarjani, Y. Hashi, J. Wu, Y. Kawazoe, Sci. Rep. RITU **A 44**, 77 (1997)

# Appendix F
# Partitioning Formula, Downfolding and Green's Functions

The partitioning formula is a useful mathematical identity used to compute the projection of the inverse of a matrix onto a given subspace. It becomes very handy in the context of Green's function calculations for problems such as transport where the transmission coefficient can be expressed exactly in terms of the Green's function matrix elements. It is also a useful tool in the context of downfolding, where again the information about the spectrum of the whole system can be included in a smaller subspace at the cost of adding an energy-dependent self-energy to the sub-Hamiltonian.

## F.1 The Partitioning Formula

For a given $(n+p) \times (n+p)$ matrix $\mathbb{m}$, let us denote its blocks $\mathbb{a}(n \times n)$, $\mathbb{b}(n \times p)$, $\mathbb{c}(p \times n)$ and $\mathbb{d}(p \times p)$ and let us define its inverse $\mathbb{M}$ by:

$$\mathbb{m} = \begin{pmatrix} \mathbb{a} & \mathbb{b} \\ \mathbb{c} & \mathbb{d} \end{pmatrix} ; \quad \mathbb{M} = \begin{pmatrix} \mathbb{A} & \mathbb{B} \\ \mathbb{C} & \mathbb{D} \end{pmatrix} \tag{F.1}$$

of blocks of similar size to the original matrix $\mathbb{m}$. Assuming the blocks $\mathbb{a}$ and $\mathbb{d}$ to be invertible, we can express the blocks of the inverse in terms of the original blocks $\mathbb{a}$, $\mathbb{b}$, $\mathbb{c}$, $\mathbb{d}$ and $\mathbb{a}^{-1}$ and $\mathbb{d}^{-1}$. Using the definition $\mathbb{m} \times \mathbb{M} = \mathbb{I}$ we obtain:

$$\mathbb{a}\mathbb{A} + \mathbb{b}\mathbb{C} = \mathbb{I} \tag{F.2}$$

$$\mathbb{a}\mathbb{B} + \mathbb{b}\mathbb{D} = 0 \tag{F.3}$$

$$\mathbb{c}\mathbb{A} + \mathbb{d}\mathbb{C} = 0 \tag{F.4}$$

$$\mathbb{c}\mathbb{B} + \mathbb{d}\mathbb{D} = \mathbb{I} \tag{F.5}$$



K. Ohno et al., *Computational Materials Science*,
https://doi.org/10.1007/978-3-662-56542-1

We can eliminate $\mathbb{C}$ from the third equation:

$$\mathbb{C} = -\mathbb{d}^{-1}\mathbb{c}\mathbb{A} \tag{F.6}$$

This is an important relation expressing the off-diagonal element of the inverse in terms of its diagonal element. Substituting back in the first equation, we can find $\mathbb{A}$ as:

$$\mathbb{A} = [\mathbb{a} - \mathbb{b}\mathbb{d}^{-1}\mathbb{c}]^{-1} \tag{F.7}$$

Likewise, after elimination of $\mathbb{B}$ from the second equation and substituting in the fourth one, we find:

$$\mathbb{D} = [\mathbb{d} - \mathbb{c}\mathbb{a}^{-1}\mathbb{b}]^{-1} \tag{F.8}$$

So that we can finally write:

$$\mathbb{M} = \begin{pmatrix} [\mathbb{a} - \mathbb{b}\mathbb{d}^{-1}\mathbb{c}]^{-1} & -\mathbb{a}^{-1}\mathbb{b}\,[\mathbb{d} - \mathbb{c}\mathbb{a}^{-1}\mathbb{b}]^{-1} \\ -\mathbb{d}^{-1}\mathbb{c}\,[\mathbb{a} - \mathbb{b}\mathbb{d}^{-1}\mathbb{c}]^{-1} & [\mathbb{d} - \mathbb{c}\mathbb{a}^{-1}\mathbb{b}]^{-1} \end{pmatrix} \tag{F.9}$$

## F.2 Downfolding and Green's Functions

The above general relations can be used in the context of Green's functions to define concepts such as self-energy, and transfer matrix relations allowing us to derive an explicit relation for the off-diagonal matrix elements of the Green's function in a tight-binding (local) representation.

For a given Hamiltonian matrix $\mathcal{H}$, let us define the corresponding retarded Green's function (GF) as $G(z) = [z\mathbb{I} - \mathcal{H}]^{-1}$. The fact that $\mathcal{H}$ is hermitian (with real eigenvalues) and the complex number $z = E + i\eta$ has a positive non-zero imaginary part $\eta$, guarantees that the Green's function is non-singular and well-defined.

Let us now assume that the Hilbert space of our system is divided into two separate subspaces called $a$ and $b$.[6] The partitioning formula allows us to write the projection of the total GF into the subspaces $a$ and $b$ in terms of the GFs of the isolated (or uncoupled) subspaces $a$ and $b$, labeled by $g_a(z) = [z\mathbb{I} - \mathcal{H}_{aa}]^{-1}$ and $g_b(z) = [z\mathbb{I} - \mathcal{H}_{bb}]^{-1}$.

$$G_{aa}(z) = [z\mathbb{I} - \mathcal{H}_{aa} - \Sigma_a]^{-1} \tag{F.10}$$

$$G_{bb}(z) = [z\mathbb{I} - \mathcal{H}_{bb} - \Sigma_b]^{-1} \tag{F.11}$$

[6] We are now switching notation from the previous section and associating $a$ and $b$ with the subspaces of $\mathcal{H}$.

$$G_{ba}(z) = -g_b(z)\mathcal{H}_{ba}G_{aa}(z) \tag{F.12}$$

$$G_{ab}(z) = -g_a(z)\mathcal{H}_{ab}G_{bb}(z) \tag{F.13}$$

where the **self-energies** $\Sigma_a$ and $\Sigma_b$ are defined as:

$$\Sigma_a(z) = \mathcal{H}_{ab}\, g_b(z)\, \mathcal{H}_{ba} \tag{F.14}$$

$$\Sigma_b(z) = \mathcal{H}_{ba}\, g_a(z)\, \mathcal{H}_{ab} \tag{F.15}$$

Physically, they represent the effect of coupling the subspace $a$ to subspace $b$, as in the GF ((F.10) and (F.11)), they appear as a $z$-dependent correction to the original isolated Hamiltonian. Representing the Hamiltonian of the coupled system, projected on the subspace $a$ by $\mathcal{H}_{aa} + \Sigma_a(z)$ is called **downfolding**. The roots of $\mathrm{Det}(z\mathbb{I} - \mathcal{H}_{aa} - \Sigma_a(z)) = 0$ with respect to $z$ are the exact same roots as the full Hamiltonian even though the matrix $\mathcal{H}_{aa} + \Sigma_a$ is of smaller size.

## F.3 Recursion Relations Satisfied by Off-Diagonal Elements of *G*

As an illustration of the use of that partitioning formula, we will consider the calculation of the off-diagonal matrix elements of the GF of a one-dimensional (1D) quantum wire. In 1D transport problems, for a wire of length $N$ unit cells, one needs to compute $G_{1N}(z)$ as the transmission is proportional to the square of this number. We can use the partitioning formula to recursively obtain $G_{1N}$ as follows:

Consider (F.13) which provides the off-diagonal $G_{ab}$ as a function of $G_{bb}$ and $g_a$, both of which can be calculated, once the related self-energies are known. For a wire of $N$ unit cells, connected to two semi-infinite leads, we can consider the cells $-\infty, \ldots, -2, -1, 0, 1$ to span the subspace $a$ and the remaining $2, 3, \ldots, N, N+1, \ldots, \infty$ cells spanning the subspace $b$, so that

$$G_{ab} = -g_1\mathcal{H}_{12}G_{bb} \tag{F.16}$$

The cells are defined in such as way that only the nearest neighbor cells have non-zero Hamiltonians and second neighbor cells are not coupled through $\mathcal{H}$. Here $g_1$ is to be understood as the surface GF of the semi-infinite system including sites 1, 0, $-1, -2, \ldots$ projected on site 1. In this geometry, it is assumed that the leads which include cells $-\infty, \ldots, 0$ and cells $N+1, \ldots, \infty$ have a periodic structure, i.e. $\mathcal{H}_{i,i+1}$ and $\mathcal{H}_{i,i}$ do not depend on $i$ for $i < 1$ or for $i > N$.

| ... | −1 | 0 | 1 | 2 | N − 1 | N | N + 1 | N + 2 | ... |
|---|---|---|---|---|---|---|---|---|---|

### F.3.1 Semi-infinite Leads: The Lopez-Sancho Method

Here we will give the recursive relations that give the self-energy due to the contact with the leads. Consider a uniform semi-infinite system where onsite matrix elements $\mathcal{H}_{i,i}$ are all equal to $\mathcal{H}_0$ and nearest neighbor hopping integrals $\mathcal{H}_{i,i+1}$ are all equal to $\mathcal{H}_{-1,0}$. The surface GF denoted by $G_{00}$ satisfies

$$G^s_{00}(z) = [z\mathbb{I} - \mathcal{H}_0 - \mathcal{H}_{0,-1}G^s_{-1,-1}(z)\mathcal{H}_{-1,0}]^{-1}$$

But because of uniformity, we must have $G^s_{00}(z) = G^s_{-1,-1}(z)$, leading to an equation which can be iteratively solved to produce the surface GF and surface self-energy:

$$G^s_{00}(z) = [z\mathbb{I} - \mathcal{H}_{00} - \Sigma^s_0(z)]^{-1}; \quad \Sigma^s_0(z) = \mathcal{H}_{0,-1}G^s_{00}(z)\mathcal{H}_{-1,0} \tag{F.17}$$

The same relations are valid for the right lead starting from cell $N+1$. $G^s_{N+1,N+1}(z) = [z\mathbb{I} - \mathcal{H}_{N+1,N+1} - \Sigma^s_{N+1}(z)]^{-1}; \ \Sigma^s_{N+1}(z) = \mathcal{H}_{N+1,N+2}\, G^s_{N+1,N+1}(z)\, \mathcal{H}_{N+2,N+1}$

The above equations have to be solved iteratively or self-consistently. In 1984, Lopez-Sancho et al. [1] presented an efficient numerical method to solve (F.17) with geometrical convergence. This method is now widely implemented in most quantum transport codes based on Landauer's formula, and we refer the reader to the original paper for its implementation.

### F.3.2 Iteration Equations Needed for $G_{1N}$

Once the surface GF are known, we can proceed to compute first $g_1$ and then $G_{bb}$. $g_1$ includes the sites from $-\infty$ to 1, so using the partitioning formula, it can be written as:

$$g_1(z) = [z\mathbb{I} - \mathcal{H}_{11} - \mathcal{H}_{10}G^s_{00}(z)\mathcal{H}_{01}]^{-1} \tag{F.18}$$

In the same way, we have

$$g_N(z) = [z\mathbb{I} - \mathcal{H}_{NN} - \mathcal{H}_{N,N+1}G^s_{N+1,N+1}(z)\mathcal{H}_{N+1,N}]^{-1} \tag{F.19}$$

Going back to (F.16) we see that for the quantum wire problem it can be written as[7]:

$$G_{1N} = g_1\mathcal{H}_{12}\, G_{2N} = g_1\mathcal{H}_{12}\, g_2\mathcal{H}_{23}\, G_{3N} = \dots \tag{F.20}$$

If we continue this relation, we arrive at the result:

---

[7] The partitioning of the wire is done by cutting between cells 1 and 2, and since $\mathcal{H}_{12}$ can only couple the cell 1 of partition $a$ to cell 2 of partition $b$, on the left we must have $g_1$, GF of the isolated or uncoupled partition $a$, and on the right $G_{2j}$, GF of the whole system $a \cup b$ projected on cells of $b$.

$$G_{1N} = \Pi_{j=1}^{N-1} g_j \mathcal{H}_{j,j+1} G_{NN} \tag{F.21}$$

For intermediate $g_j$ we should use

$$g_j(z) = [z\mathbb{I} - \mathcal{H}_{jj} - \mathcal{H}_{j,j-1} g_{j-1}(z) \mathcal{H}_{j-1,j}]^{-1}$$

Finally, for $G_{NN}$ we need the left and right self-energies:

$$G_{NN}(z) = [z\mathbb{I} - \mathcal{H}_{NN} - \Sigma_L(z) - \Sigma_R(z)]^{-1} \tag{F.22}$$

with

$$\Sigma_L(z) = \mathcal{H}_{N,N-1}\, g_{N-1}\, \mathcal{H}_{N-1,N} \tag{F.23}$$

$$\Sigma_R(z) = \mathcal{H}_{N,N+1}\, G^s_{N+1,N+1}(z)\, \mathcal{H}_{N+1,N} \tag{F.24}$$

The above equations now allow us to compute the off-diagonal elements of the full GF, $G_{1N}$ needed for the calculation of the transmission coefficient. Note that in the above treatment, the size of cells 1 to $N$ can be arbitrary without any need to invoke periodicity. The sizes of the left and right lead cells can also be independent of each other, and as a result, unlike the leads, intercell hopping matrices of the wire need not be square.

## Reference

1. M.P. Lopez-Sancho, J.M. Lopez-Sancho, J. Rubio, J. Phys. F **14**, 1205 (1984)

# Appendix G
# Statistical Analysis of Data

In what follows, we will describe standard methods to analyze statistical properties of data generated typically from a Monta Carlo or a Molecular dynamics run. Let us say we have generated a set of data $X(t)$ where the time $t$ takes $N$ discrete values $t_i = i \times \Delta t; i = 1, \ldots, N$, and we are interested in quantities such as mean, standard deviation, correlation time, and autocorrelation function of $X$. The latter is a useful quantity, related through the fluctuation-dissipation theorem (FDT) to linear response functions. Examples are diffusion coefficient, thermal conductivity, and viscosity, which are respectively related to autocorrelations of velocity, heat current, and stress tensor. These quantities can easily be extracted from MD simulations, and thus the FDT provides a way to compute the response functions from a judiciously prepared molecular simulation.

## G.1 Definitions

For a set of $N$ numbers $x_i = X(t_i)$, the mean $\overline{X}$ and variance $V(X) = \sigma_X^2$ are defined as:

$$\overline{X} = \frac{1}{N}\sum_{i=1}^{N} x_i \tag{G.1}$$

$$V(X) = \overline{(X - \overline{X})^2} = \frac{1}{N}\sum_{i=1}^{N}(x_i - \overline{X})^2 \tag{G.2}$$

It can easily be shown that $V(X) = \overline{X^2} - \overline{X}^2$. Furthermore $\sigma_X = \sqrt{V(X)}$ is called the *standard deviation*, and represents the spread of the data about the mean.

We will use these concepts for a discrete set of data or time series, but they are also applied in the context of equilibrium statistical mechanics to the "thermal average" or equilibrium expectation value of a quantity. We will represent the thermal equilibrium



K. Ohno et al., *Computational Materials Science*,
https://doi.org/10.1007/978-3-662-56542-1

average by brackets $< \ldots >$. This quantity is not exactly known as we always have a finite set of data. The mean extracted from this limited data set is an *estimator* of the true expectation value.

## G.2 Estimation

Let us now assume that the $N$ data points $x_i$ is a random sample taken from a larger population. For instance if the average age of the population of a country needs to be estimated, one will take a large enough but random sample to make an accurate estimate since collecting the data on the whole multimillion population could be very time consuming and expensive. The mean as defined above is then only an estimate of the true mean, and there is an error associated with it, which we will call $\Delta X$. To get an idea of the error in the mean, one needs to sample $N$ other data points from the full distribution, recompute the mean, and repeat this process a large number of times to see the fluctuations in the measured mean. The error in the mean is therefore the standard deviation of the means obtained from different samplings of the data. In other words, it is $\sigma_{\overline{X}}$ which can be written as:

$$\sigma_{\overline{X}}^2 =< \overline{X}^2 > - < \overline{X} >^2 = \frac{1}{N^2} \sum_{i,j=1}^{N} < x_i x_j > - < x_i >< x_j > \tag{G.3}$$

where we used the definition of mean from (G.1), and represent the true average of the whole population by $< \ldots >$. Separating the $i = j$ terms, we get:

$$\sigma_{\overline{X}}^2 = \frac{1}{N^2} \sum_{i=1}^{N} < x_i^2 > - < x_i >^2 + \frac{2}{N^2} \sum_{i=1}^{N-1} \sum_{k=1}^{N-i} < x_i x_{i+k} > - < x_i >< x_{i+k} > \tag{G.4}$$

$$= \frac{1}{N} \sigma_X^2 + \frac{2}{N^2} \sum_{k=1}^{N} (< x_1 x_{1+k} > - < x_1 >< x_{1+k} >)(1 - k/N) \tag{G.5}$$

where in the last line we used time translational invariance, and added the $k = N$ term, which is zero due to $(1 - k/N)$ term.

We deduce that for uncorrelated data, where the last term is zero, the error in the mean becomes

$$\sigma_{\overline{X}} = \Delta X \approx \sigma_X / \sqrt{N} \tag{G.6}$$

The larger the sample size $N$, the smaller the error on the estimated mean, as expected. We have used the approximation sign also because there is no access to the full data, and therefore this number represents our best estimate of the error in the mean given the single sampling performed.

## G.3 Correlations

Now we transition from general statistical concepts to statistical mechanics and MD or MC simulations of physical model systems. Response functions are related, through the fluctuation-dissipation theorem, to correlation functions which one can easily compute from a simulation. The correlation function between two observables $X$ and $Y$ is defined as:

$$C_{XY}(t) =< [X(t+t')- < X >][Y(t')- < Y >] > \tag{G.7}$$

$$=< X(t)Y(0) > - < X >< Y > \tag{G.8}$$

where the expectation value $< \dots >$ now means an equilibrium thermal average. Since at equilibrium, we can assume there is time translational invariance and the correlation function does not depend on the origin of time but only on the time difference between measurements of variables $X$ and $Y$. The term autocorrelation is used when $X = Y$, and in most cases, we will be interested in autocorrelations. If the system is *ergodic*, one can replace the thermal or "ensemble average" by a time average, meaning:

$$C_{XY}(t) = \lim_{T\to\infty}\frac{1}{T}\int_0^T (X(t+t') - \overline{X})(Y(t') - \overline{Y})dt' \tag{G.9}$$

In practice, since simulations are finite and limited in time, this is not always the case and one needs to perform both a time average and an ensemble average to obtain reliable results. By ensemble average we mean the same simulation needs to be repeated several times (as many as needed) with different initial conditions, followed by an averaging over these runs. Such runs, since independent from each other are *uncorrelated*, and one may therefore reduce the error bars in various estimates as we will shortly explain.

In practice, the run has a finite number of steps $N$ and one has to be careful in interpreting the results of a correlation calculation at large times. In the discretized case, corresponding to a real simulation, one can write the correlation function of a single run as:

$$C_{XY}(n) = \frac{1}{N-n}\sum_{i=1}^{N-n}(x_{n+i} - \overline{X})(y_i - \overline{Y}); \qquad (n \geq 0) \tag{G.10}$$

This will lead to unreliable results for large values of $n$ approaching $N$ as there will be very few terms in the sum and thus leading to large fluctuations, while if $N$ is large enough the correlations are expected to have decayed to zero as $n$ nears $N$. Ideally, $N$ should be chosen to be much larger than the *relaxation time* also called *correlation time*, the time needed for the correlations to decay. In this case, one does not need to take $n$ to values as large as $N$. Performing an ensemble average over many runs with different initial conditions will make the results smoother and decay faster to zero.

Another possibility to compute correlation functions is to use the fast Fourier transform(FFT). A theorem about the Fourier transforms (FT) states that the FT of the correlation is the product of the fourier transforms of the functions: $C_{XX}(\omega) = \frac{1}{N}|X(\omega)|^2$. Therefore one only needs to perform the FFT of the function $X(t)$, square its magnitude for each frequency $\omega$, and transform back to the time domain. The total cost for these operations is proportional to $NlogN$ whereas the direct computation of the autocorrelation is in $N^2$.

## G.4 Correlation Time

Typical correlation functions have exponential-like decay such as $e^{-t/\tau}$. In this case, the decay time $\tau$ is called the correlation time. Another meaning of the correlation time is that data separated by times larger than $\tau$ are uncorrelated or independent of each other. To obtain the correlation time in general, we use the following definition:

$$\tau = \int_0^{\infty} C_{XX}(t)/C_{XX}(0)dt \tag{G.11}$$

which is consistent with an exponentially decaying function ($\int_0^{\infty} e^{-t/\tau}dt = \tau$). Note that $C_{XX}(0) = \overline{X^2} - \overline{X}^2 = \sigma_X^2$. One property of the autocorrelation function ($X = Y$) is that, due to time translational invariance, it is even.

We see the autocorrelation function also appears in (G.5). For correlated data, it follows that the error in mean is larger than the standard deviation divided by $\sqrt{N}$. The extra term is the auto correlation $C_{XX}(k)$ integral if we take $N$ much larger than the correlation time so that the term $(1 - k/N)$ can be replaced by 1. As a result, the error in mean for correlated data ($N\mathbf{G}\tau$) can be written as:

$$\sigma_{\overline{X}}^2 = \frac{1}{N}\sigma_X^2 + \frac{2}{N}\sum_{k=1}^{N} C_{XX}(k) \tag{G.12}$$

If we go to the continuum limit, the second term becomes almost an integral. Approximating the integral from 0 to $\infty$ by the Simpson rule, we can write

$$\tau \approx \frac{1}{2} + \sum_{k=1}^{N} C_{XX}(k)/C_{XX}(0)$$

As a result the error in mean of a correlated set of data becomes

$$\sigma_{\overline{X}}^2 = \frac{1}{N}\sigma_X^2 \times (2\tau - 1) = \frac{1}{N_{eff}}\sigma_X^2 \quad \text{for } \tau \ll N \tag{G.13}$$

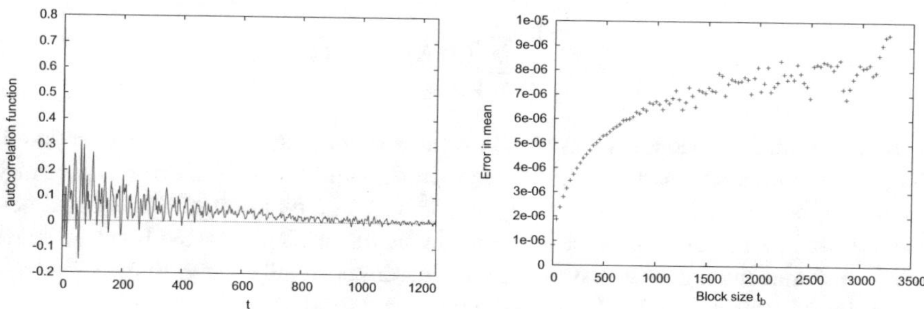


**Fig. G.1** Left: Autocorrelation of a set of MD data; Right: result of block averaging for the same set of data

We can write it as the variance divided by some effective total time $N_{eff} = N/(2\tau - 1)$, meaning that after every $2\tau - 1$ data are uncorrelated and we can perform the standard deviation from the $N_{eff}$ uncorrelated data and deduce the error on the mean from the standard formula (G.6) but with $N$ replaced by $N_{eff}$.

## G.5 Block Averaging

For a data set generated from an MD simulation, or a so-called time series, data points neighbor in time will be correlated, and the error estimation from a finite number of data points will be, as we just showed, incorrect, whereas if we were to take the same number of data but separated by a large time difference, they would be less correlated, and the error estimate will be more accurate. If we knew the correlation time, ideally we would take $N$ data points separated by $\tau$ to evaluate $\sigma_{\overline{X}}$, but the knowledge of the correlation time itself requires a relatively heavy computation. For this reason, one usually resorts to the block averaging method. It consists in separating the time series into $N_b$ blocks each of length $t_b$ such that $N = N_b \times t_b$. The data within each block $b$, $(b = 1, \ldots, N_b)$ will be averaged as (Fig. G.1):

$$< X_b >= \frac{1}{t_b} \sum_{i=1}^{t_b} x_{(b-1)t_b+i}$$

We then calculate the mean and standard deviation from these block averages. The mean of course will not depend on the block size, as ultimately all $N$ data points will be used.

$$\overline{X} = \frac{1}{N_b} \sum_{b=1}^{N_b} < X_b >= \frac{1}{N_b \times t_b} \sum_{i=1}^{N_b \times t_b} x_i = \frac{1}{N} \sum_{i=1}^{N} x_i$$

The variance will however be:

$$\sigma^2_{X_b} = \frac{1}{N_b}\sum_{b=1}^{N_b}(< X_b > -\overline{X})^2$$

If the block size $t_b$ reaches the correlation time so that data becomes uncorrelated, the error in the mean should reach an asymptotic value. Since the error is given by $\sigma^2_{\overline{X}} = \frac{1}{N}\sigma^2_X \times (2\tau - 1)$, we can plot $(2t_b - 1)\sigma^2_{X_b}/N_b$ as a function of increasing block length $t_b$, to see a saturation to the real error in the mean. Typically the function starts from $\frac{1}{N}\sigma^2_X$ and increases as the block size increases, finally saturating to $\frac{2\tau-1}{N}\sigma^2_X$. The point of saturation will also indicate the correlation time.

# Index


**A**

AB block-copolymer, 294
Ab initio, 7–8, 56, 231, 232
  molecular dynamics (MD), 7, 25, 65, 98, 117–129, 130–134, 232
  tight-binding method, 80
Abell potential, 238
Absorption spectrum, 83
Acceptance ratio method, 288
Adaptive simulated annealing, 399
Adaptive umbrella sampling, 289
Adiabatic approximation, 91, 119, 133, 137, 138
Adiabatic connection formula, 41, 42
Adiabatic nucleation, 312
Adiabatic potential surface, 117, 119
Adsorption growth, 319
Alder transition, 357
Alkyl group, 326
All-electron approach, 72–74, 89, 92, 120
All-electron mixed-basis approach, 60–62, 85, 86, 89, 124, 126–129, 130–131, 162, 164
Aluminum, 130, 164
Amorphous, 311
Anderson, 201
  localization, 19, 291
Angular momentum, 63
Anitcommutation relation, 150
Annihilation operator, 95, 143, 146, 148, 150, 153, 154
ANNN $n$-vector model, 302
ANNNXY model, 302
Antiferromagnetic coupling, 301, 302, 310
Antiferromagnetic phase, 301
Antiferromagnetic Potts model, 310
Antisymmetry of wavefunction, 15
APW method, 67, 68, 71, 72, 124
Argon, 89, 162, 164, 237
Atomic orbital, 18, 56, 59, 61, 80, 199
Atomic-Sphere Approximation (ASA), 71, 72
Atomic unit, 8
Atomization energy, 41
Auger decay rate, 175
Augmentation, 67
Augmented Plane Wave (APW), 67
Augmented Spherical Wave (ASW) method, 72, 81
Autocorrelation, 99, 100, 105, 107
Axial Next-Nearest-Neighbor Ising (ANNNI) model, 301

**B**

Ballistic deposition model, 322–324,
Band
  energy, 30–31,
  structure, 44, 45, 62
    of nickel, 62
BCC lattice, 43, 45, 47, 241, 250, 301
BCF theory, 317
BCS theory, 134
Bead jump algorithm, 297
Becke three-parameter Lee-Yang-Parr (B3LYP), 42
Becke three-parameter Lee-Yang-Parr (B3LYP) functional, 42
Becke's three-parameter exchange functional, 41
Berry phase, 47
Bethe–Salpeter Equation (BSE), 173–175
Biological macromolecule, 363
Bloch



K. Ohno et al., *Computational Materials Science*,
https://doi.org/10.1007/978-3-662-56542-1

function, 44, 48
state, 37, 44
sum, 46, 51, 56, 57
basis function, 56
theorem, 44, 45, 47, 50
Block Davidson method, 121, 396–397,
Block Davidson (BD), 121
Block spin transformation, 308
Bohr radius, 8
Boltzmann
distribution, 253, 268, 270, 279
equation, 268
weight, 278, 279, 299
Boltzmann equation, 104, 106, 107
Boltzmann theory, 103
Bond-fluctuation model, 296
Bond percolation model, 290
Born–Oppenheimer (BO)
approximation, 8, 10, 117, 119, 133, 136
surface, 117, 119, 120, 123, 125, 127
Bose–Einstein
condensation, 347
statistics, 11, 342
Bose liquid, 341
Boson, 9, 341–343, 345–347, 349, 350
Boundary scattering, 104
Branching, 344
Brownian motion, 262, 277
Brownian particle, 262
Broyden's method, 80, 397–398,

**C**

$C_2$, 126
$C_3N_4$, 164
$C_{60}$, 8, 73, 125–128, 164, 220
isomer, 127
$C_{84}$, 125
$C_{70}$, 125, 126
$C_{82}$, 125, 128
$C_{84}$, 128
Cahn–Hilliarad equation, 314
Canonical distribution, 279
Canonical ensemble, 358
Carbon, 199, 201, 203, 218, 219, 224
diamond, 85, 89
Nanotube (CNT), 98, 126
onion, 126
Carbon Nanotube (CNT), 126
Car–Parrinello method, 117–120, 121–123, 124, 125, 130–134,
Catalyst, 326
Catalytic behavior, 132
CCA model, 292, 324–331
CdS, 164
Cellular automaton, 264
Cellular method, 65
Chapman–Kolmogorov equation, 266
Character, 53
table
for $C_{3v}$, 55
for $O_h$, 55
Character system, 53
Character table, 54
Chelikowsky–Phillips potential, 239–240,
Chemical environment, 73, 88
Chemical hardness, 216
Chemical potential, 23, 24, 132, 286–287, 314
Chemical shift, 73, 89–91,
Choleski decomposition, 61, 69
Class, 50
Classical $XY$ model, 306–308
Classical Heisenberg model, 306, 307
Classical Molecular-dynamics (MD) simulation, 232
Classical particle system, 283–285,
Classical potential, 126, 231–232,
Classical spin system, 297–311,
Clock model, 307, 308
Cluster algorithm, 303
Cluster–cluster aggregation model, 324–331
Cluster expansion, 232, 241–244,
Cluster Variation Method (CVM), 232, 242, 253, 298
Coarse-grained model, 232, 242, 244
Coarse-graining, 262, 307–308, 331
Coherent Anomaly Method (CAM), 300
Coherent Potential Approximation (CPA), 134–136
COHSEX approximation, 161
Colloidal aggregation, 292, 319, 325
Commensurate–incommensurate phase transition, 302, 307
Complete Active Space Self-Consistent-Field (CASSCF), 20
Compton profile, 73
Condensation, 307, 311
Condensation-polymerization reaction, 292, 327
Conduction Band Minimum (CBM), 35, 142
Conductivity tensor, 97
Configuration Interaction (CI), 8, 19, 21, 145
Conjugacy relation, 394
Conjugate-Gradient (CG), 80, 121, 125, 207, 208, 210, 393–396

direction, 394
method, 121, 122, 207, 208, 210, 393–396
Connolly–Williams approximation, 232, 242–244
Conserved order parameter, 313
Contraction, 115, 116
Copper cluster, 202, 227
Coulomb, 365
attraction between electron and hole, 174
force
between nuclei, 120
hole, 160, 161, 169–170
integral, 16, 17, 215, 217–219
interaction, 9, 17, 18, 36, 173, 285
between ions, 204, 232, 233
electron–electron —, 10, 12, 14, 15, 17–19, 31, 160
energy, 36
operator, 17
potential, 18, 61, 63, 74, 177, 214, 215, 285
between nuclei, 31–32, 119
caused by nucleus, 10, 18, 64, 65, 92
sum, 285
system, 284, 363, 364
term, 15, 36, 214, 215, 217
Coupled Cluster (CC) theory, 8, 21
Covalent bond, 237–239
$CP^{n-1}$ model, 311
Creation operator, 95, 143, 146, 148, 150, 153, 154
Critical anomaly, 300, 308
Critical exponent, 272, 291, 292, 294, 300, 301, 307, 308, 323
Critical nucleus, 313
Critical point, 299–301, 308, 311
Critical radius, 311
Critical slowing down, 300, 315
Critical temperature, 261, 272, 299, 301
Crossover, 400
CrosswhiteBacon,EvenblyVidal,Schollwock, 350
Crystal, 43
growth, 231, 311, 313, 316–319
symmetry, 49
$Cu_3Au$, 252, 298
CuAu alloys, 252
Curie temperature, 261
Cusp condition, 177
Cusp-like anomaly, 302
Cutoff
energy, 46, 64, 65, 127, 131
radius, 63, 64, 76, 130, 200, 202, 207, 208, 211, 358

**D**

Darwin term, 74
Debye-Grüneisen model, 253
Defect, 129
Dendrite, 317, 319, 320
Dense-branching morphology, 321
Density Functional Theory (DFT), 8, 21–25, 33, 34, 36, 72–74, 76, 119, 136, 138
of electron kinetic energy, 25
time-dependent —, 24, 108–109
Density matrix, 23, 24, 75–77, 79, 80, 112–114, 138, 145
Density Matrix Renormalization Group (DMRG), 112–115
Density of States (DOS), 21, 49, 84, 132
Density matrix, 205, 207, 208, 210, 211, 216
des Cloizeaux's $c^{9/4}$ rule, 295
Detailed balance, 268, 280, 281
Deterministic model, 262
DFT+U, 38
Diamagnetic susceptibility, 87–89
Diamagnetic term, 97
Diamond, 85, 89
Dielectric constant, 83
figure for Si, 86
Dielectric response, 83–87
Diffusion, 105, 129, 130, 231, 232, 298
constant, 267, 289
equation, 267, 268, 282
Monte Carlo method, 342–346
of hydrogen in Si crystal, 130
velocity, 267
Diffusion-limited, 318
aggregation, 319
CCA model, 292, 325, 330
growth, 318
Diffusion Monte Carlo (DMC) method, 342–346
Dimer model, 308
Dipole approximation, 86
Dipole–dipole interaction, 234, 236
magnetic —, 297
Dipole moment, 234, 235
Dirac equation, 9, 74
Directed percolation model, 291
Direct method, 96
Direct term, 15
Discrete variational method, 57
Dislocation, 129

Distribution function, 265
Di-vacancy, 130
Divide-and-conquer method, 75, 77
Double bond, 127
Drude function, 98
Duality transformation, 304
DV–$X\alpha$ method, 57
DX center, 129
Dynamical matrix, 371, 377–379, 381
Dynamical Mean Field Theory (DMFT), 109–111
Dynamic critical exponent, 300
Dynamic scaling, 314
Dynamic variables, 263
Dyson equation, 135, 148, 149, 155, 158, 170

**E**
Easy axis, 297
Eden model, 322
Effective mass, 81–83
  approximation, 81, 173
  tensor, 81–83
Effective medium theory, 240
Effective one-electron wavefunction, 28
Eight-vertex model, 308
Electrical conductivity, 96–98
Electric susceptibility, 83–87, 98, 235
Electron
  correlation, 8, 16, 18–21
  gas, 18, 22–24, 26
  kinetic energy, 10, 25–28
Electron Affinity (EA), 35, 142, 164, 165, 172
Electron–hole interaction, 173
Electron-phonon
  coupling, 95
    constant, 95
  interaction, 47, 94, 95, 99, 136, 137, 142
  scattering, 99
  self-energy, 95
Embedded-atom potential, 240–241, 252, 299
Empirical, 7, 232
Endohedral fullerene, 128
Energy
  band, 45
  fluctuation, 278
  gap, 47, 49, 87
  level, 11, 28, 30, 37, 47, 52, 69, 73, 81
Enrichment algorithm, 293
Ensemble average, 271
Equation of motion, 358
Equilibrium Molecular Dynamics (EMD), 105–106
Equilibrium shape, 316
Ergodic, 269
Ergodicity, 269–272
Ergodic time, 271, 272
Etchant ballistic model, 323
Evaporation, 261
Ewald sum, 32, 71, 91, 120, 233, 285, 364
Exchange, 216, 218
  algorithm, 298
  at surface, 303
  constant $J$, 73, 297
  energy, 36
  ferromagnetic —, 300
  hole, 169–170
  integral, 16, 17
  interaction, 36, 297, 299
  nearest-neighbor —, 299–301
  of oscillator energy in Einstein model, 268
  operator, 17
  rule of dynamic variables, 282
  spatially changing —, 300
  term, 15, 18, 28, 169
Exchange–correlation, 16, 36, 204, 219
  Ceperley–Alder's —, 23, 28
  effect, 216
  energy, 22, 31
    density, 31
    functional, 26, 36
    in local approximation, 26, 28
    of electron gas, 26, 29, 32
  potential, 25–28, 73, 214
  spherical average of — hole, 26
  term, 214

**F**
Facet, 316
Fast Fourier transformation, 78, 364
Fast multipole method, 365–368
FCC
  crystal, 130
FCC lattice, 43, 45, 47, 241, 252, 298, 301
$Fe_3Al$, 130
FePt alloy, 253
  cluster, 257
Fermi
  energy, 11, 209
  level, 11, 21, 49, 62, 73
  liquid, 341

Fermi–Dirac
distribution, 127
function, 84, 127
statistics, 11, 342
Fermion, 9, 341, 342, 345, 346, 348, 350
Ferroelectricity, 96
Ferromagnet, 272
Ferromagnetic state, 272
Feynman diagram, 17
Fictitious electron mass, 118, 119
Field operator, 143
Finnis–Sinclair potential, 241, 252, 299
First Brillouin zone, 47
First principles, 7
Fixed-node approximation, 345
Fixed point, 308
FLAPW method, 71, 124
Flat surface, 316–319, 324
Fluctuation, 262, 278, 301, 307
Fluctuation–dissipation theorem, 97, 159, 282, 314
Fock
operator, 15, 17
term, 15
Fokker–Planck equation, 268
Force constant, 371, 372, 376, 378, 379, 381
Fourier
series, 45
transformation, 46
Fourier interpolation, 353–356
Fractal, 291, 300, 308, 319–331
dimension, 321, 325
pattern, 291
Frank–van der Merwe mode, 317
Free-electron limit, 47, 48, 69
Free energy, 132–133, 221, 251, 270, 282, 286–289, 314, 316
functional, 282
Frustrated antiferromagnetic phase, 298, 301
$f$-sum rule, 86–88
Full CI, 20
Fullerene, 8, 126, 128
Full-potential, 65
KKR method, 71, 130
LAPW method, 71
LMTO method, 72

**G**

$G_0W_0$, 163, 165, 167
GaAs, 89, 164
Gadolinium, 164
Galitskii–Migdal formula, 147, 153, 155
GaN, 164
GaP, 89
Gaussian-Type Orbital (GTO), 18, 57, 175
Gaussian-type potential, 240
Gear method, 357
Gel, 292
Gene, 400
Generalized Gradient Approximation (GGA), 33–346
Generalized Plasmon-Pole (GPP) model, 159, 164
Genetic algorithm, 400
Germanium, 74, 164, 203
GGA+U, 38
Giant magnetic moment, 134
Giant Magnetoresistance (GMR), 134
Gibbs–Bogoliubov inequality, 287
Gibbs-ensemble Monte Carlo simulation, 286
Gibbs–Thomson effect, 311
Glass, 311
Glass transition, 271, 315
Glauber
algorithm, 280, 298, 301
dynamics, 313
Grain boundary, 129
Gram–Schmidt orthogonalization technique, 120, 121, 123
Grand canonical ensemble, 286, 358
Greengard–Rockhlin method, 365–368
Green–Kubo (GK) formula, 99, 105, 107
Green's function, 18, 69, 75–76, 134, 135, 141–180, 343, 344, 351
method, 69
Monte Carlo method, 343
one-particle —, 17, 173
two-particle —, 173, 175
Green's Function Monte Carlo (GFMC) method, 343, 345, 346
Group, 49, 341
theory, 49–55
Gutzwiller function, 178
GW Approximation (GWA), 74, 168–170, 173–175
GW$\Gamma$, 158, 170, 171, 176, 177

**H**

Hard sphere system, 284
Harmonic approximation, 107, 221, 377
Harris functional method, 25, 81
Harrison, 200, 203
Hartree

approximation, 15, 22
operator, 15, 17
term, 15
Hartree–Fock, 19, 56, 199, 216, 346
approximation, 8, 13–22, 28, 36, 169
equation, 241
exchange energy, 36
operator, 15, 20
unrestricted —, 219
Haydock, 204, 211
HCP lattice, 301
Heat bath method, 280
Heisenberg model, 305
Heisenberg representation, 148
Heisenberg's uncertainty principle, 7
Heitler–London
picture, 19
theory, 19, 349
Helium, 87, 237
Hellmann–Feynman, 138, 199, 205, 221, 222, 359
force, 92, 120
theorem, 92, 120
Helmholtz equation, 71, 72
Herman–Skillman atomic code, 61, 66
Heyd–Scuseria–Ernzerhof (HSE), 42
Heyd–Scuseria–Ernzerhof (HSE) functional, 42
HgSe, 164
Higher-order singular value decomposition (HOTRG), 115, 116, 305, 350
High-$T_c$ superconductor, 19
High-temperature series expansion, 301
Histogram method, 286, 303
Homogeneous nucleation, 312
HOMO level, 35, 73, 87, 142
Hubbard, 214, 218
Hubbard–Stratonovich transformation, 350
Hund rule, 12
Hybrid functional, 39–43
Hydrogen Fluoride (HF), 90
Hydrogen molecule, 18
Hyperfine interaction, 73
Hyperscaling relation, 301

**I**

Ice model, 308
Idempotency, 76, 80
Imperfection, 98, 129
Importance sampling, 278–280, 340, 344, 348, 349
Impurity, 129
Impurity center, 129
Impurity scattering, 99, 104
Incomplete basis-set force, 123
Independent Electron-Pair (IEP), 21
Independent Electron-Pair (IEP) approximation, 21
Inhomogeneous electron system, 18
Inhomogeneous nucleation, 312
Interaction representation, 96
Interchange, 9
Intermetallic compound, 130
Interstitial atom, 129
Inverse Participation Ratio (IPR), 126
Inverse photoemission, 141, 142
Inversion, 51, 55
Ionization Potential (IP), 35, 41, 142, 164, 165, 172
IR, 221, 226
Iron, 130–131
Irreducible first Brillouin zone, 52
Irreducible representation, 53
Irrelevant variable, 307
Ising model, 297–305
Isothermal nucleation theory, 311
Iterative diagonalization method, 121
Itinerant-electron picture, 19

**J**

Janak's theorem, 28, 34
Jastrow function, 178, 341, 342
Jellium model, 18, 163
Johnson's sum rule, 159

**K**

Kawasaki dynamics, 282, 313
Kawasaki's (exchange) algorithm, 282, 298
Kinetic Ising model, 297
Kirkwood's method, 132, 287
Klein, 155
Kohn, 208
Kohn–Sham
equation, 25, 27, 31, 63
Hamiltonian, 27, 75, 78, 200
local-density approximation, 26
potential energy, 80
wavefuncion, 28
kondo effect, 129
Koopmans theorem, 28
Korringa–Kohn–Rostoker (KKR) method, 70, 72
Kosterlitz–Thouless transition, 307
$\boldsymbol{k} \cdot \boldsymbol{p}$ method, 81

KPZ theory, 323
Kramers–Kronig relation, 83, 84
Krypton, 89, 128, 237
Kubo
formula, 97
Kubo–Greenwood formula, 98, 107, 108

**L**

Ladder diagram, 169, 173
Lagrange multiplier, 14, 27, 79, 118, 120, 209, 243
Lagrangian, 117
λ transition, 307
Lamellar phase, 297, 316
Landauer formula, 98
Langevin equation, 282, 283
Laplace equation, 71
Lattice
animal, 292
gas model, 232, 242, 244–258, 298
BCC —, 252
FCC —, 252, 298
vector, 43
vibration, 98
Lattice Dynamics (LD) theory, 103
Layer growth, 318
LDA+U, 38
Leap-frog method, 357, 360
Legendre polynomial, 64
Lehmer method, 273
Lennard–Jones
parameter, 237
potential, 235–237, 284, 357
system, 286, 341
Li@$C_{60}$, 128
LiCl, 163
Lifetime, 103–105, 142
Lifshitz point, 302
$Li_3N$, 130
Linear Combination of Atomic Orbitals (LCAO), 18, 56, 61, 75, 199
approach, 18, 56–62, 63, 123
basis, 80, 209, 211
crystal approach, 57
density-functional approach, 56, 75
HF approach, 216
method, 237
MO approach, 57
Linear Muffin Tin Orbital (LMTO), 164, 174
ASA method, 72
method, 71, 72, 81, 92
Linear response, 81–96, 222, 223
theory, 91, 109, 156
Linearized Augmented Plane Wave (LAPW) method, 68, 77, 92, 124
Line minimization, 394
Linked-cluster diagram, 17
Linked-cluster theorem, 17
Liquefaction, 311
Liquid, 311
crystal, 310, 363
droplet, 311
molecular —, 288
quantum —, 311
supercooled —, 314
Liquid–solid phase transition, 252, 284, 311
Liquid–vapor phase transition, 311
Liquid–vapor transition, 298
Lithium, 130
Local Density Approximation (LDA), 26, 29, 30, 30, 34, 36, 37, 72, 74, 123, 131, 163, 203, 223–225, 346
Local density of states, 73, 211
Local energy, 340
Localized-electron picture, 19
Localized orbital, 37, 46, 47, 56, 73, 78, 80
approach, 56, 78
Locally self-consistent multiple scattering method, 76
Local-Spin-Density Approximation (LSDA), 26, 30, 36, 62, 74, 130, 131
Locon, 105
$L1_0$ phase, 298
$L1_2$ phase, 298
LUMO level, 35, 73, 87, 142
Luttinger–Ward
functional, 155, 168
theory, 156

**M**

Madelung
constant, 233
energy, 233, 389, 390
Magnetically ordered structure, 26, 65, 73, 297
Magnetic bubble, 297
Magnetic Kerr effect, 81
Magnetic susceptibility, 73, 87–90, 302
nonlinear —, 302
Many-Body Perturbation Theory (MBPT), 8, 141, 157, 178
Markov
chain, 266, 279, 288, 289

process, 265–268, 269, 279, 281, 289, 322
Mass–velocity term, 74
Master equation, 267–269, 281, 282
Matching condition, 63, 67, 71, 72
Mathiessen's rule, 99
Mating, 400
Matrix Product Operator (MPO), 117
Matrix Product States (MPS), 116
Maximally localized Wannier function, 47, 95
$m$-clock model, 307
Mechanical, 357, 374
Mehl, 202, 227
Melting, 231, 252, 261, 307, 313
temperature, 240, 251, 311, 313
Mermin–Wagner theorem, 306, 307
Mesoscopic system, 98
Meta hybrid GGA, 43
Metal-insulator transition, 19, 290
Metallic hydrogen, 60
Metastable, 311
Metropolis algorithm, 280, 284, 301, 340, 343, 348
Metropolis' Monte Carlo simulation, 280, 298
MgO, 164
Micelle, 294
Micelle formation, 363
Microcanonical, 101, 102
ensemble, 102
simulation, 102, 118
Microcanonical ensemble, 358
Microdomain, 297
Microemulsion, 316
Microphase separation, 316
Minimal basis, 56
Minnesota density functional, 43
Mixed-basis, 61
approach, 59–62, 63, 77, 123
pseudopotential approach, 60, 124, 130
Mixed congruential method, 274
M06-L, 43
Mobility, 313
Molecular dynamics, 199, 220, 222, 225, 359, 399
Molecular-Orbital (MO) approach, 18, 57
Møller–Plesset (MP) perturbation theory, 8, 20, 153, 169
Molybdenum, 130
Monte Carlo
integration, 276
renormalization group, 291, 302, 308
step, 284
Monte Carlo Step (MCS), 284
Mott
insulator, 19
transition, 19, 291
MP2, 21
Muffin-pan, 65
Muffin-tin, 65
approximation, 65–72
orbital, 59, 66–69
potential, 59, 65
radius, 65, 67
Muffin-Tin Orbital (MTO), 66, 71
Mullins–Sekerka instability, 317
Multicanonical Monte Carlo method, 289, 297, 302
Multiphonon process, 104
Multiple-histogram method, 286
Multiplicative congruential method, 274
Multipole, 232–235
expansion, 232, 233
field, 365, 366
interaction, 365
method, 365
moment, 366, 368
Multi-reference CI, 20
Multi-Reference single and Double Configuration Interaction (MRDCI), 20, 172, 176, 177
Multi-scale Entanglement Renormalization Ansatz (MERA), 116
Multispin coding method, 303
Multi-spin-flip method, 303
Multi-time-step method, 357
Mutation, 400
M06-2X, 43

**N**

Na@$C_{60}$, 128
Nanometer scale, 125
Nanoscale, 125
Nearly free-electron approximation, 48
Neon, 89, 237
Neumann–Ulam method, 345
Newtonian equation, 117, 119, 127
Nickel, 62, 164
NiO, 164
Non-Abelian, 50
Nonadiabatic effect, 133
Non-Equilibrium Molecular Dynamics (NEMD), 106
Non-Lehmer method, 274

Nonlinear $\sigma$ model, 305
Nonlinear susceptibility, 81
  magnetic —, 302
Non-Markov process, 266, 269, 324
Nonorthogonal, 202, 205, 214, 221, 227
Nonreversal random walk, 295
Nonsymmorphic group, 51
Normal spin orbital, 145
Nosé, 362
NTV ensemble, 284
Nuclear Magnetic Resonance (NMR), 89
Nucleation, 298, 311–316
Number density operator, 24
Numerical Atomic Orbital (NAO), 18, 61, 127, 130
Numerical integration, 276, 357
Numerical orbital, 57
$n$-vector model, 294, 300, 305, 306, 310, 311
NVE ensemble, 357, 358

**O**

Ohm's law, 97
One-electron approximation, 8, 13
One-shot GW approximation ($G_0W_0$), 163, 165, 167
$O(n)$ model, 305, 311
Open-shell, 12
Optical gap, 170, 176
Order–disorder phase transition, 244, 298
Order-$N$ method, 75–81, 207, 211, 212
Orientational glass, 311
Orientational phase transition, 310
Orthogonalized Plane Wave (OPW) method, 58, 63
Oscillator strength, 86, 87, 236
Ostwald ripening, 313
Overcompleteness, 60
Overhang, 316

**P**

Pair potential, 238, 240
Parametrization, 201, 202, 204, 221, 227
Parquet
  diagram, 178, 180
  equation, 179, 180
  theory, 178
Particle–mesh method, 364
Particle–particle method, 363
Partition function, 270
Path-Integral Monte Carlo (PIMC) method, 346–349, 349
Pauli's exclusion principle, 9, 11, 12, 15, 169
PBE0, 42
Percolation, 290–292, 293
  threshold, 290
Perdew–Burke–Ernzorhof (PBE), 33
Perdew–Burke–Ernzorhof (PBE) functional, 33, 34
Perdew–Zunger's interpolation formula, 29
Periodic boundary condition, 357
Perturbation theory, 81–96
Phase field model, 258, 313
Phase separation, 298, 313–315
Phase transition, 231, 261, 363
  commensurate–incommensurate —, 302, 307
  first-order —, 261, 308, 310, 313
  liquid–solid —, 252, 284, 311
  liquid–vapor —, 298, 311
  order–disorder —, 244, 298
  orientational —, 310
  second-order —, 261, 272, 291, 299, 300, 306–308, 311, 314
$\phi^4$ model, 283, 308, 314
Phonon, 202, 203, 221, 222, 359, 375, 376, 382
  dispersion, 91, 378
  frequency, 375, 379, 381
  lifetime, 103–107
  spectrum, 91–96, 96
3-phonon
  scattering, 103, 104
Photoabsorption spectra, 173, 175
Photoelectron, 168
Photoelectron spectroscopy, 141, 142
Photoemission, 141, 142, 168
Piezoelectricity, 96
Plane wave, 18, 46, 59, 64–66, 78
  expansion, 46, 57, 61, 63, 73, 77, 121, 123, 130, 396
Plasmon-pole
  approximation, 159, 162, 164, 175
  model, 159, 164, 172
Point defect, 129
Point group, 50
Poisson equation, 61
Polarizability, 83–87, 235, 237
Polar system, 233–235
Polymer, 363
  AB block-copolymer, 294
  chain, 264, 292–297
  network, 292, 306
Potential renormalization, 232, 244–258, 299
Potts model, 291, 305, 308, 310, 316

glass, 311
Preconditioned conjugate-gradient method, 121, 393, 395
Preconditioner, 395
Predictor–corrector, 141, 361
method, 199, 357, 362
Primitive reciprocal-lattice vector, 45, 50
Primitive translation vector, 43
Probabilistic ballistic model, 324
Projection Monte Carlo method, 350
Projector-Augmented-Wave (PAW) approach, 58, 65, 124, 130
Projector-basis technique, 77, 124
Propagator, 17, 343, 344, 346–349
Propagon, 105, 107
Protein-folding problem, 363
Proton affinity, 41
Pseudopotential, 37, 48, 58, 60, 63–65, 73, 74, 85, 89, 92, 120, 121, 123, 124, 126, 130, 164, 175, 341
nonlocal —, 63, 77, 92, 123
norm-conserving —, 63–65
separable form, 64, 123
Pseudopotential mixed-basis approach, 60, 124, 130
Pseudopotential–PW expansion method, 63, 123, 130, 164, 175
Pseudo-random numbers, 273
Pseudowavefunction, 63, 64
Pulay force, 123
PW91, 33
PW91 functional, 34

**Q**

$q$-state Potts model, 291, 305, 308, 310
Quadrupolar model, 310
Quadrupole moment, 234
Quantum
Monte Carlo method, 339–351
spin model, 349–350
Quantum dot, 98
Quantum fluctuation, 281, 282, 313
Quantum Hall effect, 134
Quantum Monte Carlo (QMC) method, 8, 23, 339–351
Quantum wire, 98
Quasi-harmonic approximation, 253
Quasiparticle (QP), 141
energy, 141, 142, 146–148, 153, 158, 164–168, 170, 175
gap, 177
equation, 147–149, 158, 160, 170
gap, 176
state, 167
theory, 28, 39, 141–180
wavefunction, 143, 146, 148, 153, 158, 170

**R**

Radial wavefunction, 66
Radio-active source, 130
Random force, 281, 282
Random-Phase Approximation (RPA), 21, 29, 158, 168, 170, 173
Random Walk (RW), 267, 271, 277, 279, 289, 293, 319
Rate-determining process, 318
Rate equation, 281
Reaction-limited, 321
CCA model, 292, 325, 330
Real space, 45
Reciprocal-lattice vector, 31, 45, 47, 48, 60, 67, 71, 90, 123
Reciprocal space, 31, 45, 61, 158
Reducible representation, 53
Relativistic correction, 74
Relativistic effect, 10, 74
Relativistic quantum theory, 9
Relativistic Schrödinger equation, 74
Relaxation, 262
time, 282, 289, 291, 296, 300, 331
Relaxation time, 95, 98, 99, 103, 106
approximation, 103
Relevant variable, 307
Renormalization group, 307
field-theoretical —, 301
Monte Carlo —, 291, 302, 308, 310
transformation, 307, 308
Replica Monte Carlo method, 302
Reptation model, 296
Repulsive, 201, 202, 204, 206, 214, 227
Residual vector, 393, 394
Response function, 97
RHEED oscillation, 317
Ring diagram, 28, 158, 168, 174
Roughening transition, 316
temperature, 316
$RP^{n-1}$ model, 310, 311
Ruderman–Kittel–Kasuya–Yoshida (RKKY) interaction, 129
Ryckaert method, 119, 121
Rydberg, 8

## S

Sankey–Niklewski formalism, 81
Scalar relativistic effect, 74
Scale invariant, 291
Scaling
  function, 314, 323
  relation, 272, 291, 301
Scanning Tunneling Microscopy (STM), 73
Schmidt decomposition, 114
Schwoebel
  barrier, 132, 318
  effect, 318
Screened Coulomb interaction, 158–161, 174
Screening, 21, 64, 207
  effect, 169
    in polymer system, 320
Segregation, 231
Self-affine, 322
Self-Avoiding Walk (SAW), 271, 277, 292–295, 306, 330
  algorithm, 294, 295
Self-consistent, 18, 22, 25, 28, 37, 68, 125, 214, 215, 218
  band calculation, 22
  field approach to Hartree–Fock, 18, 56
  potential, 93
  solution of Hartree–Fock, 18
Self-energy, 17, 39, 95, 104, 106, 135, 141, 142, 147–155, 158, 160–162
Self-Interaction, 15, 36
  Correction (SIC), 36–37
Self-Interaction- and Relaxation-Corrected (SIRC) pseudopotential, 37
Self-Interaction Correction (SIC), 36–37
Self-Interaction-Corrected (SIC) pseudopotential, 37
Semi-empirical, 56, 231–232, 238
Semi-infinite system, 302
Semi-relativistic effect, 74
Short time approximation, 343, 344
SiC(001), 164
Sign problem, 349, 350
Silica-gel, 326
Silicic acid, 326
Silicon, 85, 86, 89, 130, 132, 163, 164, 199, 201, 203, 218, 219, 225
Simple Cubic (SC) lattice, 45, 246, 301, 308
Simple sampling technique, 276–278
Simulated annealing, 126, 131, 225–227
Single bond, 127
Singular Value Decomposition (SVD), 243
Singular-Value Decomposition (SVD) theory, 114
Site percolation model, 290
Slater
  determinant, 11, 13, 14, 19, 28, 178, 341, 342
  Type Orbital (STO), 18, 57
Slater–Koster, 200, 203
Slater-Type Orbitals (STO), 18, 57
Slater–Koster
  parameter, 203
Slow dynamics, 289, 314
Smooth surface, 318
Snake model, 296
Sodium, 130, 164
Sol, 292
Sol–gel method, 325
Sol–gel transition, 327
Solidification, 231, 261, 298, 311, 313
Solid-on-Solid (SOS) model, 316
Space group, 49
Special $k$-point method, 49, 346
Spectral method, 56, 65
Spherical Bessel function, 64, 66, 68, 70
Spherical Hankel function, 72
Spherical harmonics, 66
Spherical model, 306
Spherical Neumann function, 70
Spherical wave, 71
Spin, 9, 74
  anisotropy, 74, 349
  coordinate, 9, 10, 15
  glass, 129, 302
  magnetic moment, 12
  system, 264
Spinodal decomposition, 298, 314, 315
Spinodal line, 314
Spinor, 74
Spin–orbit coupling, 10, 55, 74, 81, 83, 349
Spin-spin correlation function, 301
Spiral growth, 318
Spontaneous magnetization, 55, 299, 301
Spontaneous symmetry breakdown, 261, 272
Stability, 220, 221, 227, 228, 374
Stacking fault, 129
Staggered magnetization, 301
Star-polymer, 294, 296
Star of $\boldsymbol{k}$, 51
Steepest-Descent (SD), 120, 121, 127, 131, 207, 208, 225
  direction, 208

method, 120–123, 207, 208, 221, 282, 393
vector, 393, 394
Step free energy, 316
Step growth, 319
Step–kink–terrace mechanism, 317, 318
Stochastic model, 262
Stochastic operator, 267
Stochastic process, 263–265
Stranski–Krastanov mode, 317
Strinati scheme, 175
Sublimation, 261
Substitutional atom, 129
$SU(n)$ model, 311
Supercell, 127, 371
approximation, 127
Superconductivity, 98, 307
Superconductor, 261
Supercooling state, 311
Superfluidity, 307
Surface
adsorption, 131
fluctuation, 323
(free) energy, 316
reconstruction, 131, 258, 319
Surfactant, 315, 316
Suzuki–Trotter formula, 115, 350
Symmetry, 10
$C_{2v}$ —, 126
$C_{3v}$ —, 50, 54
$O_h$ —, 54
axis, 52
crystal —, 49
cubic —, 52, 54
group, 49, 341
icosahedral —, 125
of wavefunction, 15, 51
operator, 49–55
plane, 52
point, 52
translational —, 49, 130
Symmorphic group, 51

**T**
Tausworthe method, 274
Tempering approach, 302
Tensor network, 115–117, 303–305, 350
holographic —, 116
Tensor Renormalization Group (TRG), 115, 305, 350
Tersoff potential, 238–239
Tetrahedron approximation, 252, 253
Thermal conductivity, 99–108
Thermal diffusion, 282
Thermal expansion, 251
Thermal expansion coefficient, 251
Thermal fluctuation, 266, 272, 281, 282, 311, 313
Thermodynamic limit, 281
Thermostat, 362
Thin-film growth, 319
of diamond, 319, 323
Three-center integral, 18
Three-state Potts model, 316
Tight-binding method, 25, 56, 77, 80, 98, 126, 231, 238, 241
Time-Dependent Density Functional Theory (TDDFT), 24, 108–109
Time-dependent Schrödinger equation, 134
Time-ordering operator, 97
Time–reversal symmetry, 54
Titanium cluster, 227
Total energy, 20, 22, 25, 30–31, 33, 34, 37, 73, 78, 87, 90, 91, 117–119, 132, 133
operator, 32, 78
Transferability, 201, 202, 214
Transfer-matrix method, 299, 300, 350
Transition probability, 265
Transition rate, 267, 279, 280
Translational invariance, 43
Translation group, 50
Tri-vacancy, 130
Trotter
decomposition, 350
formula, 346
Two-center integral, 18
Two-particle Green's function, 173, 175

**U**
Ultrasoft pseudopotential, 64–65, 65, 65
Umbrella
potential, 288
sampling, 288–289
adaptive —, 289
Umklapp process, 103, 104
Unitary representation, 53
Unit cell, 357
Universal, 300, 308
Universality, 308
class, 308, 310

**V**
Vacancy, 129
Vacuum level, 142

Valence Band Maximum (VBM), 35, 142
van der Waals interaction, 30, 34
van der Waals potential, 235–237, 365
Variational degrees of freedom, 208
Variational force, 120, 123–125, 127
Variational Monte Carlo (VMC) method, 340
Variational principle, 13, 22–24, 26, 34, 178, 207
Verlet, 225, 361
  method, 119, 199, 357, 360
Vertex
  correction, 170–172, 173, 177, 178, 180
  function, 155, 158, 170, 179, 180
Virial coefficient
  second —, 237, 295
Virial theorem, 12, 32
Visco-elastic behavior, 363
Volmer–Weber mode, 317
Vortex, 307

**W**
Walker, 343
Wannier
  function, 46, 78, 95
  interpolation, 95
  representation, 95, 112
  state, 46, 178, 349
Ward–Takahashi identity, 155, 156, 167, 170, 171
Weighted-histogram method, 286
Wetting transition, 302
Whisker, 319
White noise, 314
Wick's theorem, 97
Widom's test article insertion method, 286
Wiedemann–Franz law, 99
Wigner lattice, 347
Wigner–Seitz
  cell, 45, 71, 245
  sphere, 45, 71
Wigner–Seitz cell, 245–246
Wigner solid, 29
Wulff theorem, 316

**X**
$X\alpha$ method, 28, 29
Xenon, 128, 237
X-ray, 130
  absorption, 175
  photoemission spectroscopy, 73
  photoemission spectrum, 175
$XY$ model, 305

**Z**
Zeeman energy, 297
Zero-point vibration energy, 117
ZnS, 164
ZnSe, 164
$ZrO_2$, 164

Printed in the United States
By Bookmasters